Logistic: exponential growth in a restricted environ

Mathematical model: a function when used to repre

situation

Multivariable function: one whose value depends upon two or more input variables

Natural exponential function: the exponential function whose base is e; e^x

Natural logarithm: the logarithm whose base is e; $\ln(x)$

Odd symmetry: symmetry about the origin; 180-degree rotational symmetry; $f(-x) = -f(x)$

One-to-one: characterizing a function that has exactly one input for each output

Output variable: the dependent variable in a function

Parameter: (1) in a function, a value that is constant for that particular function; (2) in a set of parametric equations, the independent variable

Parametric equations: a set of functions with the same input variable, whose outputs become the coordinates of a curve

Period of a function: the interval in which a periodic function completes its pattern once

Periodic function: one whose output values repeat at regular intervals

Piecewise function: one whose domain is divided into two or more intervals, and which is defined differently on each interval

Proportional: for two variables, related to each other through multiplication by a constant

Radian: an arc of a circle equal in length to the radius of the circle; an angle at the center of the circle subtended by such an arc, 2π radians corresponding to 360 degrees

Range: the set of possible outputs of a function

Real number: any number that can be represented as a point on the real number line

Reciprocal: the multiplicative inverse of a number; "one over" the number

Rate of change: for a function, the ratio of a change in the output to the corresponding change in the input

Root of an equation: a solution to the equation; a number that makes the equation true

Slope: for a line, the constant rate of change

Step function: a piecewise function whose outputs consist of constant-valued segments

Turning point: a point at which a function changes from increasing to decreasing, or vice-versa

Unit circle: a circle, centered at the origin, whose radius is 1 unit

Variable: a quantity that can take on more than one value

Zero of a function: an input value for which the output is zero; an x-intercept

SECOND EDITION

PRECALCULUS
Concepts in Context

Judith Flagg Moran
Trinity College

Marsha Davis
Eastern Connecticut State University

Mary Murphy
Smith College

THOMSON
™
BROOKS/COLE

Australia • Canada • Mexico • Singapore • Spain
United Kingdom • United States

THOMSON

BROOKS/COLE

Sponsoring Editor: *John-Paul Ramin*
Assistant Editor: *Lisa Chow*
Editorial Assistant: *Darlene Amidon-Brent*
Project Manager Editorial Production: *Janet Hill*
Marketing Manager: *Karin Sandberg*
Marketing Assistant: *Jennifer Gee*
Advertising Project Manager: *Bryan Vann*
Print/Media Buyer: *Kris Waller*
Permissions Editor: *Sommy Ko*
Production Service: *Hearthside Publishing Services/Anne Seitz*

Text Designer: *John Edeen*
Art Editor: *Hearthside Publishing Services*
Photo Researcher: *Terri Wright*
Illustrator: *Hearthside Publishing Services*
Cover Designer: *Marcus Fellbaum*
Cover Image: © *Arnulf Husmo/Tony Stone Images*
Cover Printer: *The Lehigh Press, Inc.*
Compositor: *Techsetters, Inc.*
Printer: *Edwards Brothers, Inc.*

For more information about our products contact us at:
Thomson Learning Academic Resource Center
1-800-423-0563

For permission to use material from this text, contact us by:
Phone: 1-800-730-2214   **Fax:** 1-800-730-2215
Web: http://www.thomsonrights.com

Library of Congress Control Number: 2003105072

ISBN 0-534-36240-0

Brooks/Cole–Thomson Learning
10 Davis Drive
Belmont, CA 94002
USA

Asia
Thomson Learning
5 Shenton Way #01-01
UIC Building
Singapore 068808

Australia/New Zealand
Thomson Learning
102 Dodds Street
Southbank, Victoria 3006
Australia

Canada
Nelson
1120 Birchmount Road
Toronto, Ontario M1K 5G4
Canada

Europe/Middle East/Africa
Thomson Learning
High Holborn House
50/51 Bedford Row
London WC1R 4LR
United Kingdom

Latin America
Thomson Learning
Seneca, 53
Colonia Polanco
11560 Mexico D.F.
Mexico

Spain/Portugal
Paraninfo
Calle/Magallanes, 25
28015 Madrid, Spain

Contents

4 EXPONENTIAL AND LOGARITHMIC FUNCTIONS   155

5 POLYNOMIAL AND RATIONAL FUNCTIONS   225

Preface

Preface to the Second Edition

In writing the second edition of *Precalculus: Concepts in Context*, we have changed neither our original goals nor our strategies for achieving those goals. We remain convinced that students learn by doing, by figuring things out for themselves (alone or in a group), and by accepting significant responsibility for their own learning. We see for ourselves that students understand better what they have successfully explained to others.

Our own classroom experiences with the first edition, along with valuable suggestions from other users, have motivated us to rewrite the text, making it easier to use, both for the student and for the instructor. Here are some of the improvements:

- There are fewer fill-ins for the student. We've attempted a better balance within each section between exposition and student writing.

- Marginal comments provide reminders and highlight key ideas.

- Every section concludes with a one-paragraph summary, *In a Nutshell*, making review easier.

- Chapter exercises follow an odd-even pattern, with solutions provided for the odds, making it easier for the instructor to select problems and for the student to gain insights from the odds that will help in doing the evens.

- Simple exponential models are introduced very early in the course, immediately following simple linear models. The formulas for both types, as well as the behavior of each, are compared and contrasted. A fuller treatment of exponential and logarithmic functions appears in Chapter 4.

- Rates of change are emphasized. Students learn to approximate instantaneous rates of change.

- Triangle trigonometry has a chapter of its own.

- The instructor has great flexibility in using the labs. Although the laboratory experience is essential in this book, very little in the body of the text depends on any specific lab. (The single exception is the *Graph-Trek* lab, which is required for an understanding of Chapter 3.) A one-semester course can use as few as two or three labs or as many as ten.

- Labs have been updated; several new labs and projects have been added to provide choice.

- Supplementary projects on sequences, series, and conic sections are included.

Once the foundation of the course (the first three chapters) is in place, the instructor may proceed immediately to Chapter 4 (Exponential and Logarithmic Functions) or to Chapter 5 (Polynomial and Rational Functions) or to Chapter 6 (Periodic Functions). These three chapters may be taught in any order. Each approach has its advantages. Chapter 4 builds directly upon the material on function transformation and inverse functions in Chapter 3 and reinforces the exponential models of Chapter 2. Teaching polynomial and rational functions immediately after function composition is the traditional order for precalculus topics, and is mandatory at some institutions. Our book supports that order, although we see such functions as less important for modeling and we postpone them until later in the semester. In our own teaching, we like to teach periodic functions immediately after function transformations so that students can experience a full range of graph shifts, stretches, and reflections in the context of sine and cosine models.

In short, the book has changed a great deal from its first edition, and yet it has hardly changed at all. We have so thoroughly revised certain sections that they will seem brand new to those familiar with the first edition. The dramatic changes in the order of topics required wholesale rewriting of large chunks of text. But we continue to stand behind the pedagogical principles of the first edition. This book still has the same flavor, although we think it tastes a little better this time.

Goals

The world, we thought at first, does not need another precalculus text.

Anyone looking at today's textbook market knows that an abundance of precalculus books is available and that many are very good. However, we have been teaching precalculus mathematics to college students for years, and we've given a lot of thought to what the course ought to entail. In attempting to integrate graphing technology and laboratory experiences with classwork, we realized that the elements and the organization of a conventional precalculus course were no longer working effectively for us and our students. In the early 1990s, we began to reexamine, not only the way in which precalculus is taught, but also what precalculus *means*.

The first edition of *Precalculus: Concepts in Context* was a completely new kind of textbook. Not only did it take a fresh look at the content of precalculus, it offered students a unique approach to learning mathematics. Starting from scratch, we constructed a course in elementary functions that emphasized the mathematical concepts needed for calculus and incorporated graphing technology as an essential learning tool. The second edition builds on that initial work and is true to its spirit.

The book is truly interactive. It engages students by asking them to respond on nearly every page, and the text is not complete until the students do their part. It involves students with their classmates through frequent activities that require collaboration.

Our goals for *Precalculus: Concepts in Context* are these:

- To present mathematics as both a symbolic language (algebra) and as a visual language (graphs), and to assist students in becoming fluent users of both languages and adept translators to and from the languages of algebra, graphs, and English.

- To enable students to recognize mathematical principles in the natural world and in everyday life.

- To engage every student actively in the process of learning and doing mathematics.

- To foster a spirit of collaboration rather than competition inside and outside the classroom.

- To use writing about mathematics as a means of promoting a deeper understanding of patterns and concepts.

- To dispel the notion that only certain people can succeed in mathematics.

- To promote the use of calculators and computers as tools for investigating mathematics, ones that reduce the time spent on numerical calculations and point-plotting and support the student in accomplishing and understanding *more* rather than less.

- To emphasize mathematical reasoning as well as numerical results and to encourage a variety of methods for reaching a conclusion.

- To persuade students that independent thinking, rather than mimicry of a worked-out example, is a problem-solving approach that will serve them best in the long run.

- To prepare students for calculus, particularly for any of the reformed calculus courses.

Computers and calculators change both the way students learn and the way instructors teach. Many ideas that once required calculus can now be understood very well at the precalculus level with the aid of a graphing utility, which we refer to as a *grapher*. There is no need, for example, to wait until calculus to introduce (informally) the rate of change of a function. On the other hand, some topics to which we used to devote much time are no longer necessary. Algebraic techniques such as synthetic division to find roots of polynomials, for instance, are not essential these days because technology gives us very good estimates of those roots. Moreover, because we can use technology to perform complicated calculations, we can situate our mathematics in authentic real-world contexts, despite their tendency to be numerically "messy." Students, instead of having to wait two or three more semesters for honest applications, get to experience first-hand the relevance of mathematics.

Adding topics to an already overpacked traditional curriculum isn't a good way to adapt to the electronic age. Many in the mathematics community see the need to change what precalculus is, not simply the way in which it is taught. We have attempted a re-vision of the precalculus curriculum, starting afresh with only those topics and techniques a student needs for a solid understanding of elementary functions and tossing out (or tossing into an appendix) everything else. We have made efforts to dovetail our changes with those of the leading calculus reform projects.

Precalculus: Concepts in Context offers a friendly, informal approach to the rigorous thinking expected in a mathematics book. (Instructors who insist upon a traditional precalculus text might be alarmed at the informal appearance of this book.) To help students feel at ease, we use humor, gentle explanations, and immediate reinforcement. We encourage sound reasoning without the formal "definition, theorem, proof" format appropriate for more advanced courses. The text sows the seeds for important calculus concepts as it develops the skills and mathematical maturity needed for calculus and for life itself. A course based on this text will succeed only with an instructor committed to the goals we have outlined above. This isn't a "teacher-proof" text.

We are thoroughly convinced of the value of learning through cooperation, and therefore we ourselves have made the writing of this book a collaborative effort. Each chapter was written by one of the three authors, reviewed by the other two, class-tested in several precalculus courses with a variety of student populations who were not shy about suggesting changes, and then completely revised by all three authors and rewritten by one of the two who had not written it originally. We encourage this sort of collaboration on a smaller scale among the students who use this text, particularly when they write and edit a group lab report.

Audience

Precalculus: Concepts in Context is written for students who have studied, but not necessarily mastered, the equivalent of two years of high-school algebra and a year of geometry. Many students who enroll in a college precalculus course have already taken a fourth year of high-school mathematics (that is, a course beyond Algebra II, which might even have been called "precalculus"), but they feel unready for calculus. Using this textbook will allow those students a fresh view of the material they saw in their earlier course and an opportunity to understand thoroughly the rules that they might simply have memorized the first time around. The approach is sufficiently different from that of a standard precalculus text that such students will not feel defeated from the beginning. ("This is the same stuff I couldn't understand last year; I won't understand it now, either.") At the same time, the text neither condescends to nor overwhelms those students who are seeing the material for the first time.

The text is suitable for a one-semester or two-quarter course. There is enough material for a longer course that stretches over two trimesters or two semesters. Specific suggestions for a variety of paths through the text can be found in the *Instructor's Guide*.

The text has been used primarily in small (15–35) class settings, but it could be used in large lecture courses as well, if students have regular opportunities to meet in smaller recitation sections with an instructor or a graduate assistant.

As its name implies, *Precalculus: Concepts in Context* is designed to prepare a student for calculus. At the same time, we recognize that precalculus is a terminal course for many students, and we have taken pains to design a course that has its own integrity and is worth studying in its own right.

Organization

The mathematical concept of function is the unifying idea of this text. We picture the book as a mathematical tapestry, in which the horizontal strands are the principal topics and the vertical strands are recurring themes.

The topics of a standard precalculus course constitute the woof strands of the tapestry. In that sense, this text is organized in a conventional manner; topics appear in the expected order. Students meet functions in general and various ways of representing them and then move to specific classes of functions, ranging from linear to exponential and periodic. Along the way, they learn about various transformations of functions and they become adept at interpreting graphs.

The major themes of the book constitute the warp strand of the tapestry:

- mathematical modeling of real-life phenomena

- distinctions between a function as an abstract mathematical entity and the same function used as a mathematical model

- rates of change

- the effects of scale

- different views of a graph

- connections among algebraic, geometric, and numerical representations of a mathematical idea

- mathematical language, or the connections between mathematical symbols and English words. ("What does it mean when a mathematician says . . . . ?")

These themes are woven through the standard topics so that they become visible at every opportunity. The interwoven themes encourage students to make connections between new material and what they already know, thus strengthening their understanding of both.

How to Use This Text

The book is written primarily for the student. We class-tested early versions of the text and rewrote every page that students found confusing. In many cases, we now employ the very phrasing that our students used when they finally grasped an idea. Here's an example: One student who had never studied functions was finding function notation frustratingly complicated. At last, she exclaimed, "Now I get it. The sentence $F(x) = 4x - x^3$ means that y depends upon x in this particular way, which we call F." The rest of the class nodded in agreement. We knew, therefore, that her sentence belonged in the text. Every student who is willing to make the effort should be able to read and understand most of the material, and therefore it should not be necessary for the instructor to cover a section in class before assigning it. Each section contains opportunities for the student to verify that he or she is on the right track before proceeding.

At the same time, there will be inevitable misapprehensions and points of confusion. Students might think they understand a topic, only to find that they cannot do the associated exercises or that they're coming up with incorrect answers. Or they will feel stuck partway through a section, unable to grasp fully the main idea. Therefore, it is essential to use class time for discussion, clarification, and expansion of the ideas presented in the reading.

Implied here is that students will actually *read* the book. The *Instructor's Guide* provides suggestions for making that outcome more likely.

Mathematics is best studied in an atmosphere that encourages collaboration, exploration, and creativity. The text lends itself well to collaborative learning by presenting some questions for which the correct response can be found via a variety of approaches or for which several different answers might be equally valid. We find it effective to involve all the students by having them discuss such questions during class in small groups (two to five students). Everyone has something to offer, and the conclusion of a group will often represent contributions from each of its members.

The laboratories are group activities on a grand scale. The instructor remains on the sidelines, allowing each group to make its own blunders and discoveries and to move at its own pace. In the process of negotiating agreements about how to interpret results, of constructing meaning from the work of the laboratory, and of organizing their conclusions for presentation in the report, group members will develop for themselves a solid understanding of the fundamental ideas of precalculus mathematics. Many instructors set aside one or two hours per week as lab time; others incorporate the lab work into their regular classes. The *Instructor's Guide* offers suggestions about how and when to assign each lab.

The text lends itself to collaborative work outside the classroom as well. Many of our students form study groups to do homework together, particularly when the homework includes exploratory exercises with a grapher. We encourage students to discuss their reading with one another and to compare with each other the responses they have written and the graphs they have drawn.

Pedagogical Elements

Precalculus: Concepts in Context uses a variety of pedagogical tools, because different students learn in different ways and because every student can benefit from several approaches to the same idea.

- *Context.* Instead of starting with abstract mathematical ideas and then looking for applications that illustrate them, we begin with the world of experience—music, commerce, psychology, natural science, and the daily news—and uncover the mathematics already present. As much as possible, we introduce each new mathematical topic by examining the idea in a context from which that topic naturally arises. A problem involving gravitational forces provides a reason to study rational functions; a news report on world population projections motivates a comparison of linear growth and exponential growth.

- *Write-in text.* Because students learn best when they are actively engaged, we have tried every trick we can think of to keep students productively occupied while they study. The responses that they are asked to write are not lengthy or especially difficult, but they serve to focus attention on the material and to help students realize whether or not they understand it. The pages of the text are incomplete until the student has written his or her responses.

- *Check Your Understanding.* Because students benefit from some immediate feedback while they are studying, we provide several opportunities in each chapter for them to try out their skills or to respond to key questions. There is space in the book for them to write their answers, and the correct answers are printed at the end of the chapter.

- *Stop and Think.* Because not every mathematical question has a short answer, because not all of the implications of a new concept will be crystal-clear from the start, and because connections always need to be made with other material, we present questions that students will not necessarily be able to answer right away, but which are worth thinking about. No space is provided for writing, and no answers are printed. These questions make good discussion-starters, especially if the students have had an opportunity to think about them before coming to class.

- *In a Nutshell*. A brief summary concludes each section, enabling the student to see at a glance the main ideas of that section.

- *Graphs and the use of a grapher*. We assume that every student either owns a graphing calculator or has frequent access to a computer with a graphing package. We encourage students to use a grapher while they study and while they're doing homework exercises. Students should use a grapher whenever it would help them to understand the material; they should not await specific instructions. As often as possible, the text shows the relationship between the graphical representation of a problem and its algebraic formulation; students should be encouraged to make their own connections as well. Students are also asked to sketch their own graphs by interpreting (rather than merely copying) the information they receive from their grapher.

- *What's the Big Idea?* This feature appears at the end of each chapter and, as its name implies, highlights just the main ideas of the chapter.

- *Progress Check*. This feature follows the "Big Ideas" and lists specific skills that each student should possess at that point in the course.

- *Key Algebra Skills*. At the end of each chapter, the essential algebra skills used in that chapter are listed with page references to the *Algebra Appendix*.

- *Exercises*. There is a wealth of exercises, more than an instructor could reasonably assign and more than a student could be expected to complete, at the conclusion to each chapter. We have deliberately grouped the exercises at the end of the chapter rather than after each section to help overcome the tendency of students to pigeonhole their math problems and to go directly to the relevant section, looking for a template example. (The *Instructor's Resource Manual*, however, indicates the sections that need to be covered before specific exercises are assigned.) Some exercises are straightforward applications of the material of the text; others require a deeper understanding or a synthesis of ideas. Exercises that would serve well as group activities or as test questions are identified in the *Instructor's Resource Manual*.

- *Projects and Explorations*. Following each set of chapter exercises are several longer activities, some of which extend the ideas of a lab but most of which stand alone. The projects can be used as homework assignments, as take-home test questions, or as mini-labs. The explorations guide students in using a grapher to understand certain types of functions. Students need not worry about making mistakes, because there's no single "correct" way to conduct an exploration. The purpose of an exploration is discovery. The *Instructor's Resource Manual* suggests ways to use each of these activities.

- *Labs*. Each chapter contains one or more major activities that serve to solidify student understanding of important concepts of that chapter. A *Preparation* section of one to three pages precedes the actual lab pages, and should be assigned in advance. The *Instructor's Resource Manual* provides detailed suggestions for organizing a precalculus laboratory and evaluating the results, as well as information indicating where in the syllabus each lab belongs.

- *Expository Writing*. We ask that students present what they have learned in each lab by writing a report in the form of an essay, complete with graphs to illustrate their discussion. Few students will have had any previous experience in doing this with mathematics, but nearly all of our students, by the end of the semester, have decided that the work that they did in the lab and wrote about in their reports was the work from which they learned the most.

Acknowledgements

Any large project such as the completion of this book is the work of many.

Our greatest thanks go to our students who, over the last several years, have worked from manuscript versions of the chapters. They spotted errors, gleefully pointing out a misspelled word or a missing negative sign. Generous with their opinions, they told us what worked for them and what didn't. They gave us concrete suggestions for improving the text, all of which we considered and some of which we used. We are grateful.

We thank our reviewers for their input.

Kenneth R. Anderson	Chemeketa Community College
Bruce W. Atkinson	Samford University
Hamid Reza Behmard	Western Oregon University
Winifred Benvenuti	Mt. Hood Community College
Gary Hagerty	Black Hills State University
Jeffery M. Lewis	Johnson Community College

We are not the only instructors to have used the text-in-progress. At Eastern Connecticut State University, Anthony Aidoo, Hortencia Garcia, Pete Johnson, Stephen Kenton, and Christian Yankov taught their classes from this book while it was still in manuscript form. For putting up with its flaws and for sharing with us their insights and experiences, we thank them. Hortencia Garcia also gave us invaluable assistance in critiquing and test-driving portions of the *Graphing Calculator Manual*.

William Dent of the University of Massachusetts Department of Astronomy reviewed our material about outer-planet light and about the work of Tycho Brahe. We appreciate his expertise.

The stunning cover is the work of Marcus M. Fellbaum of Dreamtime Graphic Design.

We have benefited enormously from a knowledgeable and hard working team at Brooks/Cole, led by our superbly responsive editor, John-Paul Ramin. Many others deserve accolades, in particular assistant editor Lisa Chow and sales representative Mike Lee. Our production group, headed up by Anne Seitz, made admirable efforts to understand and reproduce exactly what we intended on each page. Working with all of them was a pleasure.

Finally, we thank Jim Douglas and Caitlin Sporborg, who never dreamed they would see their names in a math book.

J.F.M.
M.D.
M.M.

Supplements

For the Instructor

Instructor's Resource Manual with Solutions. Since *Precalculus: Concepts in Context 2nd Edition* is not a conventional precalculus text, instructors might be reluctant to use it without ample supporting material. This comprehensive handbook contains valuable tips on using the text, including suggestions for conducting a lab period, promoting group work, squeezing the most from the class hour, making effective use of the write-in features, and fostering good writing in lab reports. It contains sample syllabi for courses of various lengths. It gives detailed section-by-section commentary, presenting the rationale for each section and concrete ways of communicating the ideas to students. (The answers to all the fill-ins are included.) A key feature of the *Instructor's Resource Manual* is several pages of specific suggestions for evaluating lab reports for each of the 14 labs. It will also contain a map that illustrates the sections to which each problem in the end-of-chapter problem sets correspond, and most importantly, it will also provide worked out solutions to all of the problems in the text. ISBN 0534378234

Test Bank. The *Test Bank* includes 4 tests per chapter as well as 2 final exams. The tests are made up of a combination of multiple-choice, free-response, true/false, and fill-in-the-blank questions. ISBN 0534378242

BCA Instructor Version. With a balance of efficiency and high-performance, simplicity and versatility, *Brooks/Cole Assessment* gives you the power to transform the learning and teaching experience. *BCA Instructor Version* is made up of two components, *BCA Testing* and *BCA Tutorial*. *BCA Testing* is a revolutionary, internet-ready, text-specific testing suite that allows instructors to customize exams and track student progress in an accessible, browser-based format. BCA offers full algorithmic generation of problems and free response mathematics. *BCA Tutorial* is a text-specific, interactive tutorial software program, that is delivered via the web (at http://bca.brookscole.com) and is offered in both student and instructor versions. Like *BCA Testing,* it is browser-based, making it an intuitive mathematical guide even for students with little technological proficiency. So sophisticated, it's simple, *BCA Tutorial* allows students to work with real math notation in real time, providing instant analysis and feedback. The tracking program built into the instructor version of the software enables instructors to carefully monitor student progress. The complete integration of the testing, tutorial, and course management components simplifies your routine tasks. Results flow automatically to your gradebook and you can easily communicate to individuals, sections, or entire courses. ISBN 0534378250

For the Student

Graphing Calculator Manual. The *Graphing Calculator Manual* provides clear, concise, and comprehensive instructions for the TI-83 Plus, 85/86, 89, 92 Plus, and the brand new Voyage 200. Each section begins with a tutorial on the basics of each calculator, such as navigating the keyboard, basic calculations, graphing, and troubleshooting, followed by text-specific instructions of the concepts covered in each chapter. Written by the authors of *Precalculus: Concepts in Context,* 2nd Edition, you can be assured of the high quality, reliability, and wonderful writing style that are hallmarks of this text. ISBN 0534378226

Student Solutions Manual. The *Student Solutions Manual* provides worked out solutions to the odd-numbered problems in the text. ISBN 0534379028

Website. http://mathematics.brookscole.com. When you adopt a Thomson-Brooks/Cole mathematics text, you and your students will have access to a variety of teaching and learning resources. This Website features everything from book-specific resources to newsgroups. It's a great way to make teaching and learning an interactive and intriguing experience.

vMentor Tutoring w/InfoTrac. With Brooks/Cole's Assessment's on-line Tutoring, your students have dynamic, flexible online tutorial resources at their fingertips. By entering a PIN code packaged with their textbook, students gain access to *BCA Tutorial,* a text-specific tutorial with step-by-step explanations, exercises and quizzes, and *vMentor,* live, one-on-one help from an experienced mathematics tutor. In addition to robust tutorial services, your students also receive anytime, anywhere access to *InfoTrac College Edition.* This online library offers the full text of articles from almost 4000 scholarly and popular publications, updated daily and going back as much as 22 years. Both adopters and their students receive unlimited access for four months.

BCA Tutorial Student Version. This text-specific, interactive tutorial software is delivered via the Web (at http://bca.brookscole.com). It is browser-based, making it an intuitive mathematical guide, even for students with little technological proficiency. So sophisticated, it's simple. *BCA Tutorial* allows students to work with real math notation in real time, providing instant analysis and feedback. The tracking program built into the instructor version of the software enables instructors to carefully monitor student progress. *BCA Tutorial Student Version* is also available on CD Rom for those students who want to use the tutorial locally on their computers and not via the Internet. ISBN 0534378269

To the Student

How to Use This Book

This book is probably quite different in style and format from other math text books you have used. Two major differences are the keys to using the book successfully.

- First, this is a book to **read.**

  You may be used to math classes where the teacher presents a technique, illustrates how to use it to solve sample problems, and then assigns very similar problems as homework exercises. Many students use the body of their text only as a reference, if at all. This book is different. It is written to and for *you*, not your instructor. We've tried to make the prose informal and clear. In the *Preface* (which is written primarily for the teacher) we suggest that students be asked to read and work through the material before it is discussed in class, and we urge that class time be used to make sure that everyone understands the reading.

- Second, this is a book to **write in.**

  You learn best when you are actively engaged in the process of learning. This text is incomplete until you supply some of the material. The responses we ask you to write are not long or particularly difficult, but they will help you realize whether or not you understand what you have read. The class period is a good time to check your write-in responses with those of your colleagues (and, if necessary, to ask your instructor) to make sure you're on the right track.

We use the term *colleagues* deliberately. We're convinced of the value of teamwork. This text is the collaborative effort of three authors, with help from other instructors who class-tested our notes and from students who served as our test drivers. Included in each chapter are one or more major investigations called *labs*, which are designed as group activities. However, we hope that you will find teamwork so valuable that you will choose to collaborate on the readings and other assignments as well as on the labs. We expect you to find that talking about the math you're learning helps to clarify your own understanding. In a group, different students contribute a variety of skills and talents. Good questions are as important as good answers; in fact, they often help others to refine their thinking and thereby improve their answers.

Although many of the problems in the body of the text have brief, definite solutions (particularly those in the *Check Your Understanding* sections), other questions are open-ended and have no single correct answer. Still others—those we designate as *Stop and Think*—are questions that you might not be able to answer right away, but that are worth thinking about and discussing with your colleagues in a study group or in class.

At first, this might not feel to you like mathematics. As one student, accustomed to cut-and-dried algebra, protested early in the semester, "This isn't math! You have to *think* about it!" Well, yes! We hope that by the end of the course you will have come to appreciate that, in fact, mathematics *does* ask you to think: not only to reason well and thoughtfully but also to explore and to try things out for yourself. One of the best ways in which a student of mathematics, or of any subject, can start a sentence is "*What if* ...?"

How to Write a Group Lab Report

We present here the advice we offer to our own students. We consider the lab experience and the subsequent writing of the report to be a major component of the precalculus course, and therefore our students have lab nearly every week. In your own course, you might have labs less frequently.

The lab report is an excellent means of gauging how well you and the other members of your group understand the material and are able to use the concepts you're learning. Each report represents the joint work and conclusions of the group, and is typed by the *scribe*. The office of scribe rotates among the group members, so that each student ends up writing approximately the same number of lab reports during the semester. Each member of the group should have a copy of the final version of the report after the instructor evaluates it. Graded lab reports form an important body of notes to which you can refer during the semester.

Each lab group will figure out effective ways of working together. The important thing is that responsibility be shared and that every member be fully apprised of what the group is doing. Any lab report that you (as scribe) submit should reflect the consensus of your group, and you should adequately review any work that another student submits in your name.

On the last page of each lab you will find specific instructions for that particular lab report. Below are some general guidelines that apply to every report.

A lab report is not a *list of answers* to all the questions posed in the lab. It is an *essay* showing what the group members have learned and what they now understand about the investigation. **Anyone should be able to read your report and understand the point of the lab.** The audience for your paper is a person who is familiar with precalculus mathematics but who has not read your textbook. You need to write in complete sentences and explain enough so the reader will understand. This does not mean that the report needs to be very long; it does mean that it must be able to stand on its own.

An effective lab report contains the following elements:

- an introduction in which you tell your reader what he or she needs to know about the lab

- the mathematical concepts involved

- an outline of your procedure

- significant findings

- conclusions.

One tendency you'll probably have at first is to read through the lab pages, highlighting the questions you find, and then go about the business of answering those questions. Fine. But that, in itself, doesn't make a lab report. Imagine yourself a tailor: you cut out the pieces of fabric and put them together with large basting stitches; then you sew the seams and remove the bastings. Your completed suit doesn't have (we hope) any basting stitches in it. Some questions on the lab pages are like basting stitches: answering them helps to shape your work and keep you from straying too far off the track. The finished product doesn't need even to mention those questions; the fact that you produced an elegant piece of work shows that you put things together correctly.

Other questions, however, do need to be answered in the lab report. How can you tell which questions are basting stitches and which are essential construction details? Experience will help, and so will the specific guidelines at the end of each lab. A question such as "What is the volume when $L = 44$?" is probably just a sample case to help you discover a pattern, and its answer ("$V = 7744$") doesn't need to be in the report unless you're using it to illustrate a concept. If, however, the lab asks, "Why does one shape yield more volume than the other?", your answer is essential to the report and should be incorporated in your discussion.

The lab instructions help to lead you, by means of questions, from specific examples to general conclusions. Your lab report, in contrast, can begin with those conclusions and illustrate them with whatever examples and graphs you need. In other words, though you do the project in the order suggested by the lab pages, you might write about it more effectively by switching things around.

Graphs can do much to clarify what you're saying. They needn't be elaborate, but axes should always be labeled and units, if appropriate, should be specified. Graphs can be woven into the text or gathered at the end of the report. There is room for creativity, because you get to decide the most effective way to present your ideas. Take the opportunity to read other groups' reports; you'll notice that no two are the same (nor should they be).

An effective presentation usually requires, in addition to graphs, two to four pages of text. If yours is longer, look carefully to see what details you might omit to tighten the paper without losing anything vital. When you truly understand an idea, you can put it into words. Writing a lab report gives you the opportunity to demonstrate the level of your understanding.

In the labs, you encounter mathematics in diverse contexts—astronomy, economics, biology, and computer-aided design, to mention just four. Mathematics is not disconnected from the world; rather, many real-world phenomena are inherently mathematical. You will almost certainly find at least one lab that engages your interest and supports your other academic work.

Introducing Functions

© John Lund/Getty Images

"I only took the regular course," (said the Mock Turtle with a sigh). "What was that?" inquired Alice. "Reeling and Writhing, of course, to begin with," the Mock Turtle replied; "and then the different branches of Arithmetic—Ambition, Distraction, Uglification and Derision."

Lewis Carroll, *Alice's Adventures in Wonderland*

In elementary school, you studied "the three R's": reading, writing, and arithmetic. Among other things, this first chapter is about how we read and write mathematics. Mathematics is in many ways a simpler and less ambiguous language than English. (Where, for instance, is the logic behind calling reading, writing and arithmetic "the three R's"?!) You already know a great deal about reading and writing mathematics and about its grammar, the algebra of real numbers. Rusty algebra will greatly hamper your mathematical communication; therefore, we have included an Algebra Appendix at the back of the book as a quick reference for the rules of this "grammar." Throughout the text, when you encounter algebraic techniques that are essential to success in precalculus and calculus, you will often see a specific reference to that appendix.

Unsure of your algebra? Make good use of the Algebra Appendix, which starts on page 467.

Like hieroglyphics, mathematics can be a very visual language. In this course, you will use technology—graphing calculators and/or computers—to depict many mathematical relationships visually. You will also learn to provide your own graphs and pictures, both quick sketches and carefully detailed representations.

Learn to use your grapher right away.

Because an electronic grapher can give us a picture of a relationship quickly and easily, our emphasis will be on *interpreting* these pictures. We will discuss which views best represent the relationship, how we can obtain specific mathematical information from a picture, and also how to recognize when the graphing utility is presenting a misleading picture. Because this book assumes access to graphing technology and a scientific calculator, we can tackle "messier" numbers and equations than you might have worked with before. Therefore, the problems we discuss, particularly those presented in the labs, can be more realistic and less artificial than ones you might have seen in the past. In this course, you will learn to use graphs to explore, measure, and analyze phenomena ranging from the severity of an earthquake to the quality of a wine vintage and from the rate of increase of an epidemic to the effect of daylight on a person's emotional state.

There might be more than one acceptable answer.

Graphing technology can free us from some of the grunt work of computation and allow us to emphasize analysis and interpretation of material. In some cases, you will select the best from several good answers; in others, you may be asked to explain why no satisfactory answer is available. ("The dog ate it," is *not* considered a satisfactory response!)

THE FAR SIDE® By GARY LARSON

"Well, here we go again. ... Did anyone here *not* eat his or her homework on the way to school?"

If you are a student who has experienced mathematics as a black or white, right or wrong discipline, you might feel uncomfortable at first with problems that have no single correct answer. Algebra, in fact, does fit the conventional "one right answer" view, so the Algebra Appendix is basically a collection of rules for manipulating symbols that represent numbers. The laboratory reports that you write, however, should be thought of as essays, and the work that you do with this text falls somewhere between the two

Get used to explaining your work!

extremes of symbol manipulation and expository writing. You will often be asked to explain your answer to a question, and *the explanation can be more important than the answer itself!* For those students who might miss the old right-wrong dichotomy, we console you with the thought that we hope to abolish the old win-lose dichotomy as well.

1.1 DEFINING FUNCTIONS

Putting your best foot forward

In a word, this text is about a certain type of mathematical relationship known as a **function.** Before we give a formal definition of the term *function*, let's investigate some examples.

We'll begin with a very down-to-earth function involving your feet or, to be more specific, the size of your feet. Our English measurement *the foot* describes a generic and generously sized appendage. In the United States, though, shoe sizes are given by numbers such as 7, $8\frac{1}{2}$, and 9. We might wonder what these numbers measure. Do you suppose they could stand for the length, in inches, of the person's foot?

Take a ruler and try a reality check. Would a female with a foot 5 inches long be likely to wear a women's size 5? How about a big-footed basketball star who wears a men's size 16? Do you imagine that his foot is 16 inches long?

A Formula Can Define a Function

The size shoe you wear is certainly *related* to the size of your foot; in fact, a shoe salesperson might say that your shoe size "is a function of" (that is, depends on) the length of your foot. A mathematician might also say that shoe size is a function of foot length, but she would mean something much more specific. In fact, she might give the following algebraic formula, which tells explicitly how to convert x, the length in inches of a woman's foot, into y, the number giving her shoe size:

$$y = 3x - 21$$

We can express this rule in English: Take a woman's foot length, measured in inches, multiply it by 3, and subtract 21 from the result. For example, if a woman's foot measures $9\frac{1}{2}$ inches in length, her shoe size is $7\frac{1}{2}$:

$$3(9\tfrac{1}{2}) - 21 = 28\tfrac{1}{2} - 21 = 7\tfrac{1}{2}$$

If you wear a size 8 shoe, what does the "8" measure? That question made us wonder whether shoe sizes (like sizes for jeans, in which 34/32 measures waist and inseam) have any meaning. What we learned from a foot-sizing chart, though, seemed entirely arbitrary: $3x - 21$ and $3x - 22.5$ have no deeper significance.

Here's one of life's little mysteries: In the United States, men's shoes are sized differently from women's shoes. Therefore, we need a slightly different shoe-size formula if we want to use it for men's shoes. If y is defined to be the size of a *man's* shoe in terms of x, the length of his foot (in inches), then

$$y = 3x - 22.5$$

Because the values of x and y vary with different customers, they are called **variables.** Because x (foot length in inches) determines y (shoe size), we call x the **independent variable** and y the **dependent variable.** If you used a calculator to compute values of this function, a value of the independent variable x would be the input, and the resulting y-value would be the output; therefore, we will sometimes speak of x and y as the **input** and **output** variables, respectively.

The statements $y = 3x - 21$ and $S = 3L - 21$ represent the same function. In each case, the value of the dependent variable is 21 less than 3 times the independent variable.

The formula $y = 3x - 21$ expresses the relationship between the two variable quantities foot length and shoe size. If, instead, we represented foot length by L and shoe size by S, we would write the same function as $S = 3L - 21$. Notice that the *relationship* that we are representing by our formula is unchanged, although we now call our independent variable L. We classify a variable as dependent or independent according to its role in the function, not according to the letter we happen to choose to represent it. (The graphing device—computer or calculator—that you use in this course will probably designate the independent variable as x or t because these are the symbols that mathematicians often use to denote the independent variable.)

The little triangle is your signal to start writing.

▷ Now use $y = 3x - 21$ to compute the y values corresponding to the input values $8\frac{2}{3}$, 9, $9\frac{1}{2}$, $9\frac{2}{3}$, 10, $10\frac{1}{6}$, and 11. Write the y-values in Table 1.1 below.

x-value	$8\frac{2}{3}$	9	$9\frac{1}{2}$	$9\frac{2}{3}$	10	$10\frac{1}{6}$	11
y-value	5	6	7.5				

Table 1.1 Table for Women's Shoe-Size Function

A table is another way of displaying a function. You just created a table of values (Table 1.1) by using the formula $y = 3x - 21$ to find y-values corresponding to the x-values supplied. The formula $y = 3x - 21$ came first, and you used it to make a table of certain selected pairs of x- and y-values. The x-values were chosen arbitrarily; we could just as easily have used a different set of x-values.

A Table Can Define a Function

In other functions, the table comes first and serves to define the function. Many functions that describe real-world relationships are determined by collecting data. The data are arranged in a table, and the pairs of values in the table are the only information immediately available.

Mathematicians and statisticians (and you, too, in some of the labs and exercises) look for a pattern in the data. In some cases, there might be a mathematical formula that models or approximates the relationship. The beauty of such a formula is that it allows us to find an output value for input values that don't happen to be in the table.

In other cases, a table might be the *only* definition of the relationship between the varying quantities. Not every function has an algebraic formula to go with it.

Look, for instance, at Table 1.2, which shows actual midnight measurements of the Connecticut River during a recent spring when the river rose dramatically and threatened to overflow its banks. The flood-watch officer recorded these values at the start of each day listed.

Date	Feet above Sea Level	Date	Feet above Sea Level
April 5	107.1	April 15	113.7
April 6	107.9	April 16	113.0
April 7	109.0	April 17	113.7
April 8	111.2	April 18	114.2
April 9	112.7	April 19	114.4
April 10	111.1	April 20	113.7
April 11	109.8	April 21	112.2
April 12	110.6	April 22	111.4
April 13	110.5	April 23	109.0
April 14	111.0	April 24	106.8

Table 1.2 Connecticut River Flood-Watch Data

STOP & THINK

Table 1.1 came from a formula, but Table 1.2 does not specify a mathematical formula for the function it gives. Such a formula, were it to exist, would allow us to determine the height of the river for times that aren't in the table, such as noon on April 20. It would have predicted the maximum height of the river and the moment when it crested, giving flood-plain residents plenty of warning for evacuating their houses and rescuing their belongings. Can you find a formula? Do you think an accomplished mathematician would be able to come up with one? Explain.

A Graph Can Define a Function

A picture is worth a thousand words

Another common method for depicting a relationship between two variables is a graph. Although the numbers in Table 1.2 give an accurate representation of the day-to-day fluctuations in the water levels, the general pattern in the rise and fall of the river is not readily apparent. The human eye and mind can usually absorb pictorial information much more easily than numeric data, so let us define the same river-level function by the plotted points on the graph in Figure 1.1.

Figure 1.1 is a graphical representation of Table 1.2.

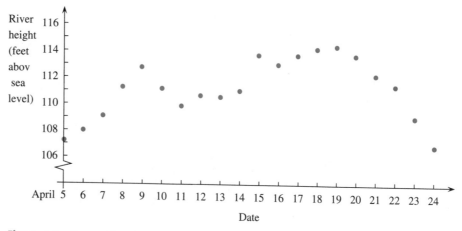

Figure 1.1 Connecticut River Flood-Watch Data

It's easy to go from the table to a graphical representation of the function. We draw two lines intersecting at right angles, called **coordinate axes.** Mathematical convention has us list the values of the independent variable (in this case, the date) along the horizontal axis and the values for the dependent variable (here, the height of the river) along the vertical axis. The table in Table 1.2 and the graph in Figure 1.1 both represent the same river-level function: For each date given, there corresponds a number, the height of the river as that day began.

Input on the horizontal, output on the vertical

In Figure 1.2, the same points have been joined by a continuous curve. Connecting the points is legitimate because the river did have a definite level at every moment, even though we don't have any information for hours other than midnight. (In fact, the flood-watch team took measurements every two hours.) Our curve, therefore, represents a reasonable guess as to what might have been happening to the river in the course of each day.

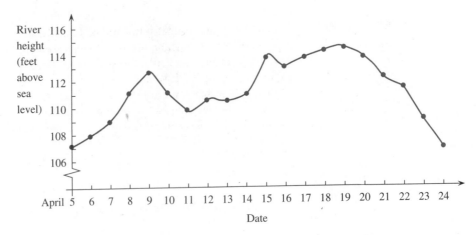

Figure 1.2   Connecticut River Level

Unlike the two shoe size functions, the river-level function cannot be expressed easily by any reasonable formula. We might find a mathematical expression to represent the points on the graph, but it would be of no use for predicting the height of the river the following week or for determining the exact moment at which the river crested.

A set of ordered pairs can generate a graph.

Let's return to our original shoe size function and use it to discuss some basic techniques available for those functions for which we *do* have an algebraic formula.

▷   On page 3, you used the formula $y = 3x - 21$ to complete Table 1.1. Now represent each x-y pair from your table as a point on the coordinate axes in Figure 1.3.

The "squiggles" in the axes indicate that a piece is missing. Because we don't need the segment between 0 and $8\frac{1}{2}$ on the x-axis or the segment from 0 to 5 on the y-axis, we've snipped them out so that the meaningful portion of the graph can be larger.

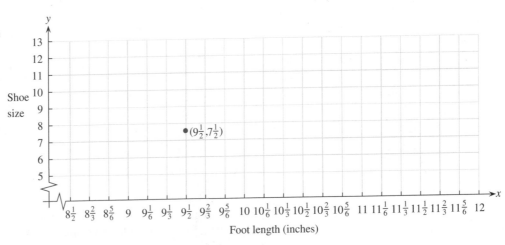

Figure 1.3   Graph of the Shoe-Size Functions

We call such pairs of values **ordered pairs.** When we list an ordered pair of values from a function, the independent variable is always the **first coordinate,** and the dependent variable is always the **second coordinate.** (We started you off by depicting the point with coordinates $(9\frac{1}{2}, 7\frac{1}{2})$, corresponding to a woman's size $7\frac{1}{2}$ shoe. Notice that this point is different from the point $(7\frac{1}{2}, 9\frac{1}{2})$; in an ordered pair, the *order* matters.)

▷   Connect the dots to see the pattern these points make.

▷   Now complete Table 1.3 for the men's shoe-size function. (Note that you will be approaching the formula from the opposite direction: You begin with a y-value and work "backward" to determine the x-value that produced it.) We have chosen some arbitrary shoe sizes; use $y = 3x - 22.5$ to find the foot length for each.

x							
y	7	$8\frac{1}{2}$	10	$10\frac{1}{2}$	$11\frac{1}{2}$	12	13

Table 1.3 Table of Men's Shoe Sizes

▷   Graph the points corresponding to these ordered pairs (x, y) on the same set of axes in Figure 1.3 but in a color different from the one you used for the women's shoe-size function.

▷   Describe the pattern formed by the points belonging to the men's shoe size function.

▷   What similarities do you see between the pattern of the first set of points and the pattern of the second set?

▷   In what way are the two patterns different?

You've just seen two types of graphs: Figure 1.1 representing points from a table of data and Figure 1.3 generated by an algebraic formula. There's another kind as well: a graph that stands alone as the sole definition of a function. Look at Figure 1.4, which defines an output variable Z as a function of an input variable u. Do you see how the graph can, all by itself, tell us which Z-value corresponds to a given u-value?

▷   What output value results from an input of 3?

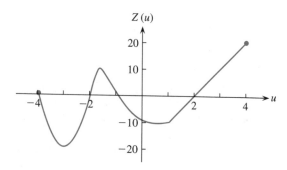

Figure 1.4 A Function Defined Solely by Its Graph

Using Function Notation

What's in a name?

To distinguish between the women's and men's shoe-size functions, let's name them w and m. The expression $w(x)$, pronounced "w of x," is standard mathematical shorthand for the output value (the y-value) that the function w assigns to the input value x:

The mathematical statement $y = w(x) = 3x - 21$ says that y depends on x in this particular way, which we're calling w.

- w is the name of the function,

- x stands for the input, and

- $w(x)$ is the value that w assigns to the input x.

If x gives the measurement of a woman's foot in inches, then we determine $w(x)$ by substituting that particular value for x in the formula $w(x) = 3x - 21$. For example, $w(9\frac{1}{2}) = 3(9\frac{1}{2}) - 21 = 7\frac{1}{2}$ or, more succinctly, $w(9\frac{1}{2}) = 7\frac{1}{2}$.

Note that $m(x)$ doesn't mean m multiplied by x.

We read that equation, or mathematical sentence, as "w of $9\frac{1}{2}$ is $7\frac{1}{2}$." It means that a woman whose foot measures $9\frac{1}{2}$ inches probably wears a size $7\frac{1}{2}$ shoe. Similarly, we pronounce $m(10\frac{1}{2}) = 9$ as "m of $10\frac{1}{2}$ is 9," and we mean that the operations performed by the function m on the input value $10\frac{1}{2}$ produce the output value 9—or, in English, that a man with a $10\frac{1}{2}$-inch foot would take a size 9.

▷   Translate the statement $m(11) = 10\frac{1}{2}$ into an English sentence.

Functions can be given any name. The name might be an entire word or just a single letter; here, we chose w and m to distinguish between the women's and men's shoe-size functions. The generic or most common designations for functions are the letters f, g, and h. Just as $y = 3x - 21$ and $S = 3L - 21$ represent the same function, so do $w(x) = 3x - 21$, $f(\square) = 3\square - 21$, and $SHOE(u) = 3u - 21$. The rule "Multiply input by 3 and then subtract 21" is what's important, not the letters chosen to write the rule.

Check Your Understanding 1.1

The answers to all "Check Your Understanding" sections in each chapter appear just before the chapter exercises. The answers to this "CYU" section begin on page 30.

1. Use an appropriate table or formula to answer these questions about the shoe-size functions m and w.

   (a) What is the value of $m(10\frac{1}{2})$? Of $w(10)$? Of $w(9\frac{1}{3})$?

   (b) If the value of the independent variable is 10, what value does w assign to y?

   (c) If the value of the independent variable is 10, what value does m assign to y?

   (d) If w determines a y-value of 10, what is the corresponding x-value?

   (e) What input value does the function m need to produce the output 10?

2. Answer the following about a function f, defined by the rule $f(\square) = \pi \square^2$.

(a) In words, what does the rule for f say?

(b) The function f gives the area of a circle whose radius is $\square$. What is $f(3)$?

(c) Write the same rule f using r as the independent variable and A as the dependent variable.

(d) Write the same rule f using x as the independent variable and y as the dependent variable.

3. Use Figure 1.4 for these questions.

(a) Give the value of $Z(0)$ and of $Z(-3)$.

(b) For what values of u does $Z(u)$ equal 0?

STOP & THINK

We can use Table 1.3 to find the value of x that corresponds to a particular y value, such as $8\frac{1}{2}$. This is not possible for the river-level data in Table 1.2. That is, we cannot use the river level to determine the date. Why not?

In a Nutshell   A function is a particular type of relationship between two variables. In the next section, we'll get more specific about which kinds of relationships qualify, but here we focused on some of the ways to define a function: a graph, a table of ordered pairs, an algebraic formula. In the case of a function defined by a formula, function notation gives us a handy way to write the function's rule.

1.2   USING FUNCTIONS TO MODEL THE REAL WORLD

Functioning continuously, functioning discretely ...

Now let's look at our shoe size function a little more closely. We used algebra to state efficiently a relationship between two variable quantities: foot length and shoe size. Our function w is a **mathematical model;** we selected certain properties of the length-to-size relationship and expressed these properties by the algebraic statement $y = 3x - 21$ or, equivalently, $w(x) = 3x - 21$.

Like most models, our equation is simpler than the reality it represents.

The Domain of a Function

We need to be careful to distinguish between the mathematical equation $y = 3x - 21$ and the situation in the shoe store.

▷   Input 2 as the value of x and use the formula to find the corresponding value of y.

The y-value you obtain is a negative number and therefore has no meaning as a shoe size. Nor can this x-value be interpreted as the length of a woman's foot. The smallest woman's shoe size is 5. Replacing y with 5 in the equation $y = 3x - 21$, we find that the corresponding x-value is $8\frac{2}{3}$ inches. A person with a smaller foot would need a child's size shoe.

▷ The largest woman's shoe size is 13. Check that it corresponds to a foot length of $11\frac{1}{3}$ inches.

> If the function models a real-world situation, the domain is determined by the context.

Thus, the x-values that make sense in our *model* lie between $8\frac{2}{3}$ and $11\frac{1}{3}$ inches. The set of all the input values that make sense for a function is called the **domain** of the function. Therefore, for our model w, the domain is the set of all real numbers between $8\frac{2}{3}$ and $11\frac{1}{3}$. (Most calculators give fractions as decimal approximations and would record these values as 8.666666667 and 11.33333333.)

> The set of all decimals, finite and infinite, is called the set of **real numbers.**

By contrast, the equation $y = 3x - 21$ has an independent mathematical existence as a function apart from its use to model women's shoe sizes. The algebra itself places no constraint on the values for x we can use as input; we can multiply and subtract *any* real numbers. Thus, as an **abstract function**, $y = 3x - 21$ has as its domain the set of all real numbers, which we will denote by $\mathbb{R}$.

Check Your Understanding 1.2

> If you're using one of the fancier calculators, such as the TI-85, you won't necessarily get an error message when you step outside the set of real numbers. If you try to take the square root of -25, for instance, you'll get $(0,5)$, which is the calculator's way of saying that the answer is an imaginary number. Imaginary numbers are part of a larger set of numbers known as the **complex numbers.** In this book, we deal only with **real numbers,** that is, numbers that have a place on the **real number line.**

1. What is the domain of the shoe-size model m if the standard selection of men's shoes ranges from size 6 to size 13?

$$9\frac{1}{2} \qquad 11\frac{5}{6}$$

2. What is the domain of the abstract function m given by the formula $m(x) = 3x - 22.5$? (That is, what values for x could you input into your calculator, multiply by 3, and subtract 22.5 from, without getting an error message?)

3. What is the domain of the river-level model given by Table 1.2 on page 4?

4. What is the domain of the river-level model given by the graph in Figure 1.2 on page 6?

Not all abstract functions have the set $\mathbb{R}$ as their domain. Two common exceptions are the reciprocal function, $g(x) = \frac{1}{x}$, and the square-root function, $h(x) = \sqrt{x}$.

5. What is the domain of g?

6. What is the domain of h?

Discrete Variables, Continuous Variables

Many of us have played "connect the dots" to graph lines and curves. In fact, this is how some of the early plotters drew the graph of a function: They started at a point, drew a line segment to the next computed point, from there to the next computed point, and so on. Today's graphing calculators and computers rapidly calculate dozens or hundreds of points for a given function and highlight the pixel *nearest* to each calculated point. If a graph is particularly steep, they might also highlight pixels in between the calculated ones so that your eye will perceive a line or a curve instead of a sequence of dots.

Calculator graphs often have a ragged appearance compared to graphs you could draw by hand. This raggedness occurs because the calculator cannot put a point except where it has a pixel. If you use a graphing calculator or computer to graph the equation $y = 3x - 21$, using as a window $-3 \leq x \leq 15$, $-30 \leq y \leq 30$, you might see a picture like Figure 1.5.

Dot
mode

Connected
mode

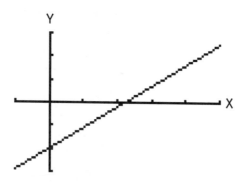

Figure 1.5 Graphing Calculator Display of $y = 3x - 21$

STOP & THINK

Even if you restrict your viewing window so that the values of the independent variable x lie between 8.666666667 and 11.33333333, the graph you see will not model the shoe-size situation satisfactorily. Why not? (*Hint*: Someone's foot measures $9\frac{3}{4}$ inches. What size shoe does she wear?)

The variable x that we use to measure women's foot length varies **continuously** between about 8 and 12 inches, limited only by the accuracy of our measuring device. By contrast, shoe sizes take on only integer or integer-plus-one-half values. We say that the dependent variable y in our model is **discrete** because consecutive y values are *separated* from one another by intervals (in this case, by $\frac{1}{2}$ unit).

This combination of a continuous input and a discrete output suggests that we should modify our model to better represent the actual shoe-size situation. We'll do just that in the next chapter. Coming up in Chapter 2: step functions.

The Range of a Function

Home, home on the range

We have defined the domain of a function as the set of input values that make sense for the function. The **range** of a function is the set of output values that the function generates from its input set or domain. The range of the abstract function[1] $w(x) = 3x - 21$, graphed in Figure 1.5 is $\mathbb{R}$, the set of all real numbers, because the graph goes on forever upward and forever downward; there is no y-value that cannot be obtained as output. But the actual shoe-size model has a *discrete, finite* range: the set of numbers $\{5, 5\frac{1}{2}, 6, 6\frac{1}{2}, 7, \ldots, 13\}$.

---

[1]As you might have detected, there is some ambiguity (which mathematicians often gloss over) associated with function notation. We said earlier that the name of this function is w and that $w(x)$ means "the value of the function w, evaluated at x." However, it is very common, though not entirely precise, to refer to "the function $w(x)$." We should really say "the function w, defined as $w(x) = 3x - 21$," but we tend to shorten the phrase by saying "the function $w(x) = 3x - 21$."

When we want the negative square root, we must write $-\sqrt{x}$. For example, $\sqrt{25}$ is 5, while $-\sqrt{25}$ is -5 (not to be confused with $\sqrt{-25}$, which isn't a real number).

The range of the abstract function $h(x) = \sqrt{x}$ is the set of all nonnegative real numbers (that is, the positive numbers and zero). There's no way to obtain a negative number as the *output* of the square-root function, because the symbol $\sqrt{x}$ denotes the nonnegative square root of the number x.

Check Your Understanding 1.3

1. What is the range of the function $H(x) = -\sqrt{x}$? Check by substituting several different values for x on your calculator. Don't forget 0.

2. What is the range of the function $\Omega(\square) = -\sqrt{\square}$?

3. Experiment with different input values such as -100, π, 0.01, -0.001, $\frac{1}{3}$, $\frac{1}{3000}$, and $\sqrt{12000}$ and try to guess the range of the abstract function $g(x) = \frac{1}{x}$.

4. What is the range of the function defined by Table 1.3 on page 7? Is the range discrete or continuous?

5. What is the range of the river-level function defined by the graph in Figure 1.1 on page 5? Is the range discrete or continuous?

6. What is the range of the river-level function defined by the graph in Figure 1.2 on page 6? Is the range discrete or continuous?

One Output for Each Input

You're my one and only

Because the concept of *function* is one of the major ideas in mathematics, mathematicians are very finicky about how they define a function. A shoe salesperson might tell a woman whose foot measures $9\frac{1}{2}$ inches that she could wear a size $7\frac{1}{2}$ or 8 shoe. (The measurement of foot length is only approximate anyway.) But a mathematician requires that a function produce *exactly one output* for each input in the domain.

Because the study of functions is the central theme of this book, here's a more formal mathematical definition:

> A **function** is a relationship between two sets, the domain and the range, such that each member of the domain corresponds to exactly one member of the range.
>
> The **domain** is the set of inputs that make sense for the function.
>
> The **range** is the set of outputs that can be expected from the function.

Remember that the function relationship can be specified by a table, a graph, a verbal statement, or a rule (often given as an algebraic formula).

For example, let F be the function that squares a number and adds 2. (F in this case is an **abstract** function; that is, we will consider it apart from any particular application.) In function notation, we can write $F(-5) = (-5)^2 + 2 = 27$ or $F(\pi) = (\pi)^2 + 2$ or $F(a) = a^2 + 2$.

▷   What is $F(0)$? $F(\sqrt{3})$? $F(-\sqrt{3})$?

For a function, each input yields exactly one output.

Your last two results should be identical. This doesn't violate the "one output for each input" requirement. In a function, two different inputs may yield the same output, but a single input cannot generate more than one output.

Check Your Understanding 1.4

These exercises test your ability to expand and simplify algebraic expressions, skills you'll need this semester and in calculus. If you have trouble, use the Algebra Appendix, pages 467–508, to brush up.

1. Write the rule for F, the function just defined, as an algebraic formula, using x for the independent variable.

Find each of the following and, if possible, simplify:

2. $F(b) =$

3. $F(x - 3) =$

4. $F(\frac{5}{x}) =$

5. $F(n^3) =$

6. $F(\sqrt{3} + 1) =$ $6 + 2\sqrt{3}$

The domain of F is the set of all real numbers, because we're allowed, mathematically, to square any number and add 2. The range, however, does not include all the real numbers.

▷   What is the smallest value in the range of F?

▷   Is there a largest value of F? Why or why not?

▷   What is the range of F?

Because no endpoints are indicated, we may assume that the graph continues indefinitely on either side in the direction in which it was headed.

Figure 1.6 is the graph of the abstract function F. Notice that there is a point on the graph corresponding to each point on the x-axis. This shows that the domain of F is the set $\mathbb{R}$. Because the point $(0,2)$ lies on the graph, 2 belongs to the range of F. Because all other points on the graph have y-coordinates greater than 2, and because there is no upper limit, the range of F is the set of all real numbers greater than or equal to 2.

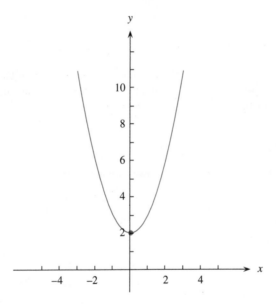

Figure 1.6 Graph of the Function That Squares Its Input and Adds 2: $F(x) = x^2 + 2$

One of the useful features of a well-drawn graph is that it allows us to determine at a glance the domain and the range of a function.

Dysfunctional relationships

Not every relationship between two sets is a function. Consider, for example, the river-level relationship given by Table 1.2, and suppose we tried to create a function using *water level* as the input variable and *date* as the output. Does knowing the level of the river allow you to determine the date? Not necessarily. Notice that the level was 113.7 feet on more than one date; an input of 113.7 would generate three possibilities: April 15, April 17, and April 20. Thus, *water level* cannot be the input variable.

Here's another example. Suppose we have a relationship whose input is the square of a number and whose output is the number that was squared to get the input. Figure 1.7 shows a graph of this relationship.

THE FAR SIDE® By GARY LARSON

I must protect Bob from the dangers of the grass.

Hey! I see something just ahead!.. Heh heh heh...

Good and evil shoes

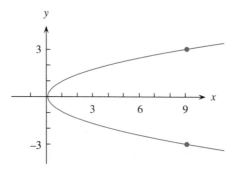

Figure 1.7 Graph of the Relationship $y^2 = x$

A number whose square is 9 is called a **square root** of 9. The symbol $\sqrt{9}$ means the positive square root, or 3; the symbol $-\sqrt{9}$ means the negative square root, or -3.

The point $(9, 3)$ is on the graph of $y^2 = x$, because 3 can be multiplied by itself to get 9. But the point $(9, -3)$ is also on the graph, because -3 is also a number whose square is 9. The graph, therefore, is showing us that an input of 9 would yield an output of either 3 or -3. But a function will not tolerate such divided loyalty. For the relationship to be a function, the input of 9 must yield precisely one output.

STOP & THINK

Precalculus textbooks traditionally present a tool known as the **vertical line test** for determining whether or not a graph can represent a function. It goes like this: If you can find a vertical line that cuts the graph in more than one place, the graph cannot represent a function; otherwise, it can. Explain why this test is valid.

Check Your Understanding 1.5

In each of the following relationships, x is considered the independent variable and y the dependent variable. Decide which are functions and which are not. Give a reason.

1.

x	y
2	0
14	9
−8	1
0	0

2.

x	y
−3	10
−1	4
−3	5
0	6

3.

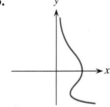

4.

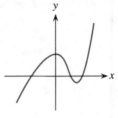

5. $y = x^2 + 16$

6. $y = \pm\sqrt{16 - x^2}$

7. Let x be the length of one side of a cube and let y be its volume.

8. Let y be the area of a triangle and let x be the length of its base.

In a Nutshell   In this section, you learned the definitions of *function*, *domain*, and *range*. You saw that the domain and the range can be limited either by mathematical impossibilities or by the nature of the situation being modeled. You learned to distinguish between continuous variables and discrete variables, and you know the difference between a relationship that is a function and one that is not.

1.3 WATCHING FUNCTION VALUES CHANGE

If you could just walk a mile in my shoes

In this section, we will focus less on function values themselves than on the various ways in which those values change and affect the functions's graph.

Some orthopedists claim that consistent running increases your foot size.

▷ Suppose a runner's foot length is 9 inches, and after jogging for a year, her foot length measures $9\frac{1}{6}$ inches. What is the corresponding *change* in her shoe size? (Note that we're asking not what size she wears, but by how much her size increases.)

▷ Undaunted, she buys a new pair of running shoes and starts training for the Boston Marathon. Her foot length increases to $9\frac{1}{3}$ inches. How much is the *additional change* in her shoe size?

▷ Her sister, who has feet 10 inches long, starts training with her, and the sister's feet increase in length by $\frac{1}{2}$ inch. What is the corresponding change in her shoe size?

▷ A supermarket tabloid reports on a long-distance runner whose foot length increased by a full inch in the course of her career. What is the resulting change in her shoe size?

STOP & THINK

Suppose the long-distance runner of tabloid fame were male, so that the function m applied instead of the function w. By how much would his shoe size have changed? Do you see a way to "read" that information directly from the formula for the function?

Constant Rates of Change

In our women's shoe-size function $y = w(x) = 3x - 21$, we called y the dependent variable because its value depended on the value of x. Now we want to investigate *how* y depends on x by examining the rate at which y changes as x changes.

Actually, this is the question you've just been considering. If you could answer the question about the long-distance runner of tabloid fame, you have discovered an important property about the rate of change of y. That question did not give the runner's foot length; no matter whether her foot grew from 9 to 10 inches or from $10\frac{1}{3}$ to $11\frac{1}{3}$ inches, she needed a shoe three sizes larger. For the function w, *a change of one unit in x always produces a change of three units in y.*

We call the ratio of the change in the y-value per unit change in the x-value **the rate of change of y with respect to x**. We write that ratio as a fraction:

$$\frac{\text{change in } y\text{-value}}{\text{unit change in } x\text{-value}}$$

For the first two runners, you should have decided that a change of $\frac{1}{6}$ in the x-value resulted in a change of $\frac{1}{2}$ in the y-value and that a change of $\frac{1}{2}$ in the x-value corresponded to a change of $\frac{3}{2}$ in the y-value.

▷   Simplify the corresponding ratios:

$$\frac{\text{change in } y\text{-value}}{\text{change in } x\text{-value}} = \frac{\frac{1}{2}}{\frac{1}{6}} =$$

$$\frac{\text{change in } y\text{-value}}{\text{change in } x\text{-value}} = \frac{\frac{3}{2}}{\frac{1}{2}} =$$

If "double-decker" fractions make you queasy, review them now in the Algebra Appendix, pages 475–477.

The two results should be equal.

The rate of change of a function is often of more interest than the actual function values. Here's how to compute the rate of change of any function $y = f(x)$ between two specific x-values, $x = a$ and $x = b$:

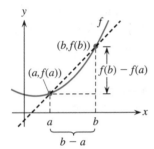

> The **average rate of change** of any function $y = f(x)$ from $x = a$ to $x = b$ is the ratio
>
> $$\frac{f(b) - f(a)}{b - a}$$

This notation is very important, so let's take a moment to sort it out.

- $f(b)$ means the y-value that the function f assigns to the input b.
  $f(a)$ means the y-value that f assigns to a.

- The order of terms in the numerator and denominator of the fraction is important! If b is the first x-value in the denominator, then $f(b)$ must be the first y-value in the numerator.

- a can be smaller than b or larger than b; *that* order doesn't matter.

When working with the shoe-size functions, you saw that the rate of change turned out to be $\frac{3}{1}$ every time. Let's use the formula for $w(x)$ to see why this happens. For any values of a and b,

$$\frac{f(b) - f(a)}{b - a} = \frac{(3b - 21) - (3a - 21)}{b - a} = \frac{3b - 3a}{b - a} = \frac{3(b - a)}{b - a} = 3$$

▷   Now you try it. Using the men's function $m(x) = 3x - 22.5$, show that the rate of change will always be 3, independent of the values of b and a.

Figure 1.8 Graph of $y = 3x - 21$

▷   You can "see" the rate of change in the graph in Figure 1.8. Pick any point (x, y) on the graph and write its coordinates. Now find the point whose x-coordinate is one unit greater. What is the new y-coordinate and how does it compare with the original y-coordinate?

▷   Start with a different point on the graph and repeat the experiment. Although the values of x and y will not be the same as before, the rate of change of y (with respect to x) is constant: three y-units for each x-unit.

Check Your Understanding 1.6

1. Determine two ordered pairs for the function $g(x) = -2x - \frac{3}{2}$.

3. Determine two ordered pairs for the function $h(x) = \frac{3}{4}x + 2$.

2. Use those two ordered pairs to find the average rate of change of g. Leave the answer as a ratio (a fraction) in lowest terms.

4. Find the average rate of change of h as a ratio in lowest terms.

Piecewise Constant Rates of Change

The Ironman Triathlon is a strenuous athletic competition in which participants complete three segments: a 4-km (2.4-mile) swim, a 180-km (112-mile) bicycle race, and a 42-km (26.2-mile) run. A strong triathlete might complete the race in 10 hours, swimming for one hour, biking for five hours, and running for four hours. Table 1.4 summarizes the results for that particular participant.

Time Elapsed (hours)	Total Distance Traveled (km)
0	0
1	4
6	184
10	226

Table 1.4 Distance versus Time for a Triathlon

Speed is a *rate*—the rate at which position changes with respect to time.

Our triathlon model will make two simplifying assumptions: first, that the athlete maintains a constant speed during any given leg of the race and, second, that no time elapses between events.

▷ For each leg of the race, calculate the speed in kilometers per hour.

▷ Using Table 1.4, draw the graph of the triathlon function on the axes in Figure 1.9. First plot points for the four ordered pairs. Then, remembering that speed between any two points is constant, connect the points with straight-line segments.

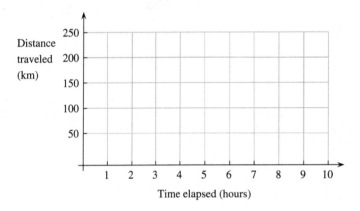

Figure 1.9 Distance versus Time for a Triathlon

Notice that each speed you calculated correlates directly with the steepness of the corresponding segment of the graph.

Just as the two shoe-size functions had a constant rate of change (three sizes per inch), the triathlon function has a constant rate for each segment: 4 km/hr for the swim, 36 km/hr for the bicycle course, and 10.5 km/hr for the marathon run.

STOP & THINK   If we allow the triathlete two minutes between events for changing clothes and grabbing some nourishment, how does the graph change?

Ever-Changing Rates of Change

The area of a square depends upon the length of its side: $A(s) = s^2$, where A is the area and s is the side length. What happens to the area if the side length increases by one unit? Well, that depends on how big the square already is. Growing from $s = 1$ to $s = 2$ increases A by three square units, while growing from $s = 3$ to $s = 4$ increases A by seven square units. As the side lengthens uniformly, the area grows at an increasing rate.

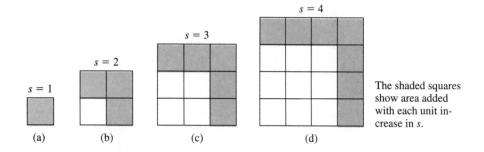

The shaded squares show area added with each unit increase in *s*.

To create the numerator, square the input values, because the function *A* is the squaring function.

We can show this mathematically:

$$\frac{A(2) - A(1)}{2 - 1} = \frac{2^2 - 1^2}{2 - 1} = \frac{4 - 1}{1} = 3,$$

while

$$\frac{A(4) - A(3)}{4 - 3} = \frac{4^2 - 3^2}{4 - 3} = \frac{16 - 9}{1} = 7$$

Because side length *s* is a continuous variable, we can compute the average rate of change of *A* for noninteger values of *s* as well.

▷   Find the average rate of change of *A* from $s = 2.3$ to $s = 3.6$.

The result should be 5.9. If you obtained something else, make sure that you divided by the quantity $(3.6 - 2.3)$.

The graph of $A(s)$ in Figure 1.10 shows why the average rates of change keep increasing as *s* becomes larger: for an increasing function, the steeper the graph, the greater the rate at which the function is changing.

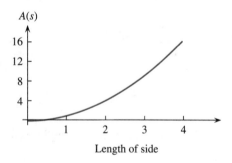

Figure 1.10   Graph of $A(s) = s^2$

One Step at a Time

A complication that we never considered in our shoe-size function is the fact that humans are not exactly symmetrical. For most people, one foot is slightly larger than the other, so their shoes are generally fit to the larger foot. You know that the use of one hand and arm dominate in all but ambidextrous people, but did you realize that one leg is usually dominant as well? Suppose that, because a particular man's right leg is dominant, his strides with that leg are, on average, longer than strides with his left. We will now investigate the effect that this stride difference could have on his path if he were wandering in a trackless waste where there were no visual clues to help him find his way.

Many (admittedly apocryphal) stories tell of a man wandering in a circle and returning to the place from which he started. One explanation could be a difference in the length of his right and left strides. In fact, if we know this difference, we can compute the radius of the resulting circle. Using mathematical language, we say the radius r is a function of the positive difference d between right and left stride lengths.

Dutton Children's Books.

Let's try to come up with a rough model of the function r. Suppose the length of a man's stride from left footprint to left footprint is 60 inches and that from right footprint to right footprint is $60 + d$ inches, as in Figure 1.11. Tightrope walkers place one foot directly in front of the other, but the rest of us have an average lateral distance between left and right steps. For our wanderer, we will assume this distance to be 8 inches.

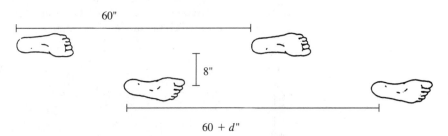

Figure 1.11   Footprints

▷ If the difference d results in the man walking in a circle, which foot would be on the inside of the circle?

Let n be the number of steps the man takes with his left foot to return to the place he started. (In mathematical shorthand, when a variable assumes integer values only, we often designate it by n rather than x.) To keep our model simple, we will assume that the number of strides with his right foot is also n.

Dutton Children's Books.

▷   Explain why the circle traced by his left foot has a circumference of $60n$ inches and why the circle traced by his right foot has a circumference of $(60 + d)n$ inches.

▷   Make a sketch here showing the two circles; they should have the same center. Label their circumferences $60n$ and $(60 + d)n$. Let r be the radius, measured in inches, of the circle traced out by his left footprints. Then the radius of the outer circle, traced out by his right footprints, is $r + 8$ inches. Label the two radii.

The ratio of C to r is constant.

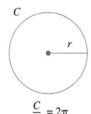

$$\frac{C}{r} = 2\pi$$

A basic fact about circles is that they all have exactly the same shape: *the ratio of the circumference to the radius of any circle is 2π.* (We examine the irrational number π in Project 1.3.) The inner circle has circumference $60n$ and radius r; the outer circle has circumference $(60 + d)n$ and radius $r + 8$. Thus,

$$\frac{60n}{r} = 2\pi \quad \text{and} \quad \frac{(60 + d)n}{r + 8} = 2\pi$$

Because both ratios are equal to the same quantity, 2π, they are equal to each other:

$$\frac{60n}{r} = \frac{(60 + d)n}{r + 8}$$

In Exercise 32, you will be asked to arrive at this formula by supplying the missing algebra. If you're rusty, this is a good time to review equations involving fractions, page 487 in the Algebra Appendix.

From this equation and a little algebra, we obtain the algebraic formula for our function:

$$r = \frac{480}{d},$$

where r and d are measured in inches.

STOP & THINK

What's the connection between this cartoon and the example you've just been reading?

THE FAR SIDE® BY GARY LARSON

"I've got it, too, Omar ... a strange feeling like
we've just been going in circles."

▷   Let's compute some ordered pairs (d,r) for our function by filling in Table 1.5. Notice
    that here we are using d to represent the independent variable, difference in stride lengths,
    and r to represent the dependent variable, radius. Since the r-values in our table are
    so large in comparison to the d-values, drawing a good graph requires a little thought
    about the scale to use on each axis. We have provided useful scales in Figure 1.12; in
    the future, you will have to plan ahead before writing numbers on the axes for a graph.

d	$\frac{1}{16}$	$\frac{1}{8}$	$\frac{3}{16}$	$\frac{1}{4}$	$\frac{3}{8}$	$\frac{1}{2}$	1	$\frac{5}{4}$
r	7680	3840						

Table 1.5   Values for the function $r = \frac{480}{d}$

Using Table 1.5, we can calculate the average rate of change of r from, for instance,
$d = \frac{1}{16}$ to $d = \frac{1}{8}$ as follows:

$$\frac{3840 - 7680}{\frac{1}{8} - \frac{1}{16}} = \frac{-3840}{\frac{1}{16}} = -61440$$

▷   It's your turn. Using your entries in Table 1.5, calculate two more average rate of changes
    of r: from $d = \frac{1}{4}$ to $d = \frac{1}{2}$ and from $d = 1$ to $d = \frac{5}{4}$.

On the number line, larger numbers are farther to the right. So, for example, $-3 > -5$, because -3 lies to the right of -5.

▷ Are these rates positive or negative? Which one is largest? (Careful!)

▷ Now plot the points (d, r) corresponding to these ordered pairs on the axes in Figure 1.12. The horizontal axis is labeled d, since d is the *independent* variable.

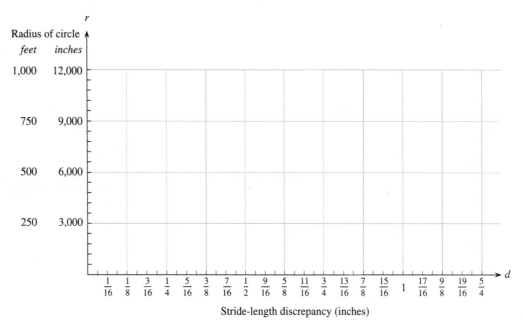

Figure 1.12 Graph of $r = \frac{480}{d}$

▷ Draw a continuous curve through the points you plotted in the preceding graph. Since the variables d and r are *continuous*, the intermediate points on the graph do have meaning. For example, the point $(\frac{5}{16}, 1536)$ lies on the graph and represents the fact that a stride difference of $\frac{5}{16}$ of an inch corresponds to a circle of radius 1536 inches.

Notice that your graph *curves* as it passes through the plotted points; a straight-line graph would not do the trick here. Do you see how the average rates of change that you just calculated gave an indication that the graph would not be straight?

The extra vertical scale helps to give meaning to the graph. Most of us have an internalized scale allowing us to picture 300 feet (100 yards) as the length of a football field, but we would have a difficult time creating a mental image of 3600 inches.

Read the graph from left to right, just as you read this sentence. As the values of the independent variable d increase from $\frac{1}{16}$ to $\frac{1}{8}$ to $\frac{3}{16}$, and so on, the corresponding dependent r-values—7680, 3840, 2560—get smaller, so the function is decreasing.

The graph in Figure 1.12 shows us that the function $r = \frac{480}{d}$ is **decreasing:** As the independent variable d increases, the r-values of points on the graph get smaller. In contrast, check the graph in Figure 1.5 on page 11. As we read the graph from left to right, the y-values of points on the graph get larger; we say that the function in Figure 1.5 is **increasing.**

Check Your Understanding 1.7

1. Would you characterize the triathlon function graphed in Figure 1.9, p. 20, as increasing or as decreasing? Write a sentence telling why.

2. What does the graph you drew in Figure 1.12 tell you about the value of r if the value of d is very close to 0 ($d \approx 0$)? (This would mean virtually no difference in stride length between the person's right and left legs.)

3. How can you find the same information by interpreting the algebraic expression $r = \frac{480}{d}$?

4. If $d \approx 0$, what does this mean about the wanderer's path?

5. Considering the function $r = \frac{480}{d}$ as a model of a physical situation in which d represents the difference in left and right stride lengths, decide on a reasonable domain for d.

6. What would be the corresponding range of values of the dependent variable r?

7. The equation $r = \frac{480}{d}$ (on your grapher, $y = 480/x$ or $f(x) = 480/x$) defines an abstract function with a much larger domain. What is the domain of this abstract function?

8. Experimenting with several different viewing windows, use your grapher to determine the range of this abstract function.

The Shape of a Graph

An increasing function can increase in three ways, as Figure 1.13 shows.

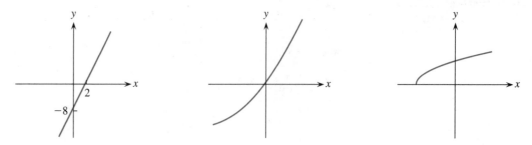

Figure 1.13   Three Increasing Functions

We can classify these functions by their type of curvature. The first graph holds no surprises: As x grows by one unit, y always grows by the same amount. (What is it?) The second graph curves upward as it rises: Each one-unit increase in x results in a successively larger increase in y as we move in the positive x direction. The third graph curves downward as it rises: Each one-unit increase in x results in a successively smaller increase in y as we move in the positive x direction. To get a quantitative feel for this, let's calculate some average rates of change.

▷ What is the constant rate of change in the first graph?

Average rate of change of
g: $\frac{g(b)-g(a)}{b-a}$

▷ The middle graph in Figure 1.13 represents a portion of the function $g(x) = x^2 + 5x$. Calculate the average rate of change of g from $x = -2.5$ to $x = -2$, from $x = 0$ to $x = 2$, and from $x = 2$ to $x = 2.1$.

▷ The third graph in Figure 1.13 represents the function $h(x) = \sqrt{x + 3}$. Calculate the average rate of change of h from $x = -2.5$ to $x = -2$, from $x = 0$ to $x = 2$, and from $x = 2$ to $x = 2.1$.

The function g increases at an increasing rate; the function h increases at a decreasing rate.

Now examine your results. The rates for the function g form an increasing sequence, translated graphically as a curve that bends continually upward (the graph gets steeper as it rises). We designate this graph as **concave up.** The rates for the function h form a decreasing sequence. The curve bends downward, becoming less steep as it rises, and we say that it is **concave down.**

The same principles apply to decreasing functions. A function can decrease in three different ways, as illustrated in Figure 1.14.

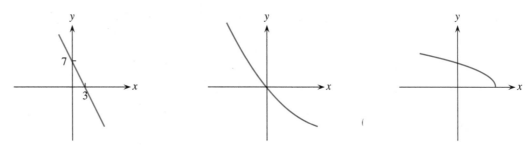

Figure 1.14 Three Decreasing Functions

▷   Can you "read" the constant rate of change for the first function in Figure 1.14? (It will be a negative number.)

▷   The middle graph in Figure 1.14 represents a portion of the function $u(x) = x^2 - 5x$. Calculate the average rate of change of u from $x = -2.5$ to $x = -2$, from $x = 0$ to $x = 2$, and from $x = 2$ to $x = 2.1$.

▷   The third graph in Figure 1.14 represents the function $v(x) = \sqrt{3 - x}$. Calculate the average rate of change of v from $x = -2.5$ to $x = -2$, from $x = 0$ to $x = 2$, and from $x = 2$ to $x = 2.1$.

The function u decreases at an increasing rate; the function v decreases at a decreasing rate.

Examine your results. You should see an increasing sequence in the rates for u (negative values, increasing toward 0), which translates graphically as a curve that curls upward, even as it heads downward. This graph, like the middle one in Figure 1.13, is *concave up*. In contrast, the rates for the function v form a decreasing sequence of negative values (negative numbers, moving away from 0). The fact that the rates decrease as x grows leads us to call this graph, like the third one in Figure 1.13, *concave down*.

Finally, a function that neither increases nor decreases is said to be **constant.** The graph of a constant function is simply a horizontal line.

Nearly every graph you will meet in this course will belong to one of the types just described or will be a combination of those types.

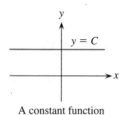

A constant function

Check Your Understanding 1.8

For each graph, give the interval(s) on which the function increases, decreases, is concave up, is concave down.

1.

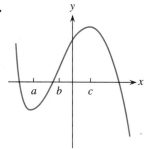

2.

3.

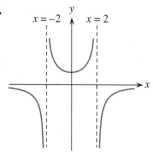

4. Calculate the average rate of change of $f(x) = \sqrt[3]{x}$ from $x = 2$ to $x = 2.1$.

5. Does the order matter? Calculate the average rate of change of $f(x) = \sqrt[3]{x}$ from $x = 2.1$ to $x = 2$. That is, put $f(2)$ first in the numerator and put 2 first in the denominator.

6. Calculate the average rate of change of $C(x) = 19$ from $x = -1$ to $x = 4$.

In a Nutshell   The average rate of change of a function f from a to b is the quotient $\frac{f(b)-f(a)}{b-a}$. If function values increase as the independent variable increases, the rates are positive and we say that the function is increasing. If function values decrease as the independent variable increases, the rates are negative and we say that the function is decreasing. If the rates of change increase as the independent variable increases, the graph of the function is concave up. If the rates of change decrease as the independent variable increases, the graph of the function is concave down.

What's the Big Idea?

- The definition of a function as a specific type of relationship between two sets: the domain of the function and the range of the function

- The use of functions to model real-world situations

- The use of the average rate of change of a function to describe the way in which a function is changing

Progress Check

After finishing this chapter, you should be able to do the following:[2]

- Explain what a variable is and make the distinction between independent and dependent variables. (1.1)

- Define a function by a table, a graph, a rule, or a set of ordered pairs. The rule is often given by an algebraic formula. (1.1)

- Use function notation, such as writing an expression for $g(t + 2)$ if g is defined by the formula $g(x) = x^2 - 2x + 5$. (1.2)

- Read information such as domain, range, and function values from a graph. (1.2)

- Distinguish between the domain and range of an abstract function determined only by the laws of algebra and the

domain and range of a mathematical model determined by what's reasonable in a given physical situation. (1.2)

- Distinguish between discrete and continuous variables. (1.2)

- Determine whether or not a given relationship is a function. (1.2)

- For a function with a constant rate of change, figure out that rate from its graph, table, or formula. (1.3)

- Compute the average rate of change of any function between two points on its graph. (1.3)

- Describe the curvature of the graph of a function. (1.3)

- Label axes and provide scales for a graph if the information is available, and draw a reasonably accurate graph of a relationship based on the information supplied. (1.3)

Key Algebra Skills Used in Chapter 1

- Using exponent rules (pages 477–479)

- Multiplying and simplifying polynomial expressions (pages 469–470)

- Working with fractions, including "double-decker" fractions (pages 473–477)

- Solving equations that contain fractions (pages 484–488)

- Writing intervals (pages 491–492)

Answers to Check Your Understanding

1.1

1. (a) $m(10\frac{1}{2}) = 9$, $w(10) = 9$, $w(9\frac{1}{3}) = 7$
   (b) 9
   (c) 7.5
   (d) $10\frac{1}{3}$
   (e) $10\frac{5}{6}$
2. (a) The function f of $\square$ is defined as π times the square of the value of $\square$.
   (b) $f(3) = 9\pi$
   (c) $A = f(r) = \pi r^2$
   (d) $y = f(x) = \pi x^2$
3. (a) $Z(0) = -10$, $Z(-3) = -20$
   (b) $-4, -2, -1, 2$

1.2

1. the set of all measurements from $9\frac{1}{2}$ inches to $11\frac{5}{6}$ inches, inclusive
2. the set of all real numbers, $\mathbb{R}$
3. the set of dates $\{$April 5, April 6, April 7, . . . , April 24$\}$

4. the entire time span from the start of April 5 through the end of April 24
5. the set of all nonzero real numbers, $x \neq 0$
6. the set of all nonnegative real numbers, $x \geq 0$

1.3

1. the set of all nonpositive real numbers (that is, the negative numbers and zero), $y \leq 0$
2. the same as 1 (same function, written differently)
3. the set of all nonzero real numbers, $y \neq 0$
4. the set $\{7, 8\frac{1}{2}, 10, 10\frac{1}{2}, 11\frac{1}{2}, 12, 13\}$, a set of discrete values
5. the set $\{$106.8, 107.1, 107.9, 109.0, 109.8, 109.9, 110.5, 110.6, 111.0, 111.1, 111.2, 111.4, 112.2, 112.7, 113.0, 113.7, 114.2, 114.4$\}$, a discrete set (You may write the elements of a set in any order.)
6. the continuous set of values ranging from approximately 106.8 to approximately 114.5, inclusive

---

[2]The section references given here point you to a good example or to the section in which the topic was introduced. They don't necessarily include all the pages on which the topic was discussed.

1.4

1. $F(x) = x^2 + 2$
2. $F(b) = b^2 + 2$
3. $F(x - 3) = (x - 3)^2 + 2 = x^2 - 6x + 11$
4. $F(\frac{5}{x}) = (\frac{5}{x})^2 + 2 = \frac{25}{x^2} + 2 = \frac{25 + 2x^2}{x^2}$
5. $F(n^3) = n^6 + 2$
6. $F(\sqrt{3} + 1) = (\sqrt{3} + 1)^2 + 2 = 6 + 2\sqrt{3}$

1.5

1. yes; each x-value is assigned one y-value
2. no; the x-value -3 is assigned two different y-values
3. no; some x-values have two or three y-values
4. yes; each x-value corresponds to one y-value
5. yes; each x-value determines one y-value
6. no; each permissible x-value (except ± 4) determines two distinct y-values
7. yes; the length of one side determines the volume
8. no; the height also plays a role in determining the area; triangles with the same base can have different heights and, consequently, different areas

1.6

1. Possibilities are $(-5, \frac{17}{2})$ and $(-\frac{3}{4}, 0)$. (There are infinitely many correct answers.)
2. $-\frac{2}{1}$, or simply -2
3. Possibilities are $(0, 2)$ and $(-2, \frac{1}{2})$.
4. $\frac{3}{4}$

1.7

1. The function is increasing because, as the independent variable (time) increases, so does the distance.
2. The value of r is enormous.

3. Dividing 480 by a very small positive number (such as 0.001) gives a very large number (in this case, 480,000 inches).
4. The wanderer's path is approximately straight—a circle of infinite radius.
5. A reasonable domain is the set of measurements greater than zero and not more than 2 inches, $0 < d \le 2$. You might choose a different upper limit.
6. The corresponding range is the set of radii greater than or equal to 240 inches, $r \ge 240$. A different domain would have a correspondingly different range.
7. The domain is the set of nonzero numbers, $d \ne 0$, because division by zero is not defined mathematically, but division by any other number is permissible.
8. The range, also, is the set of nonzero numbers, $r \ne 0$, but for an entirely different reason: the expression $\frac{480}{d}$ can never equal zero, regardless of the value of d.

1.8

1. increasing for $a < x < c$; decreasing for $x < a$ and for $x > c$; concave up for $x < b$; concave down for $x > b$
2. decreasing and concave down for all x
3. decreasing for $x < 0$, $x \ne -2$; increasing for $x > 0$, $x \ne 2$; concave up for $-2 < x < 2$; concave down for $x < -2$ and for $x > 2$
4. $\frac{\sqrt[3]{2.1} - \sqrt[3]{2}}{2.1 - 2} \approx 0.2066$
5. $\frac{\sqrt[3]{2} - \sqrt[3]{2.1}}{2 - 2.1} \approx 0.2066$ (The order is immaterial.)
6. $\frac{C(4) - C(-1)}{4 - (-1)} = \frac{19 - 19}{5} = 0$ (C is a constant function; its value is always 19.)

EXERCISES

Answers to odd-numbered exercises begin on page 517.

1. If a man and woman wear the same number shoe, who has the larger foot? How did you decide?

2. Most people know their shoe size, not their foot length.
   (a) Use the shoe-size function w or m to find the approximate length of your foot, based on the size you wear.
   (b) What is the approximate length of the foot of someone who wears a man's size 11?

3. Use the function $w(x) = 3x - 21$ to determine for what length foot a woman's shoe size and foot length are equal to the same number. Show how you found your answer.

4. Some women like to wear men's running shoes. You should be able to answer the following questions without finding any foot lengths. Note the constant difference between the two functions.
   (a) If a woman wears a size $8\frac{1}{2}$ woman's shoe, what size would she wear in a man's shoe?
   (b) If she wore a woman's size 10, what size would she wear in a man's shoe?

In Exercises 5 through 8, use guessing and pattern recognition to find a mathematical formula that could represent the relationship between x and y. Assume that x is the independent variable. (Some calculators have a stat-plot feature, which could help you to visualize the relationship.)

5.

x	y
4	17
-3	10
0	1
7	50
-7	50

6.

x	y
4	1
-3	-6
0	-3
7	4
-7	-10

7.

x	y
3	2
−3	0
0	1
−9	−2
9	4

8.

x	y
2	9
−3	−26
0	1
1	2
−1	0

9. In a newspaper or magazine, find an example of a function. Cut out the relevant article. Tell what the variables are and which one is the independent variable. Give the definition of the function, and specify the way in which it is defined (whether by a verbal statement, a table, a graph, or a mathematical formula, or by some other means).

10. The table below gives the average rainfall in Amherst, Massachusetts, as tabulated at Amherst College for each of the 12 months in 1992. Number the months 1 through 12 and draw a set of coordinate axes, labeling them carefully. Which variable should play the role of the independent variable? Justify your answer. Now plot the points corresponding to the ordered pairs from the table. Note that both variables are discrete; don't connect the points, because the in-between values have no significance. Can you find a formula for the rainfall function? Do you think one exists?

Month	Rainfall (inches)	Month	Rainfall (inches)
January	2.33	July	4.37
February	2.03	August	5.58
March	3.81	September	2.55
April	3.14	October	2.05
May	3.60	November	4.77
June	2.52	December	3.97

11. A function *Flip* takes the reciprocal of the input value.
   (a) Find $Flip(3)$, $Flip(\frac{2}{3})$, $Flip(1000)$, $Flip(-5)$, $Flip(0.01)$, $Flip(n+2)$.
   (b) Write the algebraic formula for $Flip(x)$.
   (c) For what value of x is $Flip(x)$ undefined?

12. A function SQ subtracts two from three times the square of a number.
   (a) Find $SQ(4)$, $SQ(0)$, $SQ(-5)$, $SQ(n+2)$.
   (b) Write the algebraic formula for $SQ(x)$.
   (c) Find the possible values of x if $SQ(x) = 298$.

13. $G(r) = 4r - r^2$
   (a) Find $G(\frac{1}{2})$, $G(-1)$, $G(3)$, $G(q^2)$.
   (b) Find the values of r for which $G(r)$ equals 0.
   (c) Find the values of x for which $G(x)$ equals 0.

14. $H(t) = \frac{2}{t-3}$
   (a) Find $H(1)$, $H(0)$, $H(w^2)$.

(b) Find t such that $H(t) = 1$. (Note that this is different from finding $H(1)$.)
(c) For what value of t is $H(t)$ undefined?

15. The function $Quad$ is defined by the formula $Quad(\heartsuit) = 100 + 21\heartsuit - \heartsuit^2$. Find each of the following, simplifying where possible.
   (a) $Quad(-5)$
   (b) $Quad(-x)$
   (c) $Quad(z+5)$
   (d) $Quad(z)+5$
   (e) $1 - Quad(t)$
   (f) $Quad(1-t)$
   (g) $Quad(\frac{1}{x})$
   (h) $\frac{1}{Quad(x)}$
   (i) the values of $\heartsuit$ for which $Quad(\heartsuit)$ equals zero.

16. Let G be the function defined by the rule $y = G(\diamondsuit) = 2\diamondsuit - 3\diamondsuit^2$. Find each of the following, simplifying where possible.
   (a) $G(10)$
   (b) $G(r+1)$
   (c) $G(r)+1$
   (d) $G(-t)$
   (e) $-G(t)$
   (f) $G(1 - \sqrt{x})$
   (g) the values of $\diamondsuit$ for which $G(\diamondsuit)$ equals zero.

17. In your own words, explain the difference between $y = 3x - 21$, an abstract function, and $y = 3x - 21$, a mathematical model for women's shoe sizes. Draw a graph of the line $y = 3x - 21$. Using a contrasting color, highlight the portion that represents the model.

18. For each of the following relationships, identify the variables and explain why the relationship could or could not be considered a function. If it could be a function, what is the independent variable? If the other variable were considered the independent variable, would the relationship still be a function? (Reasonable people might disagree about these; your answers will depend on how you interpret the examples, so your explanation is important.)
   (a) The telephone directory lists names and phone numbers.
   (b) The daily paper prints a horoscope for its readers based on date of birth.
   (c) The price of a gallon of gasoline changes with time.
   (d) The strength of an earthquake affects the dollar amount of damage caused.
   (e) The amount of time you can safely remain in the sun is related to the SPF number of your sunscreen.

19. The function J is completely defined by the graph that follows. Use the graph to determine the following:
   (a) $J(0)$
   (b) the value(s) of t for which $J(t) = 0$
   (c) the value(s) of t for which $J(t) < 0$

(d) the domain of J

(e) the range of J (Be careful to include in the range only those y-values that actually occur.)

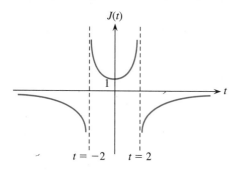

J(t)

20. The function Z is completely defined by the graph below. Use the graph to determine the following:

(a) $Z(0)$

(b) the value(s) of u for which $Z(u) = 0$

(c) the value(s) of u for which $Z(u) > 0$

(d) the domain of Z

(e) the range of Z

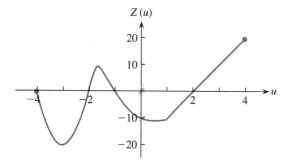

Z(u)

21. The graph of the equation $x^2 + y^2 = 9$ is a circle of radius 3, centered at the origin. (See below.)

(a) Explain why $x^2 + y^2 = 9$ isn't a function.

(b) Find two functions which together will produce the same graph.

(c) Try out the two functions on your grapher. If you get a football-shaped picture instead of a circle, switch to "square" scaling.

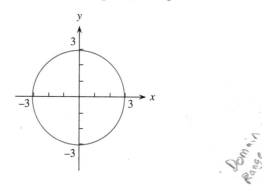

22. Write algebraic formulas for two functions which, graphed together, will produce the graph in Figure 1.7 on page 15.

23. Maybe, as a kid, you played the game of "Buzz." Players sit in a circle and count off in the usual fashion, 1, 2, 3, ..., except that if a number is a multiple of 7 or contains the digit 7, the player whose turn it is must say "buzz" instead of the actual number. Let $Buzz(n)$ be what the player should say on the nth turn. Thus, $Buzz(10) = 10$, $Buzz(21) =$ buzz, and $Buzz(97) =$ buzz.

(a) Find $Buzz(301)$, $Buzz(199)$, $Buzz(71)$.

(b) Find four other values of n for which $Buzz(n) =$ buzz.

(c) Give the domain of $Buzz$. Is it continuous or discrete?

(d) Give the range of $Buzz$. Is it continuous or discrete?

24. If a positive integer n is divided by 5, a quotient and a remainder result. For example, dividing 17 by 5 gives a quotient of 3 and a remainder of 2, and dividing 4 by 5 gives a quotient of 0 and a remainder of 4. Let REM be the function that gives the *remainder* when n is divided by 5. Thus, $REM(17) = 2$ and $REM(4) = 4$.

(a) Find $REM(59)$, $REM(3)$, $REM(110)$.

(b) Find four values of n for which $REM(n) = 3$.

(c) Give the domain of REM. Is it continuous or discrete?

(d) Give the range of REM. Is it continuous or discrete?

25. Using Figure 1.4 on page 7, find the average rate of change of Z from $u = -3$ to $u = 4$.

26. Using Figure 1.4 on page 7, find the average rate of change of Z from $u = -4$ to $u = 1$.

27. For the function $S(t) = 5t - 2t^2$, calculate the average rate of change from $t = -2$ to $t = 0$. Then calculate the average rate of change from $t = 0$ to $t = -2$, and show that your answers are the same.

28. Use algebra to show that, for *any* function f, the expression $\frac{f(b)-f(a)}{b-a}$ is equivalent to $\frac{f(a)-f(b)}{a-b}$, proving that the order in which we write the values of the independent variable is immaterial. (*Hint:* Multiply top and bottom of the first fraction by -1.)

29. The function f is defined by this graph. Give the interval(s) on which f is

(a) increasing

(b) decreasing

(c) concave up

(d) concave down

(The Algebra Appendix, pages 491–492, will help you write an interval correctly.)

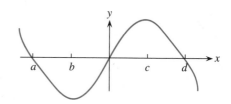

30. The function p is defined by the graph below. Give the interval(s) on which p is
   (a) increasing
   (b) decreasing
   (c) concave up
   (d) concave down

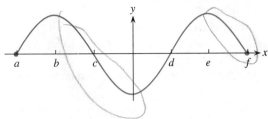

31. Sketch the graphs of $y = x^2$, $y = x$, and $y = \sqrt{x}$ together on a single set of axes, using the interval $0 \le x \le 1$. All three graphs have the same endpoints and the same average rate of change on that interval, yet their graphs are not identical. Compute the average rate of change from $x = 0$ to $x = 0.25$, from $x = 0.25$ to $x = 0.75$, and from $x = 0.75$ to $x = 1$ for all three functions; now you will see numerical differences. Tell how the rates you just calculated help to explain the shape of each graph.

32. Provide the missing algebra to show how, on page 23, we got from

$$\frac{60n}{r} = \frac{(60 + d)n}{r + 8} \qquad \text{to} \qquad r = \frac{480}{d}$$

Interpret the following graphs by writing a scenario for each. Think of a context; identify the variables; tell how the input variable produces an output with the pattern shown. (Use your imagination. If you feel stuck, get some ideas from the answers.)

33.

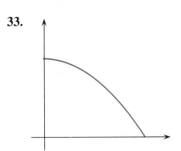

34.

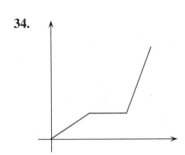

35.

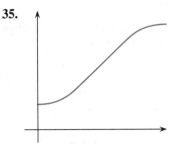

36.

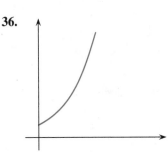

For Exercises 37–46, sketch a reasonable graph of each relationship. Choose the variables (independent, dependent), the direction of the graph (increasing, decreasing), and its general shape (concave up, concave down). Label the axes with whatever information you have. Write a sentence or two about each graph to explain what it tells about the relationship.

37. The time it takes you to travel from hither to yon is related to your driving speed.

38. The distance you can bike in one hour depends on the speed at which you ride.

39. The American Automobile Association estimates that the number of incidents of "road rage" (aggressive driving) is increasing at seven percent per year. (This means that the annual number increases, not by a fixed amount, but by a fixed percentage of the previous number. That is, each annual increase is larger than the one before.)

40. The organizers of a triathlon decide to change the order of events, having participants bike first, then swim, then run.

41. Over the years, the world record for both the men's marathon and the women's marathon has been dropping, but the women's record has been falling more rapidly than the men's. (Women's participation in the marathon has been officially sanctioned only since the late 1960s. Women had been thought too delicate to run such a long distance.) Although the winning man still finishes ahead of the winning woman, recent patterns suggest that women could catch up in a few years. (You'll need two graphs, one for the women and one for the men. Draw them on the same set of axes.)

42. The national debt is growing every year but not as rapidly as it used to grow.

43. The number of daylight hours on June 21 depends on latitude. Locations near the equator have a "longest day" of just over 12 hours (not much different from their "shortest day"). Places farther north enjoy more hours of daylight on June 21. (They pay for it, though, in the winter.) At all latitudes within the Arctic Circle, the sun does not set at all on June 21. Below the equator, the hours of daylight on June 21 decrease as you go south. At locations within the Antarctic Circle, the sun remains entirely below the horizon on that date. Draw a graph that shows the relationship between hours of daylight on June 21 and latitude, for all latitudes.

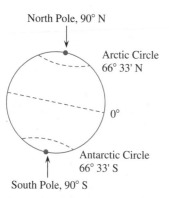

44. *The Last Laugh.* You've just heard a great joke and are eager to pass it on. At first, you're able to tell it to many people. As time passes, more and more of your acquaintances have already heard the joke. After a week or so, you can't find an audience for your joke, because everybody knows it. (Besides, you're pretty sick of it yourself.) Let *time* be one of the variables.

45. A factory worker earns a fixed hourly wage for all hours worked in a week up to 40 hours. For hours worked beyond 40, she earns time-and-a-half, that is, one-and-one-half times her regular hourly wage.

46. *Full of Hot Air.* When a balloon contains a small amount of air, a good puff of breath will cause its diameter to grow by a significant amount. When the balloon already contains a lot of air, however, adding the same amount of breath makes only a small difference in the diameter.

1.1 THE GRAPHING GAME—Exploring Linear Functions

The purpose of this activity is to help you familiarize yourself with the computer or calculator grapher that you'll be using during this course. If you're already at ease with your grapher, you might investigate some more complicated functions as well as the ones suggested here.

Adjust your viewing window so that the x- and y-axes have the same scale—so that the distance between 0 and 1 on the x-axis is the same as the distance between 0 and 1 on the y-axis. On many graphers, this adjustment is called **square scaling.**

1. Overlay the graphs of the functions $f(x) = x$, $g(x) = x + 1$, and $h(x) = x + 2$. Describe in words the pattern that you see. Sketch the graphs of these three functions on a single set of axes.

2. Overlay the graphs of the functions $f(x) = x$, $g(x) = 2x$, and $h(x) = 4x$. Sketch the graphs of these three functions on a single set of axes. Describe in words the pattern that you see. Which of the graphs rises most steeply? Which rises least steeply?

You should notice that the number multiplying the independent variable x affects the steepness of the graphs. In mathematical lingo, we call that multiplier the **slope** of the line.

3. Overlay the graphs of the functions $f(x) = x + 1$, $g(x) = 2x + 1$, and $h(x) = 4x + 1$. In what way is this set of graphs like the previous set? In what way is it different?

4. Overlay the graphs of $f(x) = x$ and $g(x) = -x$. Sketch what you see. Do these lines appear to be perpendicular? (Perpendicular lines meet at a 90° angle.)

5. Graph $f(x) = -2x$. If we read the graph from left to right, this graph slants downward; it has a negative slope. Would a line perpendicular to the graph of $f(x)$ have a negative slope or a positive slope?

6. Overlay the graph of $g(x) = 2x$. (Be sure that the x- and y-axes still have the same scales.) It should not appear perpendicular to $f(x)$. Would a line perpendicular to the graph of $f(x)$ be steeper or less steep than the line $y = 2x$?

7. Experiment with your grapher (a trial-and-error approach is fine) and find a line that appears to be perpendicular to the graph of $f(x) = -2x$. Write down the equation of this line.

8. Clear the screen and graph $f(x) = \frac{1}{4}x$ (or $0.25x$). Find a line that seems to be perpendicular to the graph of $f(x)$. Do you see a relationship between the slopes of these two lines? Is it the same relationship that existed between the two slopes in Question 7?

1.2 HOME ON THE RANGE—Domains and Ranges

Although we can use algebra to determine the domain and the range of many functions, examining the graph of a function can give us quick insight into its behavior and its limitations. Use a grapher to help you determine the domain and the range of each function, considered solely as an abstract mathematical relationship. Then examine the formula for the function to see whether your answers make mathematical sense.

1. $Q(t) = 10 - t^2$
2. $R(t) = 10 - 5t^2$
3. $y = 10 - 2x$

4. $T(x) = 5\sqrt{x}$

5. $U(x) = \sqrt{5x}$

6. $V(x) = \sqrt{5 + x}$

7. $W(x) = 5 + \sqrt{x}$

8. $f(t) = \frac{1}{t-2}$ (Remember to put parentheses around the denominator.)

9. $g(t) = \frac{1}{t} - 2$

1.3 A PIECE OF PIE—Rational and Irrational Numbers

In this project, you will see the difference between rational and irrational numbers by investigating their decimal expansions.

1. In Section 1.3, we noted that for a circle of radius r, the ratio of the circumference C to $2r$ is the irrational number π; that is, $\frac{C}{2r} = \pi$. You have probably used various approximations for π, such as $\frac{22}{7}$ and 3.1416.

  (a) Write the approximation given by your calculator. Use all of the decimal places your calculator gives.

  (b) You should see that the digit 6 is no longer in the fourth decimal place. What digit does your calculator give for the fourth decimal place? For the fifth place?

  (c) Write the ratio $\frac{22}{7}$ as a decimal by dividing 7 into 22.

  (d) You now have three different decimal approximations for the number π. All three should agree in the first decimal place. What digit is in the first decimal place (after the decimal point) in all three decimal approximations?

  (e) What is the first decimal place in which they differ?

  (f) Now, by hand, compute the decimal expansion of $\frac{22}{7}$ to 14 decimal places. This is a little tedious, but your calculator won't give an accurate enough approximation to answer this question: What is the pattern in the digits in the decimal expansion of $\frac{22}{7}$?

  (g) Use the pattern to figure out what digit is in the fifteenth decimal place without doing the division.

  (h) Without doing any more division, give the digit that you expect in the eighteenth decimal place; the twenty-fourth place; the eighty-fourth place.

When we call π an **irrational** number, we mean that *it cannot be written as a ratio of two integers.* You just saw that $\frac{22}{7}$ is a rough approximation to π, accurate to only two decimal places. In fact, you found a block of repeating digits in the decimal expansion of $\frac{22}{7}$. One of the characterizations of irrational numbers is that their decimal expansion has no pattern of repeating digits. The digits in the decimal expansion of π show no pattern whatsoever. Here are the first 40 places:

$$\pi = 3.1415926535897932384626433832795028841971\ldots$$

What digit is in the twentieth place?

2. Let's use some mathematical notation to ask the same question in a slightly different way. You know that a function is a rule establishing a relationship between two sets. You have been associating a digit with each decimal place in the expansion of π. We can think of this association as a function, which we will call *PIE*, whose value at an integer n is the digit in the nth decimal place in the expansion of π. That is,

$$PIE(n) = \text{the digit that appears in the } n\text{th place}$$
$$\text{of the decimal expansion of } \pi$$

For this function, then, $PIE(1) = 1$, $PIE(2) = 4$, and $PIE(3) = 1$, so the question about the digit in the twentieth place can be rewritten: What is $PIE(20)$?

(a) Create a table like this with values of the function PIE:

Place	Digit
0	3
1	1
2	4
3	1
4	
5	
6	
7	
8	
9	
10	
11	
12	
13	
14	
15	

(Since the whole-number part of π's decimal expansion is 3, we've added the ordered pair (0,3) to the table.)

(b) The domain of the function PIE is an *infinite* set. State the domain of the function. Is it discrete or continuous?

(c) The range of PIE is a *finite* set. Give the range. Is it discrete or continuous?

(d) Because PIE is a function, you can plot the ordered pairs (n,m) from the table you just completed, where n is the decimal position and m is the digit found there. (When a variable takes on integer values only, we often designate it by n or m rather than x or t.) Plot the points (n,m) on a set of coordinate axes. Can you see any pattern in the points?

(e) Would it make sense to connect these points? Remember that a graph is a way of conveying information. What information would be conveyed by a graph whose points were connected? What information would be conveyed by a graph of isolated points?

3. To see the difference between rational and irrational numbers, write the decimal expansions of $\frac{2}{3}$, $\frac{3}{7}$, and $\frac{3}{8}$ to 14 places. (It is actually better *not* to use a calculator but to perform the division yourself and watch what happens.) Copying what we did in Question 2 with the number π, let

- $f(n)$ be the digit in the nth decimal place in the expansion of $\frac{2}{3}$

- $g(n)$ be the digit in the nth decimal place in the expansion of $\frac{3}{7}$

- $h(n)$ be the digit in the nth decimal place in the expansion of $\frac{3}{8}$

(a) The functions f, g, and h all have the same set as their domain. What is it? (Note that for these functions, we can start counting with the first decimal place, because none of the three fractions has a whole-number part. In other words, zero is not an element of the domain of f, g, or h.)

(b) All three functions have different sets as their range. Give the range of each. In the range, you should list the *distinct* digits that appear in the decimal expansion. If 0 shows up, don't forget it.

(c) Which of the three is a *constant function*, that is, has only one y-value in its range?

Some numbers, such as $6.250000\ldots$, $-12.345634563456\ldots$, and $5.23401010101\ldots$, have decimal expansions that eventually become a block of infinitely repeating digits. Such numbers correspond to ratios of integers and are called *rational* numbers.

In others, such as $3.141592\ldots$, $2.718281828459\ldots$, and $1.010110111011110\ldots$, there is no such infinitely repeating block. The second example, $2.718281828459\ldots$, starts to fool us by repeating the "1828" part but then changes its mind and goes on to other things. (Like π, this number is renowned in mathematics. You'll meet it again in Chapter 4.) The third number, $1.010110111011110\ldots$, does exhibit a pattern; you could use the pattern to predict the digits farther down the line. This pattern, however, does not have a single block of digits repeated infinitely.

Infinite decimals, such as the three examples given, that babble on without ever becoming a block of repeating digits do not correspond to ratios of integers and are called *irrational* numbers.

1.4　A BIG MOOSETAKE

The moose populations in Vermont, New Hampshire, and Maine increased sharply during the 1980s. Meeting a moose in the wilds can be dangerous, but meeting one head-on in a car can be fatal. Even though signs warning "Moose Crossing" have been posted, moose-car collisions on northern New England roads are posing an increasing hazard for motorists (and for the moose). The table below provides data on moose-car collisions that occurred from 1980 to 1990.[3]

	Vermont	New Hampshire	Maine
1980	0	unknown	156
1984	unknown[4]	unknown	215
1985	12	49	unknown
1988	unknown	117	unknown
1990	41	170	500

1. Consider first the data for Vermont. On a separate sheet of paper, plot the number of collisions versus the year. ("Number of collisions" will appear on the vertical axis; "Year" will appear on the horizontal axis.) How many *more* collisions were there in 1990 than there were in 1985? This number represents the increase in the number of collisions over that entire five-year period. Translate this number into the average rate of change in the number of collisions *per year* for that period.

2. Now calculate the average rate of change in the number of collisions per year for the period 1980–1985.

3. Compare the average rate of change (collisions per year) for 1980–1985 and for 1985–1990. Did the rate increase, decrease, or remain the same for the two consecutive five-year periods?

4. On your plot of data, sketch the line that passes through the points for 1980 and 1985. If this line had accurately represented the relationship between the number of collisions and the year, how many collisions would you have expected in 1990? What, then, would have been the average increase per year in numbers of collisions from 1985 to 1990?

5. Make another graph using the data for New Hampshire. Do you think the points lie on a line? Check by computing the average increase per year in numbers of collisions for 1985 to 1988 and then for 1988 to 1990. Are these rates constant?

---

[3] "Comeback for Moose in New England Leads to Road Hazard," *The New York Times*, June 3, 1991.
[4] The missing numbers are not available; adequate records might not have been kept for those years. Life is like that, and we have to deal with the information we've got.

6. Finally, plot the data for Maine. Draw a line that passes through the data points corresponding to the years 1980 and 1984. Draw a second line that passes through the data points corresponding to the years 1984 and 1990. Without doing any calculations, state what happens to the average rate of change in numbers of collisions. As time passes, does that rate increase, decrease, or remain constant? Explain how you can answer this question by simply looking at your graphs.

1.5 JUST ALGEBRA

These exercises provide practice in the algebra skills called for in Chapter 1.

1. Solve each equation for the indicated variable.
   - **(a)** $ax + by + c = 0$ for x
   - **(b)** $100 = 2\pi r^2 + 2\pi rh$ for h
   - **(c)** $S = 2\pi r^2 + 2\pi rh$ for h
   - **(d)** $ax^2 + bx + c = 0$ for a
   - **(e)** $2\pi r + 2r = 100$ for r
   - **(f)** $A = \dfrac{a+b}{2} \cdot h$ for b
   - **(g)** $\dfrac{6}{x+d} = \dfrac{15}{x}$ for x
   - **(h)** $\dfrac{1}{R} = \dfrac{1}{A} + \dfrac{1}{B}$ for R

2. Simplify if possible. Then evaluate each expression for $x = -3$.
   - **(a)** $3(1 - x)^2 - 2(1 - x) + 5$
   - **(b)** $[3(x + 2)^2 - 2(x + 2) + 5] - [3x^2 - 2x + 5]$
   - **(c)** $5x^4$
   - **(d)** $-5x^4$
   - **(e)** $(5x)^4$
   - **(f)** $(-5x)^4$

Fahrenheit

Some like it hot

Preparation

After counting, one of the most basic uses of numbers is measurement. English-speaking countries have inherited a particularly arbitrary system of measurement. One example is the use of degrees Fahrenheit to measure temperature. The Fahrenheit scale was established in the early eighteenth century by Daniel Gabriel Fahrenheit, a German physicist. Fahrenheit obtained a reading for the uppermost fixed point on his scale by placing a thermometer under the armpit of a healthy man; he obtained the lowest, 32, by using an ice and water mixture to determine the freezing point of water. The number 0 approximated the temperature of an ice and salt mixture (which was widely considered to be the coldest possible temperature), so all readings on a Fahrenheit thermometer were assumed to be positive. Fahrenheit arbitrarily assigned to the upper and lower fixed points the values 96 and 32 to eliminate "inconvenient and awkward fractions."

By contrast, the Celsius scale, like our number system, is based on the number 10. The fixed points, corresponding to the freezing and boiling temperatures of water, are labeled 0 and 100, respectively. In United States, the campaign to convert to metric has been waged for decades. One problem is that to use a system of measurement effectively, a person needs to internalize the scale. We know that a room temperature of 70 degrees Fahrenheit feels comfortable but that 60 degrees requires a sweater and 50 degrees requires a coat. But if the weather forecaster predicts a high of 17 degrees Celsius, many of us would need to convert this temperature into our internalized Fahrenheit system to anticipate the weather correctly.

Many banks and businesses provide time and temperature signs with temperature given in both Fahrenheit and Celsius degrees to help us internalize the Celsius scale. Suppose that, over the course of the semester, you collected the following data from an electronic sign on a bank in your town. (Note that the sign reports only integer values for the temperatures!)

Degrees Celsius	Degrees Fahrenheit
−20	−4
−4	25
0	32
13	55
15	59
18	64
26	79
32	90

One way to understand the relationship between the two scales is to draw a graph. On graph paper, draw a horizontal axis representing degrees Celsius and a vertical axis for degrees Fahrenheit, and plot the points corresponding to the numbers in the table. Do the points make a pattern? If so, describe it.

We call the data in the table *discrete:* Each value given by the sign is separated by at least one degree from any other. However, we know that temperature varies *continuously:* For the temperature to rise from 3 degrees to 4 degrees, it must pass through all intermediate real values (even $\sqrt{11}$ and π!). Thus, we are justified in approximating our points by a continuous curve. Do this; then use your graph to estimate what Celsius temperature corresponds to 0 degrees Fahrenheit.

Draw two more graphs of the same data as follows. First, change the scale of the horizontal axis by doubling the distance you used on the original graph to represent one unit. (Leave the vertical axis unchanged.) Second, halve the distance used to represent one unit on the original graph. How do your three graphs differ from one another? Which features remain unchanged?

Bring your graphs with you to your first lab. In Lab 1A, you will be investigating the relationship between temperature measured in degrees Celsius and temperature measured in degrees Fahrenheit. You will also use your grapher (that is, your graphing calculator or a computer program) to further explore the effects of altering the scale on one of the axes of your graph.

"LET'S GO OVER TO CELSIUS'S PLACE. I HEAR IT'S ONLY 36° OVER THERE."

The Fahrenheit Lab

In your preparation for this lab, you drew three graphs of the relationship between temperature measured in degrees Celsius and temperature measured in degrees Fahrenheit. Compare your graphs with those of your lab partners to make sure you agree on how to represent the data given in the table in the preparation section. Do you all agree on what changes and what remains the same when one of the scales is altered?

In the table, both degrees Celsius and degrees Fahrenheit vary, but the variation is not random. In fact, if we know the temperature in degrees Celsius, we can find the Fahrenheit temperature. Because the temperature measured in degrees Fahrenheit is uniquely determined once we know the Celsius temperature, we can say that degrees

Fahrenheit is a *function* of degrees Celsius. Let the letter C represent the number of degrees Celsius and let F represent the number of degrees Fahrenheit. We call F and C *variables* because their values change. We would like to write the relationship between F and C explicitly using algebra. For F to be a function of C, this relationship would take the form

$$F = \text{some algebraic expression involving } C$$

An equation written in this form will show the way in which the value of F depends on the value of C, so we call F the *dependent variable* and C the *independent variable*. We want to find an equation of this form to serve as a *model* for the data in our table. (The equation is an idealization of the data; the numbers in the table might not fit the equation exactly.) That is, we want to perform the same algebraic operation on each number in the first column and obtain (approximately at least) the corresponding number in the second column.

A very simple relationship between F and C would be one in which, given a value for C, we obtain the corresponding value for F by multiplying by a fixed constant. If k represents this constant, then the equation we're looking for would be of the form $F = k \cdot C$. If the relationship between F and C is of this form, we say that F is *proportional* to C.

Here's an everyday example: The amount of sales tax is proportional to the price of an item. If a state imposes a sales tax of 7% and the item costs $22.65, the amount of sales tax is 0.07 times $22.65, or $1.59. In general, if the item costs D dollars, the amount of the sales tax is $0.07D$.

Do you think F is proportional to C? Using values from the table, along with our mathematical definition of *proportional*, construct an argument to support your answer.

Now try to construct an equation that models the relationship between F and C. We are deliberately not giving you an exact recipe for doing this. Talk with each other about how you might use the values in the table to write an algebraic relationship between the two variables. A worthwhile investigation often involves making educated guesses, trying ideas that don't work, improving those ideas, and testing your conjectures. (*Hint:* The third row of the table is very helpful.)

When you agree on an equation, test it with all the C-values in the table. Are you satisfied with the F-values it produces? Remember that not all will exactly match the ones in the table. Do you understand why some values are not exact?

What does your model predict for the value of C corresponding to $F = 0$? In the preparation section, you used your graph to estimate that C-value. Does the algebraic answer agree with your graphical estimate?

The *domain* of a function is the set of allowable values for the independent variable. We will sometimes call the independent variable the *input* variable because, if we were using a calculator to compute F, we would input -20 for C, for example, to obtain the value -4 for F.

The domain of the function defined by the table consists of eight numbers. What are they?

The domain of the function defined by the *equation* that models the data is not the same. The equation can be used to compute F given *any* value of C, so the domain of that function is $\mathbb{R}$, the set of all real numbers.

Use your grapher to graph the algebraic relationship between F and C that you found. Adjust the viewing window so that the graph looks like the first one you drew in preparation for this lab. Then answer the following questions. You will probably need to use a variety of viewing windows to observe different parts of the graph.

- What value for C gives an F-value of approximately -10? (Although Fahrenheit believed that $0°$ F was the coldest possible temperature, those of us living up north know better.)

- As C increases, what happens to F?

- As C increases, does F change more slowly than C or more rapidly?

- As C decreases, what happens to F?

- Is F always larger than C? (Are you sure?)

- Is there any temperature for which $F = C$? If so, what is this common value? (Algebra could help you with this.)

When $C = 0$, $F = 32$. When $C = 5$, $F = 41$.

- From the graph, estimate F when $C = 10$.

- How much did F change from the value it had when C was 5?

- How much do you think F will change between $C = 10$ and $C = 15$? Does your graph confirm this answer?

Complete the following: For this function, a change of _____ in the value of C produces a change of _____ in the value of F.

If the Celsius temperature goes up 1 degree, what happens to the Fahrenheit temperature?

Now look at the algebraic relationship between F and C that you postulated. Do you see a way of "reading" the information about relative changes in F and C from the equation?

From your model, you see that some integer values of C produce fractional values for F and vice versa. The electronic sign, however, displays only integer values for temperature. There are two conventions for doing this: rounding off, which replaces a rational number with the closest integer, and truncation, which drops the fractional part of a number. Which method do you think the electronic sign uses? Use specific entries from the table to justify your answer.

The Lab Report

The relationship between temperature in degrees Celsius and temperature in degrees Fahrenheit can be represented in several different ways. In this lab, you investigated three of those ways: a table, a graph, and an algebraic formula. Write a two- to three-page essay, accessible to the general reader, in which you compare and contrast those three functions. Your report should include answers to these questions:

- Is F proportional to C? Provide evidence for your answer.

- What algebraic relationship gives F as a function of C and how did you determine it?

- How does the graph of the F-C relationship respond to changes in scale? Illustrate with graphs, pointing out what changes and what remains the same.

- How can you "read" the algebraic relationship between F and C by examining its graph? That is, how does the graph reveal the two constants in the formula?

- What are some of the ways in which the function defined by the formula differs from the function defined by the table?

- Which method, rounding or truncation, does the electronic bank sign use? Show numerical evidence.

An excellent paper, rather than simply listing answers to these questions, will incorporate the answers into a discussion of the three representations of the Fahrenheit-Celsius relationship.

Follow-up Activities for the Fahrenheit Lab

1. C as a Function of F

In the Fahrenheit lab, you wrote an equation describing F as a function of C; that is, F was the dependent variable, and C the independent variable. But each Fahrenheit temperature also determines a unique Celsius temperature. We can therefore write a model in which C is the dependent variable and F the independent variable. Show the algebra that turns F as a function of C into C as a function of F.

2. Have it Your Way . . .

One group of students, working with a similar table, came up with the following model to relate Fahrenheit temperatures to Celsius temperatures: $F = \frac{20}{11}C + 32$.

1. Use your grapher to compare this function and the function that your lab group found.

2. By substituting all the values from the table on page 41, determine how well this second function, $F = \frac{20}{11}C + 32$, models the data. Would you give these students credit for their model? Why or why not?

3. Do your results suggest any conclusions about models of real-life situations?

3. Say, What?

The following passages appear on pages 301–302 of *The Time Before History* by Colin Tudge:[1]

> Twenty years could see a 7°C (44.6°F) rise in temperature: the difference between a frozen landscape and a temperate one.

And again,

> Besides, at the trough of the last ice age, 18,000 years ago, the surface of the sea in the eastern Mediterranean is known to have cooled by more than 6°C (43° F), which in ecological terms is huge, yet the elephants and their miniaturized neighbors came through those harsh times.

1. The author appears to confuse temperature changes with temperature values. Explain the mathematical misunderstanding that led to the Fahrenheit temperatures stated in these passages.

2. Edit the quoted passages to make them correct.

---

[1] Simon & Schuster, 1997.

Timing is everything

Preparation

Cecil A. DeMille has a dilemma: His roommates have given him the responsibility of taping *Jurassic Park*, which is to be shown on TV tonight. Cecil has to leave for his precalculus class shortly after the movie starts, so he won't be around to pause the machine during commercials. The entire show, then, will run for three hours, and he needs to get the whole thing on tape. That means programming the VCR before he leaves the house.

Video recording cassettes usually provide a choice of three recording speeds. At the highest speed (SP), recording quality is at its peak, but the tape will hold only two hours of video. At the intermediate speed (LP), the tape will hold twice as much but with some loss of quality. At the lowest speed (EP), a full six hours of material will fit on a single cassette, but the quality of the reproduction is less good than in either of the other modes.

Recording Mode	Tape Capacity (minutes of programming)
SP	120
LP	240
EP	360

So here's Cecil with a three-hour movie, and he must make a choice of speeds. With SP, the whole movie won't fit on one cassette. He could record the entire movie using LP or EP, but the quality would not be the highest possible, and there would be wasted tape. Thinking about how to utilize that extra tape, he has an idea: Why not start out at SP, ensuring the highest quality, and switch over to EP just in time to squeeze the last part of the movie onto the cassette.

The question is *when?* For how long can he safely record at SP without risking the loss of part of the movie? Cecil's roommates are counting on him not to lose the final 10 minutes of *Jurassic Park*.

Cecil himself is making frantic calls to electronics stores, but they're already closed. He wishes he could ask his precalculus instructor, but it's too late now.

Wait! Maybe he can use some of the math he's been learning! Just in time, he comes up with a solution. He will make the best possible recording, not lose a minute of *Jurassic Park*, and still get to class on time.

To get an idea of the time at which Cecil will switch speeds, complete the following tables.

Recording Time (minutes)	Fraction of Movie Recorded	Fraction of Tape Used
0	0	0
30	1/6	1/4
60	1/3	1/2
90	1/2	3/4
120	$\frac{2}{3}$	1

Three-Hour Program, SP Mode

Recording Time (minutes)	Fraction of Movie Recorded	Fraction of Tape Used
0	0	0
30	1/6	1/12
60	1/3	1/6
90	1/2	1/4
120	$\frac{2}{3}$	$\frac{1}{3}$
150	5/6	3/5
180	1	$\frac{1}{2}$

Three-Hour Program, EP Mode

Recording time must total 180 minutes; tape fractions should add up to one entire tape. When should Cecil switch from SP to EP?

Did you notice how helpful the tables are when you're trying to understand this problem? Their drawback, though, is that they apply only to a show that's exactly three hours long. A program of a different length would require a new set of tables. Algebra, however, will allow us to generalize the procedure.

During the lab, you will learn an algebraic solution to Cecil's problem. In the process, you will generalize the method so that it works for programs of various lengths and uses other combinations of VCR modes.

The Jurassic Lab

In preparation for this lab, each of you filled in two tables and decided on the length of time Cecil could allow the VCR to run at SP before switching over to EP to accommodate the entire length of a three-hour movie. Recall that the SP mode yields two hours of recording time on a cassette, while the EP mode gives six. Spend a few minutes comparing answers and discussing your reasoning.

Rates

This is really a problem about *rates*. The VCR has three different rates at which it records, and each specific rate determines the quantity of material that can be recorded on a single cassette.

Let's take LP, the intermediate speed, as an example. A tape recorded in LP mode holds 240 minutes of programming. Each minute of programming, therefore, occupies $\frac{1}{240}$ of the tape. That is, the tape fills at the *rate* of $\frac{1}{240}$ *tape per minute*. On the top line of the table that follows, fill in the other two rates. Include units.

Next we use these rates. If the VCR runs in LP mode for two hours (120 minutes), what fraction of the tape is filled? To obtain the answer, you *multiply* the running time by the LP rate. Now complete the table by multiplying each running time by the appropriate rate to obtain the correct fractions.

Running Time (minutes)	SP Rate $\frac{1}{120}$	LP Rate, $\frac{1}{240}\frac{\text{tape}}{\text{min}}$	EP Rate $\frac{1}{360}$
20	1/6	1/12	1/9
30	1/4	1/8	1/6
80	2/3	1/3	2/9
120	1	1/2	1/3
N			

Fraction of Tape Filled

The Three-hour Show, Revisited

Here comes the algebraic part. You will now use algebra to re-solve the problem that you did with tables in the preparation section. The algebraic expression you find will give you a big hint for answering many other questions related to Cecil's.

Let T represent the switchover time; that is, T is the number of minutes during which Cecil lets the machine run in SP mode. The total length of the program is 180 minutes. What algebraic expression represents the time remaining for the machine to run in EP mode?

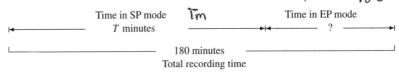

Time in SP mode — T minutes — Time in EP mode — ? — 180 minutes Total recording time

Above, you computed rates for the two modes he's using: one fraction for SP and a different fraction for EP. Use those rates, along with an algebraic expression for each of the times, to write an equation expressing the fact that the portion at SP plus the portion at EP equals one whole tape.

$$\overbrace{\text{SP rate} \cdot \text{SP time}}^{\text{SP fraction}} + \overbrace{\text{EP rate} \cdot \text{EP time}}^{\text{EP fraction}} = 1$$

Include units: Units for rate are *tape per minute*; units for time are *minutes*. Notice the way *minutes* cancels out, leaving only *tape*. Do a good job with this equation, because it's the key to the rest of the lab.

Solve that equation for T. Does your answer agree with the one you obtained from your tables? If so, you know a way of recording a three-hour movie on a standard cassette. "Big deal!" you might say. But one of the great virtues of mathematics is that it allows us to set up models and make generalizations that apply to other situations. Now we're ready to use SP and EP to handle shows of other lengths.

The x-Minute Show

The previous section should give you an idea for generalizing this procedure. We'll still let T represent the number of minutes to run the tape in SP mode, and we'll let x be the length, in minutes, of the program we want to record. T, our switchover time, depends on x; we can think of T as a *function* of x, because the value of x will determine the value of T.

Look back at that key equation you wrote for the three-hour movie. All you need to do is to replace "180" with x. You still have an equation, but it contains two variables instead of one, so you cannot solve it for a numerical answer. Instead, solve it for T in terms of x. Take your time, and don't let the fractions throw you.

In its new form, $T = \ldots$, your equation now defines T as a function of x. When a number is substituted for x, the value of T is uniquely determined. Mathematicians would write $T(x)$ rather than just plain T, emphasizing that T depends upon x in the particular way specified by the expression following the equals sign.

To check $T(x)$, evaluate it for $x = 180$ minutes (the length of Cecil's movie). It should give the same answer that you found when you solved the three-hour problem. Now try it for $x = 120$ minutes. You should obtain $T = 120$. (What does $T = 120$ mean in this context, and why does it make sense?)

If you obtained both of the above answers from your $T(x)$ function, you have probably written the model correctly.

The Graph of the Model

Sketch a graph of $T(x)$. You should see a line sloping downward from left to right. Interpret each of the following facts in the context of making a recording. (Your interpretation should go something like this: "If the movie is ____ minutes long, we need to switch from SP to EP after ____ minutes.")

- The graph passes through the point $(180, 90)$. What does this mean?

- The graph passes through the point $(240, 60)$. What does this tell us?

- The graph passes through the point $(300, 30)$. What does this say?

Here are some additional questions to consider:

> *The Godfather* is showing on public TV (no commercials!) with a running time of 171 minutes. You want a copy. When should you switch modes?

> The graph passes through the point $(60, 150)$. What, if anything, does that fact mean for our model?

> What is the x-intercept of the graph? Does it have significance for the model? If so, what does it mean?

> Where does the graph intercept the vertical axis? Does that intercept have meaning for our model? Explain.

The Domain and the Range of the Model

When you answered some of the previous questions, you probably realized that not every point on the graph has meaning in our model. As an abstract function, $T(x)$ has as its domain $\mathbb{R}$, the set of all real numbers, because any real-number value can be used as input. As a mathematical model for the VCR, though, $T(x)$ has a limited domain. What is the domain of the model? (That is, what set of x-values result in a set of $T(x)$-values that make sense in this context?) Is the domain of the model a discrete set (that is, is each x-value in the domain separated from the next one by a measurable amount) or a continuous set?

Corresponding to the domain of $T(x)$ is its range. In the case of the abstract function, the range is $\mathbb{R}$, because every real number is a possible output. In the case of the model, however, the range is limited. Given the domain you decided on, what is the range of the model? (The model's range must correspond to its domain.) Is the range discrete or continuous?

On your sketch of the graph, highlight the portion that represents the model. On each axis, write a descriptive label: *length of program (minutes)* on the x-axis and *recording time at SP (minutes)* on the T-axis.

Other Schemes

Combining the SP and the EP modes are not the only recording possibilities. Investigate the effects of using SP and LP in sequence. By adapting the procedures and the reasoning that you already used, write and graph a second mathematical model, $U(x)$, which combines SP and LP modes. Determine its domain and range.

Now combine LP and EP to produce a third mathematical model, $V(x)$. Determine its domain and range.

We have not exhausted the possibilities, but we probably have exhausted *you*. Cecil could have decided to use all three speeds in sequence. But the mathematics gets more complicated because we need an extra variable, so we will now lower the curtain on our show and raise the house lights.

The Lab Report

Suppose that you're the manager of an electronics store. Customers keep asking you and your employees for advice on how to best record movies and other long TV programs. You don't want to figure out everything from scratch each time somebody asks, and you don't want your clerks giving incorrect advice. Write a brief guide to which your clerks can refer when faced with such questions.

As an appendix to the guide, provide the mathematical background:

- Draw a graph of each of the three functions you devised.

- Highlight the portion of each graph that represents a mathematical model for a VCR taping.

- Give the domain and the range of each model, explaining what they mean in this context.

Even though the guide you write will use mathematical concepts, stick to everyday language that your clerks would have a prayer of understanding. Mathematical symbols such as $T(x)$ and terms such as *domain* and *range* belong in the appendix.

Linear and Exponential Models

2

© Benelux Press/Index Stock Imagery

In this chapter, we introduce two of the most important classes of functions used as mathematical models: linear functions and exponential functions.

We will focus on functions that can be defined by algebraic formulas. Please don't lose sight, however, of the fact that many functions do not have formulas and are defined instead by a different sort of rule, such as a table, a verbal statement, or a graph. In Chapter 1, you saw an example of such a function: the relationship between the date and the height of the Connecticut River.

One of the themes of this text is the way mathematics arises from "real life." Rather than introducing you to mathematical ideas in the abstract and then showing how they can be applied, we're more interested in plunging you into a real-world situation and helping you tease out the math that's already present there. This approach should help you discover some of the ways in which math and life fit together.

At first, it might bother you that there can be a variety of correct answers to some of the questions and that some problems can be solved in more than one way. Real life seldom comes with step-by-step instructions and an answer key. In the process of finding reasonable solutions to problems, you will learn to be open-ended and resourceful when faced with a situation that calls for quantitative reasoning. Rather than panicking because a particular answer isn't in the back of the book, you'll have the confidence to say, "My answer is a reasonable one in the given context." You'll also realize the value of having colleagues (your fellow classmates) with whom to discuss the material that all of you are learning.

A formula is only one of the many ways in which a function can be defined.

Calvin and Hobbes   by Bill Watterson

2.1   INTRODUCING LINEAR MODELS

You are already familiar with some linear models. In Chapter 1, you saw two shoe-size functions with straight-line graphs and the triathlon function whose graph consisted of three straight-line segments. In this section, we'll begin with the simplest of linear models, ones in which we know where the function's graph crosses the vertical axis (that is, the function's value at $x = 0$) and the rate at which the function values change.

A Constant Rate of Change

Most countries of the world have their own currencies, and a person traveling from the United States to another country usually needs to convert American dollars into the local currency. The terms of that transaction are determined by the exchange rate, a number that varies from day to day.

During much of the summer of 2001, the rate for the Mexican peso was approximately 9.2. This means that 100 U.S. dollars had the same value as 920 Mexican pesos and that each additional dollar was worth an additional 9.2 pesos. That rate, 9.2 pesos per dollar, is the same, regardless of the number of dollars to be exchanged. The algebraic formula

$$p = P(d) = 9.2d \qquad \text{or} \qquad y = 9.2x,$$

Using variables d and p to represent amounts in dollars and pesos, respectively, makes sense in this context. Most graphers, however, want you to use x for the independent variable and y for the dependent variable.

where d (or x) stands for the number of dollars and p (or y) the number of pesos, expresses the relationship. Its graph is a line, and its constant rate of change, 9.2, is called the **slope** of the line.

The slope of a line is a number, and that number has units: units of output per unit of input. In this case, with the input measured in dollars and the output measured in pesos, the units associated with the constant 9.2 are *pesos per dollar*.

Let's introduce a symbol, the Greek letter Δ, to stand for the amount of change in a variable. Thus, Δd means *the amount by which the number of dollars changes*, and Δp means *the amount by which the number of pesos changes*.

▷   If d increases from 80 to 90, what is the value of Δd? By how much does p increase? (That is, what is the value of Δp?)

$\Delta d = 10 \qquad \Delta p = 92$

$\Delta p_0 = 736 \qquad 828$

$\Delta p_1 = 828 \qquad \underline{-736}$

$\qquad\qquad\qquad\quad \dfrac{92}{}$

▷   If d decreases from 118 to 110, what is the value of Δd? Of Δp? (A decreasing d means a negative Δd.)

$\Delta p = 73.6$

$\Delta d = 8$

$\Delta p_0 = 1085.6$ 1085.6
$\Delta p_1 = 1012$ -1012.0

▷   Find the value of the ratio $\frac{\Delta p}{\Delta d}$ in the two preceding cases.

a) $\dfrac{92}{10} = 9.2$ b) 9.2

If you answered these few questions correctly, you understand that Δp is always equal to 9.2 times Δd, independent of the particular values of d and p. All that mattered was the *amount of change* in each variable.

$$\Delta p = 9.2\Delta d \qquad \text{or} \qquad \frac{\Delta p}{\Delta d} = 9.2$$

So, for instance, if d increases by 20 dollars ($\Delta d = 20$), p increases by

$$9.2\,\frac{\text{pesos}}{\text{dollar}} \cdot 20 \text{ dollars } = 184 \text{ pesos,}$$

and if d decreases by 40 dollars ($\Delta d = -40$), p changes by

$$9.2\,\frac{\text{pesos}}{\text{dollar}} \cdot -40 \text{ dollars } = -368 \text{ pesos,}$$

a decrease of 368 pesos (see Figure 2.1).

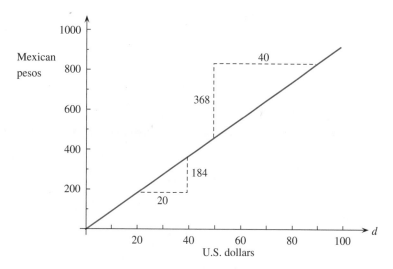

Figure 2.1   Graph of $P(d) = 9.2d$

The units for the constant 9.2 are *pesos per dollar*.

Because the ratio of Δp to Δd is constant, we say that Δp is **proportional** to Δd. The **proportionality constant** 9.2 is the slope of our currency-exchange model.

> *"My weight is always perfect for my height, which varies."*
> Erma Bombeck

You might feel as though we're beating to death this idea of proportionality. But it's very important because it's the one thing that makes a function linear.

> In any linear function, the *change* in the dependent variable is proportional to the *change* in the independent variable.
>
> $$\Delta y = m \cdot \Delta x \qquad \text{or} \qquad \frac{\Delta y}{\Delta x} = m$$
>
> for some constant value *m*, which we call the slope of the line.

Toeing the line

Figure 2.2, a picture of our two shoe-size models, shows two parallel lines. When you calculated rates of change for the functions *w* and *m*, you found that a change of 1 inch in *x* always produced a change of three shoe sizes in *y*, both for men's sizes and for women's. Therefore, both lines have a slope of *three* *y*-units for every *one* *x*-unit. Their slope is $\frac{3}{1}$, or three sizes per inch.

Parallel lines have the same slope.

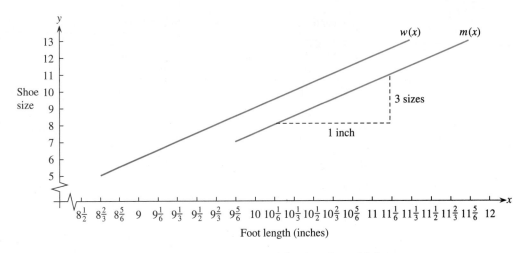

Figure 2.2 Graphs of Models $w(x) = 3x - 21$ and $m(x) = 3x - 22.5$

Check Your Understanding 2.1

CYU answers begin on page 80.

√ **1.** One week in early August 2001, the exchange rate for euros was approximately 1.12 to the dollar. Using *r* for the number of euros and *d* for the number of dollars, write a formula expressing this relationship. What units are associated with the constant in your formula?

$R(d) = 1.12d$

$r = euros$
$d = dollars$

units = "1.12 euros per dollar"

2. That same week, the exchange rate for Japanese yen was approximately 122 yen to the dollar. One year earlier, it was approximately 108 yen to the dollar.

$y = 122d$

$y = 108d$

√ **(a)** Using *y* for the number of yen and *d* for the number of dollars, write two dollars-to-yen formulas, one for each rate.

$-y(d) = 122d$

$-y(d) = 108d$

√ **(b)** Without graphing the functions, decide which graph would be steeper.

$y(d) = 122d$ because one must multiply with a larger constant

"magnitude of slope is greater"

$\frac{Dy}{\Delta x}$

3. The linear function $F = \frac{9}{5}C + 32$ converts temperatures measured in degrees Celsius into temperatures measured in degrees Fahrenheit. The value of $\frac{\Delta F}{\Delta C}$ is constant. What is that value and what are its units?

LAB

4. A linear function f has these values: $f(-1) = 17.2$, $f(3) = 1.8$. Find the slope of the line.

$(-1, 17.2) \,\&\, (3, 1.8)$

$\frac{17.2 - 1.8}{-1 - 3} \rightarrow \frac{16}{-4} \rightarrow -4 \rightarrow \boxed{m = -4}$

A Starting Value and a Constant Rate

Here we are at the heart of this section. Now that you know about slopes, all you need in order to write a linear model is a starting point for the function.

A sound investment

Jason, a freelance disc jockey, reads the following ad

> **Pioneer S-500 Pro DJ System**
> Take your music to the next level with the system that every DJ dreams about. Consisting of the model …

and decides to purchase the $3000 system. He figures that the system has a life expectancy of 10 years, after which it will be technologically obsolete. To help defray the cost, Jason can write off the depreciation of his equipment as a business expense on his income taxes each year. There are several different methods that the government allows for computing depreciation. One popular method, straight-line depreciation, is based on the assumption that equipment with a life expectancy of n years will lose $\frac{1}{n}$th of its original value each year. For Jason, this means that the $3000 system will lose $\frac{1}{10}$th of its value, or $300, per year.

Let $V(t)$ be the value of Jason's system t years after he purchases it. Then $V(0) = 3000$. Assuming that the equipment loses $300 in value each year, we can graph the function $V(t)$ immediately.

We'll start you off. On the graph in Figure 2.3, we start at the point $(0, 3000)$, then move to the right one unit (to $t = 1$) and down 300 units (to the value of $V(1)$). We mark this point on the graph.

The point at which a graph touches or crosses the vertical axis is often called its **y-intercept.**

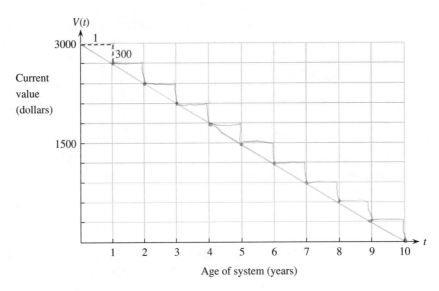

Figure 2.3 Graph of $V(t)$

▷ Plot the point $(2, (V(2))$ in the same way: Starting at $(1, (V(1))$, move right one unit and down 300 units.

▷ Continue the process until the graph has 11 points, corresponding to the 10-year life of the system.

In mathematics, the word line implies straight line.

▷ All the points should lie on the same line. Draw that line—the graph of $V(t)$.

Now that you have the graph, let's talk about the formula. You started with $V(0) = 3000$. Each time you increased the value of t by one year, you decreased the value of V by 300 dollars. Here's a formula for the linear function that says just that:

$$V(t) = 3000 - 300t$$

start here / decrease by $300 per year

A constant slope means a linear function. A negative slope means a decreasing function.

The slope of the function V is -300. Because the slope is constant, the function is linear; because the slope is negative, the function is decreasing.

▷ Use the formula for $V(t)$ to find the value of Jason's system after $2\frac{1}{2}$ years.

$V(t) = 3000 - 300t$
$V(t) = 3000 - 300(2.5) = \$2,250$

The graph of $V(t)$ crosses the t-axis at $t = 10$, a point called the t-intercept of the line. In context, $V(10) = 0$ means that after 10 years, the system is worthless. Beyond $t = 10$, all values of V are negative and have no meaning for the model.

In Section 2.2, you'll examine a nonlinear method for computing depreciation. Then you can decide which method will give Jason a better break on his taxes.

People who love people

"This Decade Will Add a Billion," declared a 1992 newspaper headline. In the accompanying article, the executive director of the United Nations Population Fund explained that the world's population was increasing by approximately a quarter of a million people every day and that she anticipated a total *increase* in population of one billion people by the end of the 1990s. On the basis of this reportedly constant growth rate, we can construct a linear model of the size of the world's population during the decade from 1990 to 2000.

A daily increase of 250,000 people translates into a yearly increase of 91.25 million (based on 365 days per year). In 1990, there were approximately 5.23 billion people in the world (that's 5230 million).

▷   Use the preceding information to complete Table 2.1.

Year	1990	1991	1992	1993	1994	1995
Population (millions)	5230	5321.25	5412.50			

Table 2.1   World Population Estimates, Based on Annual Increases of 91.25 Million

Let $P(t)$ represent the population as a function of t, the time in years since 1990. Using the data in Table 2.1, we can write $P(0) = 5230$ million persons, giving us the starting value. We already know the slope: 91.25 million persons per year.

▷   How did we determine the value of $P(1)$ from the value of $P(0)$? $P =$ millions ppl
$t = 1$ yr

$P(0) = 5230$
$P(t) = 5230 + 91.25 t$

$P(1) = 5230 + 91.25 (1) = 5321.25$

If you know the value of $P(t)$, you can determine the value of $P(t + 1)$ for any t by adding 91.25. For every one-unit increase in t, there is a 91.25-unit increase in $P(t)$.

▷   Write a formula for $P(t)$.

$$P(t) = \underline{\quad 5230 + 91.25 t \quad}$$

▷   In Figure 2.4, sketch a graph of $P(t)$.

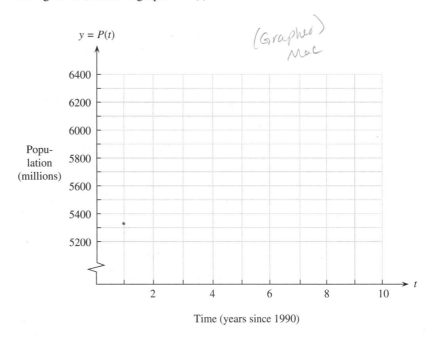

(Graphed)
Mac

Figure 2.4   Graph of $P(t)$, Linear Model for World Population Growth

STOP & THINK

> Justify this statement: According to the model shown in Figure 2.4, the increase (ΔP) in population over any time interval is directly proportional to the length (Δt) of the time interval. By what number do you multiply any Δt-value to obtain the corresponding ΔP-value? We call that number the constant of proportionality.

▷ What does the model P predict for the world population in the year 2000?

6142.5

▷ The director of the U.N. Population Fund estimated that the population would increase by one billion persons during the 10-year period from 1990 to 2000. Does model P agree with this prediction? If not, how far off is it?

▷ On the basis of this model, how many years *would* be needed for the one-billion-person increase?

Every linear function is characterized by a constant rate of change, its slope.

As you have no doubt discovered, your population prediction for the year 2000 is substantially less than the population predicted by the headlines. The linear model P is based on the assumption that as population grows, yearly increases will remain constant. Because that assumption is probably not reasonable, we will revisit this problem soon, in Section 2.2. There we'll consider a more likely model, one in which larger populations generate larger population increases.

What's my line?

Both of the previous situations (depreciation and population) were modeled by linear functions. In each case, we were able to determine a formula relating the dependent and independent variables as soon as we had a starting value for the function ($V(0)$ or $P(0)$, respectively) and its constant rate of change.

The Slope-Intercept Form of a Line

For simplicity, we will call a function's independent variable x and its dependent variable y. Then the statement "y is a linear function of x" means that the relationship can be expressed as

$$y = f(x) = b + mx, \qquad \text{where}$$

- f (or whatever other letter or letters we choose to use) names the function
- b is the value of the function when $x = 0$ (the y-intercept, or $f(0)$)
- m is the slope, the constant of proportionality linking Δy to Δx

---

You can also think of m as the multiplier of the independent variable.

The symbol m is the one that mathematicians traditionally use for slope. It probably comes from the French verb *monter*, meaning *to climb*.

By long-standing convention, mathematicians use the symbol b to stand for the **y-intercept,** the value at which a line crosses the y-axis. We can't think of any particular reason why b was chosen rather than some other symbol, but because it's the one commonly used, we'll stick with it.

The graph in Figure 2.5 shows a generic linear function. Notice that, in this particular picture, the value of b and the value of $m = \frac{\Delta y}{\Delta x}$ are both negative.

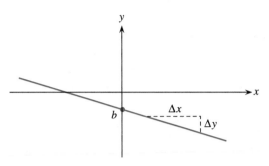

Figure 2.5   Graph of a Generic Linear Function, $y = b + mx$

Check Your Understanding 2.2

1. $PRICE(t) = 10 + 1.5t$ models the average price in dollars paid for a collectable toy t years after the toy is first marketed.

(a) What is the average price when the toy goes on the market? *$10*

(b) How much does its price increase from one year to the next? *11.5   13   14.5*

$1.50

(c) If Δt is 2, what is $\Delta PRICE$?

(d) Complete the table.

t	$PRICE(t)$
0	10
2	13
4	16
6	19
8	22
10	25

(e) Consecutive t-values in the table differ by two units. What pattern holds for consecutive $PRICE$-values?

value goes up by $1.5

? differ by ½ te 3 units

2. A baby born weighing 3.2 kg gained 0.25 kg per week.

(a) Write a linear model for the baby's weight as a function of his age in weeks.

$w(0) = 3.2$ *w = weight*
$w(t) = 3.2 + 0.25t$ *t = time (weeks)*

(b) Give a reasonable domain and range for the model. (There's no one right answer to this; use common sense and the fact that 1 kg ≈ 2.2 lb.)

(handwritten in left margin: $U(t) = 21.95t$)

Two Special Cases: Horizontal and Vertical Lines

In 1999, the Internet company America Online offered several billing options, including its Unlimited Usage plan. Under this plan, a member was charged $21.95 per month regardless of the number of hours spent online. Let's express the monthly bill, U, as a linear function of t, the number of hours the member uses in one month.

▷   What is the value of $U(0)$? ___$21.95___   What about $U(10)$? ___$21.95___

▷   When t increases from 0 hours to 10 hours ($\Delta t = 10$), what is the corresponding value of ΔU?   *0*

▷   What real number gives the slope $m = \frac{\Delta U}{\Delta t}$?   *0*

▷   Plot the points $(0, U(0))$ and $(10, U(10))$ in Figure 2.6. Connect the points and extend the line in either direction.

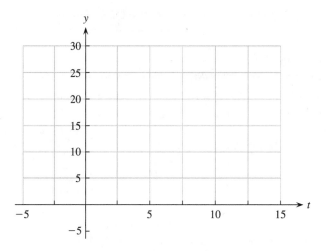

Figure 2.6   Graph of the Line $y = U(t)$

Vertical line test: If any vertical line you can draw cuts the graph no more than once, the graph represents a function.

You should have drawn a horizontal line. Notice that this graph satisfies the geometric requirement for the graph of a function: It passes the vertical line test.

▷   Write the formula for the line: $U(t) = $ _____

The function U is an example of a **constant function.** For any value of the independent variable t, the dependent variable $y = U(t)$ is equal to a constant value (here, 21.95).

The domain of $U(t)$, considered as an abstract function, is the set of all real numbers. To decide on a reasonable domain for the *model $U(t)$*, we consider how many online hours a member might possibly use over the course of a month. The domain of the model might be, for example, the set of values between 0 and 300.

The range of $U(t)$ consists of a single number, 21.95.

So far, you have graphed lines with positive slopes, negative slopes, and most recently a slope of zero ($m = 0$). Because every real number is either positive or negative or zero, there are no more possible slope values to consider. But there is one type of *line* that we haven't yet mentioned. Read on.

▷   Plot the point $(3, -2)$ in Figure 2.7. Through it, draw two lines, one horizontal and one vertical. Label the horizontal line L_1 and the vertical line L_2.

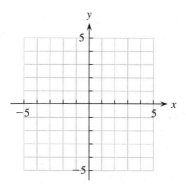

Figure 2.7 Graph of Lines L_1 and L_2

▷   Write the equation of L_1 and give its slope.

▷   The point $(3, -2)$ lies on L_2. Now pick a second point on L_2 and try to use Δx and Δy to compute its slope. What basic law of algebra does your fraction violate?

We say that a vertical line *has no slope*, because its slope cannot be defined. (This is entirely different from having a slope of zero.) Because there is no m-value for the line L_2, we cannot write L_2 in the form $y = b + mx$. The line L_2 is defined solely by its constant x-value. We indicate this by writing $x = 3$ as the equation of this vertical line.

Check Your Understanding 2.3

1. (a) Write the equation of the horizontal line through the point $(-4, 5)$.

(b) What is the slope of that line?

(c) Write the equation of the vertical line through the same point.

(d) Which of these two equations defines y as a function of x?

2. (a) Write the equation of the horizontal line through the origin. That is, write the equation of the x-axis.

(b) Write the equation of the vertical line through the origin. That is, write the equation of the y-axis.

This course places what might seem to you an inordinate amount of emphasis on (straight) lines. One reason is that lines are easy to understand and to study. If you know one point on the line and the direction in which the line is going, you know all there is to know about that line. Another reason is that we can often study a very complicated curve by magnifying a section of it so that it looks straight and then treating that section as though it were a line. There are even some functions for which the entire graph, if we look at it over a very wide interval, resembles a line. In subsequent chapters, we will study curves that look straight from close up as well as curves that look straight from far away.

In a Nutshell   A constant rate of change m, together with a starting value b, define a unique linear function $f(x) = b + mx$, whose graph is a line with y-intercept b and slope $m = \frac{\Delta y}{\Delta x}$. Every change in the output value is proportional to the corresponding change in the input value: $\Delta y = m \Delta x$. A horizontal line $y = c$ has slope $m = 0$ and represents a constant function. A vertical line $x = k$ has no slope (that is, its slope value cannot be defined); it does not represent a function.

2.2   INTRODUCING EXPONENTIAL MODELS

This section marks the debut of a new and powerful class of functions: those for which growth takes place at a fixed percentage. You'll learn the distinction between a constant growth *rate* and a constant growth *ratio*. We will interpret the word *growth* broadly; that is, we'll consider a decreasing function to have negative growth.

A Constant Growth Factor

"Kitchen Cleanup Checklist," an article in the June 1999 issue of *Better Homes and Gardens*, gave readers advice on maintaining a safe food-preparation environment in their kitchens. In answer to the question "Why bother …?," the article explained that, given the right conditions, harmful bacteria can double their numbers every 20 minutes. Furthermore, the article claimed, a single bacterium on a wet countertop might, in just eight hours, reproduce to nearly 17 million.

The doubling-every-20-minutes function b, where $b(t)$ represents the number of bacteria and t counts the number of 20-minute intervals, is not linear. Values of this function over a two-hour period, from $t = 0$ to $t = 6$, appear in Table 2.2.

Elapsed Time, in 20-minute intervals t	Number of Bacteria, $b(t)$
0	1
1	2
2	4
3	8
4	16
5	32
6	64

Table 2.2 Bacterial Counts

▷   How can you tell from the ratio $\frac{\Delta b}{\Delta t}$ that b is not a linear function?

▷   What is the ratio of any two consecutive function values (that is, output values) in Table 2.2? (Divide the later entry by the earlier entry.)

Because the number of bacteria is *doubling*, the **growth factor** for this process is 2, and the formula defining the model is $b(t) = 2^t$. Notice that the growth factor is equal to the ratio you just calculated. For every one-unit increase in t, the current $b(t)$-value is *multiplied* by 2.

▷   What would be the growth factor if the bacteria were tripling instead of doubling?

In the case of tripling, the formula is 3^t instead of 2^t, and its growth factor (also called the **base** of the function) is 3.

The exponent of an exponential function contains the independent variable.

The model $b(t)$ is an example of an **exponential function,** so named because the independent variable appears as an exponent. Figure 2.8 shows a graph of $b(t)$ over the two-hour period from $t = 0$ to $t = 6$.

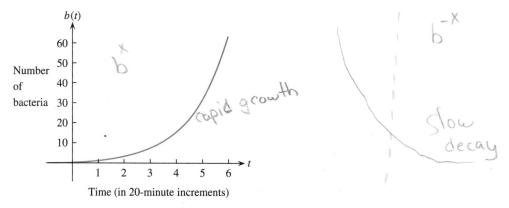

Figure 2.8   Graph of $b(t) = 2^t$

▷   The graph in Figure 2.8 is concave up. What does this shape tell you about the average rate of change of $b(t)$ as t goes from 0 to 2, from 2 to 4, and from 4 to 6? (Does the average rate increase, decrease, or remain the same?)

Could a single bacterium reproduce to almost 17 million in just eight hours? Let's use the function $b(t)$ to check the magazine's alarming statement. In eight hours, there are 24 intervals of 20 minutes. So we want the value of $b(t)$ when $t = 24$.

▷   Calculate $b(24)$. (On most graphing calculators, you will use the ^ key to raise a number to a power.) Did the author of the article do the math correctly?

Negative growth toward a noble goal

During the final decades of the twentieth century, the World Health Organization adopted as one of its goals the elimination of polio throughout the world. In the eight-year period from 1988 to 1996, cases of polio decreased by roughly 25% annually. That is, the number of cases in each successive year was approximately three fourths of the number from the previous year. The exponential model representing the incidence of polio as a function of time therefore has a growth factor (or base) of 0.75, because we multiply each year's number by 0.75 to obtain the number for the following year.

▷ Think about this: Why does a growth factor of 0.75 imply negative growth, or shrinkage?

Regardless of whether an exponential function increases or decreases, its growth factor is always positive.

We'll return to this model soon. For now, just pay attention to the growth factor: an annual decrease of 25% means a growth factor of 0.75. *Note*: Even though the number of polio cases is decreasing (the growth is negative), the *growth factor* is a positive number.

Money making money

Most banks offer compound interest on their deposits. That is, at the end of each interest period, the amount of interest is added to the current balance in the account so that at the end of the next interest period, there's a larger balance on which to compute the same percentage. Simply put, larger balances earn more interest, even though the interest rate remains the same.

A particular account offers 5% interest, compounded annually. What is the annual growth factor for this account? Someone might suggest 0.05.

▷ What do you think? Remembering that the growth factor is the number used to multiply the current value to obtain the next value, show that a growth factor of 0.05 would cause a bank balance to shrink rapidly to almost nothing.

The true growth factor in this case is 1.05. That is, to obtain the new balance, the bank multiplies the previous balance by 1 (to maintain the balance and prevent the depositor from losing money) plus 5% (to add the interest). Think of that 1.05 as representing 100% plus 5%.

The growth factor for a country that gains people at the rate of 2.5% a year is 1.025.

▷ Give the growth factor for a country that *loses* people at the rate of 2.5% a year.

STOP & THINK

You have seen that the factors 2, 3, and 1.05 represent positive growth, while the factors 0.975, 0.75 and 0.05 imply negative growth. What's the criterion for deciding whether a particular factor causes positive growth or negative growth?

A Starting Value and a Constant Growth Factor

Even with careful washing, you probably won't reduce the number of bacteria on your kitchen counter to a single bacterium. Let's see how the doubling-every-20-minutes model changes if you start with three bacteria in the puddle instead of just one. We'll call the new function d.

▷ Fill in the values missing from Table 2.3.

Elapsed Time, in 20-minute intervals t	Number of bacteria, $d(t)$
0	$3 = 3 \cdot 1$
1	$6 = 3 \cdot 2$
2	$12 = 3 \cdot 4$
3	$24 = 3 \cdot 8$
4	$48 = 3 \cdot 16$
5	$96 = 3 \cdot 32$
6	

Table 2.3 Bacterial Counts, Starting with Three Bacteria

You can't combine the 3 and the 2 in the formula $d(t) = 3 \cdot 2^t$, because order of operations requires that raising to a power be performed before multiplication.

Notice that each value of $d(t)$ is three times the corresponding value for $b(t)$, our doubling function. We express this mathematically by writing $d(t) = 3 \cdot b(t) = 3 \cdot 2^t$.

▷ Suppose your kitchen counter starts with seven bacteria. Write a formula expressing the relationship between the number of bacteria and the number of elapsed 20-minute intervals.

$$f(t) = 7 \cdot 2^t$$

▷ Imagine seven bacteria of a particularly aggressive strain that *quintuples* every 20 minutes. (Hollywood makes movies from stories like this.) Write a function to model the number of bacteria at time t, measured in 20-minute intervals.

$$f(t) = 7 \cdot 5^t$$

✱ If 2 is the growth factor, 2^t is part of the formula. If 5 is the growth factor, 5^t is part of the formula. In each model, the exponential expression must then be multiplied by the starting value: 1 or 3 or 7 in the cases above.

In general, if you begin with c bacteria and the growth factor is a, then the number of bacteria after t intervals of 20 minutes is given by the exponential function $f(t) = c \cdot a^t$.

You might wonder why we're stuck on this 20-minute interval. No, there's nothing special about that time period. We used it here because it was in the magazine article and because we didn't want to muddy the waters yet by changing the length of the interval. Soon, however, you'll find mathematical ways of expressing the same information with a time interval of your choosing.

A powerful function—one size fits all

An exponential function models any process in which function values change by a fixed ratio or percentage:

$$y = g(x) = c \cdot a^x, \qquad \text{where}$$

The growth factor, a, is always a positive number.

- c is the starting value, $g(0)$

- a is the base, or growth factor.

People who love people, exponential style

In Section 2.1, we promised to return to the world population article and show you a model that would agree more closely with the projection that the year 2000 would see one billion more people than there were in 1990. We based the linear model $P(t) = 5230 + 91.25t$ on the assumption that the same number of persons would be added to the population each year. In one decade, however, the change in $P(t)$ is 912.5, not the predicted 1000. (Remember, $P(t)$ counts people by the million, so 1000 means one billion.)

It would seem more reasonable to assume instead that the size of each annual increase depends on the current population size: Larger populations generate larger increases. We will create an alternative model, G, in which the population increases by the same *percentage* every year. But how will we determine that percentage? Keep reading.

From 1990 to 1991, population rose by approximately 91.25 million people. This increase accounts for a *relative* rise in population of $\frac{91.25 \text{ million}}{5230 \text{ million}} \approx 0.0174$, or approximately 1.74%, during that year.

The distinction between model P and model G is subtle and important. Model P has the population increasing at a constant *rate* of 91.25 million persons per year. Model G will have the population increasing by a constant *percentage*, 1.74 percent, each year. This percentage means that each person, in one year, generates 0.0174 of a new person—or, stated more sensibly, for every 10,000 persons, there will be 174 additional persons at the end of the year.

By the same reasoning that we used in the compound-interest example, we see that the growth factor for P is 1.0174 (that's 100% plus 1.74%). To get from one year's population to the next year's, we multiply by 1.0174.

▷   Use that growth factor to complete Table 2.4, rounding results to the nearest million.

Year	t	Population, $G(t)$ (millions)	Increase, ΔG (millions)
1990	0	5230	
1991	1	$5230(1.0174) \approx 5321$	$5321 - 5230 = 91$
1992	2	$5230(1.0174)(1.0174)$ $= 5230(1.0174)^2$ $\approx$	
1993	3	$5230(1.0174)^2(1.0174)$ $= 5230(1.0174)^3$ $\approx$	
1994	4		
1995	5		

Table 2.4 Estimates of World Population, Supposing an Annual Increase of 1.74%

Look at the pattern in the last column of Table 2.4. Notice that the annual population increases get larger as the size of the population grows. Compare, for example, the change in population from 1990 to 1991 with the change from 1994 to 1995. The increase in the final year should be approximately 97 million.

▷   Use the t and $G(t)$ columns of Table 2.4 to write a formula for a function of t that agrees with the pattern you observe.

$$G(t) = \underline{\hspace{4cm}}$$

▷ Use $G(t)$ to predict the world population for the year 2000, that is, when $t = 10$.

Let's set up the models side by side so that we can see what each one says.

$$P(t) = 5230 + 91.25t \qquad\qquad G(t) = 5230(1.0174)^t$$

start here
increase by 91.25 per year
start here
increase by 1.74% each year

▷ Show that model G comes closer to agreeing with the estimated one-billion increase for the decade of the 1990s than the linear model P did. (Recall that P undercounted by 87.5 million.)

You should have determined that model G is more consistent than model P with the estimated one-billion increase mentioned in the newspaper headline. But what about the statement of the U.N. official, who implied a constant growth rate when she reported the world's population growing at a quarter of a million per day? We have seen that the yearly (and hence daily) increases for model G grow larger as time passes. Was she in error? Is model G flawed? Let's do a visual comparison of the two models, whose graphs are given in Figure 2.9.

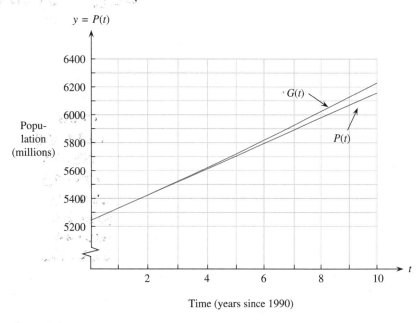

Figure 2.9 Comparing Linear and Exponential Population Growth Models

Notice that at the beginning of the decade, the predictions of the two models are close to one another. It really isn't until about 1995 or 1996 that the graph of $G(t)$ bends upward noticeably and veers away from the linear graph. Thus, for about half of the decade, the executive director's estimate of population increase would hold true, at least approximately, for both models. This points out why linear models are often used: Over a small portion of the domain, they give a good approximation even when the true function is not linear.

STOP & THINK

An update: The U.S. Census Bureau announced in mid-1999 that it expected the earth's six billionth inhabitant to arrive on October 12, 1999. In light of this new information, how do our two models look?

An even sounder investment

Remember Jason the disc jockey from page 55, who had purchased a $3000 sound system with a useful life of 10 years? Using the straight-line depreciation method, he found that he could write off 10% of the original value, or $300, each year. Financially, however, Jason would prefer to calculate the depreciation using a method that allows larger tax write-offs during the early years and smaller ones later on.

Using a second method approved by the Internal Revenue Service and called Double Declining Balance Depreciation, Jason recomputed the depreciation. By inscrutable IRS logic, a person is allowed to double the percentage from the linear method and apply it to each year's book value to determine that year's depreciation. Jason, therefore, could claim a loss in value of 20% (two times 10%) every year.

But we've been trying to model the *current value* of the equipment, not the value lost to depreciation, so we will focus on what's left each year: 80% of the value from the previous year. Our growth factor for this process is 0.80. Let $B(t)$ represent the book value of the sound system t years since it was purchased.

▷ Complete Table 2.5.

$$B(t) = 3000(.80)^t$$

Years since Purchase t	Book Value (dollars), $B(t)$	Annual Depreciation, $\Delta B \ (-\Delta B)$
0	3000	0
1	2400	600
2	1920	480
3	1536	384
4	1228.80	307
5	983.04	245.76
6	786.43	196.61
7	629.15	245.76
8	503.32	196.61
9	402.65	157.28
10	322.12	125.83
		100.67
		80.53

Table 2.5 Book Value and Depreciation over Useful Life of System

▷ Find a function of the form $B(t) = c \cdot a^t$ that produces the values in Table 2.5.

Later in this book, we will say that the *t*-axis is the **horizontal asymptote** for the function $B(t)$.

▷ Use your grapher to view the graph of $B(t)$, which should resemble the one in Figure 2.10 if you use the same viewing window. If you extend the viewing window off to the right, you will see that the function value never reaches 0, even though the curve comes closer and closer to the horizontal axis. In the real world, the equipment will become worthless and perhaps even a liability, but in the mathematical world of decreasing

Even though $B(t)$ is a decreasing function, its graph is concave up, just like the graph of the population model.

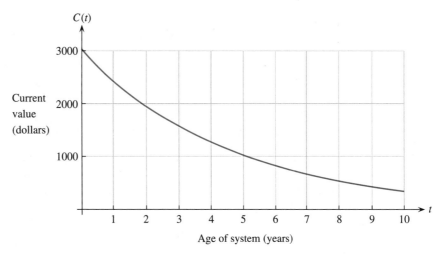

Figure 2.10 Graph of $B(t) = 3000(0.80)^t$

exponential functions, the graph remains forever above the horizontal axis. Eighty percent of *something*, no matter how small, is always something, not nothing.

This graph shows a decreasing function. The other exponential graphs you've seen in this section have shown increasing functions. Just as the *sign of the slope* determines whether a linear function increases or decreases, the *size of the growth factor* controls whether an exponential function increases or decreases. If the growth factor is between 0 and 1, the function is decreasing; if the growth factor is greater than 1, the function is increasing.

If a is the growth factor, the function increases when $a > 1$, decreases when $0 < a < 1$.

Now let's do a visual comparison of the two models. Use your grapher to view $V(t) = 3000 - 300t$ and $B(t) = 3000(0.80)^t$. The graphs intersect at 3000 on the y-axis and again several years later. We're interested in that second point.

For tax purposes, Jason wants to know the time at which the book value is the same using either method. Mathematically speaking, he needs the value of t for which $V(t) = B(t)$.

▷ Attempt to solve the equation $3000 - 300t = 3000(0.80)^t$ using algebra. What problem(s) do you encounter?

▷ Although you won't be able to solve the equation algebraically, you can find an approximate solution using the graph. Estimate the t-value at the point of intersection and call that value t^*.

▷ Carefully draw the graph of $V(t)$ in Figure 2.10. On the t-axis, mark the location of t^*.

Reminder: $\dfrac{f(b) - f(a)}{b - a}$

▷ What is the average rate of change in $V(t)$ from $t = 0$ to $t = t^*$? (Think about this for a moment. V is linear; you don't need to perform any calculations.) What is the average rate of change in $B(t)$ over the same interval?

You should obtain the same rate for each. Were you surprised? The reason is simple: The average rate of change of $B(t)$ from $t = 0$ to $t = t^*$ is the slope of the line connecting the points $(0, B(0))$ and $(t^*, B(t^*))$, and that line perfectly matches the graph of $V(t)$.

Check Your Understanding 2.4

1. A population of a city of 120,000 grows by 2.5% per year. Write an exponential function to model the size of its population t years from now.

2. A population of a city of 120,000 shrinks by 2.5% per year. Write an exponential function to model the size of its population t years from now.

3. The college population in the United States in 1995 was 14.4 million and was growing at the rate of 1.2% per year.
   (a) Write a mathematical model for the U.S. college population, measuring time in years since 1995.

   (b) How many college students does your model predict for 2006?

Calculating the Growth Factor

You've learned to write the growth factor when you know the percentage by which a function increases or decreases. And you know what to do when the exponential process is one of doubling or tripling. But most of our information doesn't come so neatly packaged. (How, you might wonder, did we know those percentages and those doubling times?) The information that we're more likely to obtain will be census data or experimental results. Let's see how we might handle those.

In 1940, the United States had 131 million inhabitants. That number had nearly doubled, to 260 million, by 1995. If we assume that the population during those years can be modeled by an exponential function, then we can take 1940 as year 0 and 131 million as our starting value and get this function:

$$US(t) = 131a^t$$

To brush up on exponents, especially fractional exponents, visit the Algebra Appendix, page 477.

But what's a? We use the fact that when $t = 55$, $US(t) = 260$, as follows:

$$131a^{55} = 260$$

$$a^{55} = \frac{260}{131}$$

Raise each side to the power $\frac{1}{55}$. (Use your calculator!)

$$(a^{55})^{1/55} = \left(\frac{260}{131}\right)^{1/55}$$

$$a \approx 1.0125$$

This means that the population was growing, on average, at an annual rate of 1.25%.

Check Your Understanding 2.5

1. China had 1.3 billion people in 1999 and was expected to grow to 1.7 billion in the next 30 years. Assuming growth at a constant percentage, find the growth factor and give the annual growth percentage.

2. A car whose value was $15,000 when purchased in 1995 had a book value of $7840 in 1999. Find the growth factor and write an exponential model for the value of the car when it is t years old.

In a Nutshell   A constant base or growth factor a, together with a starting value c, define a unique exponential function $y = c \cdot a^x$. The value of a determines whether f increases or decreases: Whenever the starting value c is positive, then $a > 1$ produces an increasing function and $0 < a < 1$ produces a decreasing function. In either case, the graph is concave up. Exponential functions serve as models of growth or decline at a constant percentage.

2.3   LINEAR MODEL UPGRADES

Sometimes our bare-bones linear model $y = b + mx$ isn't adequate, even for situations involving constant rates. In this section, we'll use what we already know from Sections 2.1 and 2.2 to create some fancier models. We begin by considering a model that starts somewhere other than at the y-axis.

When the Starting Value Isn't $f(0)$

The average girl in the United States is 102 cm tall at age 4. From then until age 13, her rate of growth is nearly constant at 6.1 cm per year. The model sounds linear, doesn't it?

▷   If *age in years* is the independent variable and *height in centimeters* is the dependent variable, what is the slope of this linear model?

If we try to use $y = b + mx$, though, we are stuck, because we don't have a value for b, the y-intercept. What's more, the y-intercept has no meaning in the model, because it would represent her height at birth, while the model applies only to the ages from 4 to 13.

There are two ways to write this model. The first is more straightforward and involves noting that the starting height (at age 4) is 102 cm, after which 6.1 cm are added every year *after the age of 4*. We write it like this:

$$y = H(x) = 102 + 6.1(x - 4)$$

where $H(x)$ is her height when she is x years old. (You can stop here, by the way. This formula tells the whole story. Multiplying it out would not improve it.)

▷ Convince yourself. Use the formula to calculate $H(x)$ for age 5 and for age 6. The height should increase by 6.1 cm each year.

Here's the second method. We start with the ordered pair $(4, 102)$, the slope 6.1, and the generic linear function:

> If a graph passes through a point, then the coordinates of the point satisfy the graph's equation.

$$y = b + mx$$
$$y = b + 6.1x \qquad \text{because } m = 6.1$$
$$102 = b + 6.1(4) \qquad y = 102 \text{ when } x = 4$$
$$b = 102 - 6.1(4) = 77.6 \qquad \text{the } y\text{-intercept}$$

Our model therefore is $y = 77.6 + 6.1x$.

▷ Show that this version of $H(x)$ is algebraically equivalent to the first one.

STOP & THINK

The average newborn is 50 cm long. Explain why $H(x) = 102 + 6.1(x - 4)$ is more informative than $H(x) = 77.6 + 6.1x$, even though they're mathematically equivalent.

The formula $y = 77.6 + 6.1x$ puts the emphasis in the wrong place (on the 77.6, a number that's meaningless for this model), when the message we really want to send is this:

$$y = H(x) = 102 + 6.1(x - 4)$$

start   add 6.1 cm per year
here   after age 4

There's no such thing as a free lunch—or a free peso

Even though one U.S. dollar might be worth 9.2 Mexican pesos (as it was in August 2001), if you actually needed to exchange dollars for pesos, you would probably encounter a service charge.

The transaction fee at one small branch bank is $15 (which is unusually high, because the bank processes exchanges through its main branch in another state). Let's see how to adjust the dollars-to-pesos model $P(d) = 9.2d$ from Section 2.1 to account for the service charge.

The bank first takes its $15 from your d dollars and then applies the exchange rate to the remainder. We'll call this adjusted model F:

$$p = F(d) = 9.2(d - 15)$$

where p, or $F(d)$, is the number of pesos received. Once again, this version tells the whole story. As you'll see, you lose information if you "simplify" the expression.

▷   Multiply the expression for $F(d)$ to put the function into slope-interc
      $b + md$.

▷   Does the value you obtained for b have any significance for this model? Why or why
      not? (Remember, b is the value of the function when $d = 0$.)

Euclid said it 2500 years ago

"Two points determine a unique line." (Euclid said it in Greek, though.) Thanks
to this truth, we can use the coordinates of any two points to write an equation for a
nonvertical line in a form known as the **point-slope form** of the line.

> **The Point-Slope Form of a Line**
> Given any two points that are not in the same vertical line, write a linear
> function in two steps:
>
> • Calculate the slope. Call it m.
>
> • Use one of the points, call it (x_1, y_1), as the starting point, and this
>   template: $f(x) = y_1 + m(x - x_1)$.

▷   Try it yourself: Write the linear function that contains the points $(-3, 7)$ and $(9, -1)$.
      (Which point are you calling (x_1, y_1)?)

▷   Do it again, calling the *other* point (x_1, y_1).

▷   Now show algebraically that both functions represent the same line by putting them both
      into the form $y = b + mx$.

STOP & THINK   Show that the point-slope formula $f(x) = y_1 + m(x - x_1)$ collapses to $y = b + mx$ when y_1 is the y-intercept.

What goes up must come down

According to the laws of physics, when a baseball is hurled straight up, its velocity is a linear function of time. If its velocity is measured, we might obtain readings similar to these:

Time Elapsed (seconds)	Velocity (ft/sec)
1.2	10
2.7	−38

A negative velocity means that the ball is falling back toward the ground.

▷   The table gives us two ordered pairs. Use them to write a linear function $v(t)$ that gives the velocity of the ball t seconds after it is thrown.

Units for this slope are
$$\frac{\text{velocity units}}{\text{time unit}} \text{ or } \frac{\text{feet/sec}}{\text{sec}},$$
sometimes written ft/sec².

The slope of this linear model is significant. Its numerical value is −32. That's 32 feet per second (in the downward direction) for every second that the ball is in motion. You might recognize this constant as an approximation of the acceleration due to earth's gravity. In a vacuum, every falling object will pick up speed at the rate of 32 feet per second per second.

▷   How fast was the ball rising at $t = 0$, the moment at which it was thrown? (Include units.)

This model needs a domain restriction on both ends. The model has no meaning for negative values of t, because the ball has not yet been thrown. It also has no meaning once the ball hits the ground. This ball happens to land after approximately three seconds (something you'll be able to figure out for yourself when you study calculus). Therefore, the domain for our model is the interval $0 \leq t \leq 3$ seconds.

▷   How fast is the ball traveling when it hits the ground? (That is, what is its velocity at $t = 3$?)

Check Your Understanding 2.6

1. You're planning a trip to Tokyo and need to change dollars to Japanese yen. The exchange rate is 122 yen to the dollar, and the bank imposes a $3 service fee. Write a function that gives the number of yen obtained if you change d dollars to yen.

2. The smallest size in women's shoes is 5, worn by a woman with feet that are $8\frac{2}{3}$ inches long. Sizes go up at the rate of three sizes to the inch. Write the shoe-size function in point-slope form and then show that it is equivalent to our model $w(x) = 3x - 21$.

3. Write a formula for the linear function that contains the points $(4, -10)$ and $(-3, -12)$.

Step Functions

Perhaps you realized back in Chapter 1 that our women's shoe-size model $w(x) = 3x - 21$ has several drawbacks, not the least of which is the fact that people often like extra toe room in their shoes. But there's another, specifically numerical, flaw. If you answered the Stop & Think question on page 11, you realized that a foot measuring $9\frac{3}{4}$ inches would, according to w, require a size $8\frac{1}{4}$ shoe, a nonexistent size. In practice, of course, the woman would ask for the next size, $8\frac{1}{2}$.

Here's the mathematical issue involved. The variable x that we use to measure foot length varies **continuously** between about 8 and 12 inches, limited only by the accuracy of our measuring device. By contrast, shoe sizes take on a limited number of values: integer or integer-plus-one-half values. We say that the dependent variable y in our model is **discrete** because consecutive y-values are *separated* from one another by intervals (in this case, by one half unit). Thus, the shoe store, which has a vested interest in a highly accurate model, probably recommends shoe sizes according to a modified model, which we'll call FIT. Table 2.6 indicates how FIT works.

Foot Length (inches)	Shoe Size
$8\frac{1}{2} < x \leq 8\frac{2}{3}$	5
$8\frac{2}{3} < x \leq 8\frac{5}{6}$	$5\frac{1}{2}$
$8\frac{5}{6} < x \leq 9$	6
$9 < x \leq 9\frac{1}{6}$	$6\frac{1}{2}$
$9\frac{1}{6} < x \leq 9\frac{1}{3}$	7
$9\frac{1}{3} < x \leq 9\frac{1}{2}$	$7\frac{1}{2}$
$9\frac{1}{2} < x \leq 9\frac{2}{3}$	8
$9\frac{2}{3} < x \leq 9\frac{5}{6}$	$8\frac{1}{2}$
$\vdots$	

Table 2.6   Partial Table of Values for Women's Shoe-Size Function FIT

The interval $8\frac{1}{2} < x \leq 8\frac{2}{3}$ denotes the infinite set of those real numbers strictly greater than $8\frac{1}{2}$ but not greater than $8\frac{2}{3}$. Notice that each of the intervals in the left-hand column has a width of $\frac{1}{6}$ unit.

The shoe store might not realize it, but FIT is an example of a **step function.** We've graphed a portion of it in Figure 2.11. An open circle signifies that the indicated endpoint of a segment is not included in the segment; a closed dot signifies that the point does belong to the segment. The graph makes it apparent why we call this type of function a step function.

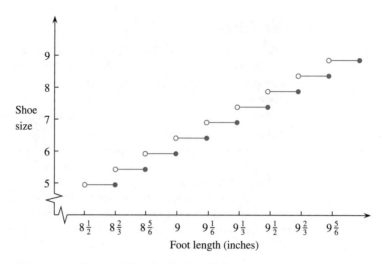

Figure 2.11 Graph of Step Function FIT

There's no shortcut for writing the formula for FIT. To express that function algebraically, we would need as many statements as there are shoe sizes, because the definition of the function changes each time x increases by $\frac{1}{6}$.

Whenever a function takes on one constant value over a specified interval, then switches to another constant value for the next interval, and so on, you have a step function, also known as a **piecewise constant** function.

Step functions appear all around us. The postage function is one such example. Current first-class postage rates are 37 cents for the first ounce plus 23 cents for each additional ounce. (These will probably have been raised by the time you read this.) Any final fraction of an ounce counts as a whole ounce. So, for instance, you can't get away with paying 20 cents for a letter weighing half an ounce, nor does a 50-cent stamp buy you any more weight than a 37-cent stamp would. Each postage fee is constant over a 1-ounce-wide weight interval. Exercise 45 asks you to draw the graph of the postage function, which looks much like the graph of FIT. Only the labels and the numbers are changed.

Piecewise Linear Functions

Step functions are a good introduction to piecewise linear functions, whose graphs are similar except that the segments aren't necessarily horizontal.

The triathlon function on page 20 is an example of a **piecewise linear** function. The graph you drew in Figure 1.9 should look like the one in Figure 2.12. Let's name the function d and see how to write its mathematical formula. We have three segments, so we'll need three statements.

▷   Find the slope (the speed of the athlete, in this case) of each segment.

The first segment starts at $d = 0$ when $t = 0$; its slope is 4. A formula for that segment is $d(t) = 4t$. Note, however, that the segment ends when $t = 1$.

The second segment starts at $d = 4$ when $t = 1$; its slope is 36. A formula for that segment is $d(t) = 4 + 36(t - 1)$. This segment ends when $t = 6$.

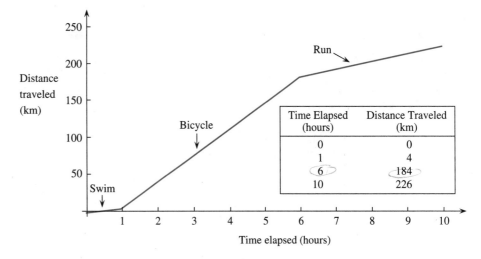

Figure 2.12   Distance versus Time for Triathlon

The third segment starts at $d = 184$ when $t = 6$; its slope is 10.5.

▷   It's your turn: Write a formula for the third segment.

After 10 hours when the race ends, the third segment also comes to an end.

This rule specifies one function, not three. Exactly one d-value corresponds to any t-value between 0 and 10.

We now have three separate formulas, all allegedly representing a single function d. To create a single function, we restrict each formula to the interval on which it applies, as follows:

$$d(t) = \begin{cases} 4t & \text{for } 0 \le t \le 1 \\ 4 + 36(t - 1) & \text{for } 1 < t \le 6 \\ 184 + 10.5t\,(t - 6) & \text{for } 6 < t \le 10 \end{cases}$$
the formula   its share of the domain

You should be ready to try a few of these on your own.

Check Your Understanding 2.7

1. A bank offers a basic checking account with the following terms: a monthly charge of \$2.50, which entitles the customer to eight free transactions, plus 75 cents for each additional transaction.

(a) Write the formula for a function $C(n)$ that gives the charge for a month in which there were n transactions. (*Hint*: You'll need two pieces.)

2. A generous employer contributes to a retirement fund an amount equal to 9% of an employee's salary for all earnings up to \$40,000 in a year and 13% of any earnings above that amount. (This sounds a little weird, but it has to do with balancing pension income and Social Security benefits.) Write the mathematical formula for a function that gives the retirement contribution $R(x)$ for a year in which an employee earns x dollars. You'll need to determine the starting amount for the second segment.

(b) Sketch a graph of $C(n)$. Think before you draw: Because there can't be a fraction of a transaction, the domain of the model is discrete, consisting only of integers. Your graph should consist of isolated points.

A Piecewise Linear and Exponential Model

On page 56, we compared two models for the current worth of a sound system. Our DJ Jason knows that it's to his advantage to claim on his income-tax return the largest allowable depreciation (and thus the lowest book value) in any given year. You found that the exponential model $B(t) = 300(0.80)^t$ gives a lower value than the linear model $V(t) = 3000 - 300t$ from $t = 0$ until their intersection at $t \approx 8.5$. Thus, for the first $8\frac{1}{2}$ years, he should use the exponential model, switching over to the linear model when it becomes more to his advantage.

The piecewise ideas we've used are equally valid here. We piece together a single function, IRS, to give the book value that Jason will use when his equipment is t years old. Its formula is

$$IRS(t) = \begin{cases} 3000(0.80)^t & \text{for } 0 \le t \le 8.5 \\ 3000 - 300t & \text{for } 8.5 < t \le 10 \end{cases}$$

The model ends at 10 years, the useful life of the equipment.

Figure 2.13 shows the graph of $IRS(t)$ and, alongside it, the two functions from which it is built. Observe that this composite graph consists simply of the appropriate segment of each of the other two graphs.

The first century philosopher Seneca, writing disparagingly of the study of mathematics, asked the following: "You know what a straight line is; but how does it benefit you if you do not know what is straight in this life of ours?" We would claim that you now have some notion about what's straight and what's not.

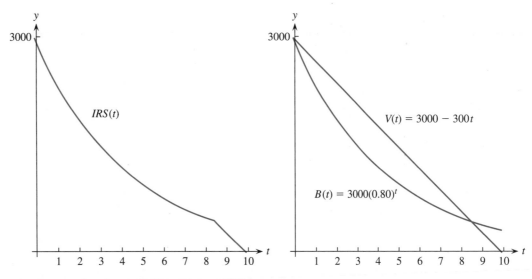

Figure 2.13 Graphs of $IRS(t)$, $B(t)$, and $V(t)$

In a Nutshell The point-slope form $f(x) = y_1 + m(x - x_1)$, gives a formula for the linear function f having slope m and containing the point (x_1, y_1). A piecewise function is a single function that is defined differently on different parts of its domain. The algebraic formula for a piecewise function contains a separate statement for each piece, specifying the portion of the domain over which that definition applies.

What's the Big Idea?

- In a linear function, the rate of change or slope, $m = \frac{\Delta y}{\Delta x}$, is constant. Linear functions model growth ($m > 0$) or decrease ($m < 0$) at a constant rate. Any one point and the rate of change determine a linear function.

- In an exponential function, the growth factor a is constant. Exponential functions model growth ($a > 1$) or decrease ($0 < a < 1$) at a constant percentage. A starting value, together with a constant growth factor or ratio, determines an exponential function.

- A piecewise function is one whose algebraic definition changes from one portion of its domain to another.

Progress Check

After finishing this chapter, you should be able to do the following:

- Recognize linear and exponential functions from their formulas. (2.1, 2.2)

- Write a formula for a linear function, given two points, or any one point and the constant rate of change. (2.1, 2.3)

- Use appropriate units for the slope of a linear model. (2.1, 2.3)

- Recognize and write equations for horizontal and vertical lines. (2.1)

- Write a formula for an exponential function, given its y-intercept and the constant growth factor. (2.2)

- Given two values of an exponential function, calculate the growth factor. (2.2)

- Given a graph, write a formula for a linear, piecewise linear, or exponential function. (2.1, 2.2, 2.3)

- Graph a piecewise function, given its formula or a description of its rule. (2.3)

- Explain the difference between the domain and range of an abstract function and the domain and range of a function used as a model. (2.1, 2.3)

Key Algebra Skills Used in Chapter 2

- Solving linear equations (pages 483–484)
- Working with exponents (pages 477–478)

Answers to Check Your Understanding

2.1

1. $r = 1.12d$; units for 1.12 are *euros per dollar*

2. (a) $y = 122d$ and $y = 108d$

    (b) The graph of $y = 122d$ is steeper, because the magnitude of its slope is greater.

3. $\frac{\Delta F}{\Delta C} = \frac{9}{5}$ degrees Fahrenheit per degree Celsius

4. −3.85

2.2

1. (a) $10   **(b)** $1.50   **(c)** 3

(d)

t	$PRICE(t)$
0	10
2	13
4	16
6	19
8	22
10	25

    (e) Consecutive *PRICE*-values differ by three units.

2. (a) $W(t) = 3.2 + 0.25t$

    (b) Possible domain: $0 \leq t \leq 26$ weeks
Corresponding range: $3.2 \leq W(t) \leq 9.7$
If the baby continued to grow at this rate, he would weigh 16.2 kg (close to 37 lb) by his first birthday, an unrealistically high weight. We've chosen to limit the model to his first six months. You might have decided differently; just be sure that the range corresponds to the domain.

2.3

1. (a) $y = 5$ **(b)** $m = 0$

    (c) $x = -4$ **(d)** the line $y = 5$

2. (a) $y = 0$ **(b)** $x = 0$

2.4

1. $y = 120000(1.025)^t$

2. $y = 120000(0.975)^t$

3. (a) $C(t) = 14.4(1.012)^t$, with population measured in *millions of students*

    (b) $C(11) \approx 16.4$; 16.4 million students

2.5

1. Solve this equation:

$$1.3a^{30} = 1.7$$

$$a^{30} = \frac{1.7}{1.3}$$

$$a = \left(\frac{1.7}{1.3}\right)^{1/30} \approx 1.009$$

This growth factor, 1.009, represents an annual increase of 0.9% (less than 1% per year).

2. $a \approx 0.85$; $V(t) = 15000(0.85)^t$

2.6

1. ~~$y = 112d - 3$~~ $y = 122(d-3)$

2. $y = 5 + 3(x - 8\frac{2}{3}) = 5 + 3x - 26 = 3x - 21 = w(x)$

3. The slope is $\frac{2}{7}$. The function is $f(x) = -10 + \frac{2}{7}(x - 4)$ if you use the point $(4, -10)$, and $f(x) = -12 + \frac{2}{7}(x + 3)$ if you use the point $(-3, -12)$. These are equivalent.

2.7

1. $C(n) = \begin{cases} 2.50 & \text{for } 0 \leq n \leq 8 \\ 2.50 + 0.75(n - 8) & \text{for } n > 8 \end{cases}$

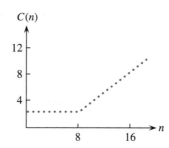

2. $R(x) = \begin{cases} 0.09x & \text{for } 0 \leq x \leq 40{,}000 \\ 3600 + 0.13(x - 40000) & \text{for } x > 40{,}000 \end{cases}$

The 3600 represents 9% of the first $40,000 of salary. Therefore, the starting point for the second segment is (40000, 3600).

EXERCISES

1. Students placed a battery-operated toy car in front of a Calculator Based Ranger (CBR), turned the car on, and used the CBR to gather time and distance data on the motion of the car. The table shows some of the data collected.

t, time in seconds since CBR began recording	d, distance in feet from CBR
0	2.500
1.8	4.084
3.6	5.668
5.4	7.252

   (a) Was the car moving toward or away from the CBR? Was its velocity approximately constant? How do you know?

   (b) Write a formula for d in terms of t. Your formula should contain two constants. What units are associated with those constants?

2. A plant nursery sells three tulip bulbs for 99 cents. For customers buying 20 or more bulbs, the nursery reduces the price to four bulbs for a dollar.

   (a) Write a formula for $c(x)$, the cost in dollars of purchasing x bulbs, if $x < 20$.

   (b) Write a formula for $c(x)$ if $x \geq 20$.

   (c) On the same set of axes, sketch graphs of your formulas from (a) and (b) over the interval $0 \leq x \leq 40$. Then color or highlight the portion of each graph that makes sense in this context.

3. Most purchases in the state of Massachusetts are subject to a 5% sales tax.

   (a) If you purchase a watch priced at $150, how much would you pay in sales tax? How much, in total, would you pay for the watch?

   (b) Let $T(p)$ represent the tax, in dollars, charged on a taxable item whose price is p dollars. Write a formula for $T(p)$.

   (c) Let $C(p)$ represent the total cost (price plus tax), in dollars, of a taxable item priced at p dollars. Write a formula for $C(p)$.

   (d) Is C proportional to p? If so, what is the constant of proportionality? If not, why not?

4. There are approximately 2.54 centimeters in 1 inch. How many inches are in 1 centimeter?

   (a) Write a formula that converts x inches into centimeters. What units are associated with the slope constant in the formula?

   (b) Write a formula that converts x centimeters into inches. What units are associated with the slope constant in this formula?

5. An athlete's resting heart rate is 50 beats per minute. When she jogs at 6 miles per hour, her heart rate is 125 beats per minute. Within certain limitations, the relationship between running speed and heart rate is approximately linear.

   (a) Write a linear model $R(x)$ that gives heart rate as a function of running speed x, measured in miles per hour. The model will have two constants. Give the units associated with each constant.

   (b) What is her heart rate, according to this model, when she sprints at 9 miles per hour?

6. A snail oozes at a uniform speed along the crossbar of a picket fence. On day zero, it is at the third picket; on day five, it has reached the tenth picket. Write a linear model of the snail's position on day n. In what units is the slope measured?

In Exercises 7–12, give a formula that could represent each graph.

7.

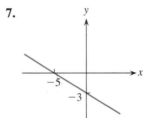

8.

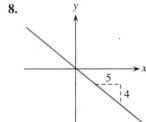

9.

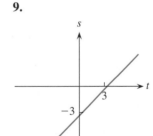

10.

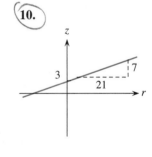

11.

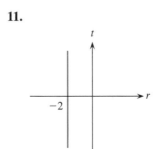

12.

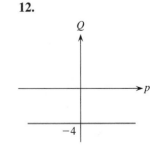

13. *I scream, you scream, . . .* Operating a business for a day costs a certain amount, known as the fixed cost, even if nothing is manufactured or sold on that day. Suppose it costs $150 per day in wages, supplies, rent, and utilities to keep an ice-cream-cone stand open, regardless of how much ice cream is sold, and that the business nets 65 cents for every ice-cream cone sold.

(a) Write a mathematical model *ICE* that gives the profits as a function of the number of cones sold.

(b) Evaluate *ICE*(0). Write a sentence explaining what it means.

(c) Evaluate and interpret *ICE*(100) and *ICE*(500).

(d) How many cones must the stand sell in order to break even for the day?

(e) Give a reasonable domain for the function *ICE*, considered as a mathematical model for this business enterprise.

(f) Give the range of the mathematical model, making sure that it corresponds to the domain you chose.

14. *"Trash has given us an appetite for art."* (Film critic Pauline Kael) The accumulation of debris in ancient cities caused the ground level to rise over the years. One well-studied case is that of Troy. A civil engineer with the U.S. Department of Commerce calculated that the elevation of the city of Troy rose at a rate of about 4.7 feet per century. The ground level of Troy in the classical epoch (about 1500 B.C.) was approximately 123 feet above sea level.

Years B.C. run *backward*. They behave on a time line like negative numbers on a number line. Although there is no "year zero" in our current system of reckoning, you can pretend, for the purposes of this exercise, that the origin of the time line is zero.

$$\longleftarrow \text{ B.C. } \quad \text{ A.D. } \longrightarrow$$

(a) If we assume a constant rate of change of elevation, how high was Troy 20 centuries ago? How high was Troy in 3000 B.C. when people first began to live there, on what was then only a rocky hump? How high would Troy be today if the rate of rise had remained constant? (The present-day elevation is actually very close to what it was in Homer's day, about 850 B.C.)

(b) Write the elevation of Troy as a linear function. Be sure to define your independent variable (not simply "time," but time since when, and measured in what units?).

You might be wondering why the city accumulated so much debris. Sanitary landfills are a relatively recent addition to our culture. In ancient times, people simply tossed their refuse out the window or even dumped it on the floor of their house. Whenever necessary, they covered it over with a layer of fresh dirt. Gradually, the floor level of the house might rise so much that the inhabitants would have to raise the roof.[1] Fires, wind, and rain also wreaked havoc on dwellings. Lacking bulldozers, the inhabitants simply leveled the wreckage and rebuilt above it.[2]

15. Could *y* be a linear function of *x*? Use the table to produce your evidence. (You need to consider all the values given.)

x	−5	−1	2	7
y	0	2	3.5	6

16. Could *y* be a linear function of *x*? Give evidence from the table.

x	−5	−1	2	7	20.6
y	0	2	3.5	6	11.9

---

[1] See *Rubbish! The Archeology of Garbage* by William Rathje and Cullen Murphy, Harper-Collins Publishers, 1992, pp. 34–35.
[2] Thanks to Robert O. Edbrooke of Buckingham Browne & Nichols School, Cambridge, MA, for providing archeological information about Troy.

17. Given the following function values, could T represent a linear function? Don't decide by drawing a graph; consider the rate at which the function changes.

$$T(2) = 11, \qquad T(5) = 6, \qquad T(12) = -4$$

18. $H(0) = \frac{66}{7}$, $H(5) = 8$, $H(12) = 6$. Can H be a linear function? If so, write its formula.

19. Two lines are parallel if they have the same slope; they are perpendicular if the slope of one line equals the negative reciprocal of the slope of the other. In other words, if line L_1 has slope m_1 and line L_2 has slope m_2, then L_2 and L_1 are parallel if $m_2 = m_1$ and perpendicular if $m_2 = -\frac{1}{m_1}$.
  (a) F is a linear function with values $F(0) = 5$ and $F(3) = 9$. Write a formula for $F(x)$.
  (b) Write an equation for the line parallel to the graph of F whose y-intercept is -2.
  (c) Write an equation for the line perpendicular to the graph of F at its y-intercept.

20. The graph shows three lines. L_1 and L_2 are parallel; L_1 and L_3 are perpendicular.

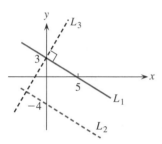

  (a) Write an equation for L_1.
  (b) Write an equation for L_2.
  (c) Write an equation for L_3.

21. In an article entitled "Is Humanity Suicidal,"[3] Edward O. Wilson wrote, "Now in the midst of a population explosion, the human species has doubled to 5.5 billion during the past 50 years. It is scheduled to double again in the next 50 years. No other single species in evolutionary history has even remotely approached the sheer mass in protoplasm generated by humanity."
  (a) On the basis of the paragraph quoted, explain why it is reasonable to assume that the author has in mind an *exponential* model for population growth.
  (b) Letting x represent the number of 50-year periods since 1993, write the formula for an exponential model that matches the information given in the article.

  (c) According to your model, how long will it take for the world's population to quadruple from its 1993 level? To grow to 16 times its 1993 level?
  (d) What is the *annual* growth factor for a doubling-every-50-years process? Does it correspond to the population growth factor we used on page 66?
  (e) Clearly, this model cannot be valid forever. Why not?

22. In 1940, the United States had 131 million people, 1.4 million of whom were college students. By 1995, the population had doubled, and the number of college students had increased tenfold.
  (a) Assuming that the U.S. population will continue to double every 55 years, write a formula $P(t)$ that gives population as a function of time, where t is the number of 55-year intervals since 1940.
  (b) Use the formula for $P(t)$ to project the U.S. population in 2006. (Be careful: What value will you use for t?)
  (c) Demographers predict 16.4 million U.S. college students in 2006. Would an exponential model based on 1940–1995 predict that value? Show how you decided.

23. Imagine that you decide to start a chain letter. You send copies to six friends (round 1) with instructions for them to send copies to six of their friends (round 2), and so forth. We'll make the improbable assumptions that no one breaks the chain and that no person receives more than one copy of the letter.
  (a) How many letters are mailed in round 3? In round 5?
  (b) Write a formula that tells how many letters are sent in round n, where $n \geq 1$. Use the formula to determine the number sent in round 10.
  (c) Suppose that the letters are sent only within the United States. How many rounds would assure that every person in the country receives a letter? (The 1999 U.S. population was estimated to be 280 million.)

24. The radioactive isotope uranium-232 has a half-life of 72 years. (This means that after 72 years, only half of the original amount remains unchanged, the other half having decayed into other radioactive substances and ultimately into a nonradioactive substance.)
  (a) Given an initial 10 grams, write a model for the amount of U-232 as a function of time, measured in half-lives. What number did you use as the growth factor?

[3] *New York Times Magazines*, 30 May 1992. Copyright ©1990/1992 by The New York Times Company. Reprinted by permission.

(b) How many half-lives will be needed for all but one eighth of the initial 10 grams to be gone? How many years is that?

(c) On the basis of this model, will there ever be a time when all of the initial amount of U-232 has decayed? If so, how long will it take? If not, why not?

(d) How would you modify the mathematical model if the initial amount were 50 grams? Explain why your answers to parts (b) and (c) would not change if the original quantity were 50 grams rather than 10.

25. There were approximately 38,000 cases of polio in 1988, a number that has been decreasing, on average, by about 25% each year.

(a) Write a model $f(t)$ that estimates the number of cases t years from 1988.

(b) How many cases does f predict for the year 2000, the year in which the World Health Organization hoped to eliminate the disease?

(c) Does your result mean that polio cannot be eliminated? Explain.

(d) How many cases does f estimate for the year 1985? (Make t negative.)

26. Dot-com, Inc., offers you a starting salary of $35,000 a year, with annual raises of $2100 if your performance is outstanding. Macronet offers $33,500 with annual raises of 7.5% as long as you receive outstanding evaluations.

(a) Write a formula for $D(t)$, your salary after t years with Dot-com.

(b) Write a formula for $M(t)$, your salary after t years with Macronet.

(c) If you plan to hold your job for at least 10 years, which position offers you more financially? What if you plan to stay in this job for only four years? (Your work, naturally, will be outstanding in either position!) These questions are more complicated than they appear. You need to consider not only the final year's salary for each position, but also the total earned over time and the fact that the same amount of extra money is worth more in the beginning than at the end, because you get to keep it longer. Reasonable people might come to opposite conclusions, so be sure to explain how you reached yours.

27. A population of 24 deer, introduced in 1990 onto an island where they have no natural predators, is growing at a rate of 29% per year.

(a) What was the deer population in 1991? In 1992?

(b) Write a mathematical model to predict the deer population t years from 1990. What is the growth factor in your model?

(c) Use the graph of your model to estimate the year in which the deer population would have reached 10 times its original size.

28. The price of a cellular phone has been steadily dropping. In 1991, one particular type sold for $330. In 1992, its price was $287.

(a) Assume that the price drops by the same *amount* each year. Write a model for the price of the cellular phone. What kind of function is your model? Use it to estimate the 1995 price of this phone.

(b) Assume, instead, that the price drops by the same *percentage* each year. Write a model for the price of the phone. What kind of function is this model? Use it to estimate the 1995 price.

(c) On the same axes, sketch a graph of each model for the years 1990 to 1999. (Be careful with years. You probably used $t = 0$ to represent 1991. What value of t, then, would represent 1990?)

(d) Which model seems more realistic to you? Explain.

29. A biology student counts 12 bacteria in a petri dish at the start of an experiment and 34 bacteria three hours later. Find the growth factor and write an exponential model for the number of bacteria t hours from the start of the experiment.

30. The aquarium in a professor's office turned green with algae, so she bought eight algae-eating snails and dropped them into the tank. Two weeks later, she counted 20 snails in the aquarium. In another two weeks, she counted 50.

(a) Find the growth factor and write an exponential model $N(t)$, where N is the number of snails and t counts the two-week periods since she first added snails to the tank.

(b) Does it make a difference whether you determine the growth factor from the 20 snails in two weeks or the 50 snails in four weeks? Explain.

(c) Recalculate the growth factor, measuring time in days, and write a model $N(x)$ for the number of snails on day x.

In Exercises 31–36, write the formula for an exponential function that could represent each graph. (Use the given information to determine each growth factor.)

31.

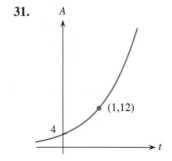

32.

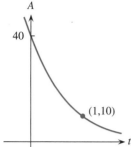

33.

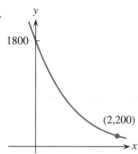

34.

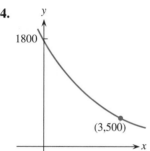

35.

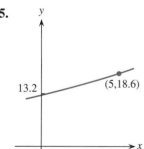

36.

In Exercises 37–40, determine whether the function specified by the table could be linear or exponential. (At least one is neither.) If it is either exponential or linear, write a possible formula for the function.

37.

x	5	10	15	20
y	100	85	70	55

38.

x	0	1	3	5
y	10	50	1250	31,250

39.

x	2	4	6	8
y	20	40	120	480

40.

x	0	3	4	6
y	120,000	7680	3072	491.52

41. Jason, flush with monetary success as a DJ, decided to start a business renting small planes. He purchased his first plane, a Saratoga II HP, for $398,900. He figured that the useful life of this plane would be 25 years.
  (a) Write a formula for the book value of his plane after t years, using straight-line depreciation.
  (b) Write a formula for the book value of his plane after t years, using the double-declining-balances method. (See page 68.)
  (c) In how many years will the book value of his plane using straight-line depreciation match its book value using double declining balances?

42. A linear function g has slope -0.6; the value of g at 2 is -5. Write a formula for $g(x)$.

43. H is a linear function; $H(12) = 7$ and $H(7) = 12$. Write a formula for $H(x)$.

44. L is a linear function; $L(-1) = 7$ and $L(2) = 3$. Write a formula for $L(x)$.

45. Let P represent the cost, in cents, of mailing a first-class envelope weighing w ounces. (You first met this function on page 76.) Assume that the first-class rates are still 37 cents for the first ounce plus 23 cents for each additional ounce. Any final fraction counts as a whole ounce.
  (a) How much postage is needed for a letter weighing $2\frac{1}{4}$ ounces? $2\frac{3}{4}$ ounces?
  (b) Draw a graph of $P(w)$ over the interval from 0 to 5 ounces.

46. The greatest integer function, denoted by $[[x]]$, has as its output the integer n such that $n \leq x < n + 1$. So, for example, $[[2]] = 2$, $[[2.7]] = 2$, and $[[-0.007]] = -1$.
  (a) What is $[[5.99]]$? What is $[[-1.8]]$?

(b) Draw a graph of the greatest integer function on the interval $-4 \leq x \leq 4$.

47. The Internet service America Online has a Light Usage Plan, according to which the user pays $4.95 per month for the first three hours and $2.50 per hour for each additional hour. The function $AOL(t)$ represents the charge in a month when the user spends t hours on line. Is this a step function? That depends. Does AOL bill for fractions of an hour, or does it charge the entire $2.50 for the final fraction of an hour (as the U.S. Postal Service charges for the final fraction of an ounce)?

(a) Sketch a graph of $AOL(t)$ for the first six hours, assuming a step function.

(b) Sketch a graph of $AOL(t)$ for the first six hours, assuming that the company bills for fractions of an hour.

(c) What is a reasonable domain for the AOL function? What is the corresponding range? (Do you need to know whether or not the company bills for fractions of an hour to answer these questions?)

48. In 1997, Shaw's Supermarket offered a Spend & Save promotional plan. A customer received Spend & Save points for each purchase of at least $30. As the promotional pamphlet explained, "Spending $47 earns 30 points, while spending $62 earns 60 points. Points are awarded only in multiples of 30."

(a) How many points would a customer receive with a purchase totaling $29.75? $91.50?

(b) Draw a graph of $P(s)$, the points earned with a purchase of s dollars.

(c) After accumulating a certain number of points, the customer could turn them in for merchandise credit. With 390 points, there was a $10 credit; with 750 points, a $25 credit; and with 1050 points, a $45 credit. Is the relationship between the amount of credit and the number of points a linear relationship? Give evidence. Can you, with the information given here, predict the merchandise credit awarded for 1290 points?

In Exercises 49 and 50, write a formula for the piecewise linear function defined by the graph.

49.

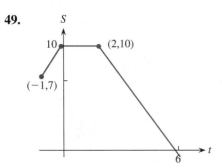

50.

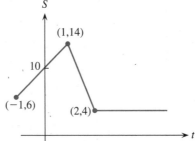

51. Lauren earns an hourly wage of $12 for her work as an emergency medical technician. If she works more than 40 hours in a single week, however, she earns $18 for each of the additional hours.

(a) Write a formula giving her weekly wage as a function of hours worked.

(b) What are a reasonable domain and range of this model? (Remember the upper limits as well as the lower.)

(c) Sketch a graph of the model. Label the axes.

52. Massachusetts does not assess sales tax on clothing, unless the item costs more than $175. In that case, the state charges a 5% tax on the amount over $175. Write a piecewise model for $C(p)$, the total cost (including tax) of an item of clothing priced at p dollars. (*Reality check*: You should pay $30 for a $30 pair of jeans but $196 for a $195 coat. Be sure $C(p)$ produces those results.)

53. An epidemiologist studying a flu epidemic has gathered data at 10-day intervals and produced this graph. This particular strain of flu last for several days, so the number of people who are ill on any given day includes those who have just succumbed as well as those who are near recovery.

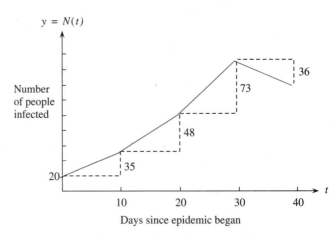

(a) How many people have the flu on day 20?

(b) During which 10-day period is the disease spreading most rapidly? How do you know? What is the approximate rate of spread (in persons per day) during that period?

(c) Write a formula for each of the four line segments.

(d) Write a formula for the piecewise-linear function $N(t)$.

(e) Use your formula to estimate $N(17)$ and $N(35)$, the number of ill people on day 17 and day 35, respectively.

54. An epidemiologist tracking the course of an outbreak of 24-hour stomach virus has gathered these data.

Days Since start of Study	Number of People Sick
0	20
10	35
20	61
30	107
40	82
50	57

(a) Plot the data from the table.

(b) These data indicate two phases of the illness. In Phase I, the illness is spreading ever more rapidly and can be modeled by an exponential function. In Phase II, the number of cases declines steadily and can be modeled by a linear function. Identify a probable time interval for each phase.

(c) For Phase I, write an exponential function. For Phase II, write a linear function. Piece the two together to create a single piecewise function $V(n)$ that gives the number of people V who have the virus on day n.

(d) On your plot of the data, sketch the graph of $V(n)$, making sure that it fits the data well.

(e) Use $V(n)$ to estimate the number of people who are ill on day 22 and on day 38.

(f) Use $V(n)$ to predict the day on which the last patient will recover.

55. In Exercise 41, you helped Jason figure out the book value of his airplane, using two different methods. Now you will help him with his income tax return. It is to his advantage to claim the larger allowable depreciation (and thus the lower book value) in any given year.

(a) From the linear and exponential book-value functions of Exercise 41, piece together a single function IRS that gives the value that Jason will report when his airplane is t years old. Write the formula for $IRS(t)$.

(b) Sketch the graph of $IRS(t)$ for the first 25 years, the expected useful life of the plane.

2.1 HOW FITTING!—The Least Squares Line

Sometimes, a mathematical model is theory-driven. For example, assuming constant annual growth leads to a linear model. At other times, the model is data-driven. That is, you have the data, and you have to decide what function best describes the pattern made by the data.

In this project, you will learn to use a calculator to produce a best-fitting line for a set of data. First, you'll create a model describing the relationship between average IQ and year.

A recent newspaper article claimed that Americans are getting smarter. Table A contains data on the average scores on IQ tests taken by Americans, data that appear to support the paper's claim.

Year	Average IQ
1932	100
1947	106
1952	108
1972	112
1978	113
1997	118

Table A

1. Plot these points on graph paper, using x to represent *years since 1932* and y for the average IQ score. (Use a zig-zag on the y-axis to break the scale and begin the labeling closer to 100, the lowest IQ score in Table A.) Draw a straight line that you think best represents and summarizes the data you plotted. (You won't be able to force the line to pass through all the points.)

2. What is the y-intercept for your line? What is its slope? Determine the linear function for the line you drew.

3. Now you'll do a little statistical analysis on the data. Begin by completing Table B. "Predicted y" means the y-value that your equation produces for the given x-value. "Residual error" is the difference between the value we actually have and the one

x	y	Predicted y	Residual error $(y - \text{predicted } y)$	(Residual error)^2 $(y - \text{predicted } y)^2$
0	100			
15	106			
20	108			
	112			
	113			
	118			

Table B

that the linear function gives. Subtract the second column entry from the third to find the residual. For the last column, square the values in the fourth column.

4. Finally, add up the last column and write the total: _____

How are we to decide whether or not this line is a good fit for the data? There are many ways of selecting a line to represent a set of data, and if you compare your line with those of other students, you'll probably find that no two are exactly alike. To avoid this ambiguity, statisticians frequently follow what is called the **least squares** criterion for selecting a single line to represent a data set.

When you calculated the residual errors for your line, you probably had some negative values and some positive values. If we added those numbers, they could conceivably sum to a value close to zero, implying hardly any discrepancy between the data and the linear model; yet the points would deviate from the line in both directions. To sidestep the issue of signs, statisticians use the *square* of each error, because the main concern is how far from the predicted value each point is rather than whether it's above or below.

The least squares criterion says to select the line for which the total of the squared residual errors is as small as possible. (This total is also known as the sum of the squared errors, or SSE.)

5. What is the SSE for your linear function?

Now let's see how well your line compares with what a statistician would consider the line of best fit. Use your calculator or statistical software to compute the least squares (also called the linear regression) equation. (Consult a graphing calculator guide or the manual for your own calculator for instructions on linear regression.)

6. Write the equation of the least squares line. Round the slope value to three decimal places and the intercept to one decimal place.

7. Complete Table C, using the least squares line for the "Predicted y."

x	y	Predicted y	Residual error (y − predicted y)	(Residual error)^2 (y − predicted y)^2
0	100			
	106			
	108			
	112			
	113			
	118			

Table C

8. Compare the SSE for the least squares line to the SSE for the linear function you wrote in Question 2. Why do you think that statisticians would prefer the line with the smaller SSE?

9. Use a grapher to view the data, the least squares line, and your original linear function. Now that you see everything, do you agree that the least squares line does a better job of capturing the overall trend of the data?

10. The newspaper article reported that scores on intelligence tests are going up at a rate of three IQ points per decade. On the basis of the least squares equation that you determined in Question 6, do you think that this is a reasonable statement? Explain.

11. Use the least squares equation from Question 6 to predict the average IQ for the current year.

2.2 SKELETON KEYS

Plan ahead, or this project will take too long! Unless your instructor is able to link calculators and transfer the data, work together. Each group member does a different equation in Question 1. Share results.

Physical anthropologists can determine much about a deceased person from his or her skeletal remains. From the femur (thigh bone) or the ulna (the inner bone of the forearm), for example, a scientist can estimate the person's height.

In this project, you will play the role of a physical anthropologist and solve the mystery of the bones. Your equations will be similar to those proposed by Dr. Mildred Trotter, a special consultant to the U.S. government during World War II, who identified dead soldiers from their skeletal remains. Forensic scientists and law enforcement agencies still use some of Dr. Trotter's formulas for estimating a person's height from the lengths of his or her bones.

The following data come from the Forensic Data Bank at the University of Tennessee. There were 29 female and 31 male skeletons represented.

	Females				Males		
Sample No.	Height (cm)	Ulna (mm)	Femur (mm)	Sample No.	Height (cm)	Ulna (mm)	Femur (mm)
1	168	258	448	1	177	279	488
2	161	227	413	2	175	272	464
3	158	237	432	3	181	290	487
4	173	265	473	4	171	268	454
5	163	236	428	5	175	271	470
6	168	244	441	6	183	281	505
7	168	246	448	7	167	264	447
8	165	246	435	8	173	276	463
9	173	266	483	9	180	288	485
10	163	250	450	10	180	278	494
11	158	232	414	11	171	260	448
12	163	245	443	12	175	272	456
13	170	253	440	13	171	258	449
14	159	233	419	14	168	262	444
15	170	253	449	15	177	266	483
16	165	252	443	16	180	281	490
17	165	249	451	17	178	278	477
18	165	240	448	18	178	271	487
19	168	248	450	19	181	·282	488
20	163	236	435	20	170	266	459
21	165	250	448	21	173	261	460
22	163	244	434	22	173	261	460
23	165	248	452	23	183	285	502
24	177	279	488	24	178	278	480
25	171	270	461	25	177	282	482
26	162	245	421	26	166	258	442
27	180	281	482	27	180	290	505
28	180	279	486	28	175	274	470
29	175	273	484	29	171	270	461
				30	172	278	449
				31	180	281	482

1. We wish to be able to predict a person's height from the length of one of his or her bones, either the femur or the ulna. Thus, *height* will be the dependent variable, and *length of femur* (or *length of ulna*) will be the independent variable. The table suggests four equations, representing the four different relationships: two for men's heights and two for women's heights.

(a) Using statistical software or a calculator, determine the regression equation (the least squares line) for each of the four relationships. Be sure to use the correct independent and dependent variable each time. Round the slope to three decimal places and the intercept to one decimal place. Write the four equations, labeling each.

(b) Use a grapher to view four plots, each one showing a different set of ordered pairs and the regression line that fits those ordered pairs. Make sure that the line appears to summarize the general trend of the plotted points.

(c) What are the units for the slope of each line?

2. Suppose that two men have femurs that differ in length by 1 cm (10 mm). By how much should we expect the men's heights to differ? (*Hint*: Use the appropriate regression line.)

3. Suppose that two women have ulnas that differ in length by 2 cm (20 mm). By how much are the women's heights likely to differ?

4. Imagine that you are called in to advise law enforcement authorities on a case in which all that remains of a person is a femur bone. With nothing more to go on, you are unsure whether the person was male or female. If the length of this bone is 470 mm, how much difference would it make if you used the regression equation for predicting the height of a man and it turned out that the bone belonged to a woman?

5. For what femur length would the regression equation for predicting a male's height give the same results as the regression equation for predicting a female's height?
(a) Show how to find the answer using algebra.
(b) Explain how to check your algebraic result using graphs.
(c) Is that result a reasonable femur length for an actual person? (Look at the data to see what's reasonable.)

6. If a man and a woman have femurs of the same length, which of them is likely to be taller? Justify your answer using graphs of the two regression lines, remembering to stay within reasonable bounds for femur lengths.

7. For what ulna length would the regression equation for predicting a female's height give the same result as the regression equation for predicting a male's height?
(a) Show an algebraic solution.
(b) Tell how to check your algebraic results using graphs.
(c) Is that a reasonable ulna length for an actual person? How do you know?

8. If a man and a woman have ulnas of the same length, which of them is likely to be taller? This situation is more complicated than the similar question in Question 6. Your answer will need two parts. Justify your answer using graphs.

9. Suppose that a student who wanted to predict women's stature from their femur lengths inadvertently used the men's equation instead.
(a) On the basis of the data we have, women's femur lengths fell in the interval from 413 mm to 488 mm. If the student used these lengths with the wrong equation to predict stature, in what interval would the predicted heights fall?
(b) Given your answer to part (a), do you think the student would notice the error? (If you have trouble visualizing metric heights, translate them into feet and inches, recalling that 1 inch is approximately 2.54 cm.)

10. A hunter discovers partial skeletal remains in the woods. Your expertise is needed to solve the mystery of these bones. Among the intact bones are a skull, a left and a right ulna, and a femur (found some distance from the other bones). From measurements of the skull, you are fairly certain that the person was male. The left and right ulnae measure 275 and 276 mm, respectively. The femur measures 474 mm.
(a) Predict the stature of the dead man.
(b) Is there any reason to think that the femur came from a different individual? Explain.

2.3   JAIL TIME—Modeling California's Prison Population

In 1980, there were approximately 24,570 inmates in California prisons. In 1981, there were approximately 29,200. In this project, you will determine six different models for the California prison population t years since 1980.

Theory-Driven Models

1. From 1980 to 1981, by how much did the number of inmates increase? Assuming a constant annual rise in the prison population of California, write a model $Y_1(t)$ to represent California's prison population t years from 1980.

2. What was the relative rise in the prison population from 1980 to 1981? Assuming a constant percentage increase, write a model $Y_2(t)$ that represents the same population.

Data-Driven Models

3. Data on California's prison population are given in the table. Use your grapher to make a **scatter plot** (a plot of just the individual points) of these data. On the same axes, overlay the graphs of $Y_1(t)$ and $Y_2(t)$. Looking at the graphs, describe any obvious problems with the models Y_1 and Y_2.

Year	Population
1980	24,570
1981	29,200
1982	34,640
1983	39,370
1984	43,330
1985	50,110
1986	59,484
1987	66,975
1988	76,170
1989	87,300
1990	97,310

4. Alter the slope of your linear model Y_1, writing an adjusted model that does a better job of representing the data as a whole. Do your best; the fit won't be great, but it will be improved. Name the revised linear model L. What is the slope of L?

5. Alter the growth factor of your exponential model Y_2, writing an adjusted model that does a very good job of representing the data. Name the revised exponential model E. What is E's growth factor?

6. Use your calculator's regression capabilities to fit a linear function to the data, and compare it to L.

7. Use your calculator to fit an exponential function, and compare it to your model E. This time, you can expect great similarity.

8. In May 1997, the California Department of Corrections projected that the prison population would reach 242,000 by the year 2006. Find out whether any of the six models (Y_1, Y_2, L, E, and the two functions that your calculator provided) you have developed here would predict something close to that number. What does this say about the models?

2.4 POWER TRIP—Exploring Exponential Functions

How can we recognize an exponential pattern? What characteristics do all exponential functions share? In this investigation, you will examine the general exponential function $c \cdot a^x$ to see the effect of changing the constant multiplier c and the base, or growth factor, a.

Changing the Base

1. First we'll find out what a does to the graph. For simplicity, let's start with $c = 1$ and experiment only with the value of a. Using a narrow viewing window such as $-2 \le x \le 2$, try several *positive* numbers for a and make comparisons. Include fractional values (numbers between 0 and 1) as well as bases greater than 1. You should see two distinctly different groups of graphs. Summarize the ways in which the size of the base affects the graph.

2. All the graphs you viewed share a y-intercept. What is it?

Whenever a curve eventually starts to resemble a line, that line is known as an **asymptote** for the graph. If you widen the viewing window, you will observe that all of the graphs tend to level off at the x-axis, either on the right or on the left. Therefore, we say that the x-axis is the **horizontal asymptote** for each of these graphs.

3. Let the base be 1 and describe the graph. Compare it with the other graphs you saw. Think about the horizontal asymptote, and suggest a reason why mathematicians do not consider 1^x an *exponential* function. (It's a legitimate function, but we don't call it "exponential.")

4. Suppose the base is 0. What sort of graph do you see? It's possible that you won't see anything at all. If that happens, try viewing the graph with the axes turned off. The graph, in fact, coincides with the right side of the x-axis. Why doesn't the function 0^x exist for $x \le 0$? Any ideas? (Along with 1^x, the function 0^x isn't considered an exponential function.)

5. Try using a negative number as the base. You'll need parentheses around it, for example, $(-3)^x$. (Without parentheses, the function is simply the opposite of 3^x.) What happens? You probably will not have much success! Substitute several values, including some fractions, for x so that you can get an idea why the grapher is unwilling to draw you a picture.

Your investigations so far should help you understand why the base of any exponential function is a *positive* number, not equal to 1.

Changing the Coefficient

6. Now hold the base constant at 2 or 3 and let c vary. Use both positive and negative values for c. Describe its effect on the graph.

7. What is the y-intercept of any exponential function $c \cdot a^x$?

8. What is the horizontal asymptote of any exponential function $c \cdot a^x$?

9. What is the range of any exponential function of the form $c \cdot a^x$ if c is positive? If c is negative?

2.5 PYRAMID POWER

In early 1997, riots erupted in the tiny Eastern European country of Albania, the poorest country in Europe. Albanian citizens were outraged over their financial losses, which were precipitated by the breakdown of several pyramid schemes. Perhaps half the families in Albania had poured their own savings, as well as contributions from family members working abroad, into these swindles. A pyramid scheme creates the illusion of success by offering a very large interest rate to investors. The money to pay the interest, however, comes not from the operations of a business but from the cash of new investors. Sooner or later, the racket must run out of new investors and collapse. Albania, just emerging from years of a repressive Communist government, had no financial regulations to restrict the operation of these investment funds and thus to protect a financially naive population.

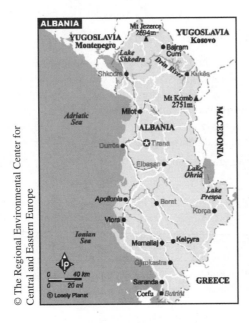

In this project, you will use what you know about exponential functions to model a representative pyramid scheme and see why it was destined for an early collapse.

One of the largest of the schemes promised gullible investors interest of 50% *per month*. That offer was particularly seductive because the banks were offering interest of only 8% *per year*, which didn't even keep up with inflation. Let's create a mathematical model of this pyramid scheme, run by a woman named Maksude Kademi, who was arrested for fraud when the pyramid tumbled.

As usual, when we model a situation from the real world, we will simplify the problem. Assume that the company accepts investments in multiples of $1000 only and that each $1000 corresponds to a different investor. Suppose she found 300 people to invest at the start; that is, the pyramid began with $300,000. Let's take as the rate of return the 50% per month that Kademi promised. (The first investors did make out handsomely and served as the best advertisers of the scheme.)

1. At the end of one month, how much interest is owed to the initial investors?

2. Assuming that Maksude Kademi wants to keep the original $300,000 for herself, she must make this interest payment by convincing other people to invest enough to cover the interest owed to the original investors. At $1000 per investor, how many *new* investors does the pyramid scheme need to rope in by the end of the first month?

3. At the end of the second month of operation, the pyramid scheme must come up with 50% interest for the original 300 investors as well as for the investors obtained during the first month. How much money does Kademi need for the interest payments at the end of the second month? How many new investors does this amount require?

4. Continue this process for two more months, keeping in mind that she has to pay interest to *all* the investors, not merely the new ones. (For now, don't worry if you get a fractional number of investors. Remember that this is only an approximation to a very messy real-world situation.) Summarize your findings in the table on page 95.

Duration of Scheme (months)	Total Amount Invested ($1000 per investor)	Interest Due at End of Month	Number of New Investors Needed
1			
2			
3			
4			

5. Your results for months 2, 3, and 4 indicate that you can obtain the number of new investors at the end of the nth month by multiplying the number of new investors in the $(n-1)$st month by the same number. What is that constant multiplier?

A characteristic of any exponential function $f(x) = c \cdot a^x$ is that an increase of one unit in the value of the independent variable x results in the multiplication of the dependent variable by the growth factor, or base, a. Therefore, your answer to Question 5 gives you the base of the exponential function that models this pyramid process.

6. Using this property of exponential functions, write a model for this pyramid scheme. The output $f(n)$ should represent the number of new investors needed at the end of month n, where $n = 1, 2, \ldots$. (Notice that n starts at 1, making this function a bit different from other exponential models, because $f(0)$ is not part of the model. So in this case, the value you obtain for c will not be the initial value for the model.) Test your model, using $n = 1, 2, 3,$ and 4.

7. Use your model to compute the number of new investors needed at the end of the twelfth month (that is, one year into the swindle). How many new investors would be needed after two years (24 months)?

Tirana investor loses his shirt.

AFP Photo/Armando Babani

The final value that you computed is the number of new investors required during just a single month, the final month of the second year. To compute the total number (ignoring fractional persons) of investors that the scheme would need to dupe over the two-year period, we would have to add all the numbers: $300 + 150 + 225 + 337 + 506 + \cdots + 1{,}683{,}411$. The total number of investors needed over the first two years is more than five million. Since the entire population of Albania is only a little more than three million, and since there were several competing pyramid schemes in operation simultaneously, you can see why the collapse was inevitable.

In the United States, pyramid schemes are illegal, but there are marketing arrangements, such as some distributorships for cosmetics, that operate on a similar exponential principle, whereby each new distributor recruits friends and acquaintances and is promised a percentage of their commissions. Unlike the Albanian pyramid companies, however, the U.S. companies actually have products, so the investors get shampoo as well as promises.

2.6   A TAXING PROBLEM—Piecewise Linear Income Tax

If you've ever filled out an U.S. income-tax return, Schedule X, copied from the instructions for Form 1040, might look familiar. The United States has a *graduated* income tax, which means that higher incomes are taxed at higher rates than lower incomes.

1. The table defines the income-tax function in 2000 for a single individual. What is the independent variable? What is the dependent variable?

Notice that the tax *rate* changes at $26,250, $63,550, $132,600, and $288,350 (few of us have to concern ourselves with some of those rates!), but that it is constant for all incomes within an interval. In other words, the function is defined differently for different income brackets. This project will guide you in drawing a graph of the income tax function and writing its algebraic formula.

[handwritten: independent: taxe rates
dependent: income]

2000 Tax Rate Schedule

Schedule X—Use if your filing status is **Single**

If the amount on Form 1040, line 39, is: Over—	But not over—	Enter on Form 1040, line 40	of the amount over—
$0	$26,250	15%	$0
26,250	63,550	$3,937.50 + 28%	26,250
63,550	132,600	14,381.50 + 31%	63,550
132,600	288,350	35,787.00 + 36%	132,600
288,350		91,857.00 + 39.6%	288,350

2. *Setting Up the Axes.*   The graph is a bit tricky, but you can save yourself time by planning ahead. First, decide the highest income you want to consider. Pick a nice round number, such as $300,000, and divide up the income axis accordingly. Calculate the tax on that top income; that figure will tell you how high your tax axis needs to go. (*Warning*: The tax for $300,000 is in the ballpark of $100,000. If your answer is closer to $200,000, you've misinterpreted the final column of the table.) Again, use round numbers for scaling the vertical axis; multiples of $5000 might be convenient.

3. *Plotting Key Points.*   Instead of randomly calculating taxes, figure the tax on the amounts where the rate changes: $26,250, $63,550, $132,600, and $288,350. Use the table for your calculation, paying particular attention to the last column. The tax rate for any income interval applies only to the income within that interval, not to the whole thing. Plot these points, as well as the tax for the highest income you selected and the tax on an income of $0.

4. *Completing the Graph.*   Now you should have six points. Should you join them? If so, how? Think about these questions and complete your graph.

5. Now you are ready to write a formula for the function. The graph should be piecewise linear, with five connected segments, each with a separate slope. (If yours isn't, you'll need to redraw it. This part should help you to figure out how.) The slopes of the segments are 0.15, 0.28, 0.31, 0.36, and 0.396 (do those numbers look familiar?), and you know the income at which each segment begins. Let TAX be the function's

name and let x be the amount of taxable income. Write a formula for $TAX(x)$. (The formula consists of five separate statements and the rule for piecing them together.)

6. Sometimes you hear people complain about getting clobbered on their income taxes because they made just a little too much money, which bumped them into a higher tax bracket. Let's investigate that complaint to see whether it's valid.

   (a) Suppose that you are single and that your 2000 taxable income was $26,250. What was your income tax?

   (b) Your neighbor, also single, had a taxable income that year of $26,300. How much more did she owe in taxes than you did? How much of her income was taxed at the higher rate?

   (c) Comment on people's misconception that earning just a few dollars too much will greatly increase their taxes.

7. Schedule Y-1 tells a married couple filing jointly how to compute their tax liability. Using similar techniques, draw the graph of this income-tax function.

8. Pick a new function name and write the algebraic formula for the function.

Schedule Y-1—Use if your filing status is **Married filing jointly** or **Qualifying widow(er)**

If the amount on Form 1040, line 39, is: Over—	But not over—	Enter on Form 1040, line 40	of the amount over—
$0	$43,850	15%	$0
43,850	105,950	$6,577.50 + 28%	43,850
105,950	161,450	23,965.50 + 31%	105,950
161,450	288,350	41,170.50 + 36%	161,450
288,350		86,854.50 + 39.6%	288,350

2.7 JUST ALGEBRA

1. Find the exact value of x in each equation.

   (a) $2x - 10 = 3x + 4$
   (b) $4(x + 7) - 2x = 8(x - 3)$
   (c) $\dfrac{6x + 8}{2} = 9x$
   (d) $1 - 3(2x + 7) = 22$

2. Solve each equation for the unknown value. Give both an exact solution and an approximate solution.

   (a) $100 = c \cdot 2.3^3$
   (b) $15 \cdot 4^{23} = c \cdot 4^{24}$
   (c) $48 = 4a^5$
   (d) $(25a^3)(10a^2) = 100a^7$
   (e) $\dfrac{18a^9}{9a^5} = (2a^3)^4$
   (f) $2a^3 = 10a^2$
   (g) $2a^5 = 10a^2$

Atmospheric Pressure

Up, Up, and Away

Preparation

All of us go around with a weight on our shoulders. In general, though, we're not aware of that particular burden, because we've always lived with it. The weight in question is that of our atmosphere, which pushes down on each of us and on every other object with a force (at sea level) of 14.7 pounds per square inch, or one *atmosphere* of pressure.

Imagine a column of air, 1 inch by 1 inch in cross section, extending from the top of your head all the way to the top of the atmosphere. That's 14.7 pounds of air pushing down on you, just at that one spot.

We are accustomed to this pressure, though, and seldom think about it unless we travel to a different altitude. Although the weight of the atmosphere varies with latitude, time, and temperature, the single most important variable is *altitude*. The greater the elevation, the less is the weight of the air above. One mountain climber received an amusing reminder of this when he ascended to an elevation of 11,000 feet, drained his plastic water bottle, replaced the cap tightly, climbed back down, got into his car, and drove to San Francisco (sea level). The empty water bottle had collapsed as though someone had stepped on it, being squeezed hard by the ambient pressure that now exceeded the pressure inside the closed bottle.

Humans seldom notice pressure differences unless the elevation exceeds about 5000 feet, but many people experience dizziness and shortness of breath, along with more serious symptoms of altitude sickness, when they go to extreme elevations without adequate preparation. (When the Olympic Games were held in Mexico City, U.S. athletes trained in Denver to accustom themselves to the high elevation.) You are probably aware, also, that airline cabins are pressurized to mimic sea-level atmosphere. Because low air pressure means a low concentration of oxygen, passengers in an unpressurized cabin might gasp for breath.

Standard air pressure at sea level is 14.7 pounds per square inch (psi), but it decreases with altitude: As altitude increases, pressure decreases. That is, the relationship between air pressure and altitude is a *decreasing* function. That function is best represented by an exponential model, but for moderate elevations, the following linear approximation works very well. In this lab, you will compare the two models.

Write the linear model $L(x) = b + mx$ that shows air pressure dropping by one half psi for every 1000-foot increase in altitude. Let the independent variable x represent altitude measured in 1000-foot increments. Remember the zero-level pressure, 14.7 psi, and be sure that you write a *decreasing* function.

A good exponential model has atmospheric pressure decreasing by $\frac{1}{30}$ of its value for every 900-foot increase in elevation.

Maybe you see an obstacle already: The linear model wants altitude in thousands of feet; the exponential model wants it in 900-foot lumps. This sort of incompatibility is commonplace whenever we function in the real world. Fear not; we can overcome this difficulty.

Write an exponential model $P(h) = c \cdot a^h$ that sets the pressure at 14.7 psi when $h = 0$ and reduces it by a factor of $\frac{1}{30}$ for every 900 feet of altitude gain. To keep the model simple, let h be measured in 900-foot increments.

Give your exponential model a reality check. The pressure at 900 feet should be about 14.2 psi. If your model reports almost no air pressure at 900 feet, rethink the value you used for a.

When you come to the lab, bring $L(x)$ and $P(h)$, your two mathematical models of atmospheric pressure.

The Air Pressure Lab

Begin by comparing with everyone in your group the two atmospheric pressure models you each wrote. If there are any discrepancies, discuss them until you can agree on one formula for $L(x)$ and another for $P(h)$.

Reconciling the Variables

Now let's work on making the models compatible so that they can be compared. The two independent variables, x and h, don't measure exactly the same thing, but they are related: When the elevation is 9000 feet, $x = 9$ and $h = 10$. In general, for any elevation E, we determine the x-value by dividing by 1000 and the h-value by dividing by 900:

$$\frac{E}{1000} = x \qquad \text{and} \qquad \frac{E}{900} = h$$

By solving this pair of equations simultaneously, you can eliminate E and end up with the algebraic relationship between x and h. Then, solve *that* equation for h.

Now rewrite $P(h)$ as a function of x by replacing the exponent h by its equivalent in terms of x. This gives you an exponential function $P(x)$ that you can compare with $L(x)$ because their independent variable is the same.

Comparing the Models

If two different functions model the same phenomenon, their graphs should be similar. Choose a window corresponding to actual terrestrial elevations, given that Mount Everest is 29,028 feet high and Death Valley is 282 feet below sea level, and that x measures the altitude in thousands of feet. Graph the two models and adjust the vertical scale so that the graphs fill the screen.

Be careful entering the formula for $P(x)$. Its entire exponent must be enclosed in parentheses.

At 5000 feet, each model should predict a pressure within 0.1 psi of the other. If yours don't agree that closely, go back and rethink the way you converted h to x.

The graphs should look almost alike at first and then diverge. Estimate the altitudes for which either model would give approximately the same result. Would you consider the linear model to be about as useful as the exponential model up to 6000 feet? 7000 feet? 8000 feet? Higher?

Measuring the Difference

Here's a neat way to measure the discrepancy between our two models: graph their difference. (On your grapher, you might be able to enter this as Y1 − Y2 instead of doing the actual subtraction.) Because we expect the differences to be small, you'll need to change the vertical dimensions of the graphing window so that you can distinguish positive and negative y-values.

Whenever we subtract a number A from a number B, a positive difference $B - A$ means that A was less than B, while a negative difference signals that A was the larger number. If the difference is zero, then A and B must have been equal. The same principle applies to functions. When you graph $L(x) - P(x)$, you'll be able to see the *intervals* on which P is greater than L and vice-versa.

Let's call the linear approximation "good" if it differs from the exponential model by no more than 0.1 psi. (The exponential model is, overall, much closer to reality.) We'll tolerate a difference of 0.1 psi in either direction; we don't care whether the error is positive or negative.

Suggestion: superimpose two horizontal lines, $y = 0.1$ and $y = -0.1$, to see the interval for which the linear model is good.

Use the graph of $L(x) - P(x)$ to determine the altitudes for which the linear model is good. On what interval does $L(x)$ overpredict the pressure? On what intervals does $L(x)$ underpredict the pressure?

The Lab Report

Your report should explain the two models of atmospheric pressure in language that is understandable to someone not in your class. In particular, explain how you reconciled the two input variables h and x and why it was necessary to do so. Describe how you compared the two models graphically, paying special attention to the graph of their difference. Discuss the pros and cons of each model, including their accuracy, understandability, and ease of use. Illustrate your report with informative graphs of the linear model, the exponential model, and their difference.

Follow-up to the Air Pressure Lab

Over What Domain is Either Model Valid?

If we use the exponential model $P(x)$ as the norm, this question is relatively easy to answer for the linear model $L(x)$. Consider the interval of altitudes on which you found $L(x)$ to be "good." Notice also that above a certain height, the linear model starts predicting negative pressures, a result that contradicts common sense. Compare $L(x)$ to $P(x)$ and decide a reasonable domain for the linear model.

The domain for the model $P(x)$ is a little more complicated, and we need additional information. $P(x)$ gives a very good estimate of the actual atmospheric pressure as long as our elevation isn't too great. What about really high altitudes? According to a Stanford University website,[1] the atmospheric pressure at 5.5 kilometers is only 50% of the surface pressure and only 10% at 16 km.

Convert kilometers to thousands of feet (1 kilometer = 3280 feet) and use $P(x)$. Do you think the model would be reliable for 5.5 km? For 16 km? Without additional information, we can only take a stab at this, but on the basis of your calculations, estimate a domain on which we might consider $P(x)$ to be valid.

---

[1] http://nova.stanford.edu/projects/mod/id-pres.html

Transforming Functions

3

In Chapter 2, you examined linear and exponential functions. By now, you should be able to graph a function of either type easily and quickly without using a grapher. In this chapter, you will expand your repertory and develop a familiarity with a variety of functions, learning to group them into families and to recognize the similarities within a family and the differences between families. Even though the exercises in the chapter depend heavily on graphing technology, they are designed to teach you several techniques that will actually allow you to rely a bit less on your grapher and more on your good mathematical sense. Your expectations of how a graph should look can be helpful in choosing the viewing window or scale that best displays the important properties of the function.

Lab 3, "Graph Trek," serves as the real introduction to this work. You should complete it as soon as possible, ideally before beginning Section 3.1.

Family Portraits

Certain mathematical relationships are used so frequently as models that you should become familiar with their graphs. We present here a portrait gallery: the "head of household" of each of seven fundamental families of functions. (The exponentials are a two-parent family.) Learn to recognize their shapes and to be able to draw a quick sketch of each. You'll refer to those shapes when you need a graph of another member of a particular family.

▷   For starters, acquaint yourself with each family by completing Table 3.1. The graphs of Figure 3.1 can guide you in determining each domain and range. We start you off in a way that emphasizes once again that the domain and the range of a function are *sets* of numbers. The "because" part asks for an algebraic explanation, and we've done a few already to give you some ideas for writing your own.

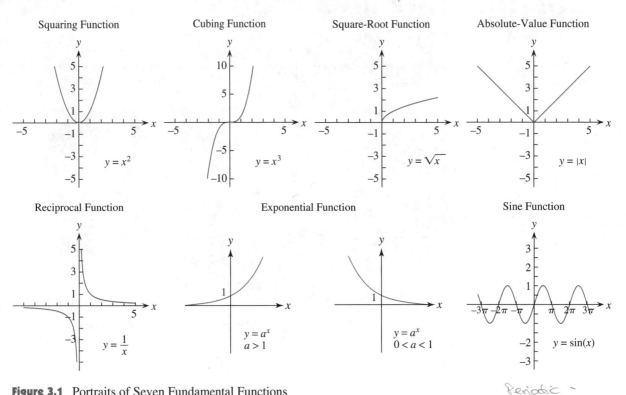

Figure 3.1   Portraits of Seven Fundamental Functions

Periodic - behavior

Function	Domain	Range
x^2	$\mathbb{R}$, the set of all real numbers, because any real number can be multiplied by itself	the set of nonnegative real numbers, because the square of any real number is greater than or equal to zero.
x^3	<u>all real numbers</u> because any real # can be multiplied by itself	<u>all real #'s</u> because all possible output values can be the result of cubing a real number. (Recall that the cube of a negative number is also negative.)
$\sqrt{x}$	<u>x ≥ 0</u> because	<u>y ≥ 0</u> because

Table 3.1   Domains and Ranges of Seven Fundamental Functions

if $a > 0$, $a^x \neq 0$
always > 0

Function	Domain	Range
$|x|$	_____R_____ because	_____$y \geq 0$_____ because
$\dfrac{1}{x}$	_everything but 0_ _____ because _all non-zero #'s_	_all non-zero #'s_ _____ because the expression $\frac{1}{x}$ can take on every real number value except zero
a^x	_____R_____ because	_____$y > 0$_____ because
$\sin(x)$	_____R_____	_____$[-1, 1]$_____

Table 3.1 (Continued) Domains and Ranges of Seven Fundamental Functions

The sine function might or might not be familiar to you. In Chapter 6, you'll learn why it looks the way it does. For now, it's enough that you recognize its shape and know that the curve oscillates forever between $y = -1$ and $y = 1$, intersecting the horizontal axis at every integer multiple of π: $0, \pm\pi, \pm 2\pi, \pm 3\pi, \ldots$.

In this chapter, we step briefly away from the physical world to develop some mathematical tools that will be useful to us soon in creating and understanding models more complicated than the ones we saw in the first two chapters. The principal context for Chapter 3 is that of mathematics itself.

The "Graph Trek" lab comes next. Section 3.1 won't make sense until you've completed that lab, which begins on page 147.

Something add to a funtion will move it vectical
Something add within a funtiom will move it horz.

3.1 TRANSFORMATIONS

Do Lab 3 ASAP.

In this section, we pose two big questions, ones that you should be able to answer after completing the "Graph Trek" lab that begins on page 147.

- What happens to the graph of a function when the *independent variable* is modified by addition or multiplication, that is, before the function itself is performed?

- What happens to the graph of a function when the *dependent variable* is modified by addition or multiplication, that is, after the function is performed?

You are most likely to understand and remember these classes of transformations if you produce the effects yourself. That's what the "Graph Trek" lab is about.

But What's It Good For?

Graph shifting, reflecting, and stretching are of more than theoretical interest. We can use transformations to help us write models. Here are several examples from your earlier work.

In Chapter 2, you learned to write a linear function, given a starting value and a constant rate of change. You also learned the point-slope form, handy when the starting value isn't the y-intercept. We could instead handle such a case neatly with a shift.

Height of a Girl   Remember the average girl (page 71) whose height at age 4 was 102 cm and who grew at the rate of 6.1 cm per year from then until age 13. A natural starting value for this model is the height we're given, 102 cm. We could write a function that says "start at 102 and grow at the rate of 6.1 inches per year":

$$y = 102 + 6.1x$$

But the function $y = 102 + 6.1x$ is awkward, because it forces us to measure her age in *years since age 4*, when we really want just to use her age. No problem; simply shift the line four years to the right:

$$y = H(x) = 102 + 6.1(x - 4)$$

Although x is still the independent variable, its definition has changed. Pay attention more to the definition of the input variable than to the letter used in representing it.

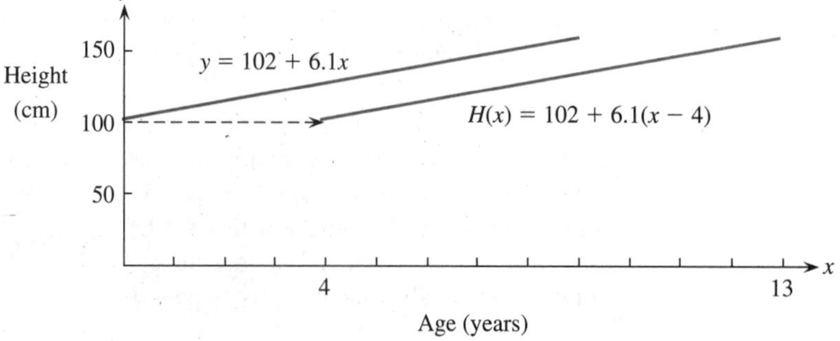

Figure 3.2   Height of Girl as Function of Her Age

Point-slope form: line through (x_1, y_1) with slope m is $y = y_1 + m(x - x_1)$.

Now we have the same point-slope version of $H(x)$ that you saw on page 71. Writing a linear function in point-slope form is like applying to the graph a lateral (horizontal) shift in the amount of $|x_1|$ units.

DJ Equipment   Recall the depreciation model (page 69). The function $B(t) = 3000(0.80)^t$ says "start at \$3000; each successive year's value is 80% of the previous." Suppose now that we were to buy some four-year-old equipment secondhand, paying \$1500. We want a model for the value of the equipment as a function of its age. (We'll assume that all DJ equipment depreciates by the same percentage.) Note that we do not know its original price. The same idea applies: Write a model using our starting value and the constant percentage; then, shift it four years to the right, because our starting value corresponds to year 4, not year 0. First we have the unshifted model

$$y = 1500(0.80)^t$$

Here, we change the definition of t.

measuring time in *years from now*. Shift four years to the right to obtain the model we want,

$$y = 1500(0.80)^{t-4}$$

measuring time in *years since equipment was new*. A little algebra can tell us the original cost of the equipment:

$$y = 1500(0.80)^{t-4}$$
$$= 1500(0.80)^t (0.80)^{-4}$$
$$\approx 1500(0.80)^t (2.44)$$
$$\approx$$

▷   You finish the job. How much was the equipment worth at time 0, when it was new?

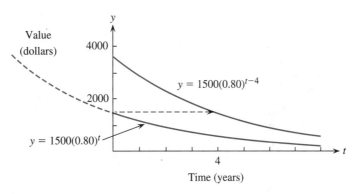

Figure 3.3 Value of Equipment as a Function of Its Age

Value Lost   Rather than the current value of the equipment, an owner might be more interested in knowing by how much it has depreciated. If $B(t)$ gives its value when it is t years old, then

$$D(t) = 3000 - B(t) = 3000 - 3000(0.80)^t$$

gives the amount lost to depreciation. In the algebraic formula for $D(t)$, do you see a reflection and a shift?

▷   Specify the transformations of $B(t)$ that produce $D(t)$: What kind of reflection? What kind of shift?

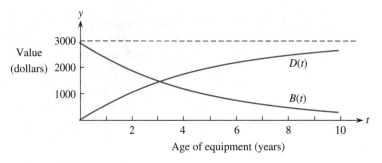

Figure 3.4   Depreciated Value and Amount Lost

Bacterial Growth   Horizontal stretching and compression becomes another handy tool in exponential models. In Chapter 2, you met a bacterium that doubles every 20 minutes. A very simple model of that process is $y = 2^t$, where t represents time measured in 20-minute intervals. But a more common time unit is *the hour* or *the minute*. If we want t to measure time in minutes, we need to stretch the graph horizontally by a factor of 20, and we write

$$y = 2^{t/20}$$

Similarly, if we want t to represent time in hours, we compress the graph horizontally by a factor of 3, writing

$$y = 2^{3t}$$

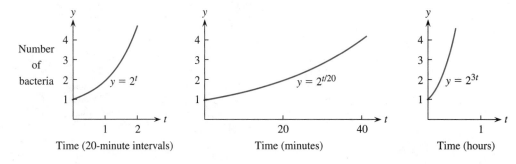

Figure 3.5   Three Models of Bacterial Growth

Check Your Understanding 3.1

CYU answers begin on page 127.

1. Describe how you could obtain the graph of $g(x) = \sqrt{x+3} - 1$ by shifting the graph of $f(x) = \sqrt{x}$. Sketch both graphs.

2. Write a formula to reflect the graph of $y = 3x - x^2$ in the x-axis.

3. Write a formula to reflect the graph of $y = 3x - x^2$ in the y-axis.

4. The solid graph represents $y = 3^x$. Write formulas for its relatives given by graphs (a), (b), and (c).

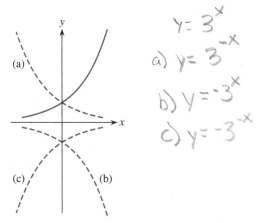

$y = 3^x$

a) $y = 3^{-x}$

b) $y = -3^x$

c) $y = -3^{-x}$

5. Use a shift to write the linear function that passes through $(-2, 10)$ with slope -3. (Make $y = 10$ the starting value and shift the line horizontally.)

6. Book circulation at the public library is growing by 6% every year. Two years ago, 75,213 books circulated. Use a shift to write an exponential model giving circulation as a function of time in years from *now*. (Your final model should have this year's circulation as its starting value.)

7. For a few years during the 1990s, Internet usage was doubling every six months. Write two models of Internet usage, using the symbol I_0 for the starting value, one with time measured in months and the other with time measured in years.

Did you have any trouble with CYU 3.1? Projects 3.2 and 3.3 provide additional help in understanding these transformations. You can do them on your own.

In a Nutshell   When the algebraic formula for any function is modified by adding a constant or multiplying by a constant, the graph is transformed in a predictable way. Adding a constant causes a shift of the graph. Multiplying by -1 (taking the opposite) causes a reflection. Multiplying by a positive constant causes a stretch or a compression of the graph.

3.2   SEQUENTIAL RELATIONSHIPS

One thing leads to another

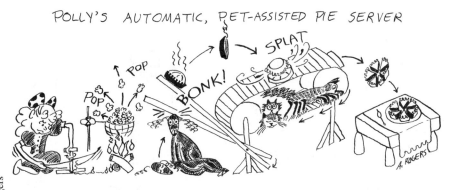

POLLY'S AUTOMATIC, PET-ASSISTED PIE SERVER

© 1993 A. Rogers

POLLY MAKES FIRE, USING PRIMITIVE DEVICE.

FIRE TRAVELS ALONG FUSE, IGNITES BLAZE, POPS POPCORN. SOUND WAKES DOG.

DOG SITS INTO BOARD. PIE TRAVELS IN ARC, LEAVES PAN ON HOOK.

PIE LANDS ON SLICER. IMPACT ALERTS CAT.

CAT CLAWS TREADMILL. PIE ADVANCES. SLICES LAND ON PLATE.

Take a good look at the illustration of Polly's invention and notice how the result of each stage gets fed into the next. If one stage were missing (if the dog, say, were out for a walk), the pie would not be served. If the stages were in some other order, something entirely different might result. Polly could end up with popcorn on the table instead of pie. In mathematics as well, there are situations in which we apply two or more functions, one after another, the output of one becoming the input of the next. In this section, we examine how to **compose** or chain together two or more functions.

Here's an example. Remember the man in Chapter 1 who wandered in a circle because the lengths of his left and right strides differed? The formula for the radius of his circular path, measured in inches, was $r = \frac{480}{d}$, where d was the discrepancy, in inches, between his left and right stride lengths. Using inches to measure d is appropriate; the difference between a person's right and left stride lengths is probably small. It is far more convenient, however, to report the radius of the path in feet. We indicated this conversion by providing a second scale, in feet, along the vertical axis for the graph of the function r on page 25. Now let's consider an algebraic method of handling the scaling issue.

The original function $r = \frac{480}{d}$ measures d in inches and produces an output, the radius r of the resultant circle, in inches. We can feed this output r into a second function, which converts inches to feet by dividing by 12. The two-step sequence starts with d in inches and ends with r in feet. Each step is a function, and the sequence is called the **composition** of the two functions:

$$d \xrightarrow{\text{multiply input's reciprocal by 480}} \frac{480}{d} \xrightarrow{\text{divide by 12}} \frac{480}{d \cdot 12} = \frac{40}{d}$$

stride-length   radius   radius
difference   (inches)   (feet)
(inches)

Composing Two Functions

Money matters

To increase sales of jewelry during 1999, the Ames department stores in Massachusetts ran numerous promotions. One week, jewelry prices were slashed by 70%. The next week, the jewelry was marked 60% off, but a sign reading "Take an additional 20% off the sale price" was posted above the "60% off" sign. Which sale is better: the first or the second? Let's use functions to find out.

Let x represent the original selling price of an article of jewelry. We'll use this variable to construct two functions: S_1, the price you would pay during the first week, and S_2, the price you would pay during the second week.

If you make a purchase during the first week, the clerk will compute the selling price by subtracting 70% from the original price. This means that you will pay only 30% of the original price. Stated algebraically,

$$S_1(x) = 0.30x$$

For a purchase during the second week, S_2 is defined by a two-step sequence. First, reduce the price by 60%:

$$x \xrightarrow{\text{reduce by 60\%}} 0.40x$$

(Remember that a 60% reduction means that you pay 40%.) Then reduce the output from the first reduction by 20%, remembering that the 20% reduction means that you pay 80%:

$$0.40x \xrightarrow{\text{reduce by 20\%}} 0.80(0.40x)$$

Stated algebraically, $S_2(x) = 0.80(0.40x) = 0.32x$.

▷ Which sale gives you the better deal?   *Second deal*

▷ Suppose the clerk takes 20% off first and then takes 60% off. Will the price you pay be any different? Why or why not?   *Yes... Different*

$$.40 \, (\cdot 80x) =$$

Baubles, bangles, and beads

You've had your eye on a handsome gold chain. This week, not only is it on sale (60% off), but you also have a coupon that will save you $10 on any jewelry purchase. Do you want to give the coupon to the clerk before or after she calculates the 60% reduction, or does it matter?

Two functions are involved here: the 60%-off function (call it $d(x) = 0.40x$) and the subtract-10 function (call it $c(x) = x - 10$). Mathematically, we're asking ourselves whether the order in which we perform the functions makes any difference. Let's try it both ways for a necklace whose list price is x dollars:

   use $120?

$$x \xrightarrow{\text{reduce by 60\%}} \underset{d(x)}{0.40x} \xrightarrow{\text{subtract 10}} \underset{c(d(x))}{0.40x - 10} \quad -2$$

or

$$x \xrightarrow{\text{subtract 10}} \underset{c(x)}{x - 10} \xrightarrow{\text{reduce by 60\%}} \underset{d(c(x))}{0.40(x - 10)} \quad 4 \quad |6|$$

▷ You decide: When do you want the clerk to subtract the value of your coupon? That is, does the order of the two functions make a difference? (If you're not sure, invent a value for the list price x.)

Compose yourself

The order of composition does appear to matter, at least in some cases. Let's try another example. Suppose we take two functions,

$$f(x) = x^3 \quad \text{and} \quad g(x) = x + 1$$

What happens when we input a number to g and then feed the output from g into f? Try this with a specific number, say, 10:

$$10 \xrightarrow{\text{add 1}} g(10) \xrightarrow{\text{compute the cube}} f(g(10))$$

We read this expression as "f of g of 10." The language implies that we apply f to the result $g(10)$.

▷ What number do you get as the final result, $f(g(10))$?

Now apply the same sequential process to an arbitrary input x:

$$x \xrightarrow{\text{add 1}} \underset{g(x)}{x + 1} \xrightarrow{\text{compute the cube}} \underset{f(g(x))}{(x + 1)^3}$$

The final result means that $f(g(x))$ is summarized in the single function $(x + 1)^3$.

What if we reverse the order of the functions? In other words, f gets the input first and the output of f is fed into g. Again, try this with one specific number, 10:

$$10 \xrightarrow{\text{compute the cube}} f(10) \xrightarrow{\text{add 1}} g(f(10))$$

▷ What is the result, $g(f(10))$? Compare this number to the one you obtained when you applied g first.

Apply the same sequence to an arbitrary input x:

$$x \xrightarrow{\text{compute the cube}} \underset{f(x)}{x^3} \xrightarrow{\text{add 1}} \underset{g(f(x))}{x^3 + 1}$$

This result means that $g(f(x))$ is summarized in the single function $x^3 + 1$.

STOP & THINK   Use algebra to expand $(x + 1)^3$ (that is, perform the multiplication necessary to remove the parentheses) and explain why the result shows that $g(f(x)) \neq f(g(x))$.

Check Your Understanding 3.2

1. Let s be the function $s(x) = \sqrt{x}$; let p be the function
   $p(x) = x + 4$; and let q be the function $q(x) = -x$.
   (a) Write an expression for the function that adds 4 to
       a number and then takes the square root. Is this
       $s(p(x))$ or $p(s(x))$?

   $\sqrt{x+4}$

   (b) What is the domain of s? Of p? Of the
       composition you chose in (a)?

   S- all real positive #'s $-4 \le x = \infty$
   P all real #'s

   (c) Write the algebraic expression for $s(q(x))$. What
       is the domain of this composition?

   $\sqrt{-x}$ all real - #'s

2. Let f be the linear function $x + 1$ and let g be the
   cubing function, x^3.
   (a) How does the composition of f with g (do g first)
       affect the graph of g? $x^3 + 1$

   $(x+1)^3$ yes

   (b) How does the composition of g with f (do f first)
       affect the graph of g?

   $f \circ g(x)$ $(2x+1)^2$

(c) One graph represents $f(g(x))$; the other
    represents $g(f(x))$. Which is which? How did
    you decide?

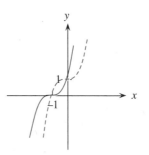

3. The conversion factor for changing U.S. dollars to
   Mexican pesos is 9.2. The bank, however, first
   deducts a $15 service charge. Express the number of
   pesos, P, obtained from d dollars as a composition of
   two functions.

4. *A yen for euros.*   On a day in August 2001, the bank
   offered 1.12 euros for each American dollar. The rate
   that day to convert euros to Japanese yen was 108.93
   yen to the euro. Name and write three functions: one
   to convert dollars to euros, one to convert euros to
   yen, and a third (their composition) to convert dollars
   to yen.

   $f(x) = 2x + 1$ $Gx(x^2)$

   $f(g(x))$ $f(x) = 2x^2 + 1$

Notation for Function Composition

In a $f \circ g$

Here we introduce a symbol, $\circ$, that is often used to indicate a chain of functions.

The notation $f \circ g$ says,
"Do g first."

> When functions g and then f are performed in sequence, the result $f(g(x))$
> is called the **composition of f with g** and is denoted by the symbol $f \circ g$.
> The symbol $f \circ g(x)$ indicates that the function g is performed first.

Similarly, the composition of g with f, denoted $g \circ f(x)$, means $g(f(x))$ and indicates
that f is done first.
    Be careful! The order in which we read the composition $f \circ g$ is not the order in
which the functions are applied to the input. The function nearest the input variable is
the one that "goes to work first."

Composition of functions is not really a new topic; only the notation is new. We've been composing functions all along without giving a special title to what we were doing. Everything you did in the "Graph Trek" lab involved composing two functions and investigating what happened to the graph.

Let's use composition notation to express some of our previous examples. If $r(d) = \frac{480}{d}$ is the radius function and $f(r) = \frac{r}{12}$ converts inches to feet, then the composition $f \circ r(d)$ does the whole job.

If $d(x) = 0.40x$ reduces prices by 60% and $c(x) = x - 10$ subtracts the $10 coupon, the composition that discounts the price by 60% and then subtracts the coupon value is $c \circ d(x)$.

Check Your Understanding 3.3

1. Name the square-root function f, the "subtract-three" function h, the "multiply-by-five" function m, the "take-the-opposite" function opp, and the "add-one" function j. Write a formula for each of the functions f, h, m, opp, and j.

    $f(x) = \sqrt{x}$

    $h(x) = x-3$

    $m(x) = x \cdot 5$

    $j(x) = x + 1$

    $opp(x) = -x$

    (d) $-\sqrt{x}$ (e) $\sqrt{5\diamond}$ (f) $\sqrt{t} - 3$

    (g) $f(x-3)$ (h) $f(5\diamond)$ (i) $f(t) + 1$

2. Use the functions you wrote in Question 1 to express each of the following transformations of f as a composition. Some of the transformations are equivalent. (This means that some of the answers will be the same.)

    (a) $\sqrt{x-3}$ (b) $5\sqrt{\diamond}$ (c) $1 + \sqrt{t}$

    $\sqrt{x-3}$ $5\sqrt{x}$

3. Why stop at two? Find three or more of the five functions which, performed in the correct order, will produce the following results. (Be sure to specify the order of composition.)

    (a) $1 - \sqrt{5x}$ (b) $5\sqrt{3-B}$ (c) $\sqrt{5\square - 3}$

The Algebra of Composition and Decomposition

There's no need to restrict ourselves to simple functions. We can compose any two or more functions to make new functions. The effects on the graph, however, are more difficult to anticipate than they were with simple additions, subtractions, and multiplications, so we'll focus here primarily on the algebra of composition.

Consider two functions, the first a relative of x^2 and the second a relative of $\frac{1}{x}$:

$$F(x) = x^2 + 5x - 1 \quad \text{and} \quad G(x) = \frac{1}{x+2}$$

To form the composition $F \circ G$, we use the output of G as input to F:

$$F \circ G(x) = (G(x))^2 + 5 \cdot G(x) - 1 = \left(\frac{1}{x+2}\right)^2 + 5\left(\frac{1}{x+2}\right) - 1$$

You can think of F as $F(\Box) = \Box^2 + 5\Box - 1$. Fill each box with a copy of $G(x)$.

Notice that $G(x)$, or $\frac{1}{x+2}$, fits *into* the F-pattern, $\Box^2 + 5\Box - 1$. Think of the composition $F \circ G$ as a process of dropping $G(x)$ into each of the boxes.

Now do the composition $G \circ F$. This time, the output of F becomes input to G:

$$G \circ F(x) = \frac{1}{(x^2 + 5x - 1) + 2} = \frac{1}{x^2 + 5x + 1}$$

Here, $F(x)$, which is $x^2 + 5x - 1$, fits into the G-pattern, $\frac{1}{\Box+2}$.

Breaking the chains

In calculus, you will often need to look at a composed function and figure out how it got that way. We can express the complicated function $V(x) = \sin(\sqrt{x+3})$, for example, as the result of adding 3 to the input value, then taking the square root, and then taking the sine of that result. In symbols, $V(x) = f \circ g \circ h(x)$, where $f(x) = \sin(x)$, $g(x) = \sqrt{x}$, and $h(x) = x + 3$.

If you need to review exponents, visit the Algebra Appendix, page 477.

The decomposition of a function is not unique, but some are more straightforward than others, and the most straightforward is usually the most helpful as well. For instance, the expression $\sqrt[3]{(x+2)^4}$ can be expressed as $f \circ g \circ h(x)$, where $f(x) = \sqrt[3]{x}$ takes the cube root, $g(x) = x^4$ raises to the fourth power, and $h(x) = x + 2$ adds 2. But it's easier to use fractional exponents and write $m \circ h(x)$, where $h(x)$ still adds 2, and $m(x) = x^{4/3}$ raises the result to a fractional power.

$\triangleright$ Show that the compositions $f \circ g \circ h(x)$ and $m \circ h(x)$ both yield $\sqrt[3]{(x+2)^4}$.

$$f(x) = \sqrt[3]{x}$$
$$g(x) = x^4$$
$$h(x) = x + 2$$

$$f(x) = \sqrt[3]{x}$$
$$f \circ g = \sqrt[3]{x^4}$$
$$f \circ g \circ h(x) = \sqrt[3]{(x+2)^4}$$

Check Your Understanding 3.4

You'll need some algebraic finesse to put a few of these into the simplified form in which we've given the answers. Fortunately, the Algebra Appendix is open 24 hours a day. See page 475 for double-decker fractions.

1. Given four functions

$$A(x) = \frac{x}{1-x} \qquad B(x) = x^2 - 3x$$
$$C(x) = |x| \qquad D(x) = 10^x$$

form the following compositions:

(a) $A \circ B(x)$

$$\frac{x^2 - 3x}{1 - x^2 - 3x}$$

(b) $A(A(x))$

(c) $D(B(x))$

(d) $B \circ D(x)$

(e) $A \circ C(x)$

(f) $C(B(x))$

2. Express the complicated functions that follow as compositions of simpler functions. Name and define the simpler functions you are using. (The answers are not unique; you might find more than one way to break things down.)

(a) $(2x - 3)^4$

(b) $\dfrac{1}{(2x - 3)^4}$

(c) $\dfrac{5}{(2x - 3)^4}$

(d) $\dfrac{\sqrt{x} - 3}{\sqrt{x} + 7}$

(e) $10^{\sin(x)}$

(f) $\sin(10^x)$

$f \circ g \circ h(x)$

$f(x) = 2x$

$g(x) = x^4$

$h(x) = x - 3$

Mind your P's and Q's

In Question 1 of CYU 3.2 (page 111), you saw that the domain of the composition $s \circ p$ was different from the domain of p and from the domain of s. Determining the domain of a composition can be tricky.

> For any two functions P and Q, the domain of the composition $P \circ Q$ is subject to two constraints:
>
> • Since Q goes to work on the input first, any input x must belong to the domain of Q.
>
> • Since P processes $Q(x)$, all values of $Q(x)$ must be acceptable to P, that is, must belong to P's domain.

Bearing in mind these constraints, let's compute the domain of $P \circ Q$, where

$$Q(x) = \frac{x - 1}{2 - x} \qquad \text{and} \qquad P(x) = \frac{4x - 1}{x + 3}$$

▷ First, determine the domain of Q.

Not every number in the domain of Q belongs to the domain of $P \circ Q$. The function P cannot process -3 (because -3 makes the denominator zero), and therefore we must eliminate from the domain of $P \circ Q$ any x for which $Q(x) = -3$.

▷ There is one value of x for which $Q(x) = -3$. Use algebra to find this number and then give the domain of $P \circ Q$. (*Hint*: Set the expression for $Q(x)$ equal to -3; solve for x. Your answer is the other value that's not in the domain of the composition.)

You should have determined that the domain of the composition $P \circ Q$ is the set of all real numbers except 2 and 2.5.

Check Your Understanding 3.5

1. Let $g(x) = \sqrt{x}$ and $f(x) = x^2$.
   (a) Evaluate $g \circ f(-3)$ and $f \circ g(4)$. Pay attention to the order of operations.

   (b) What is $f \circ g(x)$? What is the domain of $f \circ g(x)$? Be careful; the answer is not $\mathbb{R}$.

   (c) What is $g \circ f(x)$? Again, be careful. The answer is not x. Consider what happened when the input was -3.

2. Let r be the reciprocal function $r(x) = \frac{1}{x}$.
   (a) What is $r \circ r(x)$?

   (b) What is the domain of $r \circ r$? (Be careful. There's one troublesome input.)

A surprise ending

We'll consider one last composition. Let Q be defined as in the previous example,

$$Q(x) = \frac{x - 1}{2 - x}$$

and let

$$R(x) = \frac{2x + 1}{x + 1}$$

We'll begin with one specific input, 3, and calculate $Q \circ R(3)$. Follow each step carefully; you get to do the next one.

$$Q \circ R(3) = Q\left(\frac{6 + 1}{3 + 1}\right) = Q\left(\frac{7}{4}\right) = \frac{\frac{7}{4} - 1}{2 - \frac{7}{4}} = \frac{7 - 4}{8 - 7} = 3$$

Notice what happened: We started out with 3 and ended up with 3. Our output is the same as our original input. Maybe that was a fluke; let's find out.

▷ Pick any other number (not -1, though; why not?) and use it as input to the composed function $Q \circ R$. You should end up with your original number.

Algebra will help us understand what's happening. We'll compose Q with R:

$$Q \circ R(x) = Q\left(\frac{2x + 1}{x + 1}\right) = \frac{\frac{2x+1}{x+1} - 1}{2 - \frac{2x+1}{x+1}}$$

Notice how R drops into Q's pattern. Now multiply top and bottom of this monster by the expression $x + 1$, and things will immediately improve:

$$\frac{\frac{2x+1}{x+1} - 1}{2 - \frac{2x+1}{x+1}} \cdot \frac{x + 1}{x + 1} = \frac{(2x + 1) - (x + 1)}{2(x + 1) - (2x + 1)} = \frac{x}{1} = x$$

▷   Your turn: Evaluate $R \circ Q(x)$. Again, you should get x.

Inverse functions undo each other.

What's remarkable about getting x back? It means that the value of x that goes in one end is same value that comes out the other end, no matter what x-value we use and regardless of the order of the two functions. In a sense, R and Q cancel out each other's effect; each one undoes the work that the other did. We call such functions *inverses* of each other. They are the subject of Section 3.3.

In a Nutshell   To compose two or more functions means to perform them in sequence so that the output of one becomes input to the next. A composed function may have a domain restriction that's not immediately apparent unless we consider its history. To decompose a function means to show a chain of simpler functions that produces the same result.

3.3   INVERSE RELATIONSHIPS

The final example in the previous section was designed to set you up for what's coming next. There are many occasions in life, as well as in mathematics, in which we know only the outcome of a process and want to figure out its initial state.

Interchanging the Roles of the Input and Output Variables

Role reversal

A traveler from New York to Mexico City might know the amount in pesos that she wishes to bring with her. She needs to find out how much this will cost her in dollars. The bank representative tells her that the current exchange rate is 9.2 pesos to the dollar and mentions the $15 service charge.

▷   If the traveler needs 5000 Mexican pesos, how many dollars must she exchange?

Let's examine the steps involved in finding the answer:

$$p = F(d) = 9.2(d - 15) \qquad \text{Use the dollars-to-pesos function from page 117.}$$

$$5000 = 9.2(d - 15) \qquad \text{She needs 5000 pesos.}$$

$$\frac{5000}{9.2} = d - 15 \qquad \text{Divide both sides by the exchange rate.}$$

$$543.48 + 15 = d \qquad \text{Add \$15 to each side.}$$

$$558.48 = d \qquad \text{She should exchange \$558.48.}$$

Compare the two operations in the function $F(d)$ with the two operations just performed:

$F(d)$ (Finding Pesos, Given Dollars)	Finding Dollars, Given Pesos (What We Just Did)
Subtract bank fee Multiply by exchange rate	Divide by exchange rate Add bank fee

Notice that each operation in the right-hand column undoes an operation on the left side, but in the reverse order. In other words, to undo a function, we start by undoing the last operation and moving backward.

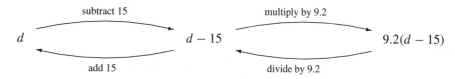

But those two operations in the right-hand column—divide by the exchange rate; add the bank fee—in themselves constitute a function:

$$d = \frac{p}{9.2} + 15$$

The function defined by the formula $d = \frac{p}{9.2} + 15$ is called the **inverse function** for the dollars-to-pesos function. It tells the number of dollars you'll pay if you want p pesos.

▷ Use the *inverse* function to determine the cost, in dollars, of 5000 pesos.

▷ Let's see a few values for our pair of functions. Fill in the missing numbers.

Number of Dollars, d	Number of Pesos, $9.2(d - 15)$
100	782
515	4,600
775	6 992

Number of Pesos, p	Number of Dollars, $\frac{p}{9.2} + 15$
782	100
4600	515
6992	775

What you can observe in these two tables is an important characteristic of any pair of inverse functions.

If an ordered pair (a,b) belongs to a function, then the ordered pair (b,a) belongs to its inverse.

We have a pair of functions that are inverses of each other. Let's rewrite them in grapher-friendly style, giving each of them x as their independent variable. Keep in mind that x in $F(x)$ represents *number of dollars*, while x in $D(x)$ stands for *number of pesos*.

$$F(x) = 9.2(x - 15) \qquad \text{and} \qquad D(x) = \frac{x}{9.2} + 15$$

Replacing the 9.2 with its fractional equivalent, $\frac{46}{5}$, gives

$$F(x) = \frac{46}{5}(x - 15) \qquad \text{and} \qquad D(x) = \frac{5}{46}x + 15$$

Notice that the slopes of these linear functions are reciprocals. As we'll see, this is no accident.

▷ Form and simplify the compositions $D \circ F(x)$ and $F \circ D(x)$, showing that the result in both cases is x.

Function F tells how many pesos you'll get for your dollars; function D counteracts the effect of F and tells you how many dollars to bring to the bank if you want to go home with a certain number of pesos. Mathematicians use the notation F^{-1} (pronounced "F-inverse") for the function that undoes the effect of F. Therefore, the function D is actually F^{-1}. We write $D(x) = F^{-1}(x) = \frac{5}{46}x + 15$.

STOP & THINK

You saw that the slopes of the lines F and F^{-1} are reciprocals; their product is 1. Can you explain why we might expect this, given that the roles of the variables have been interchanged?

When you see F^{-1}, think "inverse function for F."

The symbol F^{-1} needs a few words of explanation. If F is a *function*, then F^{-1} means "the function that undoes the work of the function F," the **inverse function** for F. The symbol F^{-1} does not mean "one-over" something! Only if, in some other context, F were being used as a *number* would the symbol F^{-1} mean the reciprocal of F, or $\frac{1}{F}$. For example, if $F = 10$, then $F^{-1} = \frac{1}{10}$. But if F is the cubing function, $F(x) = x^3$, then F^{-1} is the cube-root function, $F^{-1}(x) = \sqrt[3]{x}$.

Blow-up

Reversing the roles of the dependent and independent variables can apply to other situations as well. If you've blown up balloons for a party, you have probably noticed that a balloon seems to grow rapidly in the beginning but more slowly as it becomes larger. Consider what happens when you inflate a spherical balloon with helium. The diameter of the balloon increases as the volume of gas increases, but not linearly. In fact, the larger the diameter, the more helium you have to blow into the balloon in order to make a perceptible difference. A graph of the relationship, given by the function

$$d = \sqrt[3]{\frac{6V}{\pi}}$$

is shown in Figure 3.6. The graph is concave down—increasing at a decreasing rate—because the *growth rate* of the diameter decreases as the volume increases.

Suppose you have many balloons to inflate. To monitor helium usage, you want to know how volume depends on diameter. In other words, you need to reverse the roles of the dependent and independent variables and use the inverse function.

Each point (V,d) on the graph of d as a function of V corresponds to a point (d,V) on the graph of its inverse.

▷   From the graph in Figure 3.6, obtain approximate coordinates for four or five points. Interchanging the coordinates gives points on the graph of the *inverse* function, where volume is a function of diameter. Plot those points in Figure 3.7 and join them with a smooth curve.

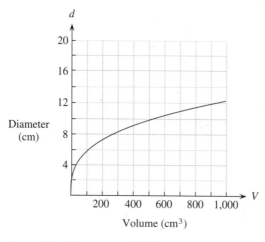

Figure 3.6   Diameter of Balloon as a Function of Volume, $d = \sqrt[3]{\dfrac{6V}{\pi}}$

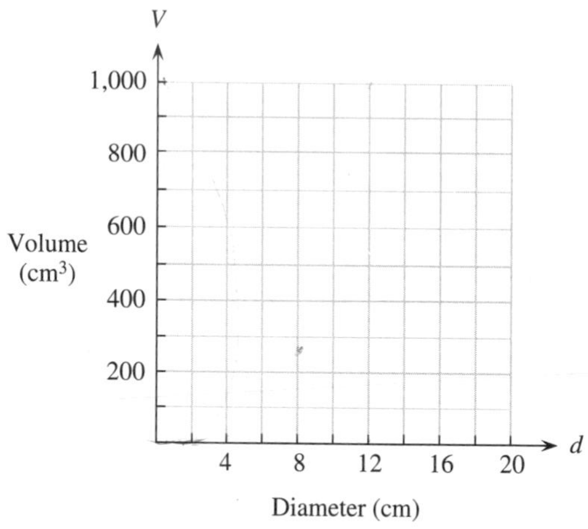

Figure 3.7   Volume of Balloon as a Function of Diameter

Let's examine how the function $d = \sqrt[3]{\dfrac{6V}{\pi}}$ is constructed. The input V is multiplied by 6 and divided by π; then the cube root of the result is computed. To determine the inverse function, we must go backward through this process.

▷   Solve the equation $d = \sqrt[3]{\dfrac{6V}{\pi}}$ for V. Pay attention to each algebraic step; you should find yourself undoing the steps that produced the function d.

If we rewrite the two functions using x as the independent variable, we have

$$d(x) = \sqrt[3]{\dfrac{6x}{\pi}} \quad \text{and} \quad V(x) = d^{-1}(x) = \dfrac{\pi}{6}x^3$$

We express them in this way to be able to plot them on the same set of axes. Don't forget, though, that x in the first function represents volume, while x in the inverse function represents diameter.

The Algebra of a Function and Its Inverse

The heart of the relationship between any function f and its inverse function f^{-1} is this:

$$f^{-1} \circ f(x) = x \qquad \text{for every } x \text{ in the domain of } f$$
$$f \circ f^{-1}(x) = x \qquad \text{for every } x \text{ in the domain of } f^{-1}$$

In plain English, each function is the other's undoing, taking us back to the original input.

This relationship gives us a recipe for finding the inverse of a function. Remember the "Surprise ending" example from page 115. Now you'll find out how we picked those two functions. Starting with the first function

$$Q(x) = \frac{x - 1}{2 - x}$$

we look for another function, $Q^{-1}(x)$, having the property that

$$Q \circ Q^{-1}(x) = x$$

To simplify the calculations, let's use the symbol y to represent $Q^{-1}(x)$, so that our equation becomes

$$Q \circ y = x \qquad \text{or} \qquad Q(y) = x$$

The Q-pattern: $Q(\Box) = \frac{\Box - 1}{2 - \Box}$

Writing an expression for $Q(y)$ is easy; just substitute y into Q's pattern to obtain the following equation:

$$Q(y) = \frac{y - 1}{2 - y} = x$$

Now solve for y:

$$y - 1 = x\,(2 - y)$$
$$y - 1 = 2x - xy$$
$$xy + y = 2x + 1$$
$$y(x + 1) = 2x + 1$$
$$y = \frac{2x + 1}{x + 1} = Q^{-1}(x)$$

If you look back at page 115, you'll recognize $Q^{-1}(x)$ as the function $R(x)$. Now you see how we found that pair of functions.

▷   It's your turn. Find the inverse of the function $R(x) = \frac{2x+1}{x+1}$ by solving the equation $\frac{2y+1}{y+1} = x$ for y. You should obtain $y = \frac{x-1}{2-x}$, which is the same function $Q(x)$ we just used, showing that the functions R and Q are truly inverses of each other.

In case you didn't obtain $y = \frac{x-1}{2-x}$, you might have an equivalent version instead. Don't give up; try multiplying your answer by $\frac{-1}{-1}$ to see whether you can make it look like the one given here.

How to Find the Inverse of a Function

- Rewrite the function in independent-dependent variable notation, that is, using x as the input and y as the output.

- Write the function's inverse by interchanging x and y. Replace y with x; replace each occurrence of x with y.

- Solve the resulting equation for y; then rename y as $Q^{-1}(x)$.

STOP & THINK
This business of "interchanging the variables" might feel like hocus-pocus, but it actually is sound mathematics. Can you justify the procedure in the preceding box?

Check Your Understanding 3.6

1. For the balloon example (page 119), evaluate $d^{-1} \circ d(x)$ and $d \circ d^{-1}(x)$. Show that the result in both cases simplifies to x.

2. Find $f^{-1}(x)$, if $f(x) = \frac{x}{2x+3}$.

$$(2y+3)x = \frac{y}{2y+3}(2y+3)$$

$$x(2y+3) = y$$

$$2xy + 3x =$$

The Geometry of a Function and Its Inverse

Any point (a,b) on the graph of f corresponds to the point (b,a) on the graph of f^{-1}.

A point (a,b) on the graph of a function f means that f turns a into b. The inverse function, then, will turn b back into a, so the point (b,a) must be on the graph of f^{-1}. Exploiting that key property will enable us to draw the graph of the inverse of a function without plotting many points.

Figure 3.8 shows a close-up of the volume function $V(x)$ together with the diameter function $d(x)$ and the diagonal line $y = x$. Because the x- and y-axes have the same scale, the $y = x$ line makes an angle of 45° with the horizontal. Imagine folding the picture along that diagonal line. Do you see the symmetrical relationship between the pair of curves?

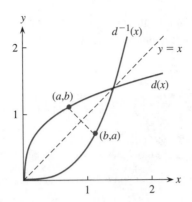

Figure 3.8 The Diameter Function $d(x)$ and Its Inverse

In the "Graph Trek" lab, you learned about pairs of graphs, $f(x)$ and $-f(x)$, that were mirror images in the x-axis and pairs of graphs, $f(x)$ and $f(-x)$, that were mirror images in the y-axis, as shown in Figure 3.9.

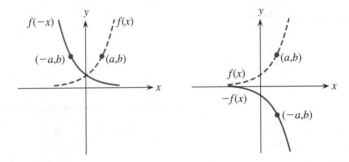

Figure 3.9 Reflection in the y-Axis; Reflection in the x-Axis

The two balloon curves have a similar mirror relationship, but in this case, the mirror is the 45° line $y = x$, as shown in Figure 3.10. That relationship of reflection in $y = x$ will be true of every function and its inverse.

Reflect the graph of f in the line $y = x$ to obtain the graph of f^{-1}.

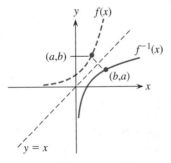

Figure 3.10 Reflection in the Line $y = x$

Under reflection in the line $y = x$, each point (a,b) is reflected onto the point (b,a). Therefore, to draw the graph of the inverse of a function f, simply reflect the graph of f in the line $y = x$.

Check Your Understanding 3.7

1. Carefully sketch a graph of the inverse of the function whose graph is shown. The intercepts should serve as guideposts: If the function has an x-intercept at $(-2,0)$, where is the y-intercept of the inverse?

(graph showing a decreasing curve through the origin, with axes labeled y vertical, x horizontal, marked at 10, −10, −10, 10)

2. The formula for the function pictured is $f(x) = -4\sqrt[3]{x+2}$. Find the algebraic formula for f^{-1}. Use your grapher to verify that its graph is the one you sketched.

STOP & THINK

How are the domain and the range of a function f related to the domain and the range of its inverse?

One-to-One Functions

All the king's horses and all the king's men . . .

A natural question to ask is whether all functions have inverses and, if not, how to decide which ones do. Recall the river-level function from Chapter 1, Table 1.2 (page 4). We asked you to consider what would go wrong if you tried to use the river-level data as input to a function that would determine the date. Look at Table 1.2 now; you'll see that April 15, 17 and 20 all have the same level, 113.7 feet. If the input to the "function" were 113.7, you would have no way to determine which date should be the output. Therefore, the "function" *isn't* really a function. Although Table 1.2 defines a perfectly valid function, river levels as a function of date, no inverse function exists.

Here's another case: Suppose you tell me that you've just squared a number (using the squaring function, x^2) and that the output is 25. Do I know the input? Well, of course you can get me on that one every time. If I say "5," you tell me your number was −5; if I say "−5," you tell me it was 5. The squaring function, in the absence of additional information, has no inverse. There's no way to determine from the output alone what the input must have been.

The problem with the function x^2 and with the river-level function is that two different inputs produce the same output. That's OK for a function, but it's not OK if we're trying to undo the function.

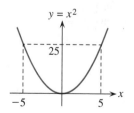

In the function $y = x^2$, two different inputs can give the same output.

For a function to have an inverse function, it cannot repeat itself; *distinct inputs must always produce distinct outputs.* If a is different from b, then $f(a)$ is different from $f(b)$. In symbols,

$$\text{If} \quad a \neq b, \quad \text{then} \quad f(a) \neq f(b).$$

We call such functions **one-to-one.** Every y has its own x.

An everyday example of a one-to-one function is the rule that assigns a social security number to each U.S. citizen: Every citizen can have a number (that makes the relationship a function), and each number corresponds to a single citizen (that makes the function one-to-one). (The Social Security function is one-to-one *in theory*. In practice, however, it's not, because a few folks have figured out how to cheat the system and obtain extra numbers.)

STOP & THINK

Precalculus texts often present a tool known as the *horizontal line test* for determining whether or not a function has an inverse. It goes like this: If you can find a horizontal line that cuts the graph of the function in more than one place, the function is not one-to-one; otherwise, it is. Why does this test work?

Horizontal line test

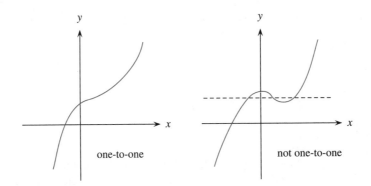

one-to-one   not one-to-one

KEEP THINKING!

If a function has an inverse function, you can find the graph of the inverse function by reflecting the original graph about the line $y = x$. The function x^2 is not one-to-one. Why is the reflection of its graph in the line $y = x$ not the graph of an inverse function?

Check Your Understanding 3.8

1. *What's your sign?* Let H be the function that assigns to a person one of the twelve signs of the zodiac based on the month of birth. Is H a one-to-one function? What does this indicate about the daily horoscope column of your newspaper?

2. Is the triathlon model on page 77 a one-to-one function? Interpret your answer in terms of distance and time.

3. Is the exponential population model on page 67 one-to-one? Interpret your answer in terms of population and time.

4. Give an example of a linear function that is one-to-one and an example of one that is not.

5. Use your grapher to view the graph of the function $y = \sin(x)$ over a wide interval such as $-8\pi \le x \le 8\pi$. Talk about flunking the horizontal line test! How often does the sine function fail the test in this interval? How many times over the entire real-number line?

Testing, testing, ...

There are times when you will simply need to check whether two functions are inverses of one another. The test is straightforward:

> When two functions, g and h, are related so that $g(h(x)) = x$ for all x-values in the domain of h and $h(g(x)) = x$ for all x-values in the domain of g, then h and g are inverses of one another.

We can say that g is h^{-1} (the inverse function for h) or that h is g^{-1} (the inverse function for g). It doesn't matter which one we designate the "function" and which one we call the "inverse." But the criteria in the box tell us how to check.

If, for example, we want to determine whether or not the two functions

$$g(x) = \frac{x}{x+2} \qquad \text{and} \qquad h(x) = \frac{2x}{1-x}$$

are inverses of one another, we need to evaluate the compositions $g \circ h(x)$ and $h \circ g(x)$ to see whether, in both cases, the end result is simply x, the original input value.

▷ Doing those two compositions is good double-decker practice; here's some space for you to work them out.

We hope you decided that g and h are, in fact, inverses of each other. If you had obtained something other than x as the result of composing two functions, you would have decided that the functions were not inverses of one another.

In a Nutshell   An inverse function undoes the work of a given function. To have an inverse function, a function must be one-to-one. If a function f has an inverse function f^{-1}, then

- $f^{-1} \circ f(x) = x$ for every x in the domain of f, and
- $f \circ f^{-1}(x) = x$ for every x in the domain of f^{-1}.

The graphs of f and f^{-1} are reflections of each other in the line $y = x$, because any point (a,b) on the graph of f has a corresponding point (b,a) on the graph of f^{-1}. The domain of f is the range of f^{-1}; the range of f is the domain of f^{-1}.

What's the Big Idea?

- Certain algebraic transformations of a function's formula (adding a constant to the independent or dependent variable, multiplying the independent or dependent variable by a constant) have specific effects on the function's graph (shifting, reflecting, stretching or compressing).

- A function's graph may exhibit symmetry about the y-axis (even symmetry), symmetry about the origin (odd symmetry), or neither of these two symmetries. A function's formula reveals its symmetry.

- When functions are performed in sequence, the result is a new function, called their composition. The order in which the functions are performed usually makes a difference in the result.

- Every one-to-one function has an inverse function. The inverse of a function undoes its effect.

Progress Check

After finishing Lab 3, Project 3.1 ("Symmetry"), and this chapter, you should be able to do the following:

- Recognize a graph as belonging to one of the function families whose portraits appear in the introduction to this chapter.

- Determine precisely how the following modifications to a function will affect its graph: adding a constant to the input or output, multiplying the input or output by a positive constant, and taking the opposite of the input or output. (3.1 and Lab 3)

- Recognize an even or odd function, given either its formula or its graph. (Project 3.1)

- Find the algebraic formula for the composition of two or more functions. (3.2)

- Work backward from a composite function to find simpler functions that produced it. (3.2)

- Determine the domain of a composite function. (3.2)

- Determine from its graph whether a function is one-to-one and thus has an inverse. (3.3)

• Given the graph of a one-to-one function, sketch the graph of its inverse. (3.3)

• Given algebraic formulas for two functions, test whether or not they are inverses. (3.3)

• Find the algebraic formula for the inverse of a function. (3.3)

• Know the relationship between the domain and range of a function and the domain and range of its inverse. (3.3)

Key Algebra Skills Used in Chapter 3

• Solving linear equations (page 483)

• Simplifying polynomial expressions (page 469)

• Multiplying polynomials (page 469)

• Adding rational expressions (page 473)

• Simplifying rational expression, including "double-decker" fractions (pages 474–479)

• Solving fractional and radical equations (pages 487–488)

Answers to Check Your Understanding

3.1

1. The graph of g is a shift of the graph of f three units left and one unit down.

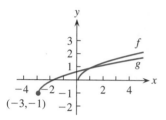

2. $y = -(3x - x^2) = x^2 - 3x$

3. $y = 3(-x) - (-x)^2 = -3x - x^2$

4. Graph (a) represents $y = 3^{-x}$; (b) is $y = -3^x$; (c) is $y = -3^{-x}$.

5. $y = 10 - 3(x + 2)$

6. $y = 75213(1.06)^{t+2} = 75213(1.06)^t(1.06)^2$
$= 84509(1.06)^t$

7. $y = I_0(2)^{t/6}$ (time in months); $y = I_0(2)^{2t}$ (time in years)

3.2

1. (a) $s(p(x)) = \sqrt{x + 4}$.
   (b) The domain of s is the set of nonnegative numbers, $x \geq 0$; the domain of p is the set $\mathbb{R}$ of all real numbers; the domain of $s(p(x))$ is the set $x \geq -4$.
   (c) $s(q(x)) = \sqrt{-x}$. Its domain is the set $x \leq 0$, the nonpositive numbers.

2. (a) Doing the cubing function before the linear function shifts the x^3-shape one unit vertically in the positive direction (i.e., up!).
   (b) Doing the linear function before the cubing function shifts the x^3-shape one unit to the left.
   (c) One way to tell them apart is to observe that the dashed curve has its "wiggle" (inflection point) on the y-axis, indicating that it represents a vertical shift of the x^3-shape, so it must be $x^3 + 1$. Similarly, the solid curve has its inflection point

on the x-axis, so it represents a horizontal shift and must be $(x + 1)^3$.

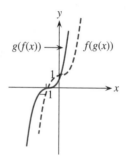

3. A function to subtract \$15 is $f(d) = d - 15$. A function to multiply by 9.2 is $g(d) = 9.2d$. $P(d) = 9.2(d - 15)$ is the composition $g(f(d))$.

4. If d is the number of dollars and r is the number of euros, then $r(d) = 1.12d$ converts dollars to euros; $Y(r) = 108.93r$ converts euros to yen; the composition $Y(r(d)) = 108.93(1.12d) = 122d$ (rounded to the nearest yen) converts dollars directly to yen. (Feel free to use other letters.)

3.3

1. $f(x) = \sqrt{x}, h(x) = x - 3, m(x) = 5x$, $opp(x) = -x$, and $j(x) = x + 1$. (The variable does not have to be x.)

2. (a) $\sqrt{x - 3} = f(h(x)) = f \circ h(x)$
   (b) $5\sqrt{\Diamond} = m \circ f(\Diamond)$
   (c) $1 + \sqrt{t} = j(f(t))$
   (d) $-\sqrt{x} = opp \circ f(x)$
   (e) $\sqrt{5\Diamond} = f \circ m(\Diamond)$
   (f) $\sqrt{t} - 3 = h(f(t))$
   (g) $f(x - 3) = f(h(x))$
   (h) $f(5\Diamond) = f \circ m(\Diamond)$
   (i) $f(t) + 1 = j(f(t))$

3. (a) $1 - \sqrt{5x} = j \circ opp \circ f \circ m(x)$

(b) $5\sqrt{3 - B} = m(f(opp(h(B))))$ (That was tricky! Be sure you understand how this works.)

(c) $\sqrt{5\square - 3} = f \circ h \circ m(\square)$

3.4

1. (a) $A \circ B(x) = \dfrac{x^2 - 3x}{1 - x^2 + 3x}$

(b) $A(A(x)) = \dfrac{\frac{x}{1-x}}{1 - \frac{x}{1-x}} = \dfrac{x}{1 - 2x}$

(c) $D(B(x)) = 10^{x^2 - 3x}$

(d) $B \circ D(x) = (10^x)^2 - 3(10^x) = 10^{2x} - 3(10^x)$

(e) $A \circ C(x) = \dfrac{|x|}{1 - |x|}$

(f) $C(B(x)) = |x^2 - 3x|$

2. (a) $(2x - 3)^4 = f \circ g(x)$, where $g(x) = 2x - 3$ and $f(x) = x^4$. (Several other compositions are possible, but this is the most straightforward.)

(b) $\dfrac{1}{(2x - 3)^4} = d \circ f \circ g(x)$, where $g(x) = 2x - 3$, $f(x) = x^4$, and $d(x) = \frac{1}{x}$. More simply, $\dfrac{1}{(2x - 3)^4} = r \circ g(x)$, where $g(x) = 2x - 3$ and $r(x) = x^{-4}$.

(c) $\dfrac{5}{(2x - 3)^4} = m \circ r \circ g(x)$, where $g(x) = 2x - 3$, $r(x) = x^{-4}$, and $m(x) = 5x$.

(d) $\dfrac{\sqrt{x} - 3}{\sqrt{x} + 7} = u \circ v(x)$, where $v(x) = \sqrt{x}$ and $u(x) = \dfrac{x - 3}{x + 7}$.

(e) $10^{\sin(x)} = D \circ c(x)$, where $c(x) = \sin(x)$ and $D(x) = 10^x$.

(f) $\sin(10^x) = c \circ D(x)$, where $D(x) = 10^x$ and $c(x) = \sin(x)$.

3.5

1. (a) $g \circ f(-3) = g(9) = 3$ (Notice how a negative input produces a positive output.) $f \circ g(4) = f(2) = 4$

(b) $f \circ g(x) = x$, $x \geq 0$ (Do you understand why negative values are not in the domain?)

(c) $g \circ f(x) = \sqrt{x^2} = |x|$

2. (a) $r \circ r(x) = \frac{1}{1/x} = x$, $x \neq 0$.

(b) The domain of $r \circ r$ is the set of nonzero real numbers.

3.6

1. $d^{-1} \circ d(x) = d^{-1}\left(\sqrt[3]{\dfrac{6x}{\pi}}\right) = \dfrac{\pi}{6}\left(\sqrt[3]{\dfrac{6x}{\pi}}\right)^3$

$= \dfrac{\pi}{6} \cdot \dfrac{6x}{\pi} = x$

$d \circ d^{-1}(x) = d\left(\dfrac{\pi}{6}x^3\right) = \sqrt[3]{\dfrac{6\left(\frac{\pi}{6}x^3\right)}{\pi}} = \sqrt[3]{x^3} = x$

2. The equation $x = \dfrac{y}{2y + 3}$ gives the inverse. Solve for y:

$$x(2y + 3) = y$$
$$2xy + 3x = y$$
$$2xy - y = -3x$$
$$y(2x - 1) = -3x$$
$$y = \dfrac{-3x}{2x - 1} = \dfrac{3x}{1 - 2x} = f^{-1}(x)$$

3.7

1.

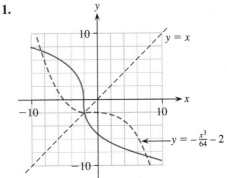

2. Solve for y: $x = -4\sqrt[3]{y + 2}$. The inverse function is $y = -\frac{x^3}{64} - 2$.

3.8

1. H is not one-to-one; everyone born between May 22 and June 21, for example, is a Gemini. The sign alone does not single out an individual, and it would be remarkable for one prediction to suit those hundreds of millions of people.

2. Yes. Distance traveled uniquely determines elapsed time.

3. Yes. If we know the population, we can determine the year.

4. Any nonconstant linear function $y = mx + b$ is one-to-one. A constant function such as $y = -3$ is not.

5. The graph of the sine function intersects some horizontal lines 16 times between $x = -8\pi$ and $x = 8\pi$. It intersects the x-axis 17 times. Over the whole real line, the sine function repeats itself infinitely many times.

EXERCISES

1. Each graph belongs to one of the seven function families pictured on page 102. Identify the family and write an algebraic formula to represent the graph.

(a)

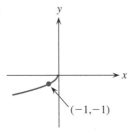

$(-1,-1)$

(b)

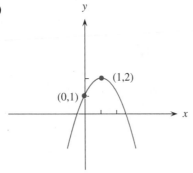

$(1,2)$ $(0,1)$

(c)

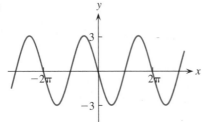

-2π 2π 3 -3

(d)

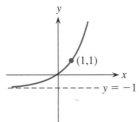

$(1,1)$ $y=-1$

(e)

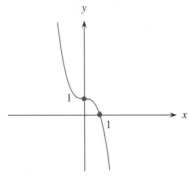

1 1

(f)

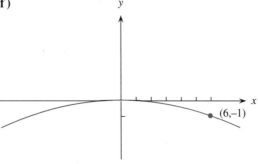

$(6,-1)$

2. Each graph belongs to one of the seven function families pictured on page 102. Identify the family and write an algebraic formula to represent the graph.

(a)

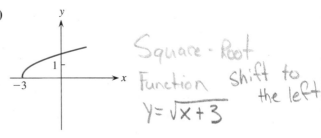

-3 1

Square-Root Function shift to the left
$y=\sqrt{x+3}$

(b)

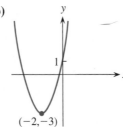

1 $(-2,-3)$

Squaring Function
$y=(2+x)^2-3$
Shift left an down

(c)

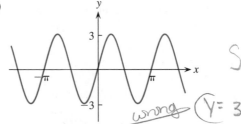

$-\pi$ π 3 -3

Sine Function
wrong $\boxed{y=3\sin(x)}$

(d)

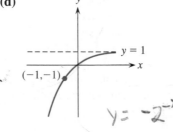

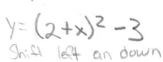
$y=1$
$(-1,-1)$

Exponential Function

$y=-2^{-x}+1$

(e)

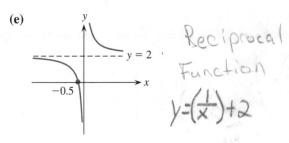

Reciprocal Function

$y = \left(\frac{1}{x}\right) + 2$

(f)

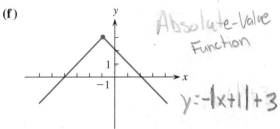

Absolute-Value Function

$y = -|x+1| + 3$

3. Let $f(x) = \frac{1}{x^2}$. Write the formula for the function that results when the graph of f is
   (a) shifted three units right
   (b) shifted three units up
   (c) reflected in the x-axis
   (d) reflected in the y-axis
   (e) compressed toward the x-axis by a factor of 3
   (f) compressed toward the y-axis by a factor of 3

4. Let $g(x) = 10^x$ and repeat the steps of the preceding exercise.

5. On a particular day, the exchange rate for converting U.S. dollars to euros is 1.12 euros to the dollar. One bank does not charge its customers for the conversion; another bank imposes a $3 service charge.
   (a) For each bank, write a mathematical model for converting dollars to euros.
   (b) The second model can be considered a horizontal shift of the first. Explain this.
   (c) The second model could instead be considered a vertical shift of the first—in which direction and by how much?

6. A store has a promotion one week in which all jewelry is marked "60% off." The following week, it invites its customers to "take an additional 15% off a 50% markdown price."
   (a) Write functions that compute the reduced prices under each of the two sales. Show how the formula for the second sale arises from the composition of two functions.
   (b) Which sale offers the better deal to the customer? Explain.
   (c) Adjust one of the percentages in the second sale so that it yields the same savings as the first sale.

7. Let $f(x) = x - 1$ and let $g(x) = x^2$.
   (a) Find $f \circ g(5)$ and $f \circ g(x)$.
   (b) Find $g \circ f(5)$ and $g \circ f(x)$.

(c) Does the order of composition make a difference?
(d) Sketch the graphs of $g(x)$, $g \circ f(x)$, and $f \circ g(x)$. How does composing f with g change the graph of g? How does composing g with f change the graph of g?

8. Use algebra to explain why, for the graph of $\frac{1}{x}$, a horizontal stretch by any factor of k, $k > 1$, is equivalent to a vertical stretch by the same factor. (Be careful; for a horizontal stretch, multiply the independent variable by a fraction.)

9. We can consider kx^2, $k > 1$, to be a vertical stretch of the graph of x^2 by a factor of k. What horizontal compression produces the same graph?

10. A vertical shift of three units, applied to the graph of $y = -\frac{3}{2}x + 3$, causes the graph to coincide with the graph of $y = -\frac{3}{2}x$. What horizontal shift produces the same effect? A good sketch of the two lines will help you figure this out.

11. The population of Sinefield grows by 2.2% per year, while that of Triangle decreases by 1.8% each year. Sinefield's population in 1990 was 28,170; Triangle held a census in 1995 and counted 5987 people.
   (a) Write a population model for each town, using 1990 and 1995, respectively, as the starting year.
   (b) Reconcile the models so that they both start in 1990.

12. The population of Parallel doubles every 30 years. Its current population is 50,000.
   (a) Write a mathematical model for the population as a function of time from now, measured in 30-year periods.
   (b) Use an appropriate stretch to rewrite the function so that time is measured in *years*.

13. Use the given partial graph to create two complete graphs.
   (a) Make a graph symmetric about the y-axis.
   (b) Make a graph symmetric about the origin.

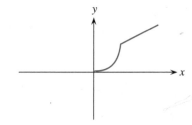

14. Classify the capital letters of the alphabet—A, B, C, D, E, F, G, H, I, J, K, L, M, N, O, P, Q, R, S, T, U, V, W, X, Y, Z—according to the following categories:
(a) letters symmetrical about their vertical axis
(b) letters symmetrical about their horizontal axis
(c) letters that have 180° rotational symmetry
(d) letters with none of the above types of symmetry

15.

$y = f(x)$

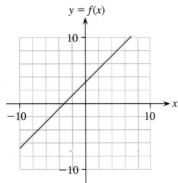

$y = g(x)$

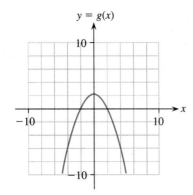

(a) With the given graphs as your only guide, estimate these values: $f(g(-4))$, $f(g(-2))$, $f(g(0))$, $f(g(2))$, and $f(g(4))$

(b) Use your results to sketch a graph of $f(g(x))$. To what family does this composition belong?

16. (a) With the given graphs as your only guide, estimate these values, one of which is not defined: $r \circ h(-1)$, $r \circ h(0)$, $r \circ h(1)$, $r \circ h(2)$, $r \circ h(3)$, $r \circ h(4)$, $r \circ h(5)$, and $r \circ h(6)$.

(b) Use your results to sketch a graph of $r \circ h(x)$. To what family does $r \circ h$ belong?

$y = h(x)$

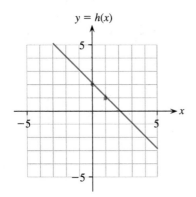

$y = r(x)$

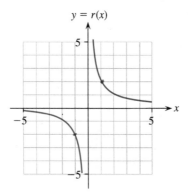

17. Most of the following functions can be broken down into a chain of two or more simpler functions. For those that can be readily decomposed, name and define the simpler functions and express the complicated function as a composition of those simpler functions. If a function cannot be readily expressed as a composition, say so.
(a) $F(x) = (25x + 7)^{10}$
(b) $G(x) = \dfrac{10^x}{10^{2x} - 4}$
    (*Hint:* 10^{2x} is the square of 10^x.)
(c) $H(x) = (1.05)^{-x}$
(d) $J(x) = 2^x \sin(x)$
(e) $K(x) = \sqrt{1 - (\sin(x))^2}$
(f) $L(x) = \dfrac{3^x}{x^3 + 1}$

18. Most of the following functions can be broken down into a chain of two or more simpler functions. For those that can be readily decomposed, name and define the simpler functions and express the complicated function as a composition of those simpler functions. If a function cannot be readily expressed as a composition, say so.
(a) $a(x) = \sin(2^x)$
(b) $b(x) = \dfrac{1}{\sin(x^2)}$
(c) $c(x) = 2^{-x^2}$
(d) $d(x) = 5^{2x} + 3 \cdot 5^x - 10$ $x^2 + 3x + 10$?
    (*Hint:* 5^{2x} is the square of 5^x.)
(e) $f(x) = (x^2 - 3)(\sqrt{x} + 2)$ not a composition
(f) $g(x) = 7x^4 - 4x^{-3}$ no

19. If you're not careful when you compose functions, you might lose some vital information. Suppose $f(x) = \sqrt{x}$ and $g(x) = x^2 - 3$.
(a) Find a formula for $g \circ f(x)$.
(b) Use that formula to evaluate $g \circ f(-1)$.
(c) What is the domain of f? Is -1 a legitimate input to f?
(d) If -1 cannot go into f, it certainly cannot go into $g \circ f$! What is the domain of $g \circ f$?
(e) Sketch the graph of $g \circ f$.

20. Let $h(t) = \frac{1}{t}$ and $j(t) = \frac{1}{t+3}$.
   (a) Find a formula for $h \circ j(t)$.
   (b) Give the domain of $h \circ j$. Pay attention to the domain of j.
   (c) Find a formula for $j \circ h(t)$.
   (d) Give the domain of $j \circ h$. Don't forget the domain of h.

21. Let $g(t) = \sqrt{t}$ and $h(t) = \frac{1}{t}$.
   (a) Find $g(h(t))$ and $h(g(t))$. Does the order of composition matter? (Do they have the same graph?)
   (b) The graph of either composition bears some resemblance to that of $y = \frac{1}{t}$, but it has only one branch. Explain why.

22. Many people would say that the operations of squaring and of taking the square root undo one another, giving you back your original number. In this problem, we examine that assumption. Let $f(x) = \sqrt{x}$ and $g(x) = x^2$.
   (a) Evaluate $f(g(x))$ for any $x > 0$. Do the same for $g(f(x))$. So far, do f and g act like inverses of each other?
   (b) Let's find out for sure. Recall that for f and g to be inverses, the statement $g(f(x)) = x$ must be true for every x in the domain of f, and the statement $f(g(x)) = x$ must be true for every x in the domain of g. Think about the domain of f, and explain why the $g(f(x)) = x$ part does work. Sketch a graph of $g(f(x))$.
   (c) Now explain why $f(g(x))$ is not the same as x for every x in the domain of g. Numerical examples might help. Notice that g's domain includes negative numbers.
   (d) You already know a name for the composition in part (c), $\sqrt{x^2}$. What is it? Sketch its graph.

23. This table gives values for the triathlon function, page 77.

Time Elapsed, t (hours)	Distance Traveled, $d(t)$ (km)
0	0
1	4
6	184
10	226

   (a) Evaluate $d^{-1}(4)$ and interpret its meaning.
   (b) Explain the difference in meaning between $d(6)$ and $d^{-1}(6)$.

24. An athlete finds that, when she jogs at 6 miles per hour, her heart rate is 121 beats per minute. When she sprints at 10 miles per hour, her heart rate is 165 beats per minute. Within certain limitations, the relationship between running speed and heart rate is linear.
   (a) Write the linear model $R(x)$ that gives heart rate as a function of running speed x, measured in miles per hour.
   (b) According to this model, what is her resting heart rate?
   (c) Find $R^{-1}(x)$.
   (d) Find $R^{-1}(100)$ and interpret its meaning.
   (e) Give a reasonable domain and range for the model R and for the inverse function R^{-1}.

25. Given $f(x) = \frac{1}{x} + 3$, find $f^{-1}(x)$. Sketch both graphs.

26. Given $g(x) = \frac{1}{x+3}$, find $g^{-1}(x)$. Sketch both graphs.

27. Find a formula for the reflected graph when $y = \sqrt{x+2}$ is reflected
   (a) in the x-axis
   (b) in the y-axis
   (c) in the line $y = x$

28. Find a formula for the reflected graph when $y = \frac{3}{x-2}$ is reflected
   (a) in the x-axis
   (b) in the y-axis
   (c) in the line $y = x$

29. Which two functions are inverses of each other? Show how you decided.

$$f(x) = \frac{x - 21}{3}, \qquad k(x) = 3x - 21,$$

$$h(x) = \frac{x + 21}{3}, \qquad l(x) = \frac{1}{3}x + 21$$

30. **(a)** Which two functions are inverses of each other? Show how you decided.

$$g(x) = \frac{1}{x - 1}, \qquad j(x) = \frac{x + 1}{x},$$

$$m(x) = \frac{x - 1}{x}, \qquad r(x) = \frac{x}{x - 1}$$

   (b) Show that one of the four functions is its *own* inverse.

31. Refer to the depreciation model on page 69. Does $B(t)$ have an inverse function? If so, interpret the expression $B^{-1}(1000)$ and approximate its value. If not, explain clearly why no inverse function exists.

32. It might surprise you that a 20% price markdown, followed by a 20% markup, does not return an article to its original price.

(a) Give the formula $p(x)$ for the 20% markdown function and the formula $r(x)$ for the 20% markup function.

(b) By finding $r \circ p$, show that p and r are not inverses.

(c) How *would* you undo a 20% markdown? Find $p^{-1}(x)$.

33. Reflect the graph of $y = x^2 + 3$ across the line $y = x$ and sketch the result. Does the new graph represent a function? Can you find formulas for two separate functions that, glued together, produce the new graph you drew?

34. Turn back to the discrete model for shoe sizes, *FIT*, whose graph is on page 76. Explain why the function *FIT* does not have an inverse. Use a numerical example to show why we cannot use this model to determine the *exact* length of a woman's foot on the basis of the size shoe she wears.

35. On page 119, you found the volume of a balloon as a function of its diameter as $V(d) = \frac{\pi}{6} x^3$. By substituting $2r$ for d, show that this function is equivalent to the standard formula for the volume of a sphere. (Check this in the Algebra Appendix, page 507.)

Shifts
outside = up + down   y-axis
inside = left + right   x-axis

Reflections
outside = flip   x-axis
inside = flip   y-axis

3.1 SYMMETRY

SYMMETRY ABOUT THE *y*-AXIS (EVEN SYMMETRY)

> "Beauty is truth, truth beauty,—that is all
> Ye know on earth, and all ye need to know."
> John Keats, *"Ode on a Grecian Urn"*

Beauty is often associated with symmetry. We would all agree that the shape of a Grecian urn is symmetrical. When the curve representing the graph of a function $f(x)$ is placed symmetrically on the coordinate axes, we say it is **symmetric about the *y*-axis** because if you folded this paper along the *y*-axis, the two halves of the graph would lie on top of each other. (Try it!)

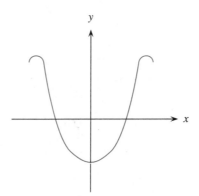

In this investigation, you will learn an algebraic technique for deciding, even without seeing the graph, whether or not a function is symmetric about the *y*-axis.

1. Here are the graphs of four more functions, $f(x)$, three of which are symmetric about the *y*-axis. Which three? Now, for each of these functions, draw the graph of $f(-x)$. (In the "Graph Trek" lab, you learned how to draw the graph of $f(-x)$, given the graph of $f(x)$. Recall that $f(-x)$ is *not* an upside-down version of $f(x)$!)

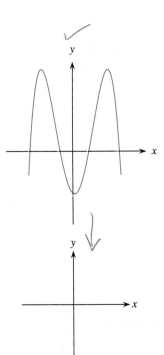

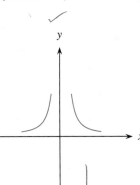

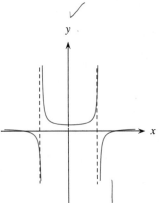

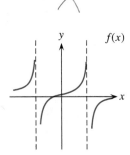

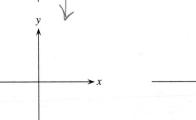

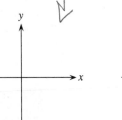

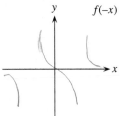

What do you observe about three of the graphs you just drew—something that isn't true of the other one?

2. Here are the formulas for the four functions pictured. For each one, write the formula for $f(-x)$ and compare it to the formula for $f(x)$. What do you observe?

   (a) $f(x) = -x^4 + 10x^2 - 9$ $\rightarrow -(-x)^4 + 10(-x)^2 - 9$

   (b) $f(x) = \dfrac{1}{x^2}$ $\rightarrow \dfrac{1}{(-x)^2}$

   (c) $f(x) = \dfrac{1}{4 - x^2}$ $\left(\textit{Warning}: \text{Remember, this really means } \dfrac{1}{4 - (x^2)}.\right) \rightarrow \dfrac{1}{4 - (-x)^2}$

   (d) $f(x) = \dfrac{x}{4 - x^2}$ $\rightarrow \dfrac{(-x)}{4 - (-x)^2}$

3. Recall from the "Graph Trek" lab how the graphs of $f(x)$ and $f(-x)$ are related. In your own words, explain why you should expect the algebraic results you just obtained, simply by looking at the pictures.   Inside △ . reflects on y-axis, ⌣
   and falls into a specific graph family

The algebraic formula itself reveals whether or not the graph of a function will be symmetric about the y-axis.

> **Symmetry about the y-Axis**
> Whenever $f(-x) = f(x)$ for every possible value of x (that is, they have the same algebraic formula), the graph of f is symmetric about the y-axis.

4. Test the following algebraically to predict whether or not their graphs will be symmetric about the *y*-axis. Check your results by looking at their graphs.

(a) $7 - 3x^2$

(b) $x^6 - 40x^2 + 5$

(c) $x^6 - 40x^3 + 5$

(d) $\dfrac{x}{x^2 - 4}$

(e) $\dfrac{x^2}{x^2 - 4}$

(f) $\dfrac{x}{x^3 - 4x}$ Simplify the fraction you write as much as possible.

5. Do you have an idea why this kind of symmetry is called *even*? (*Hint*: Consider the exponents in functions 4(a) and (b).)

need even exponents

Symmetry about the Origin (Odd Symmetry)

© Anglo-Australian Observatory,
Photograph by David Malin

The curve here mimics the symmetry of the spiral galaxy in the photo. It is placed on the coordinate axes symmetrically, but its symmetry is different from the even symmetry of the Grecian urn curve in the previous section. One way to describe its symmetry is to imagine placing a pin through the graph at the origin and then spinning or rotating the paper 180° around the pin, as a pinwheel (or a galaxy) rotates about its center. The new position of the graph is identical to its old position. We say such a graph is **symmetric about the origin.**

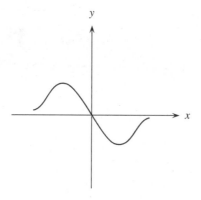

There is another algebraic technique for deciding from a function's formula whether or not the function has origin symmetry. Before learning that trick, though, you will perform several reflections.

6. Here are the graphs of four functions, $f(x)$, three of which are symmetric about the origin. Which three? Below each one, draw the graph of $f(-x)$. Below that, in the third row, draw the graph of $-f(x)$. (Be sure to use the *original* $f(x)$ each time.)

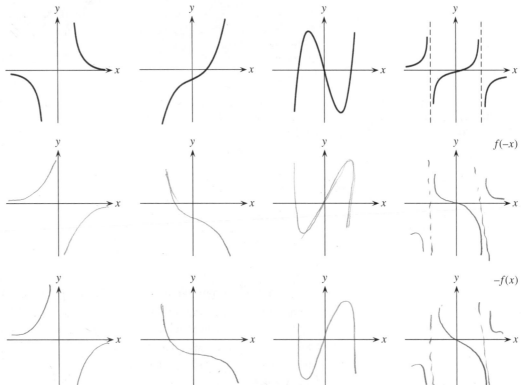

the reflections across the y- and x- axes are the same

Examine the eight graphs you just drew. What do you notice?

7. Each time that you drew the graph of $f(-x)$, you did so by reflecting the graph of $f(x)$ about one of the axes. Each time that you drew the graph of $-f(x)$, you reflected the graph of $f(x)$ about the other axis.

    (a) The graph of $f(-x)$ is the reflection of the graph of $f(x)$ in the ____y____ axis.

    (b) The graph of $-f(x)$ is the reflection of the graph of $f(x)$ in the ____x____ axis.

8. Here are the formulas for the four original functions. For each one, write the formula for $f(-x)$ and the formula for $-f(x)$. Can you, for each function, make the two expressions identical using the rules of algebra? For one of the four functions, that will not be possible.

    (a) $f(x) = \dfrac{1}{x^3}$

    (b) $f(x) = x^3 + 2x - 3$

    (c) $f(x) = x^3 - 4x$

    (d) $f(x) = \dfrac{x}{4 - x^2}$

9. In your own words, explain why you could have expected such results simply by looking at the four given graphs.

The formula for a function is a giveaway for this type of symmetry, too.

> **Symmetry about the Origin**
>
> Whenever $f(-x)$ is algebraically equivalent to $-f(x)$ (that is, when we can make them the same using rules of algebra), the graph of $f(x)$ will be symmetric about the origin.

For help in writing the opposite of a fraction, see the Algebra Note on page 139.

10. Test the following algebraically to predict whether or not their graphs will show origin symmetry. Check your results by looking at their graphs.
    (a) $7x - 3x^3$
    (b) $x^5 - 25x^3 + 100x$
    (c) $x^5 - 25x^3 + 100$
    (d) $\dfrac{3}{x - 3}$
    (e) $\dfrac{x}{x^2 - 4}$
    (f) $\dfrac{x^2 - 4}{x}$

11. Does a graph with origin symmetry necessarily pass through the origin? Explain.

12. Suppose the graph has origin symmetry *and* a y-intercept. Does it pass through the origin? (*Hint*: Could the graph pass through $(0,2)$ instead? What would go wrong?)

13. Why do you suppose this type of symmetry is called *odd*? (*Hint*: Consider the exponents in functions 10(a) and (b).)

Testing for Symmetry

Let's summarize this investigation of symmetry. If we know the formula for a function, we can test for y-axis or origin symmetry by evaluating $f(-x)$.

> **The Symmetry Test**
>
> • If $f(-x)$ is algebraically equivalent to $f(x)$, the graph has symmetry about the y-axis (also known as "even" symmetry).
>
> • If $f(-x)$ is the opposite of $f(x)$ (that is, if $f(-x)$ and $-f(x)$ are algebraically equivalent), the graph has symmetry about the origin ("odd" symmetry).
>
> • If $f(-x)$ is something else, the graph has neither type of symmetry. (The graph might possibly be symmetric about a different line or a different point, but we do not call the function "even" or "odd.")

Symmetry Exercises

Apply the symmetry test to the following functions. Predict whether their graphs will have y-axis symmetry, origin symmetry, or neither. (In every case, start by computing $f(-x)$.) Then check your results by looking at the graphs.

1. $x^4 + \pi$

2. $x^2 + \pi x$

3. $\sqrt{x}$

4. $\sqrt[3]{x}$

5. $x^{2/3}$

6. $2 - x + x^3$

7. $\sqrt{x^4 - 9}$

8. $\dfrac{x}{x - 3}$

9. $\dfrac{x}{3 - x^2}$

10. $\dfrac{x^2 - 9}{x^2 + 2}$

11. $\dfrac{\sqrt[3]{x}}{x^3}$

12. $\dfrac{x^3 - 9}{x^2 + 2}$

13. $\dfrac{x^2 + 3}{x}$

14. $\dfrac{x^2 - 4}{x^2 - 2}$

15. $\dfrac{x^3}{x^2 - 2}$

Algebra Note

Associated with every fraction are three signs:

- The sign of the numerator
- The sign of the denominator
- The sign of the entire fraction

Changing any *one* of the signs changes the sign of the fraction. Changing any *two* signs does not alter the value of the fraction.

For example, the opposite of $\frac{2}{3}$ is $-\frac{2}{3}$ or $\frac{-2}{3}$ or $\frac{2}{-3}$ but not $\frac{-2}{-3}$.

Similarly, the opposite of $\dfrac{x - 3}{x^2 + 4}$ is $-\dfrac{x - 3}{x^2 + 4}$ or $\dfrac{-x + 3}{x^2 + 4}$ or $\dfrac{x - 3}{-x^2 - 4}$.

3.2 VERTICAL AND HORIZONTAL EFFECTS

This project will help you organize what you learned in the "Graph Trek" lab. After finishing the lab, complete Tables A and B. To guide you, we've provided some partial descriptions.

Operation	Effect on Graph of f
Adding a constant: $f(x) + c$	If $c > 0$, If $c < 0$,
Taking the opposite (multiplying by -1): $-f(x)$	Reflect the graph vertically across the x-axis. Portions below the x-axis are now above it, and portions above the x-axis are now below it.
Multiplying by a positive constant: $k \cdot f(x)$	For $k > 1$, stretch the graph away from the _____-axis by a factor of _____. For $0 < k < 1$, compress the graph toward the _____-axis by a factor of $\frac{1}{k}$. (Because k itself is a fraction, the compression factor $\frac{1}{k}$ is *greater* than 1.)

Table A Transformations of the Graph of a Function f When the *Dependent Variable* Is Modified

Operation	Effect on Graph of f
Adding a constant: $f(x+c)$	Shift the graph $|c|$ units horizontally, to the _____ if $c > 0$, to the _____ if $c < 0$.
Taking the opposite (multiplying by -1): $f(-x)$	
Multiplying by a positive constant: $f(k \cdot x)$	For $k > 1$, compress the graph toward the _____-axis by a factor of _____. For $0 < k < 1$, stretch the graph away from the _____-axis by a factor of $\frac{1}{k}$. (When k is a fraction, the stretch factor $\frac{1}{k}$ is *greater* than 1.)

Table B Transformations of the Graph of a Function f When the *Independent Variable* Is Modified

For a vertical change, modify the output. For a horizontal change, modify the input.

Have you noticed that all operations affecting the independent variable produce their graphical effects in the horizontal direction, while all operations affecting the dependent variable affect the graph in the vertical direction? This should not be too surprising, because the independent variable is measured horizontally and the dependent variable is measured vertically.

In the "Graph Trek" lab, you grouped the transformations into three categories: shifts, reflections, and stretches or compressions. Tables A and B encourage you to organize them according to the variable being affected. This gives you two different ways to understand and remember the effects.

3.3 COMPANION POINTS

In the "Graph Trek" lab, you saw what happens to entire graphs under various transformations. In this project, you will track the location of each point on a graph as the graph undergoes a transformation.

Shifts

Consider the function $f(x) = x^2$ and the vertical shift $g(x) = f(x) + 4 = x^2 + 4$. You expect the y-intercept to move from 0 to 4, but did you realize that the y-value of *every* point increases by four units? Try a few input values and watch what happens.

x	x^2	$x^2 + 4$
-2		
0		
3		

If (a,b) is a point on the graph of $f(x)$, then $(a, b+4)$ is a point on the graph of $f(x) + 4$. We say that the points (a,b) and $(a, b+4)$ are **companion points** under the vertical shift $f(x) + 4$. In the example above, $(-2, 4)$ and $(-2, 8)$ are companion points.

Notice that the first coordinate remains the same; a vertical shift affects only the second coordinate.

Take the same function, $f(x) = x^2$, and the horizontal shift $h(x) = f(x - 3) = (x - 3)^2$. Here, we subtract 3 from the input *before* the squaring takes place.

x	$x - 3$	$(x - 3)^2$
-2		
-1		
0		
1		
2		
3		
4		
5		

This time, the x-intercept moves from 0 to 3; it moves three units to the right, and so does every other point on the graph. The companion point for any point (a,b) on the graph of $f(x)$ under the transformation $f(x - 3)$ is the point $(a + 3, b)$. In the case of $f(x) = x^2$ and $f(x - 3) = (x - 3)^2$, the points $(-2,4)$ and $(1,4)$ are companion points. The second coordinate stays the same, because a horizontal shift affects only the first coordinate.

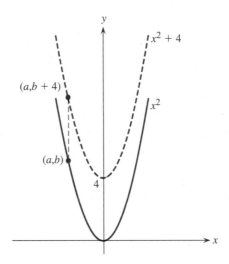

Vertical Shift of x^2, Four Units Upward

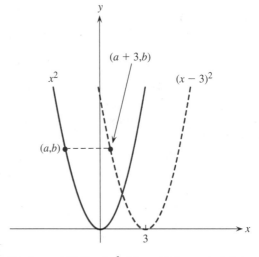

Horizontal Shift of x^2, Three Units to the Right

1. Using $f(x) = x^2$, give the companion points for the points $(0,0)$, $(2,4)$, $(-3,9)$, and (c,c^2) under the transformation $f(x) - 5$.

2. Using $f(x) = x^2$, give the companion points for the points $(-3, f(3))$, $(0, f(0))$, and $(c, f(c))$ under the transformation $f(x + 5)$.

3. Change the function to $g(x) = \frac{1}{x}$ and find the companion points for the points $(-1, g(-1))$ and $(3, g(3))$ under the transformation $g(x) + 2$.

4. Redo Question 2 using the function $g(x) = \frac{1}{x}$ instead.

5. The men's shoe-size model $m(x) = 3x - 22.5$ can be thought of as a vertical shift of the women's model $w(x) = 3x - 21$. By what amount? In which direction? The point $(10, 9)$ is on the graph of $w(x)$. What is its companion point on the graph of $m(x)$ under that vertical shift? Interpret that companion point in terms of foot length and shoe size.

6. From a different point of view, the model $m(x)$ can be considered a horizontal shift of $w(x)$. By what amount? In which direction? What is the companion point for $(10,9)$ under that horizontal shift? What does it mean in terms of foot length and shoe size?

7. Sketch $\sqrt{x}$ and $\sqrt{x} - 1$. The points $(0,0)$, $(1,1)$ and $(9,3)$ are on the first graph. Find their companion points on the second graph.

8. Sketch $\sqrt{x}$ and $\sqrt{x + 3}$. Find the companion points for $(0,0)$, $(1,1)$ and $(9,3)$ on the second graph.

9. What are the companion points for $(0,0)$, $(1,1)$ and $(9,3)$ on the graph of $\sqrt{x + 3} - 1$?

Reflections

When a graph is reflected in the x-axis (by taking the opposite of the whole function), each of its original points acquires a companion point with the same first coordinate and a new second coordinate which is the opposite of what it had been. Under a reflection in the y-axis (when the opposite of the input variable is used), the first coordinate is replaced by its opposite while the second coordinate remains the same.

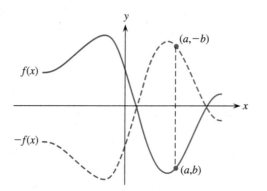

Reflection in the x-Axis

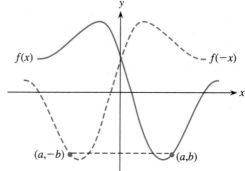

Reflection in the y-Axis

10. Sketch a graph of the function $f(x) = \sqrt{x + 3} - 1$. Don't plot many points, but do locate the two intercepts and the endpoint correctly. (See Question 9.)

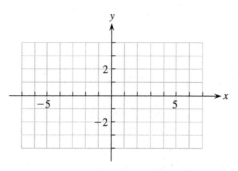

11. Complete the table and, in contrasting colors, draw the reflections $y = -f(x)$ and $y = f(-x)$ for the function $f(x) = \sqrt{x+3} - 1$.

Point on $f(x)$	Companion Point on $-f(x)$	Companion Point on $f(-x)$
$(-3,-1)$		
$(-2,0)$		
$(0, \sqrt{3} - 1)$		
$(6,2)$		

Stretches and Compressions

When a graph is stretched or compressed vertically (by multiplying the entire function), the second coordinate of each point changes: it is multiplied by the stretch factor or divided by the compression factor. Similarly, if the graph is stretched or compressed horizontally (by multiplying the input), only the first coordinate of each point changes, being multiplied by the stretch factor or divided by the compression factor.

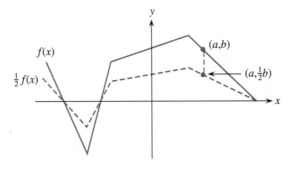

Vertical Compression by a Factor of 2:
the x-intercepts stay put.

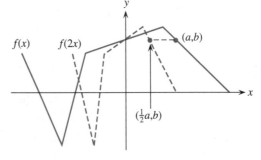

Horizontal Compression by a Factor of 2:
the y-intercept doesn't budge.

12. To see how this works, make another copy of the graph of $f(x) = \sqrt{x+3} - 1$. Complete the following table and, in contrasting colors, draw the graphs of $y = 3f(x)$ and $y = f(2x)$:

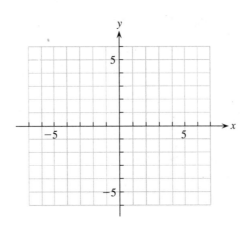

Point on $f(x)$	Companion Point on $3f(x)$	Companion Point on $f(2x)$
$(-3,-1)$		
$(-2,0)$		
$(0, \sqrt{3} - 1)$		
$(6,2)$		

You don't need to use this point-by-point approach when it's time to shift or reflect or stretch a graph. Your goal should be to move beyond point-plotting to an understanding of each graph as an entire object that can be manipulated in various ways. But companion points can be *your* companions as you move toward that goal.

3.4 ABSOLUTELY!—Absolute Value in Functions

By now, you should be getting the idea that success in graphing depends in large part upon learning a few basic shapes and then being able to recognize patterns. This assignment will teach you how to toss absolute values into the mix.

$|f(x)|$, The Absolute Value of a Function

The absolute value of a quantity is always nonnegative. The absolute value of an entire function, therefore, will have to be nonnegative as well. Keep this in mind as you investigate these functions.

1. Have your grapher draw both $f(x) = x^2 - 4$ and $g(x) = |x^2 - 4|$ on the interval $-3 \le x \le 3$. Sketch what you see, using two different colors and showing clearly where the graphs coincide and where they differ. Explain what the absolute value did to the graph of the second function. Why did some portions of the graph change, while others remained the same?

2. In the same manner, graph $f(x) = 4 - x^2$ and $g(x) = |4 - x^2|$. Compare your results with what you saw in the first set of graphs. Are you surprised? Explain why two of the graphs are identical.

3. For each function given, sketch a graph. Then write the formula for $g(x) = |f(x)|$. On the same axes in a different color, make a graph of $g(x)$. Check your work with your grapher.
   (a) $f(x) = x - 2$
   (b) $f(x) = \frac{1}{x}$
   (c) $f(x) = \sin(x)$

4. If the graph of some function $KIWI(w)$ looks like the graph below, sketch the graph of $|KIWI(w)|$.

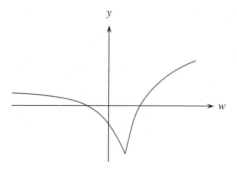

5. Write a sentence or two summarizing what you have learned about taking the absolute value of an entire function. (In the next section, you will learn what happens to a function when the absolute value is applied first.)

$f(|x|)$, The Function of an Absolute Value

Something different happens if we apply the absolute value to the *independent* variable. The expression $f(|x|)$ implies that, regardless of whether the input variable is negative or positive, the function will use only the *magnitude* (a nonnegative quantity) of that variable. For example, if x happens to be -5, the value to be computed will be $f(|-5|)$, or $f(5)$.

6. Let's see how this affects a graph. With your grapher, draw the graph of $f(x) = x^2 - 2x$ on the interval $-4 \le x \le 4$. Write the algebraic formula for $g(x) = f(|x|)$ and have your grapher overlay its graph on that of $f(x)$. Sketch what you see, using two colors and showing clearly where the graphs coincide and where they differ.

7. Substitute another function for $f(x)$ and view its graph together with the graph of $f(|x|)$. Do this for several different choices of $f(x)$ until you feel certain of the effect of using absolute value in this way. Describe what happened and explain why the graph of $f(|x|)$ looks the way it does.

8. You should observe that some of the outputs are still negative (that is, below the x-axis). Explain why, even though we're dealing with absolute values, we can still end up with negative values for the *function*. Why didn't that occur in the last section, when you took the absolute value of an entire function?

9. Here are some practice problems. For each function given, sketch a quick graph. Then write the algebraic formula for $g(x) = f(|x|)$. Use your drawing of $f(x)$ to make a graph of $g(x)$ on the same axes in a different color. Check your work with a grapher.
   (a) $f(x) = x - 2$
   (b) $f(x) = \sqrt{x + 1}$
   (c) $f(x) = \sin(x)$

10. Using the graph of $KIWI(w)$, sketch the graph of $KIWI(|w|)$.

11. Explain what you did in each case to the graph of $f(x)$ to produce the graph of $f(|x|)$.

12. What type of symmetry do you observe in any $f(|x|)$ graph?

3.5 BEAUTIFUL COMPOSITIONS—Linear Transformations

What happens to the graph of any function $f(x)$ when you compose $f(x)$ with a linear function or compose a linear function with $f(x)$? That's what you will discover in this investigation.

1. Start with two simple functions: the quadratic function $f(x) = x^2$ and the linear function $g(x) = x + 3$.
   (a) First compose f with g. Write a formula for $f \circ g(x)$, that is, $f(g(x))$. Now graph $f(g(x))$ and $f(x)$ together. What happened to the graph of f when it was composed with g?
   (b) Next compose g with f. Write a formula for $g \circ f(x)$, also known as $g(f(x))$. What is the effect of this composition on the graph of f?
   (c) What are the domain and range of f? Of $f \circ g$? Of $g \circ f$? How does each composition affect the domain and the range?
   (d) If you were to compose any other quadratic function with any other linear function, would the result necessarily be a quadratic? Suppose you composed the functions in the opposite order. Do you think the result would still be quadratic?

2. Repeat the steps in Question 1 using $f(x) = \sqrt{x}$ and the same linear function. When you get to part (d), ask yourself whether the composed functions will necessarily be members of the square root family of functions.

3. Now it's time to generalize. Let $f(x)$ be any function and let $g(x) = x + c$ any linear function with slope 1. Answer the following, supporting your answers with several examples. In your examples, please consider both positive and negative values for c.
   (a) How does the composition of f with g affect the graph? The domain? The range?
   (b) How does the composition of g with f affect the graph? The domain? The range?

4. All of the linear functions so far have had a slope of 1. Now use $f(x) = \sin(x)$ and $g(x) = 3x$.
   (a) Write a formula for $f \circ g(x)$ and one for $g \circ f(x)$. Now graph $f(x)$ together with the two compositions.
   (b) What happened to the graph of f when you composed f with g? What happened when you composed g with f?
   (c) Which of these compositions affected the range of f?

5. Suppose $f(x) = \sqrt{x-1}$. This time, let $g(x)$ be $\frac{1}{2}x$.
   (a) Write a formula for $f \circ g(x)$ and one for $g \circ f(x)$. Now graph $f(x)$ together with the two compositions.
   (b) What happened to the graph of f when you composed f with g? What happened when you composed g with f?
   (c) Which of these compositions affected the domain of f?

6. It's time for the full generalization. Let $f(x)$ be any function and $g(x) = mx + b$ be any linear function. Answer the following, providing several examples as supporting evidence.
   (a) What will happen to the graph of $f(x)$ when you compose f with g?
   (b) What will happen to the graph of $f(x)$ when you compose g with f?

Graph Trek

To boldly go where no lab has gone before

Preparation

Before you begin this lab, let's be sure that you understand function notation. Given a specific formula for a function, say, $f(x) = 3x^2 - 2x + 5$, write the formulas for the following:

$$f(x) + 1, \quad f(x + 1), \quad -f(x), \quad f(-x), \quad f(2x), \quad 2f(x)$$

Repeat this exercise, changing the formula for $f(x)$ to $3\sqrt{x-1}$. Then do it again using the function $f(x) = 10^x$. Check your answers with the ones given below.

Function notation is very efficient because it allows us to express the algebraic rule for a mathematical function economically, using only a few symbols. This very economy of expression, unfortunately, can also be a source of confusion. Be sure you understand, for example, the difference in meaning between $f(x) + 1$ and $f(x + 1)$, and between $-f(x)$ and $f(-x)$.

In the "Graph Trek" lab, you will see that the graph of each function you just wrote is a simple variation of the graph of the original function—a parabola or a square root curve or an exponential curve—and you will learn exactly how to get each new graph from the old one without plotting more than a couple of points.

Imagine yourself a research mathematician. The purpose of the lab is to understand what happens to the graph of any function $f(x)$ when you apply a **transformation** to the function—that is, when you change its algebraic formula in one of the ways you did above. You will have three research tasks during the lab:

- Collecting information from your grapher
- Deciding the effect of the given transformation (backed up by examples)
- Determining why the effect occurred.

As background for this research, you will need to be familiar with the general shapes of our seven fundamental functions. Review their graphs now on page 102.

Answers

$f(x)$	$3x^2 - 2x + 5$	$3\sqrt{x-1}$	10^x
$f(x) + 1$	$3x^2 - 2x + 6$	$3\sqrt{x-1} + 1$	$10^x + 1$
$f(x + 1)$	$3(x+1)^2 - 2(x+1) + 5$	$3\sqrt{x}$	10^{x+1}
$-f(x)$	$-3x^2 + 2x - 5$	$-3\sqrt{x-1}$	-10^x
$f(-x)$	$3x^2 + 2x + 5$	$3\sqrt{-x-1}$	10^{-x}
$f(2x)$	$12x^2 - 4x + 5$	$3\sqrt{2x-1}$	10^{2x}
$2f(x)$	$6x^2 - 4x + 10$	$6\sqrt{x-1}$	$2 \cdot 10^x$ (not 20^x)

The Graph Trek Lab

Pay particular attention to the title of this lab. The lab instructions are designed to help you discover and understand some mathematical patterns. They aren't meant to inhibit your own creative investigations, however. The beauty of graphing technology is that it allows us to try out our ideas quickly. ("What would happen if . . .?") If your group sees something interesting or puzzling, discuss it with one another. The individual explorations of each lab group are an important component of this lab and ought to be included in the report.

Shifts: $f(x) + c$ and $f(x + c)$

Question 1: How does adding a constant change the graph?

Here's our first question: What happens to the graph of a function when we add a constant to either the dependent or the independent variable? To go about this systematically, let's start by adding to the dependent, or output, variable. For example, to use the function $f(x) = \sin(x)$, you would view its graph along with the graphs of functions such as $\sin(x) + 4$, $\sin(x) - 1$, and $\sin(x) + 0.5$,

Now choose another function $f(x)$ and view its graph along with those of several functions of the form $f(x) + c$, where c is a positive or negative constant.

You should see that each new graph has exactly the same shape as the original but that it has shifted to a new position. Describe the relationship between each of the new graphs and that of your chosen function. What does the constant do? Why do you think this happens?

Repeat the above procedure with several different types of functions—quadratic, exponential, linear, square root—and see whether the effect of adding a constant to the output is always the same.

As soon as all of you feel confident that you can predict the appearance of each transformed graph, give yourselves a little test. Here is a function $y = CUP(t)$, defined solely by this graph:

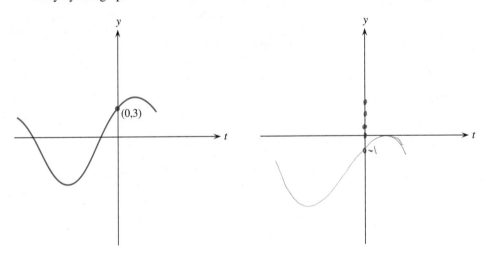

Draw the graph of $y = CUP(t) - 4$. Label its y-intercept.

So far, you've been adding a constant to the entire function, the dependent variable. Suppose instead that you were to add a constant to the independent variable; that is, you first add the constant to the input variable and then perform the function. What do you think would happen to the original graph? From your work in the preparation section, you know that the order of operations makes a difference in the *formula* for the function. (Compare the formulas for $f(x) + 1$ and $f(x + 1)$ on page 147.) Now you will see that the order of operations also makes a difference in the *graph* of the function.

Make a prediction about the graph of $y = (x + c)^2$ for some positive or negative constant c. What will be its relationship to the graph of $y = x^2$?

Now choose a specific value for c. Graph x^2 and overlay the graph of $f(x + c)$, that is, of $(x + c)^2$. Is this what you expected?

Explain to each other *what* happened and *why* you think that effect occurred. (The "why" is a lot harder than the "what," but don't avoid thinking about it.) How did your choice of a value for c affect the basic graph of x^2 ?

Test your conjecture by trying other values, both positive and negative, for c.

Is the behavior that you've witnessed peculiar to parabolas, or will other functions respond in the same way? Check it out by changing the basic function and adding different constants to the independent variable *before* performing the operation. You might, for instance, compare $\sqrt{x - \frac{1}{2}}$ to $\sqrt{x}$ and $\frac{1}{x+\pi}$ to $\frac{1}{x}$. Are you able to come up with a rule for what happens?

The grapher is nothing but a very fast point plotter, calculating the points for each new graph without reference to the previous one. *You*, however, can recognize a pattern that enables you to get a new graph from the old one with a minimum of effort. Do several different examples until each of you understands this pattern and can describe, in general terms, what happens to the graph of a function when you add a constant to the independent variable.

Now generalize about the relationship between the graph of some function $BALE(a)$ and that of $BALE(a + c)$. Suppose the graph of $y = BALE(a)$ looks like this:

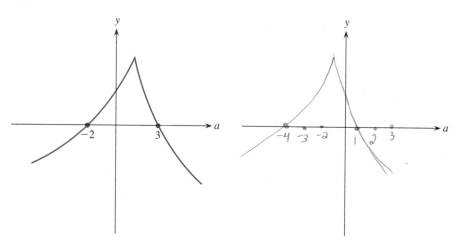

Draw the graph of $y = BALE(a + 2)$. Label its a-intercepts.

It's time to check your progress.

- Can you write a function that shifts the graph of $\sin(x)$ two units down? One unit to the left? $y = \sin(x) - 2$ $y = \sin(x - 1)$
- Can you sketch the graph of $(x - 2)^2 - 4$ by shifting the graph of x^2 two ways?
- To shift the graph of $\frac{1}{x}$ to the right, which would you use: $\frac{1}{x} + 1$, $\frac{1}{x} - 1$, $\frac{1}{x+1}$, or $\frac{1}{x-1}$?

Reflections: $-f(x)$ and $f(-x)$

Question 2: How does a negative sign change the graph?

Here's our second question: What happens to the graph of a function if we take the opposite of either the input or the output?

Again, we'll begin with the output, taking the opposite of an entire function after the function's job has been accomplished. Sketch the graph of $y = -x^2$. (Note that $-x^2$ is different from $(-x)^2$. Be sure you know why.) With the graphs of x^2 and $-x^2$ on the screen at the same time, describe the relationship between the two.

Mathematicians look for patterns, and you might expect to find the same relationship between the graphs of any pair of functions $f(x)$ and $-f(x)$, that is, a function and its opposite. Check this out with several other functions and their opposites. Remember that the opposite of a function is simply the function itself preceded by a negative sign. Some suggestions follow, but make up your own as well. Each person in the group should try two or three. Compare your results.

$$\sqrt{x}, \quad 2^x - 1, \quad x^3, \quad x^2 + 3x - 4$$

(Remember to take the opposite of the *entire* function.)

You should observe that the portion of the original function that lay above the x-axis ended up below it and vice-versa. Can you explain *why* this effect occurred?

We say that the x-axis is the **axis of reflection** for the transformation $-f(x)$.

Now it's time to generalize. Suppose the graph of some function $y = POINT(u)$ looks like this:

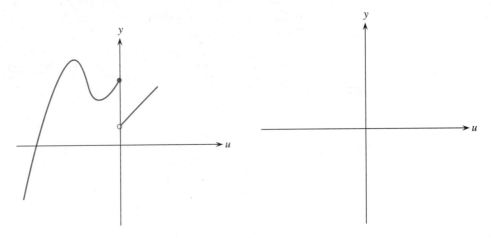

Sketch and label the graph of $y = -POINT(u)$.

Now for the independent variable. Suppose you look instead at $f(-x)$, that is, at the original function when the input variable is replaced by its opposite, before the function goes to work. Try this with the function $f(x) = \sqrt{x}$. What's the formula for $f(-x)$? What's the relationship between the graphs of $f(x)$ and $f(-x)$? If you don't see two separate graphs on the screen, pick a different window. Compare these two graphs with the graph of $-\sqrt{x}$. Try to get all three graphs on the screen simultaneously. The position of the negative sign makes a difference! Explain.

You should now see one graph in the first quadrant, one in the second, and one in the fourth. Can you complete a symmetrical design by putting a graph into the third quadrant as well? You'll have to figure out its algebraic formula first.

Try graphing $f(-x)$ for several other functions, perhaps the same ones that you used for learning about $-f(x)$. Keep doing examples until you can predict with confidence how the graph of $f(-x)$ will look.

You should observe that whatever was on the left side of the y-axis ends up on the right, and vice-versa. (Why do you think this happened?) The y-axis is the axis of reflection for the transformation $f(-x)$.

What happens if you start with $f(x) = x^2$ or $f(x) = |x|$? Why?

$y = (-x)^{\frac{1}{2}}$

Here's another picture of $POINT(u)$.

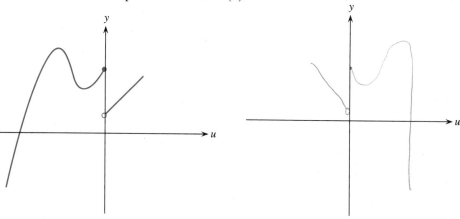

Sketch and label the graph of $y = POINT(-u)$.

At this point, you understand a lot about shifts and reflections.

- Can you write a function to reflect the graph of $x^2 + x$ in the x-axis? In the y-axis?
- Can you write a function to reflect the graph of $\sqrt{x + 1}$ in the x-axis? In the y-axis?
- Which of these is a reflection of the graph of 3^x in the y-axis: 3^{-x} or -3^x?

Stretches and Compressions: $k \cdot f(x)$ and $f(k \cdot x)$

Question 3: How does multiplication change the graph?

Now for our final question: What happens to the graph of a function when we multiply either the dependent or the independent variable by a constant?

Thus far, you have slid graphs into new positions and flipped them over an axis, but you have not changed their shapes. Now you will learn, via multiplication, to make a graph taller or shorter, wider or skinnier. (These terms are only metaphorical. The graph of x^2, for instance, already extends infinitely far in the positive direction; it cannot become "taller." But we can do something to the function to make the graph stretch upward more steeply.)

Once again, let's begin by modifying the output. Find out what happens to the graph of $\sin(x)$ when the whole function is multiplied by a positive constant. Try several different constants, including at least one value between 0 and 1.

Describe to one another the effects you see. What features of the graph remain the same? What features have changed? The original sine function had a maximum value of 1 and a minimum value of -1. What do you observe about the maximum and minimum values of the transformed graphs?

Try this with other functions, maybe $\sqrt{x - 2}$, 2^x, $x^2 + x$, or others of your choosing. Be sure that you multiply the entire function, not just one of the terms, by the various constants. Also, be careful with the order of operations. Three times 2^x is $3(2^x)$, not 6^x.

Some functions have certain points that stay put under the transformation. On which axis are those points located?

Here's an interesting graph: $x^3 - x$. In a window about four units wide and four units high, you should see the graph make two small loops, one up and one down. Can you make the loops three times as tall? Can you make them half as tall?

Here's a function $y = FOND(u)$ defined solely by its graph:

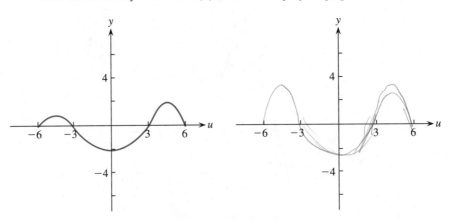

Draw the graph of $y = 3 \cdot FOND(u)$. Label its y-intercept.

You're on the home stretch! (Bad pun—sorry.) You still need to consider what happens to the graph when you multiply the input. Maybe you felt surprised a few pages ago when you saw that adding 2 to the input moved the graph *left* instead of right. The effect of multiplying the input variable might also surprise you.

Let's try a few. Compare $\sin(2x)$ and $\sin(3x)$ to the basic $\sin(x)$. Is this what you expected? Since a "2" on the *outside* of a function served to stretch the graph vertically, we might think that a "2" on the *inside* would stretch the graph horizontally. What actually happened?

We can think of the "2" and the "3" as factors that serve to speed up the wave, causing it to oscillate twice or three times as fast as a basic sine wave. A more general interpretation, one that applies to all functions and not only sine waves, is that the graph is being squeezed horizontally toward the y-axis by a compression factor of 2 or 3.

How would you stretch the sine wave horizontally, making it oscillate more slowly? Test your conjecture—does it work? If you want to stretch the graph horizontally by a factor of 3, by what constant should you multiply the independent variable?

Investigate this effect with other functions. If you use 2^x, you'll be graphing transformations such as 2^{3x} and $2^{0.4x}$. (Your grapher will want those exponents in parentheses.) There is one point on the graph of 2^x that doesn't move under this transformation. What is it? Go back to $x^3 - x$ (that's the one with two loops) and make the loops twice as wide, then four times as narrow (one fourth as wide—same thing). Pay attention to the multipliers as you make this happen.

Are you ready to generalize? Here's $FOND(u)$ again.

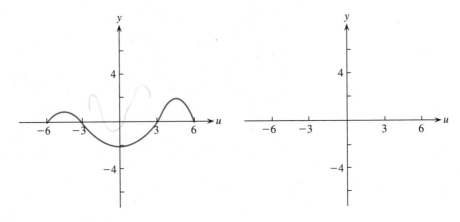

Draw the graph of $y = FOND(3u)$. Label its u-intercepts.

The Lab Report

What happens to the graph of a function when a constant is added to the output or to the input? (Consider negative as well as positive constants.) What happens to the graph of a function when the output or the input is replaced by its opposite? What happens to the graph of a function when the output or the input is multiplied by a positive constant? *elongation*
Your report should answer these questions fully but concisely.

- Summarize what you learned about the graphs of $f(x) + c$, $f(x + c)$, $-f(x)$, $f(-x)$, $k \cdot f(x)$, and $f(k \cdot x)$ in relation to the graph of any function $f(x)$.

- Illustrate with representative sketches of a few well chosen functions. Be sure that the graphs that form part of your report are sufficiently general to make the patterns clear.

- An excellent paper will strive to answer the question "why" for each transformation.

- Include any interesting or puzzling observations your group may have made.

Although this lab was long, the lab report need not be.

—increase range but maintains domain
—increase in frequency w/ same range

Exponential and Logarithmic Functions

4

© Yu Gang/Sovfoto/Eastfoto/PictureQuest

When you approximated the growth factor for the exponential model of U.S. population on page 71, your calculation involved the root $\left(\frac{260}{131}\right)^{1/55}$. Even a $12 calculator can produce that result in an instant. Imagine life before our electronic tools were invented. How on earth would you find a useful numerical approximation for such an expression?

Until the Scottish mathematician John Napier discovered logarithms, that evaluation would have been virtually impossible. Napier devoted 20 years of his life, from 1594 to 1614, to the painstaking creation—by hand, of course!—of detailed tables of powers. Those tables enabled anyone who learned to use them to perform impressive feats of arithmetic by treating the numbers involved as *powers of 10*. Those powers, known as logarithms, transformed multiplication and division into addition and subtraction, exponentiation into multiplication, and root-finding into division. In other words, logarithms simplified each mathematical operation by one level. For the next 360 years, Naperian logarithms were the calculational tool of choice for students everywhere, either in the form of log tables or as logarithmic scales on a slide rule. (A slide rule is pictured on the following page.)

The slide rule, which was found in the pocket of every engineer in 1970, is now a museum piece, while books of log tables lurk in dusty corners of used-book stores. The world uses ever more sophisticated calculators and computers for its increasingly elaborate calculations. But logs live on!

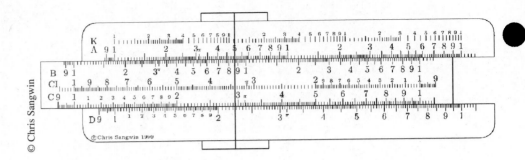

© Chris Sangwin

In this chapter, you'll see the value of a logarithmic scale for plotting certain types of data, and you'll learn to use logs in solving equations that until now you had no way of solving algebraically.

4.1 EXPONENTIAL FUNCTIONS

A gymnastics team practices its balance-beam routine, improving month by month. At first, scores increase rapidly; as more time passes, additional efforts result in smaller gains. Over time, a graph of the team's average score for the event might look like Figure 4.1, as each gymnast strives for a perfect 10.

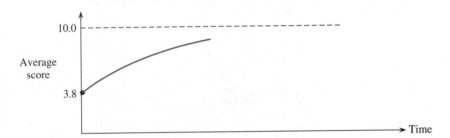

Figure 4.1 Average Balance-Beam Score versus Elapsed Time

A mathematics student pours himself a mug of steaming coffee and sits down to work on his lab report. Struggling to explain exponential growth in understandable English, he forgets to drink the coffee. The coffee cools, losing heat rapidly at first and then more slowly as the liquid approaches room temperature. If the room remains at 20° Celsius, the graph in Figure 4.2 could represent the temperature of the coffee over time.

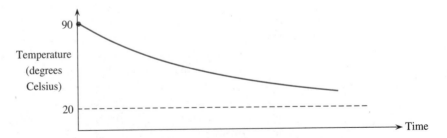

Figure 4.2 Coffee Temperature versus Elapsed Time

Both of these situations can be modeled by functions from the exponential family. The coffee-temperature graph is a decreasing exponential function that has been shifted up 20 units. As Figure 4.3 shows, the gymnastics-score graph is obtained from a decreasing exponential function that has been reflected in the horizontal axis and then shifted up 10 units.

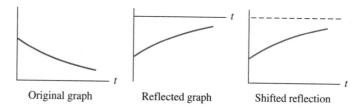

Original graph   Reflected graph   Shifted reflection

Figure 4.3 Transforming a Decreasing Exponential Function

In this section, you will learn to adapt the basic exponential formula to new situations by using the transformations of Chapter 3. First, though, we will do a brief review of exponents, justifying our use of any real number as an exponent, and we will summarize the properties of exponential functions.

Characteristics of Exponential Functions

On all fours

Recall that a basic exponential function can be written in the form $f(x) = c \cdot a^x$, where c is the y-intercept of the function f (its value when $x = 0$) and a is a positive number called the *base* or the *growth factor* of the function. Let's examine a very simple exponential function, one whose starting value, c, is 1 and whose base is 4: $f(x) = 4^x$. You'll use properties of exponents that you learned in algebra. Here they are again—be sure they're in your data bank.

For more help with exponents, turn to page 477 of the Algebra Appendix.

$a^0 = 1$, if $a \neq 0$

a^{-p} means $\frac{1}{a^p}$

$a^{1/p}$ means $\sqrt[p]{a}$ if the root exists

$a^{q/p}$ means $\left(\sqrt[p]{a}\right)^q$ if the root exists

So, for example, $4^{1.5} = 4^{3/2} = (4^{1/2})^3 = 2^3 = 8$ and $4^{-2} = \frac{1}{4^2} = \frac{1}{16}$.

▷   Without the aid of a calculator, complete Table 4.1 to get the coordinates of some points on the graph of 4^x.

x	−2.0	−1.5	−1.0	−0.5	0.0	0.5	1.0	1.5	2.0
4^x									

Table 4.1 Some Points on the Graph of $f(x) = 4^x$

▷   Use the points from Table 4.1 to plot a graph of $f(x) = 4^x$ in Figure 4.4.

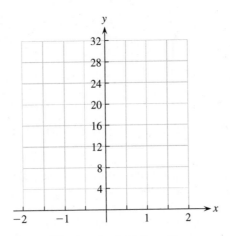

Figure 4.4 Graph of $f(x) = 4^x$

Check Your Understanding 4.1

CYU answers begin on page 190.

1. Consider the function $h(x) = 8^x$. Without using a calculator, evaluate $h(2)$, $h(0)$, $h(-1)$, $h(\frac{2}{3})$, and $h(-\frac{4}{3})$.

2. Graph $h(x) = 8^x$ and $f(x) = 4^x$ simultaneously on your grapher. How are the graphs similar?

How are they different?

Trading places

Now we'll use an exponential function whose base is between 0 and 1: $g(x) = \left(\frac{1}{4}\right)^x$.

▷   Complete Table 4.2 without using a calculator.

x	-2.0	-1.5	-1.0	-0.5	0.0	0.5	1.0	1.5	2.0
$\left(\frac{1}{4}\right)^x$									

Table 4.2 Some Points on the Graph of $g(x) = \left(\frac{1}{4}\right)^x$

Stop for a moment to compare the entries in Tables 4.1 and 4.2; you should observe that corresponding entries are reciprocals of each other. If we plot the points from both tables, we obtain the graphs in Figure 4.5.

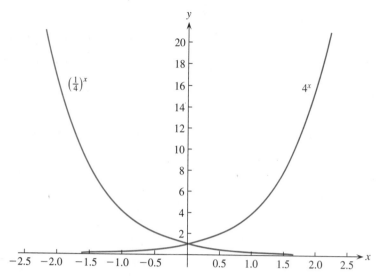

Figure 4.5 Graphs of $g(x) = \left(\frac{1}{4}\right)^x$ and $f(x) = 4^x$

Each of the graphs in Figure 4.5 is the reflection in the y-axis of the other graph. Does it surprise you that using the *reciprocal* of the base produced a *reflection* of the graph? This effect follows naturally from the way exponents work; let's see why it happens. As a reminder, here are some fundamental rules.

> **Laws of Exponents**
>
> **1.** $a^{p+q} = a^p \cdot a^q$
>
> **2.** $a^{p-q} = \dfrac{a^p}{a^q}$
>
> **3.** $a^{pq} = (a^p)^q$
>
> **4.** $(ab)^p = a^p b^p$
>
> **5.** $\left(\dfrac{a}{b}\right)^p = \dfrac{a^p}{b^p}$

Recall from Chapter 3 that the transformation $f(-x)$ creates the y-axis reflection of the graph of $f(x)$.

▷ If $f(x) = 4^x$, write the algebraic formula for $f(-x)$. Then use what you know about exponents to prove that it is equivalent to $g(x) = \left(\frac{1}{4}\right)^x$. (*Hint*: The fifth law of exponents tells us that $\left(\frac{1}{4}\right)^x$ is the same as $\frac{1^x}{4^x}$, or $\frac{1}{4^x}$.)

Check Your Understanding 4.2

1. By hand, sketch a graph of $y = \left(\frac{1}{3}\right)^x$. Don't overdo it; three or four well-chosen points will allow you to draw an excellent graph. Be sure to use points on both sides of the y-axis.

2. Give a formula for the reflection of $f(x) = \left(\frac{1}{3}\right)^x$ about the y-axis. Sketch and label its graph.

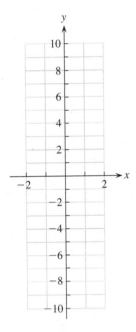

3. Give a formula for the reflection of $f(x) = \left(\frac{1}{3}\right)^x$ about the x-axis. Sketch and label its graph.

The functions 4^x and $\left(\frac{1}{4}\right)^x$ represent the class of simple exponential functions $y = a^x$, whose graphs you've met before. We'll use them to spell out the important properties of this function class. Refer to their graphs in Figure 4.5.

Domain   The domain of 4^x is $\mathbb{R}$. Any real number can serve as an exponent: $4^3, 4^{-5}$, $4^{2/7}$, and even 4^π have mathematical meaning. (4^π is close to $4^{3.14}$, which is $4^{314/100}$; it's even closer to $4^{3.14159}$. That is, we can get as close to the value of 4^π as anyone wishes, simply by using more decimal digits of π. The approximate value of 4^π is 77.88, between 4^3 and 4^4 as we should expect.)

See the Algebra Appendix, page 478, for a refresher on negative and fractional exponents.

Range   The range of 4^x is the set of all positive real numbers. Take a look: as we choose x-values farther and farther out to the right, 4^x-values get bigger and bigger. They go right through the ceiling—any ceiling you want—if the exponent is large enough.

x	0	1	2	3	4	5	6
4^x	1	4	16	64	256	1024	4096

▷   How big an integer does x have to be to make 4^x bigger than a million?

Here's how mathematicians express the fact that the values of 4^x grow without limit as x increases:

$$\text{as } x \to \infty, \qquad 4^x \to \infty.$$

We read this sentence, "As x approaches infinity, 4^x approaches infinity." The symbol ∞ does not represent a *number*; rather, it says that 4^x has no upper bound.

Well, what about negative outputs from 4^x? Could the function ever have zero as its output? Look what happens as the x-values march off to the left (values are rounded to 5 places):

x	0	-1	-2	-3	-4	-5	-6
4^x	1	0.25	0.0625	0.01563	0.00391	0.00098	0.00024

No matter how small the output of 4^x becomes, it never reaches 0; 4^x is always positive and its graph remains above the x-axis.

If $a > 0$, then $a^x > 0$ for any real number x.

A general principle is at work here: **raising a positive quantity to any real-number power always produces a positive result.** Keep this fact in mind, because it can be very handy in solving equations. See Exercises 5 and 6.

Asymptote   Look at the left side of the graph of 4^x. As x decreases, the graph becomes nearly flat, looking more and more like the x-axis itself. We say that the x-axis is the **horizontal asymptote** for this function; that is, the x-axis is the horizontal line that the curve eventually resembles. Symbolically,

$$\text{as } x \to -\infty, \qquad 4^x \to 0.$$

You saw this behavior numerically: for negative x, 4^x is a positive fraction, closer and closer to zero as x takes on smaller values.

▷   How small an integer does x have to be to make 4^x less than one one-millionth (0.000001)?

Figure 4.6 shows a visual summary of the global behavior of 4^x.

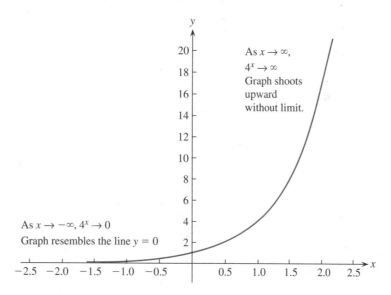

Figure 4.6 Global Behavior of the Exponential Function $f(x) = 4^x$

The behavior of $\left(\frac{1}{4}\right)^x$ is like that of 4^x, but in reverse: The function approaches the x-axis asymptotically on the right side. In symbols,

$$\text{as } x \to \infty, \qquad \left(\frac{1}{4}\right)^x \to 0.$$

Review concavity, pages 26–28.   **Concavity**   The graph of 4^x is concave up. As it rises, it bends upward; the function increases at an increasing rate. Increasing rates mean upward concavity, independent of whether the graph itself is rising or falling.

Check Your Understanding 4.3

1. Give the domain and the range of the function $\left(\frac{1}{4}\right)^x$.

2. Translate into English: as $x \to -\infty$, $\left(\frac{1}{4}\right)^x \to \infty$. What does this say about the function $\left(\frac{1}{4}\right)^x$?

3. What is the horizontal asymptote of $\left(\frac{1}{4}\right)^x$, and how would you express that in symbols?

4. What type of concavity does the graph of $\left(\frac{1}{4}\right)^x$ have?

STOP & THINK

Why can't -4 be the base of an exponential function? Why do you suppose mathematicians don't consider a function such as $H(x) = (-4)^x$ a member of the exponential family? (*Hint:* Evaluate H for several values of x, including some fractions.)

We've used 4^x and $\left(\frac{1}{4}\right)^x$ as representatives, but the features you've just learned for those two functions hold for every exponential function a^x. Figure 4.7 summarizes what you should know about a^x.

Important properties of exponential functions

Domain: $\mathbb{R}$
Range: positive reals
Asymptote: x-axis
y-intercept: 1
Increasing for $a > 1$
Decreasing for $0 < a < 1$
Concave up

$f(x) = a^x$
$a \geq 1$

$g(x) = a^x$
$0 < a < 1$

Figure 4.7 Graphs of $y = a^x$

Modeling with Exponential Functions

Coffee time

The coffee graph

Now back to the coffee. Figure 4.2 shows a decreasing exponential function with a vertical shift. Let $h(t)$ stand for the temperature of the coffee t minutes after it is poured. We have to shift the basic exponential function $c \cdot a^t$ up 20 units, so the function will have this form:

$$h(t) = c \cdot a^t + 20$$

Now we want to find what c and a are. We'll need the starting value and one more bit of information, just as we did in Chapter 2. Use the fact (from the graph in Figure 4.2) that $h = 90$ degrees Celsius when $t = 0$:

$$h(0) = 90 = c \cdot a^0 + 20$$

If you solve this equation for c, you'll get 70. You might have expected c to be 90, because the initial temperature of the coffee is 90 degrees Celsius, but in this function, c is the initial *difference in temperature* between the coffee and the room. Here's our model so far:

$$h(t) = 70a^t + 20$$

To determine a, we need another bit of information. Suppose that after 10 minutes, the temperature of the coffee is 68 degrees Celsius.

$$h(10) = 68 = 70a^{10} + 20$$

If you forget how to do this, see pages 70–71.

▷ Solve this equation for a; you should obtain $a \approx 0.963$.

The complete formula for the graph in Figure 4.2 therefore is $h(t) = 70(0.963)^t + 20$. That's the graph of $70(0.963)^t$ shifted up 20 units. The asymptote for h is the horizontal line $y = 20$.

A balancing act

Determining the gymnastic-score function is a similar process, but we must incorporate a flip as well. If $S(t)$ represents the average score t weeks from when practice began, the function will have this form:

$$S(t) = -c \cdot a^t + 10,$$

indicating a basic exponential graph that is first reflected in the horizontal axis and then raised 10 units.

To specify c and a, we need more information. Figure 4.1 shows that $S(0) = 3.8$. Let's use that fact; this time, you do the work. Keep going until you have a value for c.

▷ $S(0) = 3.8 =$

▷   Check that your *c*-value equals the *difference* between the initial average and a perfect score.

      An additional fact will enable you to find the growth factor *a*. Suppose that after six weeks, the team's average score has risen to 5.7.

▷   Use $S(6) = 5.7$ to obtain an approximate value for *a*.

      The complete function for the graph in Figure 4.1 is $S(t) = -6.2(0.94)^t + 10$ or, equivalently, $S(t) = 10 - 6.2(0.94)^t$. That's the function $6.2(0.94)^t$, reflected in the *t*-axis and shifted up 10 units. The line $y = 10$ is the horizontal asymptote.

Check Your Understanding 4.4

A kiddie pool is placed in a shady spot and filled with water, whose initial temperature is 55 degrees Fahrenheit. The air is warm—92 degrees Fahrenheit—and the water in the pool heats up.

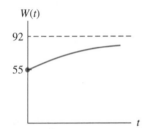

1. Write the general form for an exponential function $W(t)$ to model the temperature of the water after t minutes. Don't solve anything yet; use letters for any unknown constants.

2. Using the initial temperature, find one of the constants.

3. If, after 20 minutes, the temperature of the water is 64 degrees, find the growth factor for the model.

4. Write the complete function $W(t)$.

In a Nutshell   The domain of every exponential function a^x is $\mathbb{R}$; its range is the positive reals. The graph of a^x is concave up and has the *x*-axis as its horizontal asymptote. Shifts, reflections, and stretches or compressions can make a^x into a model for a variety of situations.

4.2   THE NATURAL EXPONENTIAL FUNCTION

What is this thing called e?

Suppose we want to know how fast a cup of coffee is cooling at a particular instant or how rapidly a stockpile of radioactive plutonium is decaying at a specific moment. We'd like a mathematical model that would tell us its own growth rate. Simply by observing the value of the function at a particular point, we'd know how rapidly it is growing at that point.

Such a function does exist. It is the exponential function e^x, where e represents a number whose approximate value is 2.718. Like the more familiar π, e is an irrational number: a nonrepeating, nonterminating decimal. Unlike π, its meaning cannot be briefly explained, but we will give you a glimpse of what's unique and natural about e and why mathematicians love it. The number e pops up frequently in mathematics, often in improbable places. It was first recognized in tables of compound interest, but it also arises naturally in statistics, probability, economics, psychology, physics, and a host of other disciplines.

Here's one way to calculate e. Add up the following numbers:

$$1 + \frac{1}{1} + \frac{1}{1 \cdot 2} + \frac{1}{1 \cdot 2 \cdot 3} + \frac{1}{1 \cdot 2 \cdot 3 \cdot 4} + \frac{1}{1 \cdot 2 \cdot 3 \cdot 4 \cdot 5} + \frac{1}{1 \cdot 2 \cdot 3 \cdot 4 \cdot 5 \cdot 6} + \cdots$$

The table shows where the numbers are headed. If you keep going forever, you'll have e exactly. In real life, of course, we have to stop somewhere, and the best we can achieve is a decimal approximation for e. The prolific Swiss mathematician Leonhard Euler (pronounced "oiler"), working in the eighteenth century (without a calculator, remember!) introduced the symbol e and calculated its value to 23 decimal places.

Number of Terms	Sum so Far
1	1.00000000
2	2.00000000
3	2.50000000
4	2.66666667
5	2.70833333
6	2.71666667
7	2.71805556
8	2.71825397

> The function $g(x) = e^x$ is, mathematically speaking, the most important of all exponential functions, so much so that mathematicians refer to it as the **natural exponential function,** or even simply *the* exponential function.

When you study calculus, you will understand e better, but there is one remarkable property of the function e^x that you can learn now. (A student recently told one of the authors that she was e-centric. We hope that you, too, will become a little e-centric.)

Approximating an Instantaneous Rate

Average rate of change of a function g: $\frac{g(b)-g(a)}{b-a}$

In Chapter 1, you learned to find the average rate of change for a function g over an interval $a \leq x \leq b$. The average rate of change gives the slope of the line joining the points $(a, g(a))$ and $(b, g(b))$. In this section, we will use that formula in a new way: to approximate the rate of change of a function *at one particular point*, rather than over an interval.

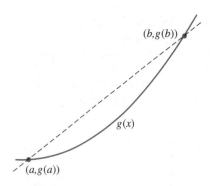

Figure 4.8 Average Rate of Change of Function g

Suppose, for example, that we want the instantaneous rate at which the coffee (from Section 4.1) was cooling 10 minutes after it was poured, when its temperature was 68° C. The temperature of the coffee is given by the function $h(t) = 70(0.963)^t + 20$; we need the slope of that curve at $t = 10$.

Let's consider $(10, h(10))$ to be the target—the point at which we want the rate of change of h. Because the graph of h is smooth and unbroken, it is nearly linear in the immediate vicinity of the target point, and the slope of that "line segment" is a good approximation to the **instantaneous rate of change** of the function h at the point $(10, h(10))$. We choose two points close to the target to calculate a slope. For convenience, we will use $(9.99, h(9.99))$ and $(10.01, h(10.01))$. Here goes!

An instantaneous rate of change gives the slope of the curve at a single point.

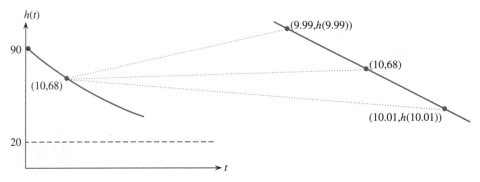

Figure 4.9 Graph of $h(t)$ with Magnified View

▷   Calculate the average rate of change of h from $x = 9.99$ to $x = 10.01$.

You should find that at $t = 10$, the rate of change of h is approximately -1.8102. That is, ten minutes after being poured (when its temperature is 68° C), the coffee is losing heat at the rate of about 1.8 degrees per minute.

▷   Why should we expect this rate to be negative?

Now we'll use the same technique on the function e^x and discover something exceptional about this function.

▷   Ask your grapher to draw $y = e^x$. If you zoom in repeatedly anywhere on the curve (at the y-intercept, for instance), the graph will appear to straighten out. Try it.

How to approximate an
instantaneous rate of
change for e^x

Let's estimate the slope of the graph of e^x, its instantaneous rate of change, at the y-intercept. Our strategy again will be to use two points on the curve, one just to the left of our target, $(0, e^0) = (0, 1)$, and one just to its right, for the slope calculation. The points $(-0.01, e^{-0.01})$ and $(0.01, e^{0.01})$ will work nicely because on that narrow interval, the curve is nearly linear.

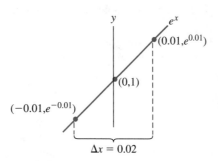

Figure 4.10 Magnified View of e^x

Find a point on either side
of the target. Then do $\frac{\Delta y}{\Delta x}$.

▷ With a calculator, evaluate the fraction

$$\frac{e^{0.01} - e^{-0.01}}{0.01 - (-0.01)} \quad \text{or, more simply,} \quad \frac{e^{0.01} - e^{-0.01}}{0.02}.$$

The result should be $1.00001666\ldots$, a number very close to 1. If you obtained something different, check your calculator technique. Use parentheses and perform a single sequence of steps, pressing "enter" only at the end. Accuracy is really important here, and you'll lose accuracy if you round off any numbers in midstream. Rather than writing down intermediate results, allow the calculator to hold them in its memory.

The slope of e^x at $x = 0$ seems to be about 1. Notice that the y-value of e^x at $x = 0$ is also 1. The slope of e^x at $x = 0$ equals e^0. This is no coincidence!

How to fill in the table

▷ Complete the table by first writing a three-decimal-place approximation of each e^x-value, and then finding the instantaneous rate of change at the point. Round slope values to three places. Using $x = -1$, for example, your slope calculation could have this setup:

$$\frac{e^{-0.99} - e^{-1.01}}{-0.99 - (-1.01)} \quad \text{or, more simply,} \quad \frac{e^{-0.99} - e^{-1.01}}{0.02}.$$

You do not have to use this exact window. *Any* two points on the graph of e^x that are very close to $(-1, e^{-1})$ will do. You might use $(-0.999, e^{-0.999})$ and $(-1.001, e^{-1.001})$ instead.

x	e^x	Slope of e^x
-1		
0		
1		
2		

The slope of e^x at $x = c$ is equal in value to e^c.

Notice that the slope value and the function value are remarkably close. Their similarity is because of this mathematical fact: For the function e^x, the y-value at any point equals the slope of the function at that point. The rate at which e^x grows is equal, at every moment, to its value at that moment; e^x lets us know how fast it's going.

The preceding statement gives a hint as to why the family of exponential functions whose base is e is the family of choice for every process involving natural growth.

Figure 4.11 shows that the graph of e^x lies between the graph of 2^x and the graph of 3^x, closer to 3^x than to 2^x because the value of e is between 2 and 3 and is closer to 3 than to 2.

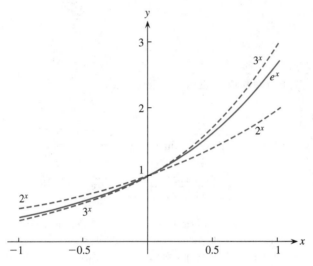

Figure 4.11 Graphs of 2^x, e^x, and 3^x

Check Your Understanding 4.5

1. Give the domain, the range, the y-intercept, and the horizontal asymptote for the function $F(x) = e^x$.

2. Use a slope calculation to approximate the instantaneous rate of change of e^x at $x = -2$. Check that your result is nearly equal to e^{-2}.

3. Without performing any calculations, give the exact instantaneous rate of change of the function e^x at $x = -1.9$ and at $x = \pi$. Then find a decimal approximation for those two values.

4. Using an interval that is 0.02 unit wide, approximate the instantaneous rate of change of the function 4^x at $x = -2$ and at $x = 2$. Compare your first result with 4^{-2}; does it match? Compare your second result with 4^2; does it match?

Base-*e* Exponential Functions

In CYU 4.5, you saw further evidence that e^x measures its own rate of change, and you saw that 4^x, a different exponential function, does not. The value of 4^0 is 1, but the slope of 4^x at $x = 0$ is more than one. Wouldn't it be handy if every exponential function could have *e* as its base? Stay tuned.

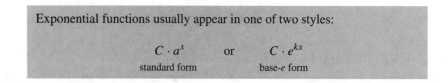

Exponential functions usually appear in one of two styles:

$$C \cdot a^x \qquad \text{or} \qquad C \cdot e^{kx}$$

standard form   base-*e* form

Every exponential function can be written in either form, and each form has certain advantages. Standard form allows us to determine at a glance the growth factor. Base-*e* form gives us quick information about the rate of change. (This advantage will become more apparent to you when you study calculus.) Therefore, we want to be able to go back and forth between the two forms. Let's see how to write 4^x and $\left(\frac{1}{4}\right)^x$ in base-*e* form.

▷ With a calculator, verify that $e^{1.386} \approx 4$ and that $e^{-1.386} \approx \frac{1}{4}$.

Laws of Exponents, page 159 ▷ Which exponent rule allows us to write $\left(e^{1.386}\right)^x$ as $e^{1.386x}$?

The function 4^x is approximately equivalent to the function $e^{1.386x}$. Similarly, $\left(\frac{1}{4}\right)^x \approx e^{-1.386x}$. And, as you'll learn in calculus, finding the instantaneous rate of change for $e^{1.386x}$ or for $e^{-1.386x}$ is nearly as easy as it is for e^x. Why 1.386? Because it's the power of *e* that (approximately) equals 4, as shown in Figure 4.12.

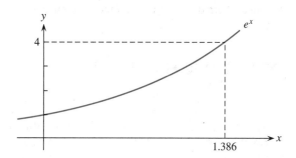

Figure 4.12 Solving $e^k = 4$ Graphically

Now you'll work in the other direction, converting a function from standard form to base-*e* form.

▷ By approximating $e^{0.01725}$, rewrite the function $G(t) = 5230e^{0.01725t}$ in standard form. You should obtain the function from page 67 that we used to model world population growth.

▷ With a grapher, overlay the graphs of $5230e^{0.01725t}$ and $5230(1.0174)^t$. On the interval $0 \le t \le 10$, you should find the graphs indistinguishable.

Again, where did that 0.01725 come from? Starting with $5230(1.0174)^t$, we had to express 1.0174 as a power of e: $e^k = 1.0174$. The constant k was whatever exponent was necessary to make that happen, and the number 0.01725 did the trick. In other words, we had the *output* of the exponential process, and we needed the corresponding *input* value (that is, the exponent itself). This should remind you of inverse functions: finding the input when you know the output. Trial-and-error would work, as would a good graph (as in Figure 4.12), but an algebraic solution to the equation $e^k = 1.0174$ requires a new set of tools, known as logarithms.

Get ready to meet the Logarithmic Family!

Check Your Understanding 4.6

1. Rewrite the function $B(t) = 3000e^{-0.22314t}$ in standard form.

2. What is the growth factor for the function $B(t)$? Is $B(t)$ an increasing or a decreasing function? (You might remember $B(t)$ from Chapter 2 as the value of Jason's sound system.)

3. In standard exponential form, a function decreases if its growth factor is between 0 and 1. In base-e form, what indicates a decreasing function?

In a Nutshell   The number e is the most commonly used base for exponential functions. It serves as a "common currency" for all members of the exponential family.

The instantaneous rate of change of a function is the slope of its graph at a particular point. We can approximate the slope of a curve at a point with the average-rate-of-change formula, choosing two points that are very close to our target.

For the function e^x, the slope agrees[1] with the value of the function (the y-value) no matter where on the curve you look. In other words, e^x and its growth rate are one and the same.

4.3 LOGARITHMIC FUNCTIONS

The powers that be

In this section, we introduce the family of logarithmic functions. If you've had an unpleasant experience with logs, you might be unenthusiastic at the prospect of visiting them again. But logarithmic functions are simply the inverses of exponential functions. You need only reflect. (Pun intended. If you missed it, review Section 3.3.) What's more, they provide a handy tool for solving exponential equations such as $e^x = 4$.

A Logarithmic Scale

A *New York Times* article examining the potential risk from asteroids hitting the earth contained the graphic display shown in Figure 4.13.

---

[1]Working with electronic devices, we are able to obtain only *approximations* for the actual function values and growth rates. The actual values are irrational numbers; a calculator provides decimal approximations. Given appropriate electronic tools, we can obtain those values to whatever degree of accuracy we desire.

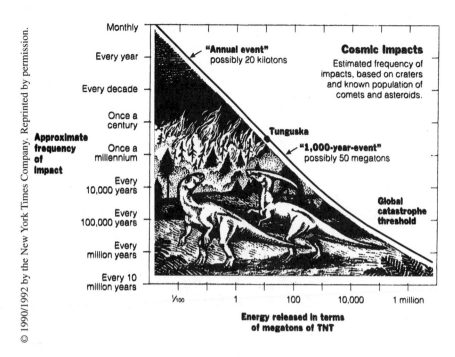

Figure 4.13 Frequency of Cosmic Impacts

Focus on the horizontal scale, which we've reproduced in Figure 4.14, isolated from the rest of the display. The scaling on that number line is quite different from any that you've seen previously in this text. Until now, we've held to the uniform-scale convention that equal intervals on an axis must represent the same quantity—the numbers associated with consecutive tick marks must differ by a constant amount. In Figure 4.14, however, the difference in value between consecutive tick marks is not constant. The difference between 1 and 10 is 9, for instance, while the difference between $\frac{1}{10}$ and 1 is only $\frac{9}{10}$. The numbers on this scale increase as powers of 10.

Figure 4.14 Horizontal Scale from Figure 4.12

STOP & THINK
Why was the horizontal scale in Figure 4.13 laid out in that way? In conventional scaling, equal spacing between tick marks represents equal amounts. If conventional scaling were adopted for the horizontal axis of Figure 4.13, how would the display change?

The scaling in Figure 4.14 makes it easy to locate numbers that are powers of 10, but not so easy to locate other values. Where, for instance, would you place 40 or $\frac{1}{2}$ relative to the other numbers on the line? Clearly, 40 belongs somewhere between 10 and 100, but should it be placed closer to 10 or to 100? Similarly, $\frac{1}{2}$ is between $\frac{1}{10}$ and 1 but not halfway; where, precisely, does it belong on this scale? To answer these questions, we need to do some work with exponents.

We will rescale the numbers in Figure 4.14 by replacing each value by its corresponding power of 10. Thus, we replace 100 by 2 (because $10^2 = 100$) and 10^5 by 5. In other words, we will focus on the **order of magnitude** of each of the numbers.

▷   Create a power scale by writing the corresponding powers underneath the numbers in Figure 4.14. You will write only the exponents.

▷   Does the power scale that you have just written conform to conventional scaling? That is, are consecutive scale numbers separated by a constant amount? If so, what is that constant separation?

Now return to an earlier question: Where do we place 40? The power scale that you just created will help us to decide. We need to know *what power of 10* gives 40. (Do you see why?) This is an *inverse* problem: We know the output of 10^x but not the input. Although the equation $10^x = 40$ doesn't have an integer solution, we know that this power must be more than 1 and less than 2, and we can estimate it with the graph of 10^x in Figure 4.15 or with a grapher.

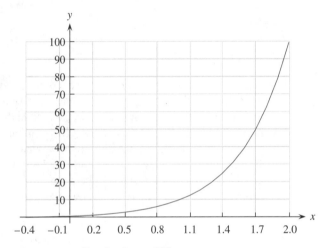

Figure 4.15   Graph of $y = 10^x$

▷   Use Figure 4.15 to verify that 1.6 would be a good choice for the power of 10 that produces 40.

▷   Check this on your calculator. (Use the 10^x key.) Is $10^{1.6}$ close to 40?

Because 1.6 is six tenths of the way between 1 and 2, 40 should be placed approximately six tenths of the way between 10 and 100 on the horizontal axis of Figure 4.14. Figure 4.16, a magnified portion of the same horizontal scale, shows where to place 40 in relation to 10 and 100.

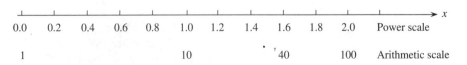

Figure 4.16   Magnified Horizontal Scale

Check Your Understanding 4.7

1. Use the graph of 10^x in Figure 4.15 to estimate the power of 10 that gives 5 and the power of 10 that gives 80.

2. Use those answers to locate 5 and 80 on the arithmetic scale in Figure 4.16.

STOP & THINK   Notice that the numbers on the horizontal scale in Figures 4.13 and 4.14 never reach zero. Why not?

The Base-10 Logarithm

Log on

When you replaced the original scale numbers in Figure 4.14 with their corresponding powers of 10, you were writing the **common logarithm** or **base-10 logarithm** of those numbers. The base-10 logarithm of 40 turned out to be approximately 1.6, because $10^{1.6} \approx 40$.

▷   What is an approximate value for the common, or base-10, logarithm of 80? (If you completed CYU 4.7, you won't need a calculator.)

The statement $\log_{10}(40) \approx 1.6$ means that $10^{1.6} \approx 40$.

At this point, we'll introduce some notation for the common logarithm. We write $\log_{10}(40)$ (pronounced "log, base 10, of 40") for the common logarithm of 40. More generally, $\log_{10}(x)$ means *the power of 10 that yields the quantity x*.

The *output* of a logarithmic function is an *exponent*. Thus, the statement $y = \log_{10}(x)$ is equivalent to the statement $10^y = x$. $\log_{10} x$ can also be written as $\log x$.

To find the inverse of $y = g(x) = 10^x$, interchange x and y: $x = 10^y$ then solve for y: $y = g^{-1}(x) = \log_{10}(x)$.

Does that last sentence ring a bell? If you had been given the function $y = g(x) = 10^x$ and were asked to find its inverse, you would probably have known enough to start out by writing $x = 10^y$, but then you would have been stuck, because you had (up until this moment) no algebraic method of solving for y. Now you do!

How about the graph of $y = \log_{10}(x)$? As you learned in Section 3.3, an easy way to draw the graph of an inverse function is to reflect the graph of the original function (in this case, $y = 10^x$) in the diagonal line $y = x$. When you do so, you'll see a picture like the one in Figure 4.17.

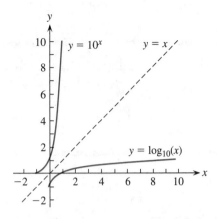

Figure 4.17 Graphs of $y = 10^x$, $y = \log_{10}(x)$, and the Line of Reflection $y = x$

▷ Using square scaling, reproduce the graphs of Figure 4.17 on your grapher. The key for the base-10 logarithmic function will probably be labeled simply "log."

Check Your Understanding 4.8

Section 4.3 contains many short CYU exercises. If you want to understand logs, don't skip any!

1. Write each statement in exponential form. The first one is done as a sample.
   (a) $\log_{10}(0.001) = -3$, because $10^{-3} = 0.001$

   (b) $\log_{10}(10000) = 4$, because $10^4 = 10,000$

   (c) $\log_{10}(\sqrt[3]{100}) = \frac{2}{3}$, because $10^{\frac{2}{3}} = \sqrt[3]{100}$

   (d) $\log_{10}(0.5) \approx -0.3$, because $10^{-.3}$

2. Write each statement in logarithmic form. The first one is done as a sample.
   (a) If $10^{x+3} = 7$, then $\log_{10}(7) = x + 3$

   (b) Because $10^{-1/2} = \frac{1}{\sqrt{10}}$, then $\log_{10}\left(\frac{1}{\sqrt{10}}\right) = -\frac{1}{2}$

   (c) Because $10^0 = 1$, then $\log_{10}(1) = 0$

   (d) If $10^{2t} = 450$, then $\log_{10}(450) = 2t$

3. Evaluate each of the following without using a calculator. (In each case, ask yourself what exponent is needed to convert the base, 10, into the number in parentheses.) The first one is done as a sample.
   (a) $\log_{10}\left(\frac{1}{100}\right) = -2$

   (b) $\log_{10}(0.00001) = -5$

   (c) $\log_{10}(10) = 1$

   (d) $\log_{10}(10^{13}) = 13$

   (e) $\log_{10}(\sqrt[5]{10}) = .2$

4. Use a calculator to find an approximate value for each of the following.
   (a) $\log_{10}(\frac{1}{2}) \approx -.301$
   (b) $\log_{10}(2) \approx .301$
   (c) $\log_{10}(20) \approx 1.301$
   (d) $\log_{10}(200) \approx 2.301$
   (e) $\log_{10}(400) \approx 2.602$

STOP & THINK

Look for numerical patterns in your answers to Question 4. What relationships do you see among any of the five logarithms? Can you account for any patterns? You might want to try out your conjectures with the base-10 logs of other numbers, such as 8 or 32 or 40. (*Hint*: A logarithm is an exponent; therefore, the numbers your calculator supplies are exponents.)

The Family of Logarithmic Functions

We've begun our discussion of logarithms by presenting the base-10 logarithm, because it's frequently used in science, is associated with scientific notation, and is the easiest log to understand. But 10 is not the only base for logarithms.

Look back at the graphs of $y = a^x$ in Figure 4.7 on page 169. We defined an exponential function for every positive base except 1. For bases greater than 1, the function a^x is increasing; for bases between 0 and 1, the function a^x is decreasing. In either case, a^x is a **one-to-one function** and has an inverse function.

If you need a refresher, find inverse functions on pages 116–126.

The inverse function of any exponential function a^x is the corresponding logarithmic function, $\log_a(x)$ (pronounced "log, base a, of x"). Every base that has an exponential function has a corresponding logarithmic function as well, and they are inverses of each other. Given any *positive* input x, the function $\log_a(x)$ gives as its output *the power of a that yields x*.

STOP & THINK   Why must the input x in $\log_a(x)$ be positive?

$\log_{10}(1) = 0$

$10^0 = 1$

~~any number~~ You can't get a negative answer from an exponential function.

The quantity $\log_a(x)$ is the *power* to which you raise a to get x. In other words,

The statement $\log_a(x) = y$ is equivalent to the statement $a^y = x$.

These two facts are important. You'll use them often.

▷ $\log_a(1) =$ _____ because _____, and

$\log_a(a) =$ _____ because _____ for every real number a.

Warning: To avoid confusion and error, be very careful when you write logarithms. In the expression $\log_{10}(x)$, the 10 is a *subscript* giving the base of that logarithmic function, and it should be written slightly below the line on which "log" and "x" are written. *Don't let the 10 drift up*, or you could end up with $\log(10x)$ or $\log(10^x)$, both of which are mathematically very different from $\log_{10}(x)$.

Check Your Understanding 4.9

1. Rewrite as an exponential statement.
   (a) $\log_{16}(2) = \frac{1}{4}$ (b) $\log_2\left(\frac{1}{4}\right) = -2$

   $16^{\frac{1}{4}} = 2$ $2^{-2} = \frac{1}{4}$

2. Without using a calculator, evaluate each expression.
   (a) $\log_3(9)$ (b) $\log_9(3)$ (c) $\log_2(32)$

   $= 3$ $9^x = 3$

   (d) $\log_2\left(\frac{1}{32}\right)$ (e) $\log_9(27)$

Like exponential functions, logarithmic functions form a family. Figure 4.18 shows a generic log graph for bases greater than 1, paired with the graph of an increasing exponential function. Log bases between 0 and 1 are scarcely ever used; therefore, we've omitted that graph.

In inverse functions, the roles of x and y are interchanged.

If you compare the characteristics of log functions with those of exponential functions, you'll see that every attribute of the independent variable becomes a feature of the dependent variable and vice-versa.

For $\log_a(x)$,
Domain: positive reals
Range: $\mathbb{R}$
Asymptote: y-axis
x-intercept: 1
Increasing and concave down
for $a > 1$

For a^x,
Domain: $\mathbb{R}$
Range: positive reals
Asymptote: x-axis
y-intercept: 1
Increasing and concave up
for $a > 1$

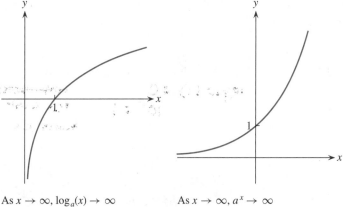

As $x \to \infty$, $\log_a(x) \to \infty$ As $x \to \infty$, $a^x \to \infty$
As $x \to 0$, $\log_a(x) \to -\infty$ As $x \to -\infty$, $a^x \to 0$

Figure 4.18   Graphs of $y = \log_a(x)$ and of $y = a^x$ for $a > 1$

The Natural Logarithm

The most frequently used logarithmic functions are the inverses of 10^x and e^x. We've already discussed $\log_{10}(x)$, the inverse of 10^x. Now we'll investigate the base-e logarithm, the inverse of e^x. Using standard log notation, we would write this function as $\log_e(x)$. Mathematics, however, reserves a special notation for the base-e logarithm, **ln(x)**, and a special title, the **natural logarithm**.[2]

> For any positive number N, the expression $\ln(N)$ means the base-e logarithm of the number N, or the power to which e must be raised to yield N.

---

[2]The symbol *ln* is an abbreviation for the term, *natural logarithm*, but because it was used first in France, the initials appear in the order in which they would appear in French, where the adjective follows the noun. That's one reason for calling the function "ln" instead of "nl."

▷ Use the graph of e^x in Figure 4.19 to guide you in drawing the graph of $\ln(x)$.

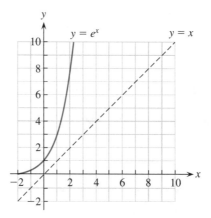

Figure 4.19 Graphs of $y = e^x$, $y = \ln(x)$, and $y = x$

▷ Use Figure 4.19 to estimate natural logarithm values corresponding to the x-values in Table 4.3. The first two are done for you.

x	0.1	0.5	1.0	1.5	2.0	3.0	4.0
$\ln(x)$	-2.3	-0.7					

Table 4.3 Estimates of $\ln(x)$ for Selected Values of x

▷ With a calculator, verify your Table 4.3 entries.

▷ The domain of e^x is $\mathbb{R}$ and its range is the set of positive real numbers. What are the domain and the range of the function $\ln(x)$?

▷ The graph of $y = e^x$ has $y = 0$ (the x-axis) as its horizontal asymptote. What kind of asymptote does $y = \ln(x)$ have? What is its equation?

You've just answered questions that help to relate the natural exponential function to the natural logarithmic function. Briefly, here's the connection:

> The statement $\ln(N) = p$ is equivalent to the statement $e^p = N$.

On page 169, you recognized 1.386 as an approximate solution to the equation $e^x = 4$. Now you know that the exact solution is $x = \ln(4)$. In other words, $\ln(4)$ is the exponent that you place on e to obtain 4; $e^{\ln(4)} = 4$. That exponent happens to be approximately 1.386.

The solution to $e^x = N$ is $x = \ln(N)$; $\ln(N)$ is the power of e that produces N.

Similarly, you verified that 0.01725 is an exponent k such that $e^k \approx 1.0174$. The exact value of that exponent k is $\ln(1.0174)$; $\ln(1.0174)$ is the power to which you raise e to get 1.0174.

Natural logarithms make it easy to write any exponential model in base-e form. So, for instance, the world population model $G(t)$ becomes

$$G(t) = 5230(1.0174)^t = 5230 \left(e^{\ln(1.0174)}\right)^t \approx 5230 e^{0.01725t}$$

Note that the expression $e^{\ln(1.0174)}$ replaces the number 1.0174.

Check Your Understanding 4.10

1. Rewrite each statement in logarithmic form. Use "ln." (Yes. Get used to it. You can think "$\log_e$" but you should write "ln.")
   (a) $e^{-1/2} = \frac{1}{\sqrt{e}}$

   (b) $e^0 = 1$

   (c) $e^{0.0023t} = V$

2. Rewrite each statement in exponential form.
   (a) $\ln\left(\frac{1}{e^2}\right) = -2$ (b) $\ln(e) = 1$

3. Without a calculator, find the exact numerical value of each expression.
   (a) $\ln(e^4)$ (b) $\log_{10}(10000)$

   (c) $\ln(\sqrt[3]{e^2})$ (d) $\ln\left(\frac{1}{e^3}\right)$

4. Find the *exact* value of x. (You should use the symbol "ln" or "$\log_{10}$" in your answer.) Then give a three-place decimal approximation.
   (a) $e^x = 10$ (b) $e^{x+3} = 10$

   (c) $10^{2x} = e$ (d) $\ln(x) = 4$

   (e) $\ln\left(x^3\right) = 4$

5. Rewrite $B(t) = 3000(0.80)^t$ in base-e form.

Figure 4.20 allows you to compare three different logarithmic graphs. Notice how the size of the base determines the steepness of the graph.

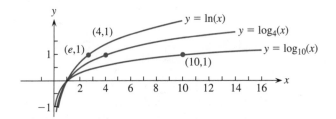

Figure 4.20 Graphs of $\log_{10}(x)$, $\log_4(x)$, and $\ln(x)$

▷   How do you know that the point $(4,1)$ must be on the graph of $\log_4(x)$?

We emphasize that the logarithmic function is a legitimate function in its own right. In the next section, you will learn various transformations of the generic log graph, and you will learn to manipulate logs in clever ways without necessarily referring back to the exponential function from which the logarithms have come.

In a Nutshell   A logarithm is an exponent. The logarithmic function $\log_a(x)$ is the inverse of the exponential function a^x. Any positive number except 1 can serve as the base of a logarithmic function, but the two most commonly used bases are 10 and e. Logs give us an algebraic tool for solving equations in which the unknown appears as an exponent.

4.4 LOGARITHMIC TRANSFORMATIONS

Power play

Because log functions are inverses of exponential functions, the output of a logarithmic function is an exponent. This little fact will allow us to do some amazing feats of algebra and will give us algebraic techniques for solving problems that until now required trial-and-error or graphical approximations. Here we go!

Rules for Logarithms

Logarithms have their own set of rules, each log rule having a counterpart in the exponent rules from Section 4.1. To make the relationship clearer, we list the three basic Laws of Logarithms and leave space for the corresponding Law of Exponents from page 159. To help yourself make sense of the rules, keep in mind that *the entire expression* $\log_a(x)$, not the a alone nor the x alone, represents an exponent. In every case, a is assumed to be a positive number, not equal to 1, and M and N represent positive quantities.

Laws of Logarithms	Laws of Exponents
1. $\log_a(M) + \log_a(N) = \log_a(MN)$	1. _____
2. $\log_a(M) - \log_a(N) = \log_a\left(\frac{M}{N}\right)$	2. _____
3. $p \cdot \log_a(N) = \log_a(N^p)$	3. _____

▷   Beside each rule for logs, write the corresponding exponent rule from page 159.

In the exercises, you'll have a chance to verify the log rules. For now, though, let's see how we can exploit them. Logarithms enable us to replace a higher-level operation, such as division or exponentiation, with a lower-level operation, like subtraction or multiplication. This strategy is important for calculus, and we can make good use of it right here as well.

Your calculator will verify that $\log_{10}(2) \approx 0.301$. Here are a few examples of log magic using just that fact and the three log laws.

- To find $\log_{10}(8)$, we recognize 8 as 2^3:

$$\log_{10}(8) = \log_{10}(2^3)$$
$$= 3\log_{10}(2) \quad \text{(by Law 3)}$$
$$\approx 3(0.301) = 0.903$$

- To find $\log_{10}(5)$, we think of 5 as $\frac{10}{2}$:

$$\log_{10}(5) = \log_{10}\left(\frac{10}{2}\right)$$
$$= \log_{10}(10) - \log_{10} 2 \qquad \text{(by Law 2)}$$
$$\approx 1 - 0.301 = 0.699$$

- Finally, $\log_{10}(200)$ is easy if we rewrite 200 as $2 \cdot 100$:

$$\log_{10}(200) = \log_{10}(2 \cdot 100)$$
$$= \log_{10}(2) + \log_{10}(100) \qquad \text{(by Law 1)}$$
$$\approx 0.301 + 2 = 2.301$$

Why should we care? After all, a calculator will give values for any of these logs at the push of a button—no arithmetic needed. Is there a reason for us to act as though a calculator weren't available? Yes, there is. If we're working with variables, a calculator might not be enough, as the following example shows.

Would you know how to sketch a graph of $y = \ln\left(x^3\right)$? That is, could you guess the relationship between its graph and that of $y = \ln(x)$? That relationship is much easier to see if we use Rule 3 to write the function in a different form:

$$\ln\left(x^3\right) = 3\ln(x),$$

showing that $\ln\left(x^3\right)$ is a vertical stretch, by a factor of 3, of $\ln(x)$.

▷ Using the graph of $\ln(x)$ in Figure 4.21 as a guide, carefully sketch the graph of $\ln\left(x^3\right)$. Be sure your graph shows the correct x-intercept and vertical asymptote. Label the graph you draw.

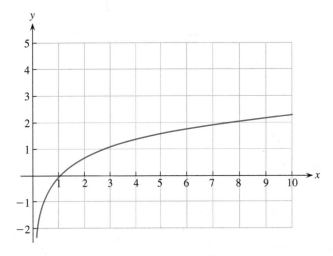

Figure 4.21 Graph of $y = \ln(x)$ and Some of Its Relatives

Check Your Understanding 4.11

1. Rewrite the expression $\ln(8x)$ as the sum of two logs, one of which is a constant. (*Hint*: The natural log of a constant is also a constant.)

2. Use the rewritten version to sketch a graph of $y = \ln(8x)$ on the axes of Figure 4.21. Label it. What is the precise relationship between the graph you drew and the graph of $y = \ln(x)$?

3. Use one or more log rules to rewrite the formula $y = \ln\left(\frac{1}{x}\right)$ as a simple transformation of the graph of $y = \ln(x)$. Add that graph to Figure 4.21; label it. What is the precise relationship between the graph you just drew and the graph of $y = \ln(x)$?

4. Rewrite as a single logarithm with a coefficient of 1: $\frac{1}{2}\ln(x) - \ln(5)$.

5. Rewrite as a sum or difference of simpler logs: $\log_{10}\left(0.001x^3\right)$.

STOP & THINK   How would you draw the graph of $h(x) = \ln\left(x^2\right)$? You might suppose that it would look like $\ln\left(x^3\right)$, only less steep, but there's a catch. (*Hint*: Consider both the domain of h and the symmetry you should expect in the graph.)

Two Important Identities

The fact that a^x and $\log_a(x)$ are each other's inverses leads to a pair of useful identities.

Remember: $\log_a(a) = 1$ ▷   In one or two steps, simplify the expression $\log_a(a^x)$ as much as possible.

If you did that correctly, you just verified the first identity:

$$\log_a(a^x) = x$$

So, for example, $\log_5\left(5^{12}\right) = 12$. Do you see the log-base-5 function undoing the 5-to-a-power function?

▷   Simplify $\log_{10}\left(10^{-2.4}\right)$ and check your answer with a calculator.

The second identity puts the functions in the opposite order:

$$a^{\log_a(x)} = x$$

The left side of the second identity simplifies automatically when we recall that $\log_a(x)$, by definition, means *the power to which we raise the base a to obtain x*. By raising a to that power, we must get x.

So, for example, $5^{\log_5(678)} = 678$. Here, the 5-to-a-power function undoes the log-base-5 function. Be sure you understand these additional examples as well:

$$\log_7\left(7^{29-\sqrt{q}}\right) = 29 - \sqrt{q} \qquad \text{by the first identity, and}$$

$$10^{3\log_{10}(t)} = 10^{\log_{10}(t^3)} \qquad \text{by the third law of logs}$$

$$= t^3 \qquad \text{by the second identity}$$

We can replace any positive number N by $e^{\ln(N)}$. They are equivalent.

The second identity also implies that any positive number N can be rewritten as $e^{\ln(N)}$, a fact that comes in handy in calculus.

STOP & THINK   Below are graphs of $\log_a(a^x)$ and $a^{\log_a(x)}$. One graph is identical to the line $y = x$; the other isn't. Why don't these two functions have the same graph?

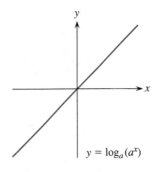

$y = \log_a(a^x)$

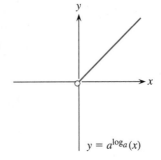

$y = a^{\log_a}(x)$

Using Logs to Solve Equations

Now we have the algebraic equipment necessary to answer a whole flock of interesting questions which, until now, required graphical solutions or trial-and-error estimates. Here are some.

As easy as falling off a log

The population of Vietnam in the year 2000 was approximately 79 million persons and was growing at the rate of roughly 3.3 percent per year. At that pace, in what year will the population reach 100 million?

A function to model the population, in millions of persons, is $V(t) = 79(1.033)^t$. The question asks for the value of t for which $V(t)$ equals 100. Let's begin.

$$79(1.033)^t = 100$$

$$(1.033)^t = \frac{100}{79}$$

Oops! We have a problem here—a logjam?: The variable t appears as an exponent, and we need to get it out of that position if we're to solve the equation. Watch this nifty trick. First we take the natural logarithm of each side of the equation:

$$\ln(1.033)^t = \ln\left(\frac{100}{79}\right)$$

Then apply the third law of logs:

$$t \ln(1.033) = \ln\left(\frac{100}{79}\right)$$

The t is no longer an exponent! Solve for t by dividing both sides by $\ln(1.033)$:

$$t = \frac{\ln\left(\frac{100}{79}\right)}{\ln(1.033)} \approx 7.26$$

The model predicts 100 million people about three months into the year 2007.

In taking the logarithm of each side of the equation, we *changed to a logarithmic scale*. (This is exactly what you did in Section 4.3 when you rescaled a horizontal axis by replacing each number by its corresponding power of 10.) Because the logarithmic function is one-to-one, we can always move back and forth between a conventional scale and a logarithmic scale. Any base will work, but we ordinarily use one of the two bases, e or 10, that our calculators provide.

If the unknown is an exponent, try logs.

Notice the beauty of this process! The logarithmic function allows us to yank the exponent t down onto level ground where we can solve for it with ordinary algebra.

> In general, whenever we have to solve an equation in which the variable appears in an exponent, logs are the tool of choice.

Compounding the problem

Banks and investment plans remind us about the power of compounding. Suppose one investment of $1500 grows at 5% per year while another investment of $1000 grows at 7% per year. Because the second investment grows faster, it will eventually catch up to the first. In how many years will that happen?

First, we write a function for each investment: $1500(1.05)^t$ and $1000(1.07)^t$. Then we set them equal to each other and solve for t:

$$1500(1.05)^t = 1000(1.07)^t$$

$$\frac{1.05^t}{1.07^t} = \frac{1000}{1500} \qquad \text{Isolate the factors containing } t.$$

$$\left(\frac{1.05}{1.07}\right)^t = \frac{2}{3} \qquad \text{Apply the fifth law of exponents.}$$

$$\ln\left(\frac{1.05}{1.07}\right)^t = \ln\left(\frac{2}{3}\right) \qquad \text{Take the natural log of each side.}$$

$$t \ln\left(\frac{1.05}{1.07}\right) = \ln\left(\frac{2}{3}\right) \qquad \text{Apply the third law of logs.}$$

$$t = \frac{\ln\left(\frac{2}{3}\right)}{\ln\left(\frac{1.05}{1.07}\right)} \qquad \approx \qquad 21.489$$

The investments will have equal value after about $21\frac{1}{2}$ years.

Notice that we don't use a calculator until the final step. Our final answer will be most accurate if we do not approximate any values along the way. (Remember that the calculator provides decimal approximations rather than exact values.)

Things are a little different if the investments offer interest that is compounded *continuously*. In that case, surprisingly, we use the natural base *e*. Compound interest, in fact, was the context in which *e* was first recognized and defined. You'll learn more about how *e* relates to compound interest in Project 4.1.

The formula for continuously compounded interest

> The general formula for the worth of an investment after t years of continuous compounding is
>
> $$M(t) = M_0 e^{rt},$$
>
> where M_0 is the initial investment and r is the interest rate written as a decimal.

For our two investments, then, we write $1500e^{0.05t}$ and $1000e^{0.07t}$. Let's find out when these investments have equal value.

$$1500e^{0.05t} = 1000e^{0.07t}$$

$$\frac{e^{0.05t}}{e^{0.07t}} = \frac{1000}{1500} \qquad \text{Isolate the factors containing } t.$$

$$e^{0.05t - 0.07t} = \frac{1000}{1500} \qquad \text{Apply the second law of exponents.}$$

$$e^{-0.02t} = \frac{2}{3} \qquad \text{Simplify each side.}$$

$$-0.02t = \ln\left(\frac{2}{3}\right) \qquad \text{Rewrite the statement in logarithmic form. (Take the natural log of each side.)}$$

$$t = \frac{\ln\left(\frac{2}{3}\right)}{-0.02} \quad \approx \quad 20.273$$

In our previous example, the smaller investment needed $21\frac{1}{2}$ years to catch up to the larger. Here, the smaller investment needs less time—just over 20 years. If you have a choice, go for continuous compounding!

Breaking a logjam

Frequently, our information comes as a pair of census facts. For instance, the city of Atlanta, Georgia, and its surrounding urbanized area experienced rapid growth during the late twentieth century. The region had 1.86 million people in 1990 and 2.40 million in 1997.

Let's write a base-*e* population model, $A(t) = C \cdot e^{kt}$, with these data. If 1990 is our zero year, then $A(0) = 1.86$ and the model, so far, looks like this:

$$A(t) = 1.86e^{kt}$$

where k is a growth constant we need to find. Using the 1997 population, we have

$$A(7) = 1.86e^{k(7)} = 2.40$$

Solve the equation for k:

$$1.86e^{7k} = 2.40$$

$$e^{7k} = \frac{2.40}{1.86}$$

$$7k = \ln\left(\frac{2.40}{1.86}\right)$$

$$k = \frac{\ln\left(\frac{2.40}{1.86}\right)}{7} \approx 0.0364$$

The completed population model is $A(t) = 1.86e^{0.0364t}$.

Check Your Understanding 4.12

1. Simplify each expression as much as possible.
   (a) $\ln\left(\frac{e}{x}\right)$

   $\ln(e) - \ln x$

   $1 - \ln(x)$

   (b) $\ln(\ln(e^e))$

   $\ln(e)$

   1

   (c) $10^{\log_{10}(z^3)}$

   z^3

   (d) $10^{7 \cdot \log_{10}(z)}$

   $10^{(\log_{10}(z))^7}$

   $(10^{\log 10z})^7 = z^7$

2. Solve the equation $79(1.033)^t = 100$ from page 182 again, using base-10 logarithms this time.

   $t = \frac{\log\left(\frac{100}{79}\right)}{\log 1.033}$ $\log 1.033^t = \frac{100}{79} \log$

   $\log(1.033) \cdot \log(t) = \log\left(\frac{100}{79}\right)$

3. Use logs to find the exact value of x, then give a three-place decimal approximation.
   (a) $3^x = 40$

   $\log 3^x = \log 40$ $x = \frac{\log 40}{\log 3}$
   $x \cdot \log 3 = \log 40$

   (b) $(0.04)^{2x-1} = 5$

   $2x - 1 \cdot \log(6.04) = \log(5)$

   (c) $3000(0.80)^x = 1000$

4. Find the doubling time for a population that grows at 3.3% annually.

   $2 = 1 \cdot (1.033)^t$

   $2 =$

5. Determine the annual growth percentage for the Atlanta model, $A(t) = 1.86e^{0.0364t}$.

In a Nutshell   Logarithms simplify mathematical operations: To multiply two expressions, we add their logs; to divide them, we subtract their logs; to find a root or raise to a power, we multiply logs. Many equations in which a variable appears as an exponent can be solved easily with the help of logarithmic rescaling.

4.5 LOGISTIC GROWTH

Exponential population models, because of their ever-increasing rates of change, cannot hold true forever. If the population of Vietnam or of metropolitan Atlanta continued to grow at a constant percentage, there would eventually be people on every square inch of the earth's surface. Similarly, if AIDS were to continue to spread exponentially, humankind would become extinct. In the real world, outside forces and internal limitations collaborate to inhibit growth.

When a population grows exponentially, its growth rate at any instant is *proportional* to its size at that time. The larger the current population, the larger its rate of growth. The first graph in Figure 4.22 illustrates this: As time passes, the function values increase and the curve becomes increasingly steep.

Exponential Growth in a Restricted Environment

The graph on the left in Figure 4.22 represents a model that ignores real-world limitations. In actuality, however, increasing population density means increasing competition for available resources, causing growth to slow. In other words, there comes a time when *the growth rate begins to decrease*. This is not to say that the *population* decreases; the population continues to grow, but its rate of growth is smaller each year than it was the year before. The *S*-shaped graph on the right in Figure 4.22, known as a **logistic**[3] **growth** model, represents a model that takes into account population density and other growth-limiting factors.

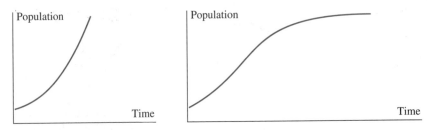

Figure 4.22 Exponential and Logistic Growth Curves

The logistic curve shows a population that increases slowly at first and then more rapidly. Initially, the curve resembles the shape of an ordinary exponential graph. After the population reaches a certain size, however, its growth slows. When the population approaches a constant value, called the **carrying capacity** of the environment, it virtually stops growing.

Let's reword the previous paragraph using rate-of-change language, as it applies to our logistic graph:

▷   Complete the following by choosing the appropriate verb: *increases, decreases,* or *approaches zero.*

When the population is small, it _____ and its rate of growth

_____.

After the population reaches a certain size, it still _____, but its rate

of growth _____.

---

[3]The word *logistic* comes from the Greek word *logizein*, to calculate. *Logarithm* has its roots in two Greek words: *logos*, word or reason, and *arithmos*, number. Although the two terms are similar, they are unrelated in meaning. Logistic growth doesn't have much to do with logarithms.

As the population approaches its carrying capacity, it remains virtually constant

and its rate of growth _____.

▷   On the logistic curve in Figure 4.22, find the point at which the growth rate appears to
be greatest, that is, the steepest part of the curve. Mark the spot with an X.

   Notice that the graph is concave up to the left of the X and concave down to the
right. The curvature of the graph tells us whether the growth rate of the function is
increasing or decreasing.
   We call the point you marked "X" a **point of inflection;** it is a point on the curve
at which the concavity changes. At that point, the *rate of growth* is neither increasing
nor decreasing.
   A logistic growth model has this general formula:

$$g(t) = \frac{A}{1 + Be^{-kt}}, \qquad \text{where } A, B, \text{ and } k \text{ are positive constants}$$

This fairly complicated formula involves a base-e exponential function, along with constants A, B, and k, which govern the initial value of the population, the carrying capacity,
and how quickly the population approaches the carrying capacity. The constant A is particularly informative, giving us the carrying capacity directly.
   Calculus will help you understand where this formula comes from, but you can use
a grapher now to investigate the graph of a logistic model and to understand its features.

A Logistic Population Model

A 1996 U.N. study produced the figures and projections shown in Table 4.4 for the
population of the world. The study predicts that the number of humans will stabilize in
the twenty-third century at about 10.7 billion.

▷   Using 1804 as the zero year, complete the x-column in Table 4.4.

Calendar Year	Year, x	Population (billions), $P(x)$
1804	0	1
1927		2
1960		3
1974		4
1987		5
1999		6
2011·		7
2025		8
2041		9
2071		10

Table 4.4 U.N. Population Projections

▷   In Figure 4.23, carefully plot the ordered pairs $(x, P(x))$ and connect them with a smooth curve. What you draw should closely resemble the S-shaped curve in Figure 4.22, suggesting that a logistic model will fit the data very well.

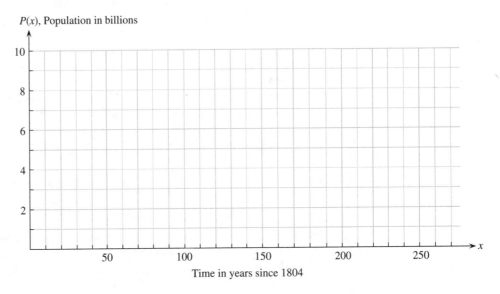

P(x), Population in billions

Time in years since 1804

Figure 4.23   Graph of U.N. Population Projections

Statistical software produced a good mathematical fit for the curve in Figure 4.23:

$$H(x) = \frac{9.68}{1 + 818e^{-0.03458x}} + 1$$

Notice how this formula agrees with the general logistic formula $\frac{A}{1 + Be^{-kt}}$, with an upward shift of one unit (one billion persons).

▷   With a grapher, plot the ordered pairs $(x, P(x))$ from Table 4.4. Then overlay the graph of $H(x)$ and admire the close fit. (Be careful with syntax: Enclose the entire denominator in parentheses as well as the entire exponent.)

Now we'll see how the growth rate changes over time by calculating some instantaneous rates of change.

You've computed rates this way before. See page 166.

The quotient $\frac{H(0.01) - H(-0.01)}{0.02} \approx 0.0004$ billion per year gives an excellent approximation to the instantaneous rate of change of H at $x = 0$. In English, this says that world population in 1804 was growing at the rate of 400,000 people per year.

Similarly, $\frac{H(123.01) - H(122.99)}{0.02} \approx 0.0244$ billion per year does a good job for the rate at $x = 123$. (In 1927, the rate was 24.4 million people per year.)

▷   Now you complete Table 4.5 by approximating the remaining rates. With a graphing calculator, you can save time as follows. In the function list, define Y1 using the $H(x)$ formula. Then, on the home screen, evaluate (Y1(156.01)−Y1(155.99))/.02. By modifying that entry, you can quickly obtain all the rates.

Precise numbers are less important than the pattern: The slope increases for about 195 years, peaking at the end of the twentieth century, after which the slope decreases, even though the population itself continues to climb.

x	$H(x)$, rounded to the nearest billion	Slope of $H(x)$
0	1	0.0004
123	2	0.0244
156	3	
170	4	
183	5	
195	6	
207	7	
221	8	
237	9	
267	10	

Table 4.5 Logistic Population Model and Its Growth Rates

Check Your Understanding 4.13

1. Use a grapher to examine $H(x)$ in the twenty-third century ($397 \leq x \leq 497$). What is the approximate rate of change of H on that interval? (You shouldn't need a calculator to answer this.)

2. Does the graph of $H(x)$ support the U.N. prediction of a population that stabilizes in the twenty-third century at 10.7 billion people? Explain.

If the logistic family of functions were merely a bunch of fancy formulas with pretty graphs, we wouldn't have bothered to include them in this text. But because they are so useful as models of natural, economic, and social phenomena, we decided to introduce them. Here is a sampling of functions that can be represented by logistic models:

- Newly diagnosed cases of AIDS in the United States

- Solar oven use in rural Lesotho

- The percentage of magazine ads containing website addresses

- The percentage of the population that has heard a rumor (or a joke, or an announcement).

In each case, *time* is the independent variable.

An interesting fact about logistic models is that they have *two* horizontal asymptotes: one at the bottom ($y = 0$, unless the graph has a vertical shift) and one at the top, or carrying capacity.

In calculus, you will learn to express logistic growth more simply, in terms of the carrying capacity and the changing rate of change.

In a Nutshell   In the real world, no exponential growth model is valid indefinitely. Over the long term, many processes that initially looked exponential experience a slowing of growth and a leveling-off behavior. Logistic functions represent this type of behavior: increasing for all x, concave up until the inflection point, concave down from there on, leveling off at the carrying capacity.

What's the Big Idea?

- The irrational number e is the *natural base* for exponential growth and decline. The growth rate (slope) of $y = e^x$ at every point is equal to its y-value at that point.

- Logarithmic functions are inverses of exponential functions. The input of an exponential function is an exponent; the output of a logarithmic function is an exponent. The graph of an exponential function has a horizontal asymptote; the graph of a logarithmic function has a vertical asymptote.

- Logarithmic functions help to solve equations in which the unknown appears in an exponent.

- No exponential growth model can represent reality indefinitely. Logistic functions can model exponential growth in a restricted environment.

Progress Check

After finishing this chapter, you should be able to do the following:

- Apply the laws of exponents to algebraic expressions. (4.1)

- Write exponential models. (4.1)

- Approximate an instantaneous rate of change, the slope of a curve at a point. (4.2)

- Convert exponential functions from standard form to base-e form and vice-versa. (4.2, 4.3)

- Use a logarithmic scale to rescale data that vary widely in magnitude. (4.3)

- Translate exponential statements into logarithmic statements and vice-versa. (4.3)

- Sketch graphs of logarithmic functions. (4.3)

- Use the definition and the properties of logarithms to simplify expressions containing logarithms. (4.3, 4.4)

- Use logarithms in solving equations. (4.3, 4.4)

- Given a logistic model, find and interpret its point of inflection and its carrying capacity. (4.5)

Key Algebra Skills Used in Chapter 4

- Working with exponents (page 477)

- Working with logarithms (page 481)

Answers to Check Your Understanding

4.1

1. $h(2) = 64$, $h(0) = 1$, $h(-1) = \frac{1}{8}$, $h(\frac{2}{3}) = 4$,
   $h(-\frac{4}{3}) = \frac{1}{16}$

2. Both graphs hug the x-axis on the left, pass through the point $(0,1)$, and rise steeply on the right. Both graphs are entirely above the x-axis. On the right side, the graph of 8^x is steeper than the graph of 4^x.

4.2

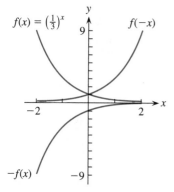

$f(x) = \left(\frac{1}{3}\right)^x$

1. $f(x) = \left(\frac{1}{3}\right)^x$
2. $f(-x) = \left(\frac{1}{3}\right)^{-x} = 3^{-x}$
3. $-f(x) = -\left(\frac{1}{3}\right)^x$

4.3

1. domain: $\mathbb{R}$; range: positive reals
2. "As x approaches negative infinity, $\left(\frac{1}{4}\right)^x$ approaches infinity." As the exponent becomes more and more negative, the function values increase without limit.
3. the x-axis; as $x \to \infty$, $\left(\frac{1}{4}\right)^x \to 0$
4. The graph is concave up (decreasing at an increasing rate).

4.4

1. $W(t) = 92 - c \cdot a^t$
2. $W(0) = 92 - c \cdot a^0 = 55$, so $c = 37$
3. $W(20) = 92 - 37a^{20} = 64$; therefore, $a = \left(\frac{28}{37}\right)^{1/20} \approx 0.986$
4. $W(t) = 92 - 37(0.986)^t$

4.5

1. domain: $\mathbb{R}$; range: positive reals; intercept: $y = 1$; asymptote: $y = 0$
2. $\dfrac{e^{-1.99} - e^{-2.01}}{0.02} \approx 0.13533754$; $e^{-2} \approx 0.13533528$
3. $e^{-1.9}$ and e^π, respectively; approximately 0.15 and 23.14
4. Slope of 4^x at $x = -2$ is approximately 0.0866, but $4^{-2} = \frac{1}{16} = 0.0625$. Slope of 4^x at $x = 2$ is approximately 22.1814, but $4^2 = 16$.

4.6

1. $B(t) = 3000(0.80)^t$
2. The growth factor is 0.80—less than 1, making B a decreasing function.
3. A negative k-value indicates a decreasing function.

4.7

1. $5 \approx 10^{0.7}$, so the *power* required is 0.7; $80 \approx 10^{1.9}$, so the power is 1.9.
2.

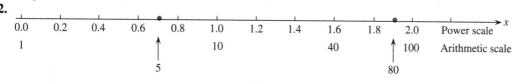

4.8

1. (b) $10^4 = 10000$
   (c) $10^{2/3} = \sqrt[3]{100}$
   (d) $10^{-0.3} \approx 0.5$
2. (b) $\log_{10}\left(\frac{1}{\sqrt{10}}\right) = -\frac{1}{2}$
   (c) $\log_{10}(1) = 0$
   (d) $\log_{10}(450) = 2t$
3. (b) -5
   (c) 1
   (d) 13
   (e) $\frac{1}{5}$
4. (a) -0.301
   (b) 0.301
   (c) 1.301
   (d) 2.301
   (e) 2.602

4.9

1. $16^{1/4} = 2$
2. $2^{-2} = \frac{1}{4}$
3. (a) 2
   (b) $\frac{1}{2}$
   (c) 5
   (d) -5
   (e) $\frac{3}{2}$ (because $9^{3/2} = 27$)

4.10

1. (a) $\ln\left(\frac{1}{\sqrt{e}}\right) = -\frac{1}{2}$
   (b) $\ln(1) = 0$
   (c) $\ln(V) = 0.0023t$
2. (a) $e^{-2} = \frac{1}{e^2}$
   (b) $e^1 = e$
3. (a) 4
   (b) 4
   (c) $\frac{2}{3}$
   (d) -3
4. (a) $x = \ln(10) \approx 2.303$
   (b) $x = \ln(10) - 3 \approx -0.697$
   (c) $x = \frac{\log_{10}(e)}{2} \approx 0.217$
   (d) $x = e^4 \approx 54.598$
   (e) $x = e^{4/3} \approx 3.794$
5. $B(t) = 3000\left(e^{\ln(0.80)}\right)^t \approx 3000e^{-0.22314t}$

4.11

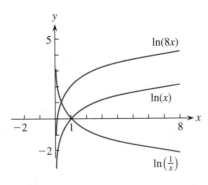

1. $\ln(8x) = \ln(8) + \ln(x) \approx 2.08 + \ln(x)$
2. The graph of $y = \ln(8x)$ is the graph of $y = \ln(x)$ shifted up 2.08 units.
3. $y = \ln\left(\frac{1}{x}\right) = \ln\left(x^{-1}\right) = -1\ln(x) = -\ln(x)$. (You might have used the division rule for logs instead.) The graph of $y = \ln\left(\frac{1}{x}\right)$ is the graph of $y = \ln(x)$ reflected in the x-axis.
4. $\ln\left(\frac{\sqrt{x}}{5}\right)$
5. $-3 + 3\log_{10}(x)$

4.12

1. (a) $1 - \ln(x)$
   (b) 1 (c) z^3
   (d) z^7 (First, rewrite $7\log_{10}(z)$ as $\log_{10}(z^7)$.)
2. If we replace each occurrence of "ln" by "log," the procedure is identical and the result is the same.
3. (a) $x = \log_3(40) = \frac{\ln(40)}{\ln(3)} = \frac{\log_{10}(40)}{\log_{10}(3)}$ (your choice)
   ≈ 3.358
   (b) $x = \frac{1}{2}\left(\frac{\ln(5)}{\ln(0.04)} + 1\right) = 0.250$ (exactly)
   (c) $x = \frac{\ln(1/3)}{\ln(0.80)} \approx 4.923$
4. Solve $(1.033)^t = 2$ for t; $t = \frac{\ln(2)}{\ln(1.033)} \approx 21.35$ years
5. $e^{0.0364} \approx 1.037$; approximately 3.7%

4.13

1. On the interval [397, 497], the rate of change of H is approximately zero because the graph of $H(x)$ is nearly horizontal.
2. Yes. The function values on that interval are all approximately 10.68, and the function has virtually stopped changing.

EXERCISES

> Properties of exponents, page 157; laws of exponents, page 159

1. Let f be the function $f(x) = 3 \cdot 2^x + 1$.
   (a) Without a calculator, evaluate $f(0)$, $f(-2)$, $f\left(\frac{3}{2}\right)$, and $f\left(-\frac{1}{3}\right)$. Some answers contain radicals.
   (b) State the domain and the range of f.
   (c) Sketch a graph of $f(x)$.
   (d) On the same axes, sketch and label a graph of $f(-x)$.
   (e) Write an algebraic formula for $f(-x)$, and explain why the reflected graph you drew is the same as the graph of $3\left(\frac{1}{2}\right)^x + 1$.

2. Let g be the function $g(x) = 3 \cdot 2^{x+1}$.
   (a) Without a calculator, evaluate $g(0)$, $g(-2)$, $g\left(\frac{3}{2}\right)$, and $g\left(-\frac{1}{3}\right)$. Some answers contain radicals.
   (b) State the domain and the range of g.
   (c) Sketch a graph of $g(x)$.
   (d) On the same axes, sketch and label a graph of $g(-x)$.
   (e) Write an algebraic formula for $g(-x)$, and explain why the reflected graph you drew is the same as the graph of $6\left(\frac{1}{2}\right)^x$.

3. Each table represents values of a function. Decide whether or not the function could be exponential or linear. If yes, write an algebraic formula that could represent the function; if it is neither exponential nor linear, explain why not.

(a)

x	y
-2	$\frac{2}{9}$
-1	$\frac{2}{3}$
0	2
1	6
2	18
3	54
4	162
5	486

(b)

x	y
-2	12
-1	3
0	0
1	3
2	12
3	27
4	48
5	75

4. Follow the directions for the preceding exercise.

(a)

x	y
-2	$-\frac{7}{3}$
-1	$-\frac{5}{3}$
0	-1
1	$-\frac{1}{3}$
2	$\frac{1}{3}$
3	1
4	$\frac{5}{3}$
5	$\frac{7}{3}$

(b)

x	y
-2	$\frac{3}{4}$
-1	$\frac{3}{2}$
0	3
1	6
2	12
3	24
4	48
5	96

5. A positive number raised to any real-number power is also a positive number; that is, if $b > 0$, then $b^x > 0$. In particular, b^x can never equal zero. This fact will help us solve equations involving exponential terms. The strategy is to collect all nonzero terms on the left side of the equation and rewrite that expression in factored form. If any factor is exponential, then that factor does not contain a solution to the equation. Solve the following equations.

(a) $x^2 \cdot 10^{3x} - 10^{3x} = 0$

(b) $(2x - 1)3^{x^2-1} = 0$

6. Solve each equation.

(a) $x^2(x - 5)5^{-x} = 0$

(b) $2x(10^x) = 10^{x+2}$

In Exercises 7–10, write the formula for an exponential function that could represent each graph. (*Hint*: First identify any shifts, reflections, or stretches and code them into the formula. Then use a known point to determine the function's growth factor.)

7.

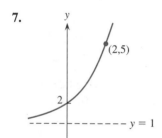

8.

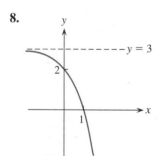

9.

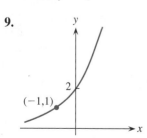

10.

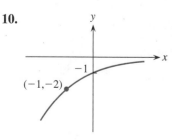

11. The function 1^x isn't considered exponential. Review the properties of exponential functions (pages 160–162) and describe all the ways in which $y = 1^x$ fails to measure up.

12. The function $y = 0^x$ doesn't belong to the exponential family either.

(a) What are the domain and the range of $y = 0^x$? (Be careful with the domain! Ask your calculator to evaluate 0^0 and 0^{-1}. What happens?)

(b) Notice the bind we're in if we try to assign a value to 0^0: Zero raised to a nonnegative power should still be 0, but any other number raised to the zero power is supposed to be 1. The expression 0^0 is mathematically undefined because it confronts us with a dilemma; we can't call it 0, and we can't call it 1. Try making the base very close to zero and keeping the exponent zero: 0.001^0, 0.00001^0, and so on. What happens? Now try making the exponent very close to zero and keeping the base zero: $0^{0.001}$, $0^{0.00001}$, and so on. Suppose both the exponent and the base are very close, but not equal, to zero. Is the result close to 0 or close to 1? Write a few sentences summarizing your investigation.

13. Sam, a sprinter, has a resting pulse of 54. In the 200-meter dash, his heart rate climbs to 160 beats per minute. As soon as he stops running, his heart rate begins to drop rapidly. If, 20 seconds after he reaches the finish line, his pulse is 92, write an exponential function $H(t)$ to model his heart rate t seconds after he stops. (Remember that his heart rate does not drop below 54.) Sketch a graph of $H(t)$.

14. Esperanto, the most successful of the so-called planned or artificial languages, was devised in the late nineteenth century by L. L. Zamenhof, a Polish doctor. Intended as a second language in which speakers of different tongues could communicate readily, Esperanto has a small basic vocabulary of approximately 500 root words, from which many other words can be created by using simple rules. Suppose a school offers a course in Intensive Esperanto, ESP 101, that meets daily during the entire academic year. Elena enrolls, already knowing 20 words of Esperanto when she begins the course. By day 5, she has learned 80 additional words for a total of 100 (of the basic 500). Her progress follows a characteristic learning curve—an upside-down exponential function. Write a mathematical model $ESP(t)$ to represent the number of the 500 basic vocabulary words Elena knows on day t of the course. Sketch a graph of $ESP(t)$.

15. Copy a graph of $y = e^x$. Without the help of a grapher, sketch on the same axes the graphs of the following functions, labeling the y-intercept and the horizontal asymptote. Describe how each graph is related to the graph of e^x.
    (a) $3e^x$
    (b) $e^x + 3$
    (c) e^{3x}

16. Follow the directions for the preceding exercise.
    (a) $5 - e^x$
    (b) $e^{0.5x}$
    (c) $0.5e^{-x}$

17. You know that for e^x, the value of the function at any point equals its rate of change at that point. In this exercise, you will determine whether or not that property changes when e^x is multiplied by a constant. Choose three A-values—$A > 1$, $0 < A < 1$, and $A < 0$—and approximate the instantaneous rate of change of Ae^x at any two points on each curve. (Altogether, you will calculate six rates.) Then write the value of Ae^x at each of those points and compare each function value with its corresponding rate. If they match closely, you may conclude that a vertical stretch or compression or a reflection in the x-axis does not destroy the special slope-giving property of e^x. Be sure to present your evidence along with your conclusion.

18. A horizontal shift of e^x also preserves its slope-giving property. Examine two functions e^{x+c}, choosing two c-values: one positive and one negative. For both of these transformed functions, provide evidence, either numerical or graphical, that the slope of e^{x+c} at a point equals its y-value at that point. Numerical evidence means approximating the instantaneous rate of change at a minimum of two points on each curve and comparing each rate to its y-value. Graphical evidence would be a convincing argument, with sketches, explaining why a horizontal shift should not alter the relationship between the y-value and the slope. When you finish, use the laws of exponents to show that e^{x+c} is mathematically the same kind of creature as Ae^x, which is why they have the same property.

19. Remembering that e raised to any power is never 0, solve these equations.
    (a) $e^x + xe^x = 0$
    (b) $2xe^{-x^2} - 2x^3e^{-x^2} = 0$
    (c) $\dfrac{3e^x - (3x - 1)e^x}{e^{2x}} = 0$

20. Solve these equations.
    (a) $2xe^x + x^2e^x = 0$
    (b) $e^{3x} + 3xe^{3x} = 0$
    (c) $\dfrac{2x^2e^{x^2} - e^{x^2}}{x^2} = 0$

21. Make a copy of the scale in Figure 4.14 on page 171. Carefully locate on it the numbers 0.05, 0.5, 50, and 5000.

22. Refer to the display in Figure 4.13 on page 171. We have already discussed the horizontal scaling. Now look at the vertical scale.
    (a) Does the vertical scale follow the conventional pattern (equal spacing representing equal amounts)? Explain.
    (b) Where would you place the frequency "once every 500 years," closer to "once a century" or to "once a millennium"? Give a reason.

23. Evaluate each of the following without a calculator.
    (a) $\log_2(64)$ $\log_2(64)$
    (b) $\log_{16}(64)$
    (c) $\log_3\left(\frac{1}{27}\right)$
    (d) $\log_{27}(3)$
    (e) $\ln\left(e^{3.45}\right)$

24. Evaluate each of the following without a calculator.
    (a) $\log_8(2)$
    (b) $\log_4(8)$
    (c) $\log_{12}(1)$
    (d) $\log_{10}(0.0001)$
    (e) $\ln\left(\sqrt[3]{e}\right)$

25. Copy a graph of $y = \ln(x)$. Without the help of a grapher, sketch on the same axes the graphs of the following functions, identifying the vertical asymptote and any intercepts. Describe how each graph is related to the graph of $\ln(x)$.
    (a) $\ln(x + 2)$
    (b) $\ln(2x)$
    (c) $-2\ln(x)$

26. Follow the directions for the preceding exercise.
    (a) $\ln(x) + 2$
    (b) $\ln\left(\frac{e}{x}\right)$
    (c) $\ln(x^2)$ (Careful! There's more to this than meets the eye.)

In Exercises 27–30, write the formula for a logarithmic function that could represent the graph. (*Hint*: First identify any shifts, reflections, or stretches and code them into the formula. Then use a known point to determine the value of the base.)

27.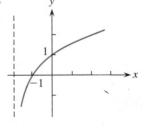

28.

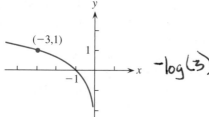

$-\log(3)$

29.

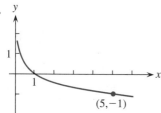

30.

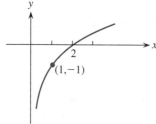

31. Give the formula for a function $y = f(x)$ that could represent this table.

x	y
$\frac{1}{9}$	-4
$\frac{1}{3}$	-3
1	-2
3	-1
9	0
27	1
81	2

32. Give the formula for a function $y = f(x)$ that could represent this table.

x	y
2000	-3
200	-2
20	-1
2	0
0.2	1
0.02	2
0.002	3

33. Given $f(x) = \ln(1 + 2x)$, find $f^{-1}(x)$.

34. Given $g(x) = \ln(3x - 2)$, find $g^{-1}(x)$.

35. Given $h(x) = 3e^{0.2x}$, find $h^{-1}(x)$.

36. Given $j(x) = 10e^{x-1}$, find $j^{-1}(x)$.

37. Find both the exact value of x and a three-place decimal approximation.
   (a) $e^{3x} = 10$
   (b) $10^{2x+1} = e$
   (c) $\ln(x) = 3$
   (d) $\ln(x^6) = 3$

38. Show that this statement is true: Every exponential function $f(x) = C \cdot a^x$ can be considered a transformation (some combination of reflections, horizontal or vertical stretches, horizontal or vertical compressions) of the graph of e^x. (*Hint*: First rewrite $C \cdot a^x$ in base-e form.)

39. Use log laws and properties to simplify each expression as much as possible.
   (a) $\ln\left(\frac{x}{e}\right)$
   (b) $\log_{10}(0.001x)$
   (c) $e^{3\ln(x)}$

40. Use log laws and properties to simplify each expression as much as possible.
   (a) $\ln(\ln(\ln(e^e)))$
   (b) $\log_2\left(2^{\pi+1}\right)$
   (c) $e^{-10\ln(x)}$

41. Rewrite each expression as a single logarithm with a coefficient of 1.
   (a) $2\log_{10}(x) + \frac{1}{2}\log_{10}(3x)$
   (b) $\ln(2t) - 4\ln\left(\frac{1}{t}\right)$
   (c) $\ln(400) + 0.95t\ln(10)$

42. Rewrite each expression as a sum or difference of simpler logarithms, writing your answer in a form that doesn't involve products, quotients, or powers.
   (a) $\log_{10}(4x^3)$
   (b) $\ln\left(\sqrt{\frac{x}{e}}\right)$
   (c) $\ln(500(1.07)^x)$

43. Solve each equation exactly. (The symbol "log" or "ln" or the number e should appear in your answer.) Then calculate a three-place decimal approximation.
   (a) $10^x = 6$
   (b) $6^x = 10$
   (c) $10^{4x-3} = 470$
   (d) $8 \cdot 3^x = 5 \cdot 2^x$
   (e) $\dfrac{2x\ln(x) - x}{(\ln(x))^2} = 0$

44. Solve each equation exactly. (The symbol "log" or "ln" or the number e should appear in your answer.) Then calculate a three-place decimal approximation.
   (a) $2^x = 0.96$
   (b) $0.96^x = 2$
   (c) $2 \cdot 10^{3x+1} = 85$
   (d) $8 \cdot 3^{-x} = 5 \cdot 2^x$
   (e) $\ln(x) + 1 = 0$

45. A population is growing at an annual rate of 2%. In how many years will the population double? Quadruple? Show an algebraic solution.

46. Country A, with a population of 5.24 million, is growing at an annual rate of 1.3%, while Country B, with 3.16 million people, is growing at an annual rate of 2.1%. In how many years will Country B have more people than Country A? Show an algebraic solution.

47. Change of base. *Count Base-e and all that jazz.* Your calculator has only two logarithmic functions: base-e and base-10. Suppose that you wanted to approximate a logarithm with another base—say, $\log_5(37)$. This means *the power of 5 that yields 37.*
 (a) Solve $5^x = 37$ exactly. Your result should be the quotient of two logs.
 (b) Let's generalize this. To find $\log_a(N)$, where N is any positive number and a is any acceptable base, you will solve the equation $a^x = N$ for x. Do so, first using natural logs and again using base-10 logs, and you will derive the change-of-base formula:

$$\log_a(N) = \frac{\ln(N)}{\ln(a)} = \frac{\log_{10}(N)}{\log_{10}(a)}$$

Use either version of this formula to express any logarithm in terms of one of the two logs on your calculator.

48. The preceding exercise implies that all log graphs are equivalent to a transformation of the natural-log graph. Explain how the graph of $h(x) = \log_a(x)$ can be obtained from the graph of $\ln(x)$ by a vertical stretch or compression (along with a reflection, if $0 < a < 1$).

49. *Verifying the first rule of logarithms.* In this exercise, you'll prove that logarithms turn multiplication of numbers into addition of their logs, or that $\log_a(M) + \log_a(N) = \log_a(MN)$, for any positive numbers a, M, and N ($a \neq 1$).
 (a) Let p represent $\log_a(M)$ and let q represent $\log_a(N)$. (This means that p and q are exponents.) We have two equations:

$$p = \log_a(M) \qquad \text{and} \qquad q = \log_a(N)$$

Rewrite each equation in exponential form.
 (b) Now multiply the left sides by each other and the right sides by each other to produce a new equation. Use a law of exponents to rewrite this equation so that the symbol a appears only once.
 (c) Translate that equation into logarithmic form again. It should tell you that the log of the *product* of M and N is the *sum* of p and q. But

what are p and q? Replace those symbols by what they stand for, and you should have a statement of the first rule of logarithms.

50. *Verifying the second rule of logarithms.* By adapting the method of the preceding exercise, prove that logarithms turn division of numbers into subtraction of their logs, or that $\log_a(M) - \log_a(N) = \log_a(\frac{M}{N})$.

51. *Verifying the third rule of logarithms.* In this exercise, you'll prove that logarithms turn exponentiation of numbers into multiplication of their logarithms, or that $p \cdot \log_a(N) = \log_a(N^p)$.
 (a) Let q represent $\log_a(N)$. This gives us the equation $q = \log_a(N)$. Rewrite that equation in exponential form.
 (b) Raise both sides of the exponential equation to the power p. You should still have an exponential statement, but one side is raised to the power pq and the other side is raised to the power p.
 (c) Rewrite this equation in logarithmic form, using a as the base.
 (d) Finally, replace q by what it stands for, and you should have a statement of the third rule of logarithms.

52. The formula $M(x) = \log_{10}\left(\frac{x}{0.001}\right)$ gives the relationship between the seismographic reading x and the Richter scale earthquake magnitude $M(x)$. Another version of the Richter function is $M(x) = \log_{10}(x) + 3$. Use algebra to show that the two formulas are equivalent, justifying each step with an appropriate log rule.

53. In 1960, when many major colleges and universities in the United States still excluded female students, there were 298 colleges just for women. During the next several decades, that number dropped steadily as many of those women's colleges became coed or closed their doors. Here are some figures on women's colleges.

Year	Colleges
1960	298
1990	93
1994	84
2001	65

 (a) Plot the points from the table. The pattern should resemble exponential decline.
 (b) Using the exponential regression feature of a graphing calculator, obtain a model to approximate the data. (For convenience, measure time in *years since 1940.*)

(c) Use the model to estimate the number of U.S. women's colleges in 1980.

(d) According to the model, what is the annual percentage decrease in the number of women's colleges?

(e) According to the model, in what year would we expect the number of women's colleges to drop to 40?

54. The price of a first-class stamp rose more or less exponentially during most of the twentieth century. In 1960, a stamp cost 4 cents; in 2003, it cost 37 cents.

(a) Using the 1960 price and the 2003 price, write a base-e exponential model to approximate the price of a first-class stamp.

(b) Use the model to predict the year in which a stamp will cost a dollar.

(c) As dramatic as the price rise seems, the cost of a stamp has only kept pace with inflation. In 1920, a first-class stamp (which sold for 2 cents!) represented about two minutes' salary for the average U.S. worker; the cost of a stamp in 2003 was still the equivalent of two minutes worth of work. Rewrite your model from part (a) in standard form and determine the average annual inflation percentage for the first-class stamp.

55. Osteoporosis is a major health concern for women. Unless preventive measures are taken, a postmenopausal woman is likely to experience mineral loss in the spongy bone of her spine at the rate of 5% per year. At that rate, in how many years would she have lost half the mineral content of that bone?

56. Recent studies by anthropologists at the University of Michigan show that for the past 10,000 years, the average size of a human tooth has been decreasing at the rate of 1 percent every 1000 years. (The shrinkage rate was smaller before humans discovered that cooking their food made it easier to chew. Evolution seems to have been on our side during those epochs when we humans needed bigger teeth.) If we suppose that the trend for shrinking teeth continues, when can our descendants expect to have teeth that are only 75% as large as ours?

57. The radioactive isotope uranium-232 has a half-life of 72 years. (This means that, after 72 years, only half of the original amount remains unchanged, the other half having decayed into other radioactive substances and ultimately into a nonradioactive substance.)

(a) Given an arbitrary initial quantity A_0 of U-232, write a mathematical model in base-e form that predicts the amount of U-232 remaining t years later.

(b) By rewriting the model in standard form, determine the annual percentage decay rate of U-232.

58. A remarkable property of every exponential function of the form $y = C \cdot a^{kx}$ is that its logarithm is linear; the new function $L(x) = \ln(y) = \ln\left(C \cdot a^{kx}\right)$ is actually a linear function, whose slope m and y-intercept b are determined by the constants C, a, and k. In other words, a logarithmic scale turns an exponential function into a linear function.
Pick any specific exponential function (you choose values for C, a, and k) and have your grapher draw the graph of $\ln(y)$. You should see a line. Using algebra and the laws of logarithms, rewrite your particular $\ln(C \cdot a^{kx})$ expression in the form $L(x) = \ln(y) = b + mx$. Then give the exact slope and y-intercept of that line.

59. Suppose that x and the natural log of y are linearly related by the formula

$$\ln(y) = 3x + \ln(2)$$

Find a formula not involving logarithms that expresses y as a function of x.

60. Suppose that the natural log of x and the natural log of y are linearly related by the formula

$$\ln(y) = 3\ln(x) + \ln(2)$$

Find a formula not involving logarithms that expresses y as a function of x. (Note the difference between this exercise and the preceding one.)

61. Suppose that x and the natural log of y are linearly related as shown in this graph.

(a) Express $\ln(y)$ as a function of x.

(b) Express y as a function of x.

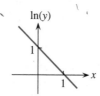

62. During the 1980s, the course of the AIDS epidemic in the United States exhibited a logistic growth pattern. The data in the table show the number of newly diagnosed cases, in thousands, each year from 1981 through 1991, with $x = 0$ representing 1981.

x	Thousands of Cases
0	1
1	1
2	3
3	6
4	12
5	19
6	28
7	34
8	39
9	41
10	42

A logistic model for these data is the function

$$A(x) = \frac{43}{1 + 64.4e^{-0.79x}}$$

(a) Use the graph of $A(x)$ to find the approximate coordinates of the point of inflection on the logistic curve. According to this model, when was AIDS spreading most rapidly in the United States? At that time, at what rate was the number of new cases increasing? (Find the slope of the curve at the inflection point.)

(b) According to this model, what is the maximum number of new cases that the country might expect in any one year?

63. During the period from 1982 to 1992, AIDS spread among women in the United States with alarming rapidity. This table shows the number of new cases reported each year, with $x = 0$ representing 1982. At first glance, the pattern appears exponential. Let's look more closely.

x	Cases, $W(x)$
0	2
1	7
2	13
3	31
4	84
5	151
6	226
7	330
8	503
9	726
10	972

(a) In true exponential growth, the ratio of cases in any two consecutive years should be equal. Calculate the ratios $\frac{W(x+1)}{W(x)}$ for $x = 5, \ldots, 9$ (five separate ratios). After the first three, which look roughly equal, you should see a decrease.

(b) That decreasing ratio suggests a logistic model. Here's one that fits the data:

$$W(x) = \frac{2432.5}{1 + 141e^{-0.458x}} - 22$$

Use the graph of $W(x)$ to find the approximate coordinates of the point of inflection on the logistic curve. According to this model, when was AIDS spreading most rapidly among U.S. women? At that time, what was the rate, in new cases per year, at which the disease was spreading?

(c) Estimate the maximum number of new cases that can occur in any one year, according to this model.

64. A graph of the national debt of the United States for the final three decades of the twentieth century looks dramatically logistic. This mathematical model

$$D(x) = \frac{6.3}{1 + 104.3e^{-0.209x}} + 0.3$$

represents the size of the national debt, in trillions of dollars, x years from 1970. Use $D(x)$ to answer the following.

(a) In 1970, at what rate was the debt increasing? Express your result in billions of dollars per year. (Note that the function values give *trillions* of dollars.)

(b) In what year did the debt increase most rapidly?

(c) What was the rate of increase at that time?

65. Telephone answering machines have been around for more than a hundred years. In 1900, the Danish inventor Valemar Poulsen patented the Telegraphone, "a device to record telephone messages in the absence of the called party." They became popular, however, only in the late twentieth century. The citizens of Logtown weren't among the early enthusiasts or the most avid adopters of answering machines, which at first they considered rude. Nonetheless, by the year 2000, nearly everybody had one, and a person without one was thought to be inconsiderate. The table, giving the percentage of home phones with answering machines, shows their increasing popularity in Logtown.

Year	Percentage
1980	5
1985	19
1990	52
1995	83
2000	95

A function to model the table is this logistic curve:

$$y = \frac{100}{1 + 19e^{-0.3x}}$$

where y stands for the percentage of answering machines x years after 1980.

(a) Use the model to determine the year in which answering-machine coverage increased most rapidly in Logtown.

(b) By what year should we expect virtually every phone in Logtown to have an answering machine. (You may settle for 99%.)

66. The growth of a particular population of wild animals is represented by this logistic model:

$$P(t) = \frac{3500}{1 + 210e^{-0.52t}}$$

where $P(t)$ is the number of animals in the population at time t, measured in years from the present.

(a) Estimate the number of animals after five years and after eight years.

(b) Find the *average* rate of growth of the population from year 5 to year 8.

(c) Approximate the growth rate of $P(t)$ at $t = 5$ and at $t = 8$. Compare these two growth rates with the average rate of growth over the interval $[5, 8]$.

(d) At what time does the growth rate of the population appear to be greatest?

(e) For this species of animal, what is the carrying capacity of the environment?

4.1 COMPOUND INTEREST

Too much of a good thing is wonderful.   Mae West

In this project, we investigate compound interest and the effect of frequent compounding. Much to our surprise, we encounter that amazing number, e.

1. Suppose you put $1000 into an investment that pays 8% compounded annually. This means that at the end of the year, 8% of the current balance is added to your account. How much will you have in the account after one year? After 10 years? After t years?

2. Suppose instead that the 8% interest is compounded quarterly. This means that at the end of every three months, 2% (that is, one fourth of 8%) of the current balance is added to the account. How much will you have after one year? After 10 years? After t years?

3. Your calculations in Questions 1 and 2 should convince you that quarterly compounding is financially better for you than annual compounding. Investigate the effect of *monthly* compounding for one year (one twelfth of 8% of the current balance will be added each month), and compare your results with the previous ones.

4. "Is there any limit to the benefits of frequent compounding?" you might wonder. Let's find out. Investigate the effects of compounding weekly (divide the interest rate by 52 and do 52 compoundings), daily, and hourly. In each case, your calculations should follow this formula:

$$1000 \left(1 + \frac{0.08}{n}\right)^n$$

where n is the number of compoundings per year. Suppose the interest were compounded every minute? Would you have more money at the end of the first year than you would with hourly compounding? Record your results in the table. (Don't do "continuously" yet.)

Frequency of Compounding	Balance after One Year
Once a year	
Quarterly	
Monthly	
Weekly	
Daily	
Hourly	
Each minute	
Continuously	

5. On the basis of your investigations, what is the value of the expression $\left(1 + \frac{0.08}{n}\right)^n$ as n becomes large without bound?

6. Some banks offer what they call *continuous compounding*. This does not mean that they actually work nonstop all year calculating the interest on your account, but that

(handwritten notes in left margin):
$1000 \cdot .08^x + 1000 \cdot .9^x$
$1000 \cdot 1.08^x$

of moose in New Hampshire, with respect to time, for the years from 1991 until the moose have disappeared from the region. Your formula should be written as a piecewise-linear function. Be sure that it gives the correct results for 1991 and 2001.

11. Sketch the graph of this function. Provide scales and labels.

12. According to this model, during what year will the moose disappear from New Hampshire?

4.4 STEEP LEARNING CURVES

Two college students, Dmitri and Nico, participate in a psychology experiment in which they attempt to learn as many nonsense words as they can. Their learning process can be modeled by these functions:

$$D(t) = 6\sqrt{t} \quad \text{and} \quad N(t) = 40\left(1 - e^{-0.089t}\right)$$

where $D(t)$ and $N(t)$ represent the number of nonsense words Dmitri and Nico, respectively, know after t minutes.

1. How many words does each student know after four minutes? On average, how many words per minute has each one learned?

2. To determine the rate at which each is learning when four minutes have elapsed, approximate the instantaneous rate of change of $D(t)$ and of $N(t)$ at $t = 4$. Give the rates in *words per minute*.

3. Graph the two models on the same screen and use the graphs to answer these questions.

   (a) After three minutes, who knows more words? At that moment, who is learning faster? How do you know?
   (b) After eight minutes, who knows more words? Who is learning faster? How do you know?
   (c) After 40 minutes, who knows more words? Who is learning faster? How do you know?
   (d) During which period of time does Nico know more words than Dmitri?

4. Use algebra to determine how long it takes each student to learn 25 words, giving the answers to the nearest minute.

5. Given unlimited time, what is the maximum number of words Nico will ever learn. Tell how you decided.

6. Write a short paragraph comparing and contrasting the learning styles of these two students, based upon your investigation of the graphs. Include a description of the changing rate at which learning takes place. Use language that is accessible to people who aren't studying mathematics.

4.5 EXPLORING LOGISTIC GROWTH

Section 4.5 gave the general formula for a logistic growth model:

$$g(t) = \frac{A}{1 + Be^{-kt}}, \quad \text{where } A, B, k > 0$$

Use a grapher to investigate the effects of the various constants, changing one of them at a time until you have an idea what each one does. All of the graphs should have the same fundamental S-shape, but you might have to adjust the viewing window several times in order to see it. Write a summary of your conclusions. You will probably have to be satisfied with partial results.

4.6 I HEAR THE TRAIN A-COMIN' . . .

This project is a follow-up to Lab 4C, "Earthquakes"

The human ear is sensitive to a wide range of sound intensities. Sound becomes audible after it reaches an intensity of 10^{-16} watt/cm^2 and becomes painful to the ear at about 10^{-4} watt/cm^2 (the threshold of pain).

1. Compare the sound intensity for the faintest audible sound to the sound intensity at the threshold of pain. How many times as great is the latter?

In the "Earthquakes" lab, you saw that measurements of earthquake strength given by seismograph readings, x, became more manageable when converted to a logarithmic scale, the Richter scale. Specifically, this relationship was given by the function

$$M(x) = \log\left(\frac{x}{0.001}\right)$$

Similarly, converting sound intensity into a base-10 logarithmic scale produces numbers that are easier to use. The sound intensity level $\beta(x)$, measured in decibels, is related to the sound intensity x, measured in watts/cm^2, by the formula

$$\beta(x) = 10\left(\log\frac{x}{x_o}\right)$$

where $x_o = 10^{-16}$ watt/cm^2 (a reference intensity corresponding to the faintest sound that the average person would be able to hear).

2. What is the sound intensity level in decibels for the faintest sound that can be heard by the average person? What is the decibel level for the threshold of pain?

3. The sound intensity level of a quiet whisper is about 20 decibels, while that of an ordinary conversation is about 65 decibels. Compute the sound intensities, measured in watts/cm^2, for the quiet whisper and ordinary conversation. Compare the two numbers you just computed. How many times greater is the larger?

4. A difference in sound intensity level of 1 decibel is generally taken to be the smallest increase in volume that the average human ear can detect. How much of an increase in watts/cm^2 will be required to raise the sound level of a 20-decibel whisper by 1 decibel? How much of an increase will be required to raise the level of a 65-decibel conversation by 1 decibel?

5. The sound of a train passing over a trestle measures about 90 decibels. Suppose a particular train caused a noise of 91 decibels. Would the increase in watts/cm^2 corresponding to a change from 90 to 91 decibels be greater or less than the increase required to go from 65 to 66 decibels? Explain.

6. Examine the graph of $\beta(x)$. As in the "Earthquakes" lab, you will need to use several different intervals, ones that are reasonable in this context, in order to get a complete picture of this logarithmic curve. What happens to the graph as the sound intensity x gets very close to zero? How does the basic shape of this graph compare to the graph of $M(x)$ in the "Earthquakes" lab? Comment on the similarities between the formulas for $M(x)$ and $\beta(x)$.

4.7 JUST ALGEBRA

1. Solve each equation for the unknown value. Give an exact solution and, where appropriate, an approximate solution.

   (a) $1000 = 500(1 + r)^{15}$

   (b) $500 = 1000 \left(1 + \frac{r}{4}\right)^{8}$

   (c) $95a^{1/2} = 65a^{7/2}$ (*Warning*: Don't divide both sides by $a^{1/2}$, or you will lose a solution.)

   (d) $H = M \cdot r^{-3}$ (Solve for r.)

2. The exact solution to each of the following equations involves a logarithm. Find both the exact solution for t and a three-place decimal approximation.

   (a) $100 \cdot 3^{t} = 200$

   (b) $450 \cdot 1.03^{t} = 175 \cdot 1.08^{t}$

   (c) $\dfrac{200}{1 + 100e^{-0.4t}} = 1000$

3. Determine a formula for the inverse of each of the following functions.

   (a) $f(x) = 0.2x^{3/5} - 1$

   (b) $g(x) = x^{-1}$

   (c) $h(x) = \log_3(x + 2)$

   (d) $j(x) = 5^{x-2} + 6$

Bridge over troubled water . . .

Preparation

In the 1980s, the United States had the dubious distinction of being the first country to experience the devastation of the AIDS epidemic. In fact, 10,805 of the 13,170 cases of AIDS reported to the World Health Organization through 1984 occurred in the United States.[4] Distress over the tragedies of individual cases was compounded by the realization on the part of public health officials that a mathematical model for the spread of the epidemic was an exponential function. In an essay in the *New York Times* (April 19, 1987), the noted biologist Stephen Jay Gould wrote, "The exponential spread of AIDS underscores the tragedy of our delay in fighting one of nature's plagues." Writing in 1987, Gould had sufficient evidence to support his claim, as we will see in this lab.

To appreciate the urgency of the situation described by Gould and others, we need to understand the nature of exponential growth. The simplest example is a trick problem that you might have seen before:

> If you place a penny on the first square of a checkerboard, two pennies on the second square, four on the third square, and eight on the fourth, and if you continue doubling the number of pennies on each subsequent square, how many pennies are on just the last (64th) square?

Write the answer using an exponent and then give the dollar value of the pennies. (*Note*: If you could stack the pennies, you'd have a stack about as high as the universe is wide!)

Fill in the following table, in which the second column gives the number of pennies on the square whose number is in the first column.

Number of Square	Number of Pennies
1	
2	
3	
4	
5	
6	

Graph the points determined by the table. This is an example of *discrete* data; the only values that are meaningful for both the *x*- and the *y*-coordinates are positive integers. Now, to more easily perceive the pattern of the data and to help determine the mathematical function that can model the data, connect the points with a curve as smoothly as possible.

---

[4]J. M. Mann and D. J. M. Tarantola (Editors), *AIDS in the World II*, Oxford University Press, 1996.

Letting y be the number of pennies on the xth square, write an equation giving y as a function of x. The function you determined is an *exponential* function. In this function, where does the independent variable appear?

Make four more graphs, one for each data set given in the AIDS lab. Bring all the graphs with you to the lab.

The AIDS Lab

The home page on the World Wide Web for the Centers for Disease Control in Atlanta shows graphs of death rates in various sectors of the U.S. population. In 1992, HIV infection became the leading cause of death among men aged 25–44 years; in 1994 it became the third leading cause of death among women in that same age group. (HIV is generally considered to be the virus that causes AIDS.) Many cases of AIDS in the heterosexual population have been associated with intravenous drug use. Table 1, however, gives the much smaller numbers of cases of AIDS among women in the United States ages 13 and older who were neither intravenous drug users nor sexual partners of users of intravenous drugs.[5]

Year	Cases Reported
1982	2
1983	7
1984	13
1985	31
1986	84
1987	151
1988	226
1989	330
1990	503
1991	726
1992	972

Table 1 AIDS in U.S. women (cases not associated with intravenous drugs)

Plot the points corresponding to this table. Let the integers $0, 1, 2, \ldots, 10$ on the horizontal axis correspond to the years $1982, \ldots, 1992$ and let the numbers on the vertical axis represent the number of AIDS cases reported in this population of women.

Draw a smooth curve that approximates the data. (Don't attempt to hit all the points. In particular, expect the y-intercept of the curve to be more than 2.) A well-fitting curve will be a sort of *average* for the data, with some points above the curve and others below it. In this particular case, the too-high points should be clustered in the middle, while some much-too-low points should appear on the right side.

Now use the curve to estimate the number of cases reported in 1993.

[5]W. A. Rushing, *The AIDS Epidemic*, Westview Press, 1995, p. 109.

In the exponential function that we used to model the coins on a checkerboard, the *base* was 2, because the number of coins *doubled* with every square. Compare with your partners the formulas each of you wrote for the checkerboard problem. When you are in agreement, use your grapher to draw your checkerboard function. Experiment with viewing windows until the graph on the screen matches the one you drew.

Now let's return to the AIDS statistics in Table 1. The curve you drew should have the same general shape as the graph for the checkerboard problem. Therefore, you can find a function of the form $f(x) = C \cdot a^x$, where C and a are positive constants to be determined, to represent that curve.

If your grapher has a curve-fitting command, ask it to fit an exponential function to the data in Table 1.[6] If your grapher doesn't have that capability, use your hand-drawn curve to determine good approximations for C and a. Here's how:

- Estimate the y-intercept, which should be greater than 2 and not necessarily an integer. Use that as the C-value. $C \approx 3$

- Estimate the coordinates of one other point (x_1, y_1). (A point near $x = 9$ or $x = 10$ would be best.) Don't use a value from Table 1; use the curve you drew to represent those data. $x_1\ 9.5 = y_1 = 840$

- Solve $y_1 = C \cdot a^{x_1}$ for a to complete the model. The a-value should be a number that is more than 1 but less than 2. $840 = 3 \cdot a^{9.5}$

Now you have a function

$$f(x) = C \cdot a^x,$$

expressed in standard exponential form. Its base, or growth factor, is the number a, and its constant coefficient is the C-value you determined.

Graph the exponential model $f(x)$ together with the data points. A detail that you observed earlier—that the data points on the right side are below the exponential model—suggests that the overall trend of the data is about to change.

We might need something more sophisticated than an exponential function if we are to model this population for years beyond 1992. We'll return to this idea later in the lab and again in Section 4.5, "Logistic Growth." For now, though, let's stick with the model we have.

To appreciate the impact of the growth factor of an exponential function, you will now compare your $f(x)$ and the checkerboard function, whose growth factor is 2. (The checkerboard function, unless you did some algebraic fiddling, is not in standard form because its exponent is not a simple unadorned x. This fact won't hinder our comparison.)

In the checkerboard problem, each time the input variable increased by one square, the output (number of pennies) was multiplied by 2. For the exponential model $f(x)$, each time the input variable increases by one year, the output (number of cases) is also multiplied by a specific amount. What is that amount?

The coefficient C for the women's AIDS model is larger than the starting value of 1 for the checkerboard model. This causes the AIDS model to give larger output values at first (for small values of x) than the checkerboard function does. However, no matter how great the difference in initial values, the function with the larger base will eventually overtake the function with the smaller base. Which model has the larger base?

---

[6]Some software produces a base-*e* function instead of one in standard form. If yours does that, convert the model to standard form, $f(x) = C \cdot a^x$, asking your instructor for help if you haven't yet learned to do this.

Find the first integer value of x for which the checkerboard model is greater than the AIDS model. Both of these curves become incredibly steep as x grows. If you have trouble reading the graphs, you could instead evaluate each function for successive values of x and notice how the checkerboard model catches up to the AIDS model.

The first point that you plotted for the AIDS data was $(0,2)$, although that point does not lie on the graph of the model $f(x)$. What is the value of the model when $x = 0$? When $x = 1$? All points that you plotted have integer coordinates, but the function that models the data assigns noninteger values of y to integer values of x. Remember that a model only approximates the data. If you did the Fahrenheit lab, you created a model of the temperature sign data that didn't give the exact data points either. In this lab, we'll look at other reasons that a model might fit the data imperfectly and then over only a limited time interval.

Now let's look at another set of data. In an article by R. Steinbrook in the *Los Angeles Times* from January 5, 1990, "Slower Spread of AIDS in Gays Seen Nationally," the information in Table 2 on the number of AIDS cases reported in Los Angeles was given.

Date of Diagnosis	Cases Reported
Jan–June '83	116
July–Dec '83	154
Jan–June '84	197
July–Dec '84	269
Jan–June '85	415
July–Dec '85	503
Jan–June '86	668
July–Dec '86	773
Jan–June '87	952
July–Dec '87	933
Jan–June '88	955
July–Dec '88	943
Jan–June '89	967

Copyright ©1990 by the *Los Angeles Times*. Reprinted by permission.

Table 2 AIDS in Los Angeles

To get a feeling for these data, plot the points corresponding to Table 2, numbering the horizontal axis 0, 1, 2, . . . to count the time intervals, with 0 representing the period January–June 1983, and number the vertical axis to count AIDS cases reported in Los Angeles during that period. You will see that in this case, not all of the points appear to lie along an exponential curve. To create a model for this situation, we will need to splice two functions together. Here's a start. Using your grapher to help you, draw the graph of the exponential function

$$j(x) = 121(1.313)^x$$

on the same axes where you plotted the points from the table.

For what time period is j *not* a good model? For those x-values, a much simpler function will serve to model the data. Give such a function and the interval on the x-axis where it is appropriate. Now use both formulas to write a *single* piecewise function that could represent the entire period 1983–1989. Be sure to indicate which formula goes with which x-values.

The article in the *Los Angeles Times* stated that "public health officials cite several possible causes for the slowdown, including the adoption of safer sexual practices by many gay men to prevent infection." In this situation, changing behavior altered the basic conditions contributing to the epidemic and, fortunately, made the exponential model obsolete for these data. (Go back and check the date of the article by Stephen Jay Gould that introduced this lab.)

There is another reason why an exponential function can model a real-world situation over only a brief period. Since exponential functions grow so rapidly, they exhaust the population very quickly. Nothing in life can continue to grow indefinitely at an ever-increasing rate.

A common pattern for an epidemic is exemplified by the Table 3, which represents data from the entire U.S. population. We rounded values from *The AIDS Epidemic*[7] to the nearest thousand cases.

Year	Cases Diagnosed (in thousands)
1981	1
1982	1
1983	3
1984	6
1985	12
1986	19
1987	28
1988	34
1989	39
1990	41
1991	42

Table 3   AIDS in the United States

Plot the points corresponding to this table on a graph where the integers 0, 1, 2, ..., 10 on the horizontal axis correspond to the years 1981, ..., 1991 and the numbers on the vertical axis represent AIDS cases (in thousands) diagnosed in the United States that year. Draw a smooth curve that approximates your data. Color the part of the graph that looks like an exponential curve. After approximately what year does the exponential function no longer provide a good model? (By the way, in what year did Stephen Jay Gould write the essay mentioned in the lab preparation?)

It is more difficult to write a function for this graph, and we won't ask you to try! The points follow, approximately, a pattern that mathematicians call **logistic growth;** you'll see a way to model logistic growth in Section 4.5.[8]

Although the epidemic has slowed in the United States, it continues to rage unchecked in other parts of the world, particularly in sub-Saharan Africa and southeast Asia. An article in the July 5, 1997, issue of *The Economist* states that the number of South Africans infected with HIV is 2.4 million, three times as many as in the United States. Ironically, the article suggests that the ending of apartheid has hastened the spread of the disease by opening the South African economy to greater trade and migration flows from the north.

---

[7]Rushing, op. cit., p. 111.
[8]Some calculators will fit a logistic function to a set of data. If yours does, request a logistic model for these data. You'll be amazed by the close fit.

Table 4 gives the number of AIDS cases in South Africa as reported in *AIDS in the World II*. (These are cases of AIDS, not simply of HIV infection, and include only those reported to the World Health Organization. Many cases of the disease go unreported.) Graph this set of points as you did the others in this lab. Label the *x*-axis 0, 1, 2, . . . , 9 to correspond to the years 1985–1994 and the vertical axis to count the number of AIDS cases reported. Which of the other three graphs in this lab does it most nearly resemble?

Year	Cases Reported to WHO
0 1985	9
1 1986	34
2 1987	48
3 1988	94
4 1989	176
5 1990	304
6 1991	393
7 1992	658
8 1993	1267
9 1994	2774

Table 4 AIDS in South Africa

If your grapher has a curve-fitting command, ask it to approximate the data with an exponential function. You should get a model similar to

$$h(x) = 14.437(1.777)^x$$

The growth factor, or base, of this standard-form model tells the amount by which the output from a given year is multiplied to obtain the output for the following year. What is that growth factor?

We can think of that growth factor (the base of the function) as a percentage. The base, 1.777, can be written as 177.7%, or 100% + 77.7%, indicating that AIDS in South Africa is growing, on average, at 77.7% per year! Look again at the first model, $f(x)$, and figure the annual percentage growth of AIDS in that population of U.S. women.

The Lab Report

Write a report explaining to a student who's not taking your precalculus course why Stephen Jay Gould could state (in the same article quoted in the lab preparation), "The AIDS pandemic . . . may rank with nuclear weaponry as the greatest danger of our era." In your report, describe the nature of exponential growth and tell how the base of the exponential function governs the growth of the function values. Place Gould's essay in historical context by noting the year in which it was written and relating it to the information conveyed by each of the four graphs. Then describe the limitations inherent in an exponential model, referring to the data for the United States and for Los Angeles. Give the model you found for the U.S. women's data and the complete function that you wrote to model the Los Angeles data. Finally, discuss the alarming nature of the women's AIDS data and the statistics from South Africa, noting their similarities and what they indicate about the stage of the epidemic in these two population groups. Include graphs of the four sets of AIDS data and of the three mathematical models. On each graph, note the pre-1987 and post-1987 portions.

Don't put all your eggs in one basket!

Photo Courtesy Texas Parks & Wildlife 2003, Bill Reaves

Preparation

The Kemp's ridley sea turtle (sometimes called the Atlantic ridley sea turtle) is the most endangered of the world's sea turtles. In recent decades, this once-abundant turtle has faced threats ranging from entanglement in the nets of shrimp trawlers to the killing of the adults for meat and leather to rampant poaching of its eggs (believed by some to be an aphrodisiac). The females and their nests were easy prey for poachers because 95% of the nesting occurs on a single beach at Rancho Nuevo in Tamaulipas, Mexico. Even after the turtle received protection in Mexico and was listed as an endangered species under U.S. law, its numbers continued to decline. By 1978, scientists feared that the turtle might become extinct unless strong conservation measures were implemented. While Kemp's ridley populations are still only a small fraction of what they were in the 1940s, the joint effort between Mexican and U.S. scientists and conservationists might have turned the tide (no pun intended) for the Kemp's ridley sea turtle.

Your task in this lab will be to examine data on Kemp's ridley sea turtles and to form mathematical models describing first their decline and later their rebound. You can determine for yourself how close the turtles were to extinction. Then, on the basis of more recent data, you can predict when the turtle might be upgraded from "endangered" to "threatened" under the Endangered Species Act.

Before beginning the investigation, you'll learn to use base-10 logarithms to transform an exponential function into a linear one. This knowledge will help you to understand a linear model proposed by a Mexican scientist studying these turtles. We'll start with a couple of purely algebraic examples to help illustrate the concept.

Recall the standard form for an exponential function: $y = c \cdot a^x$. For our first example, we arbitrarily choose $c = 3$ and $a = 1.2$. Write a formula for the function:

$$y = \underline{\hspace{5in}}$$

Now use that function to complete Table 1.

x	y	$\log(y)$
0		
1		
2		
3		
4		
5		
6		

Table 1

Next use Table 1 to create two separate graphs: the points for y versus x on one set of axes and the points for $\log(y)$ versus x on the second set. Connect the points in each graph with a smooth curve. Your first graph should have the familiar upward-bending shape of an exponential function. But your graph of $\log(y)$ versus x should look like a line. Determine the approximate equation of this line. In other words, in the equation that follows, fill in the vertical intercept and the slope:

$$\log(y) = \underline{\hspace{3cm}} + \underline{\hspace{3cm}} x$$

After rounding, your result should be $\log(y) = 0.477 + 0.0792x$.

Now suppose you had started with this *linear* function and wanted to work your way back to an exponential function of the form $y = c \cdot a^x$. You can accomplish this by using rules of exponents, as follows:

$$\log(y) = 0.477 + 0.0792x$$
$$y = 10^{0.477 + 0.0792x} \qquad \text{Rewrite as an exponential statement.}$$
$$= 10^{0.477} \cdot 10^{0.0792x} \qquad \text{(Which exponent rule permits this?)}$$
$$= 10^{0.477}(10^{0.0792})^x \qquad \text{(By which law of exponents?)}$$

Use your calculator to compute $10^{0.477}$ and $10^{0.0792}$ to two decimal places and see that the final result matches the original exponential function, $y = 3 \cdot 1.2^x$.

Now it's your turn. Select your own exponential function, using other values for c and a. Write the formula for the function and use it to complete Table 2.

x	y	$\log(y)$
0		
1		
2		
3		
4		
5		
6		

Table 2 $y =$ _____

Plot the points for $(x, \log(y))$. They should lie on a line. By finding the vertical intercept and calculating the slope, write an equation for that line, $\log(y) = b + mx$. Then, following the procedure of the preceding example, work your way back to your original exponential function, $y = c \cdot a^x$.

The examples you just completed illustrate two general conclusions:

- If a function is exponential, its logarithm is linear.
- If the logs of a set of y-values form a linear pattern, the y-values themselves will form an exponential pattern.

In this lab, you'll learn how scientists and statisticians make use of these mathematical facts.

The human eye is very good at detecting a linear pattern in plotted points. You know a line when you see one. If the data follow a curved pattern, however, determining the precise type of curve is difficult. Although it might be exponential, it could be something else instead. Our eyes alone can't give us enough help in deciding. How do you think logarithms might help a scientist to determine whether a given set of data could be appropriately modeled by an exponential function?

When you get to lab, share your exponential functions and the corresponding graphs of $\log(y)$ versus x with the members of your group. Compare your answers to the question in the preceding paragraph.

The Turtles Lab

It is impossible to estimate the size of the Kemp's ridley population directly. The adult males and the "kids" stay at sea. Only the adult females come ashore to lay their eggs. Some females nest every year, others every two or three years. Since most of the nesting is done on one beach, annual population estimates generally focus on the number of nesting females at Rancho Nuevo.

Alarming Decline, 1947–1986

A 1947 aerial film of an *arribada*, as a mass nesting is termed, showed an estimated 40,000 turtles nesting at Rancho Nuevo on a single day. By 1968, however, the female nesting population at Rancho Nuevo was reduced to 5000 for the entire season (April to July). Table 3 provides additional data.

Year	x	Nesting Females, y	$\log(y)$
1947	0	40,000+	
1968		5,000	
1970		2,500	
1974		1,200	
1986		621	

Table 3 Estimates of the Nesting Female Population at
Rancho Nuevo

Plot the points corresponding to Table 3. Let the integers 0, 1, 2, ..., 39 on the
horizontal axis correspond to the years 1947, 1948, 1949, ..., 1986 and let the numbers
on the vertical axis represent the population estimates. (For plotting purposes, treat
40,000+ as 40,000.) Because it is difficult to see whether or not this pattern is exponential, compute and plot the base-10 logs of the population numbers. Check that the pattern
looks approximately linear. This indicates that an exponential model is appropriate for
the data. With a graphing calculator or statistical software, fit an exponential function
to these data, and sketch that curve on your data plot.

According to the *Recovery Plan for the Kemp's Ridley Sea Turtle*,[9] this turtle has
experienced one of the most drastic population declines recorded for any animal. Use
your model to estimate the annual percentage decrease for the female nesting population.
If no outside forces intervene to change the pattern, in what year do you predict that the
Kemp's ridley turtle would, for all practical purposes, be extinct? (Be prepared to defend
your interpretation of "for all practical purposes.")

Gradual Losses, 1978–1985

In 1978, a binational Kemp's Ridley Working Group was formed, comprising Mexican
and U.S. university researchers and several agencies: Instituto Nacional de la Pesca,
the U.S. Fish and Wildlife Service, the National Marine Fisheries Service, the National
Parks Service, and the Texas Parks and Wildlife Department. Their efforts resulted in
increased protection at the nesting beach, promotion of the development of a second
nesting beach (at Padre Island National Seashore in Texas), and an experimental "head
start" program in which hatchlings were raised in captivity for their first year to decrease
infant mortality. Despite these efforts, the Kemp's ridley population continued its decline
late into the 1980s, as shown by Jack B. Woody's[10] data in Table 4.

---

[9]U.S. Fish and Wildlife Service and National Marine Fisheries Service, *Recovery Plan for the Kemp's
Ridley Sea Turtle* (Lepidochelys kempii), National Marine Fisheries Service, St. Petersburg, FL,
1992.

[10]Jack B. Woody, "Kemp's Ridley Continues Decline," in *Marine Turtle Newsletter*, Vol. 35: 4–5,
1985.

1	2	3	4	5
Year	x	Number of Known Nests	Nesting Females (estimated)	Log of Column 4
1978	1	924	711	
1979		954	734	
1980		868	668	
1981		897	690	
1982		750	577	
1983		746	574	
1984		798	613	
1985		673	518	

Table 4 Population Estimates for 1978–1985

Once again, we have a handy way of deciding whether or not an exponential model is suitable for this set of data. Complete columns 2 and 5 of Table 4. Then plot column 5 versus x. If the result appears approximately linear, then an exponential model would be a good choice.

Dr. Rene Marquez (Instituto Nacional de la Pesca) decided that it did appear linear and developed this logarithmic model for the size of the nesting population:

$$\log(N(x)) = 2.89 - 0.0195x$$

where $N(x)$ is the expected number of nesting females and x is the time in years since 1977. (Marquez used an x-value of 1 to represent the year 1978.) Add a graph of Marquez's model to your plot. The line should resemble the pattern formed by the points.

Now use the method that you learned in the lab preparation to transform Marquez's model into an exponential model. In other words, determine a formula for $N(x)$. From Table 4, plot column 4 versus x; then overlay the graph of $N(x)$. Does this exponential model appear to fit the data reasonably well?

According to your exponential model N, what is the annual percentage rate of decline for the female nesting population? Compare this percentage with the percentage decline for 1947–1986, and decide whether the conservationist's efforts had any positive effect on the Kemp's ridley population. If the nesting population continues to follow the trend of the model N, in what year would the Kemp's ridley turtle be, for all practical purposes, extinct? Compare this prediction with your earlier prediction, which was based on Table 3.

Woody counted nests, not turtles. He estimated the number of nesting females by dividing the number of observed nests by his estimate of the average number of nests per female. Use Table 4 to determine the average he must have been assuming.

More recent research suggests that the average number of nests per female is closer to 2.3. Would this revised assumption increase or decrease the estimates of the nesting population?

Modify your exponential population model $N(x)$ to reflect the assumption of an average of 2.3 nests per female. How does this modification affect the percentage rate of decline for your population model?

A Vigorous Comeback, 1988–1997

With the Kemp's ridley turtle population still in decline, environmentalists turned their attention to another major cause of sea turtle mortality: drowning in the nets of shrimp trawlers. For nearly 15 years, the National Marine Fisheries Service worked on the development of turtle excluder devices (TEDs), which allow turtles caught by accident to swim free, unharmed. Because at first (1983) the use of TEDs was voluntary, the effort failed miserably until the U.S. National Marine Fisheries strengthened the regulations. After December 1, 1994, all trawlers in U.S. waters from Virginia to Mexico were required to use TEDs year-round. Improvement in nest counts from 1988 to 1997 is apparent in Table 5.

Year	Number of Known Nests
1988	842
1989	878
1990	992
1991	1155
1992	1286
1993	1229
1994	1560
1995	1943
1996	2080
1997	2387

Table 5 Nest Counts from a Variety of Sources

Plot the points corresponding to Table 5. Code the years by letting $x = 0$ correspond to 1988 and $x = 9$ correspond to 1997. Then fit an exponential model to the data you have plotted. How well does your model appear to represent the pattern in the data? On the basis of your model, what is the annual percentage rate of increase in the number of observed nests? In the population of nesting females? (To answer the last question, does it matter whether the average number of nests per female is 2.3 or the Woody's estimate of 1.3?)

Under the Endangered Species Act, the Kemp's ridley sea turtle will be upgraded from "endangered" to "threatened" once the population of nesting females reaches 10,000. On the basis of your model and Woody's assumption of the average number of nests per female, predict when the Kemp's ridley sea turtle will be upgraded.

The Lab Report

Use mathematics to tell a story of the Kemp's ridley sea turtles. What is the purpose of using the logs of the data? Explain (that is, justify mathematically) why graphing $\log(y)$ versus x and getting a linear pattern indicates that the original data y versus x has an exponential pattern. Include scatter plots and graphs of models corresponding to the data in Tables 3, 4, and 5. Determine the annual percentage change for all your models. Explain how modifying your models by a constant multiple affects those percentage rates. Discuss your predictions for the turtle's extinction, taking care to explain your interpretation of "extinct for all practical purposes," and your prediction for upgrading the turtles' status to "threatened."

Follow-up to The Turtles Lab

One Model or Several?

In the "Turtles" lab, you determined several separate models describing the population over time of the Kemp's ridley sea turtle. Would it make more sense to combine all the data from Tables 3, 4, and 5 into a single data set and to fit a model to the combined data? Explain why or why not.

Shake, rattle, and roll

Preparation

IZMIT, Turkey (CNN) - As rescue workers fought desperately through the night to free victims from layers of twisted concrete and steel, shocked survivors of Tuesday's 7.8-magnitude earthquake in northwestern Turkey struggled to come to terms with the scope of devastation and human loss.

U.S. geologists said the quake registered a magnitude of 7.8, the strongest ever to hit western Turkey. But they later acknowledged its itensity might be downgraded as low as 7.4 as additional measurements are taken into account.

—"Turkish Earthquake Kills More Than 2,000,"

Web posted at 10:42 P.M. EDT, August 17, 1999
http://www.cnn.com/WORLD/europe/9908/17/turkey.quake.07/index.html

The map shows the region referenced in the CNN report. It pinpoints Izmit, the city closest to the earthquake's epicenter.

© tagent@turner.com

Earthquakes are dramatic events that attract a great deal of media attention. Coverage of earthquake disasters includes the death toll, reports of destruction, and estimates of economic impact, as well as specific information on the earthquake. Location of the earthquake's *epicenter* (the spot on the earth's surface directly above the region where ground movement first begins) along with its size as measured on the Richter scale are two pieces of information contained in nearly every report.

Have you ever wondered about the technical details contained in such news reports? For example, how did scientists determine that the epicenter for the August 1999 earthquake in Turkey was just outside of Izmit? How much stronger in terms of ground movement is a quake that measures 7.8 on the Richter scale than a quake measuring 7.4?

Modern seismographs located around the world (hundreds are in California alone) record a number of different types of earth movements. Two types of motions that can be detected are P-waves and S-waves, both of which radiate out in rings from the epicenter. P-waves cause back-and-forth movements, whereas S-waves result in side-to-side motion. These two types of waves travel at different rates, the P-waves arriving first, followed by the S-waves. Just as the time delay between thunder and lightning can be used to estimate distance from a storm, the time delay between the arrival of the P-waves and the S-waves can be used to determine the distance from the seismograph to the epicenter.

A seismograph, then, can measure both the strength of a quake and its distance from the seismographic station, but it cannot by itself determine the direction from which the P-waves and the S-waves are arriving. To know that, geologists need information from more than one station. Here's a geometric method for finding the epicenter of an earthquake.

The map on the following page has five seismographic locations. Assume that an earthquake generates the following data (distance units unspecified):

Location	Distance from Epicenter
1	2 units
2	6
3	7
4	6
5	7

Using a compass (the circle-drawing kind), locate the epicenter of this hypothetical earthquake on the map. (If you do not have a compass, you can make one from a piece of string and a pencil. Tie the string to the pencil. From the pencil, mark off a length of string matching the distance from a particular location. Hold the mark at that location and pull the pencil away until the string is taut. Put the point of the pencil on the paper, keep the string taut, and draw an arc. Every point on the arc is the desired distance from the location.) What was the minimum number of seismographic locations needed to pinpoint the epicenter of the earthquake?

All earthquakes are compared to what is called a zero-level earthquake—that is, one that would produce readings of 0.001 millimeter on a seismograph located 100 miles from the epicenter. (Thus, an earthquake with a seismograph reading of 0.02 is 20 times as strong as the zero-level earthquake, because $\frac{0.02}{0.001} = 20$.) The following is a hypothetical list of seismographic readings. Assume that the readings have been adjusted for distance so that the numbers can be compared with one another.

$$0.002, \quad 15.4, \quad 0.03, \quad 2005, \quad 0.5, \quad 432$$

Try to plot these six readings on a number line. What difficulty do you encounter?

The magnitude, $M(x)$, of an earthquake as measured on the Richter scale is computed by taking the base-10 logarithm of the ratio of the seismograph reading to the zero-level reading, 0.001. You can express this relationship algebraically as follows:

$$M(x) = \log\left(\frac{x}{0.001}\right)$$

where x is the seismograph reading in millimeters (adjusted for distance from the epicenter).

Compute the Richter scale magnitude of each hypothetical earthquake from the list of seismograph readings given above. Now plot these Richter scale magnitudes on a number line. Why are these numbers easier to plot than the first set?

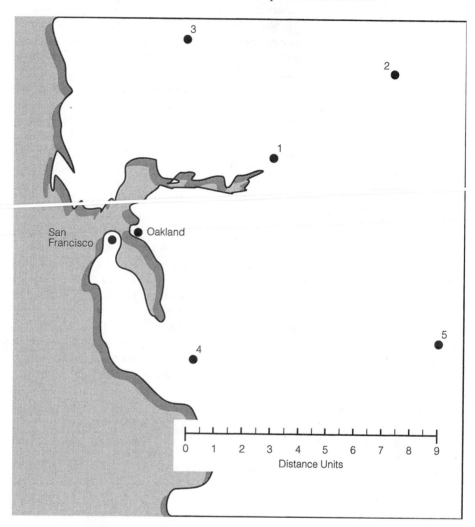

The Earthquake Lab

Begin this lab by comparing with each other the work you did in the preparation section. Did you all center the earthquake at the same location? How many seismographic locations did you need? Discuss the difficulties you encountered in plotting the raw seismograph readings. How did your conversion of these readings into their Richter scale equivalents make the data easier to represent on a number line?

Because Richter magnitude is not well understood by the general public, reporters sometimes explain that an earthquake measuring 6 on the Richter scale is ten times as powerful as an earthquake measuring 5 on the Richter scale and that an earthquake that measures 7 is 100 times as powerful as an earthquake that measures 5. Examine the algebraic formula for $M(x)$ and explain how the factors of 10 and 100 were determined.

If two earthquakes have Richter numbers of 7 and 3, respectively, how many times as strong as the second one is the first one?

Graph the function $M(x) = \log\left(\frac{x}{0.001}\right)$. The screen should show a curve that lies entirely on the positive side of the y-axis. Why does the function not exist for $x \leq 0$?

What about the output values? Can $M(x)$ itself ever be negative? If so, for what values of x does this happen? How large, in theory, can the function $M(x)$ grow? Answering these questions will demand some serious searching, with many adjustments to the viewing window. It is difficult to obtain a complete view of the graph of $M(x)$ without taking several different "snapshots." Don't rule out using algebra as a tool to help you here.

As an abstract function, what is the domain of $M(x)$? What is its range?

The graph shows a function that *increases* for all values of x. The exponential growth functions in the AIDS lab were increasing functions, also, but they increased in a different manner, bending upward as they rose. We say that such exponential functions *increase at an increasing rate*. What comparable statement could we make about this logarithmic function?

Now you will decide on a reasonable domain and range for the mathematical model. Do you think that the portion of the graph where $M(x)$ is negative is relevant to the function M used as a mathematical model for earthquake strength? Consider the fact that the smallest earthquake that humans would be likely to notice measures approximately 2.5 on the Richter scale. Very sensitive seismographs, however, can detect local tremors whose magnitudes are -2.0 or even lower.

The following table, listing some major earthquakes of this century, can give us an idea for a reasonable upper bound for the model.

What portion of the graph of $M(x)$ do you think actually represents the model? Give the domain and the range for the model by setting reasonable bounds for Richter scale readings and corresponding bounds for the seismograph readings.

Some Recent Major Earthquakes

Year	Place	Richter Number	Deaths	Year	Place	Richter Number	Deaths
2000	Indonesia	8.0	103	1979	Colombia-		
1999	California	7.1	0		Ecuador	7.9	800
1999	Taiwan	7.6	2,400	1978	Iran	7.7	25,000
1999	Turkey	7.4	17,118	1977	Romania	7.5	1,541
1997	Iran	7.1	2,400	1976	Turkey	7.9	4,000
1992	Indonesia	7.5	2,500	1976	Philippines	7.8	8,000
1992	California	7.4	63	1976	China	7.8	200,000*
1990	Iran	7.3	29,000	1976	Guatemala	7.5	22,778
1990	Philippines	7.7	440	1970	Peru	7.7	66,794
1985	Mexico	8.1	9,500	1970	Turkey	7.4	1,086
1983	Turkey	7.1	1,300	1968	Iran	7.4	12,000
1980	Italy	7.2	4,800	1964	Alaska	8.4	131
1980	Algeria	7.3	4,500	1962	Iran	7.1	12,230

*Official estimate; may have been considerably higher.

Seismologists do not always agree on the precise magnitude of a given event. Initial reports on the August 1999 quake in Turkey gave it a Richter number of 7.8 and mentioned that it was nearly as powerful as the terrifying 1906 San Francisco earthquake. By the following day, scientists were downgrading the intensity to 7.4 as they considered additional measurements. Use your graph of $M(x)$ to compare the higher reading to the lower reading. How many times as strong as a 7.4 event is one with a Richter measurement of 7.8?

Quakes measuring 7.0 and above are considered "major." How many times as strong as a quake measuring 7.0 is one measuring 7.4 (like the one in Turkey)?

Stop and look carefully at your results for the two comparisons you just performed. You should get the same answer both times. Can you find a pattern relating the Richter numbers you were asked to compare? A solid understanding of the mathematics involved will not only help you to appreciate how a logarithmic scale operates, but also points the way to a very fast method for comparing the strengths of any two earthquakes from their Richter magnitudes. Don't move on until you understand this shortcut.

Several strong quakes that did not make the table because their magnitudes were less than 7.0 nevertheless resulted in tens of thousands of casualties. In Morocco in 1960, for instance, 12,000 people died in an earthquake of magnitude 5.8. Does there appear to be any correlation between the number of deaths and the Richter magnitude?

Two articles appeared on the same page of *The Boston Globe* on August 16, 1991. "Nuclear weapon tested in Nevada" reported the detonation of a nuclear weapon 1600 feet underground. The test caused the ground to sway, registering 4.4 on the Richter scale. "Quake put at 3.0 hits central Pa." reported on a light earthquake with epicenter 10 miles from State College, Pennsylvania. That earthquake, which measured 3.0, was the strongest to hit this area since 1944. Please comment on these reports. You might use your newfound ability to compare two quakes in your comments.

Earthquake measurement is only one example of logarithmic scale. Measuring the loudness of sound in decibels and the acidity of rain on the pH scale are two other examples. In Lab 8A, you will meet the Bordeaux Equation, in which the output Q is a logarithm (that is, an exponent) and needs to be put into the form e^Q to yield a meaningful number. In that lab, you will see bunched results (Q-values) become nicely spread out when they are converted to e^Q-values. In this lab, you saw input data (ground movement measurements) that were too spread out but that yielded meaningful numbers (Richter magnitudes) after logarithmic scaling.

The Lab Report

Describe how you used geometry to locate the epicenter of an earthquake and how you determined the minimum number of seismographic stations needed to pinpoint its location. Explain to someone who is not in this course how the Richter magnitude is determined from the seismograph reading and how it describes earthquake strength. Be sure to include reasons why we would want to use a logarithmic scale in this case. Support your reasons with specific examples. Show how to use Richter magnitudes to compare the strengths of two earthquakes. Illustrate your discussion with one or more graphs of the Richter function $M(x)$ and indicate the portion of the graph that gives reasonable bounds for Richter earthquake magnitudes.

Finally, report any other observations your group made. Did there appear to be a relationship between the magnitude of an earthquake and the number of resultant deaths? (Include specific examples and speculate as to other factors, besides Richter magnitude, that might contribute to the differences in death tolls.) What were your reactions to the two *Boston Globe* items from August 1991?

Polynomial and Rational Functions

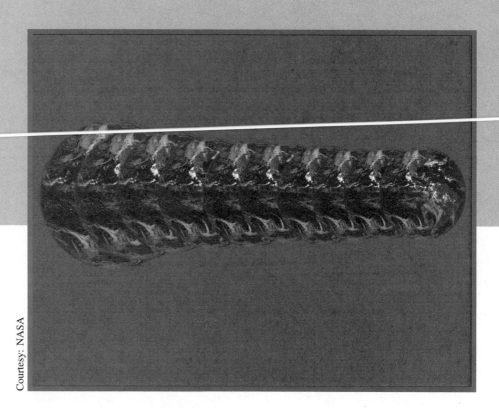

Courtesy: NASA

Although this text emphasizes linear and exponential functions more than any other kinds, those two types alone aren't adequate models for all situations. A linear or exponential function either increases or decreases; it doesn't do both. Life, on the other hand, gives us many instances in which something at first increases, then decreases, or vice-versa. For situations like these, we might turn to polynomial models.

Our study begins with the simplest type—polynomials of degree 2—which are called quadratic functions.

5.1 QUADRATIC FUNCTIONS AND MODELS

Humpty-Dumpty sat on a wall

For many years, the Worchester Polytechnic Institute in Massachusetts held an egg-drop contest. Each team of engineering students designed a container for a raw egg so that it could be dropped from a second-story window of Stratton Hall and reach the ground intact. The winning entry in 1995 was an egg encased in concrete. An interesting but less successful entry was a giant inflated gumby toy. The egg was taped to the head, and the gumby was dropped feet first. Unfortunately, the gumby flipped over in mid-flight and landed on its head, with a predictable outcome.

To win an egg-drop contest, a team must design a container that allows the egg to survive its great fall. A key question for contestants is this: How fast will the package be traveling when it crashes? We'll use a mathematical model of the egg's descent to determine its velocity on impact.

An object falling under the force of earth's gravity falls faster and faster as time passes; the rate at which it drops is not constant. The distance the object falls during the first 0.1 second is less than the distance it falls during the second 0.1 second, which is, in turn, less than the distance it falls during the third 0.1 second, and so on. We will write a mathematical model for the position of a falling egg.

Let $E(t)$ represent the height (in feet) of the egg t seconds after it is released. The quadratic model

$$E(t) = -16t^2 + 20$$

is a good mathematical representation of its vertical position from the time that it is dropped until the moment it hits the ground.

Each term in that quadratic function has a physical explanation. The $-16t^2$ represents the effect of acceleration (or increase in velocity) due to gravity; it has a negative sign because the force operates to *decrease* the height of the egg. The 20 is simply the initial height, 20 feet.

Before the students can find out how fast it's going when it lands, they need to predict *when* it will land. That is, they must solve the equation $E(t) = 0$.

▷   Why is $E(t) = 0$ the equation to solve?

▷   Help them out. After how many seconds will the container hit the ground?

You should obtain two answers, but only the positive one is relevant here because the domain of the *model* begins with $t = 0$, when the egg is dropped. (The domain of the model ends with the number you just found, because the egg does not continue to fall once it has landed.)

Impact Velocity—An Instantaneous Rate of Change

Figure 5.1 shows a graph of the egg-drop model, $E(T)$. Knowing that the egg started to fall at $t = 0$ and landed at $t = \frac{\sqrt{5}}{2} \approx 1.118$, the team can calculate the average rate at which it fell by dividing the total distance the egg falls by the time its fall takes:

Average rate of change of a function $f(x)$ on an interval $a \le x \le b$: $\dfrac{f(b) - f(a)}{b - a}$

$$\frac{E(1.118) - E(0)}{1.118 - 0} \approx \frac{0 - 20}{1.118} \approx -17.9 \text{ feet per second}$$

the negative sign indicating that the height was *decreasing* with time. The slope of the dashed line in Figure 5.1 is -17.9.

That rate, however, is only an average velocity over the entire duration of the fall and doesn't tell the engineers what they really need to know: the impact velocity. Observe that near $t = 1.118$, the graph of $E(t)$ is considerably steeper than the dashed line. This shows that the egg was falling much faster at the end of its flight than at the beginning. We need to determine the slope of the curve *at* $t = 1.118$; that slope is the impact velocity.

If you have studied either Chapter 4 or Chapter 6 already, you are familiar with this strategy.

In calculus, you'll learn a way to determine the exact value of an instantaneous rate of change. Here, we will use a technique that gives a very good approximation. The idea is to choose a brief time interval near the time of impact and to calculate the average rate of change on that interval. For our purposes, a t-interval of width 0.001 will serve well. We'll evaluate E at two moments: at $t = 1.118$ (approximate time of impact) and 0.001 second earlier.

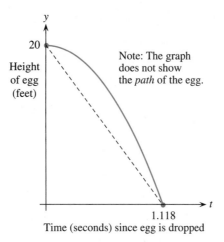

Figure 5.1 Graph of Egg-Drop Model

▷   View the graph of $E(t)$ on the interval $[1.117, 1.119]$. It should appear linear, similar to what you see in Figure 5.2.

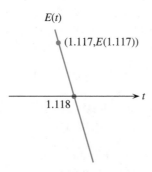

Figure 5.2 Estimating Impact Velocity

Here's the setup:

$$\frac{E(1.118) - E(1.117)}{1.118 - 1.117} = \frac{-16(1.118)^2 + 20 - (-16(1.117)^2 + 20)}{0.001}$$

$$\approx -35.76 \text{ feet per second}$$

The egg was traveling at approximately 35.76 feet per second when it hit the ground. (That's over 24 miles per hour.)

There is no need for you to evaluate $E(1.118)$ or $E(1.117)$ explicitly; let your calculator handle the computations internally. The final result will be more accurate if you do not round off intermediate results.[1]

---

[1]You might be tempted to use 0 in place of $E(1.118)$, since 1.118 is our estimate for the t-intercept. Don't. $E(1.118)$ is very close to zero, but it isn't exactly zero, and that tiny difference matters very much on this extremely narrow time interval.

Check Your Understanding 5.1

CYU answers begin on page 254.

1. Write a function to model the height of an egg container, as a function of time, if it is dropped from a fourth-floor window (approximately 50 feet).

2. Find a good approximation for the impact velocity of the egg in Question 1.

3. Explain why the result is negative. (If yours *isn't*, check your slope setup.)

Suppose the egg, instead of being *dropped* from the second-story window, is thrown directly upward from that height. The quadratic model gains a new term, one that represents the initial upward velocity of the container. If the initial velocity is 19 feet per second, then the model for the egg's height becomes

$$H(t) = -16t^2 + 19t + 20$$

Notice that the coefficient of the linear term $19t$ is positive because the egg is tossed *up*. If, instead, it were hurled toward the pavement, that velocity coefficient would be negative.

How fast is it going when it hits the ground? Will it be falling faster than the egg that was simply dropped? Let's use Figure 5.3 to compare the graphs of the two models.

The *path* of each egg is vertical. The graphs show height versus time.

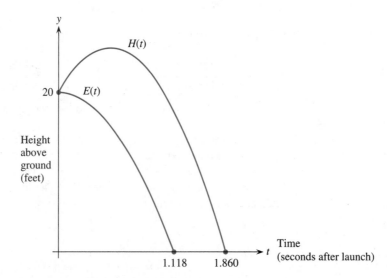

Figure 5.3 Two Egg-Drop Models

The procedure for finding the impact velocity is the same: Find out when it lands; estimate the slope of the curve at that point on the graph.

▷   Use the quadratic formula or the "zero" or "root" function of a graphing calculator to verify that this egg lands at $t \approx 1.860$.

▷   Evaluate the slope quotient $\dfrac{H(1.860) - H(1.859)}{0.001}$ to estimate the impact velocity.

You should determine that this egg is traveling a bit faster than the first one when it lands. At the t-axis, the graph of $H(t)$ is a tiny bit steeper than the graph of $E(t)$.

Parabolas: Intercepts, Axis of Symmetry, and Vertex

Quadratic function:
$f(x) = ax^2 + bx + c$

Both egg-drop models belong to the class of **quadratic** functions $y = ax^2 + bx + c$, in which a, b, and c are constants, $a \neq 0$. Every quadratic function has a term in which the variable is squared; the word *quadratic* comes from the Latin word for *square*. The graph of every quadratic function is a graceful curve known as a **parabola.**[2]

Quadratic functions are important in modeling and are straightforward to analyze mathematically. In the graph of a parabola, two bits of information are especially significant:

- The coordinates of the **vertex** (the highest or lowest point)

- The intercepts

With a grapher, these are easy to find to any desired degree of accuracy. However, they are almost as easy to find algebraically as well. Let's use the second egg-drop model, $H(t) = -16t^2 + 19t + 20$, as an example.

▷   Find the y-intercept, $H(0)$. You should obtain the initial height.

You already know that one of the two t-intercepts is approximately 1.860. To find its *exact* value, use the quadratic formula to solve the equation $-16t^2 + 19t + 20 = 0$.

What? You forget the quadratic formula? The Algebra Appendix is at your service; see page 484.

▷   Try it. Be sure you still know how to apply this piece of wisdom.

$$\frac{-b \pm \sqrt{b^2 - 4ac}}{2a} \qquad \frac{-19 \pm \sqrt{19^2 - 4(-16)(20)}}{2(-16)} \qquad \frac{+19 \pm \sqrt{1641}}{-32}$$

$$\frac{19 \pm \sqrt{1641}}{32} \qquad 1.86$$

The quadratic formula also gave a second t-intercept, a negative number. That negative t-value is an intercept of the abstract function but not of this model. Figure 5.4 shows $H(t)$ and its two t-intercepts.

Finding the vertex is even easier. Every parabola is balanced on either side of an **axis of symmetry.** As Figure 5.4 shows, this parabola's axis is the vertical line $t = \frac{19}{32} \approx 0.594$, the average of the two t-intercepts. The axis of symmetry gives the first coordinate of the vertex, so the highest point of this parabola is located at $\left(\frac{19}{32}, H\left(\frac{19}{32}\right)\right)$. These are messy numbers, perhaps, but what's a calculator for, anyway?

▷   Find the y-value of the vertex. $-16\left(\frac{19}{32}\right) + 19\left(\frac{19}{32}\right) + 20$

---

[2]The word parabola comes from the Greek *para*, beside, and *ballein*, to throw. Perhaps the ancient Greeks also held egg-drop contests.

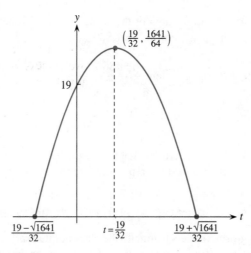

Figure 5.4 Graph of $H(t) = -16t^2 + 19t + 20$

The graph of every quadratic function $y = ax^2 + bx + c$ is symmetric about the vertical line $x = -\frac{b}{2a}$, its axis of symmetry. The vertex or turning point of the parabola lies on the axis of symmetry as seen in Figure 5.5. Therefore, plug the x-value $-\frac{b}{2a}$ into the function to obtain the highest or lowest value of the function.

The number $-\frac{b}{2a}$ tells *where* to find the greatest or least value; $f\left(-\frac{b}{2a}\right)$ *is* that value.

The vertex of a parabola $f(x) = ax^2 + bx + c$ is located on the axis of symmetry,

$$x = -\frac{b}{2a}$$

The vertex is the point

plug in the points

$$\left(-\frac{b}{2a}, f\left(-\frac{b}{2a}\right)\right)$$

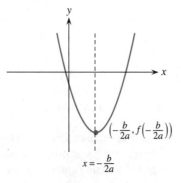

Figure 5.5 Graph of $f(x) = ax^2 + bx + c$

Zeros, Roots, and Factors

By the **zeros** of a function f, we mean the x-intercepts of the graph of $f(x)$, which are the **roots** of the equation $f(x) = 0$. If we're lucky, a quadratic expression will factor neatly, and we can determine the zeros at a glance. Here's an example:

$$q(x) = 10x^2 + 11x - 6 = (5x - 2)(2x + 3)$$

Finding the zeros means solving the equation $q(x) = 0$. One solution is given by $5x - 2 = 0$, the other by $2x + 3 = 0$.

▷   You do the math. What are the two zeros of $q(x)$?

$$x = \frac{2}{5} \qquad x = -\frac{2}{3}$$

Going backward, we can create a quadratic function if we know its zeros. One quadratic function whose zeros are -4 and $\frac{2}{3}$ is

$$y = (x + 4)\left(x - \frac{2}{3}\right)$$

Another is

$$y = (x + 4)(3x - 2) = 3(x + 4)\left(x - \frac{2}{3}\right)$$

Yet another is

$$y = -199(x + 4)(3x - 2)$$

There's no limit to the number of quadratic functions with a given pair of zeros.

STOP & THINK

Can a quadratic function have exactly one zero? (That is, can its graph have exactly one x-intercept?) If so, how?

Optimization

Optimization means finding the best value of a particular mathematical model—the highest, the cheapest, the shortest, the fastest, depending on context. With quadratic models, that means locating the vertex of the parabola.

Feeling run down?

Too much or too little air pressure can adversely affect the life of an automobile tire, so the owner's manual of a car gives the recommended pressure. To come up with these recommendations, a manufacturer performs a quality-control test to measure the life of a tire for various amounts of air pressure. The quadratic model

$$L(x) = -1106 + 69.7x - 1.06x^2 \qquad -1.06x^2 + 69.7x - 1106$$

comes from one such study for a particular type of tire. The dependent variable L represents the life of the tire (measured in thousands of miles driven), and x represents tire pressure (measured in pounds per square inch, or psi).

$$x = -\frac{b}{2a} \qquad x = -\frac{69.7}{2(-1.06)}$$

Let's use algebra to figure out the optimal pressure for this type of tire.

Even though the quadratic term of $L(x)$ appears at the end of its formula rather than first, a is still the coefficient of the x^2 term.

▷   Find the equation of the axis of symmetry of $L(x)$.

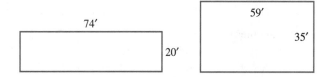

▷   Find the coordinates of the vertex of the parabola.

32.88, 1150.88

▷   What is the optimal pressure for these tires? (Which coordinate did you use?)

▷   How many miles, according to this model, can we expect the tire to last with optimal inflation? (Pay attention to units; L is measured in thousands of miles.)

Notice that each coordinate of the vertex has relevance to the tire situation.

Don't fence me in

In the preceding optimization example, you were handed the function. Sometimes you have to create one yourself, as in the following situation.

Suppose Jack has 188 feet of fencing to make a rectangular enclosure for his cow. He could fence in a rectangle 20 feet by 74 feet ($20 + 20 + 74 + 74 = 188$) or one that is 35 feet by 59 feet ($35 + 35 + 59 + 59 = 188$).

▷   Find the area of each field.

Although each has a perimeter of 188 feet, one field is larger than the other. Jack wants as large an area as possible. If one dimension of a rectangle has length x and the rectangle's perimeter is 188 feet, you can write an algebraic expression for the other dimension.

▷   Label the other three sides of this rectangle; be sure that the sides add to 188.

x

▷   Write an algebraic expression for the area.

Here comes the parabola part. Your answer should be a quadratic expression that is equivalent to $A(x) = 94x - x^2$. $A(x)$ gives the area, A, in square feet, as a function of x, the length of one side. The axis of symmetry of $A(x)$, where the vertex lies, is $x = 47$. The coordinates of the vertex are $(47, A(47)) = (47, 2209)$. The largest area, then, is 2209 square feet.

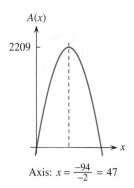

$A(x)$

2209

x

Axis: $x = \dfrac{-94}{-2} = 47$

▷   Now calculate the other dimension of the rectangle. You should see that it, too, is 47 feet. The largest rectangle, given a perimeter of 188 feet, is a *square*. (A square is a rectangle whose four sides are equal.)

Check Your Understanding 5.2

The length of time it takes Madeline to drive to or from work depends on the hour of the day. The model $g(t) = \frac{5}{3}t^2 - 10t + 45$ represents her commute; $g(t)$ is the duration in minutes of a trip begun t hours after 8 A.M. The model is valid between 8 A.M. and 6 P.M. only.

1. If Madeline leaves home at 8 A.M., how long will her trip take?

2. If she leaves work at 6 P.M., how many minutes will she need?

3. For the shortest possible trip, what time should she depart?

4. Leaving at that hour, how long a drive will she have?

The function $g(t) = \frac{5}{3}t^2 - 10t + 45$ from CYU 5.2 has no real-number zeros. As Figure 5.6 shows, the graph sits entirely above the horizontal axis. This lack of t-intercepts makes perfect sense: A t-intercept would indicate a time of day when Madeline's drive took no time at all. Watch what happens when we attempt to solve the equation $g(t) = 0$. By the quadratic formula,

$$t = \frac{10 \pm \sqrt{(-10)^2 - 4\left(\frac{5}{3}\right)(45)}}{2\left(\frac{5}{3}\right)}$$

Oops! These aren't real numbers. (Why not?) Both the algebra and the graph tell us that the function has no real zeros.

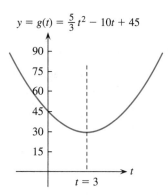

$y = g(t) = \frac{5}{3}t^2 - 10t + 45$

Figure 5.6

▷   Explain why the numbers $\dfrac{10 \pm \sqrt{(-10)^2 - 4\left(\frac{5}{3}\right)(45)}}{2\left(\frac{5}{3}\right)}$ are not real.

Quadratics Functions as Transformations of x^2

Every quadratic function $g(x) = ax^2 + bx + c$ is related to the basic quadratic function x^2 by some combination of the transformations you learned in the "Graph Trek" lab. The transformations become apparent when the function appears in the following form:

You, too, can express any quadratic in $a(x - h)^2 + k$ form; see Project 5.3, "Completing the Square."

$$g(x) = a(x - h)^2 + k$$

|a| > 1 stretch U
|a| < 1 compress U
a > 0 up concavity min
a < 0 down concavity max

Each parameter helps to describe the parabola $a(x - h)^2 + k$: ✓ (h,k) is vertex

- $|a|$ is its vertical stretching factor (stretch if $|a| > 1$, compress if $0 < |a| < 1$)
- The sign of a gives its orientation: opening up if $a > 0$, down if $a < 0$
- h gives the horizontal shift; $x = h$ is the axis of symmetry
- k gives the vertical shift
- (h,k) is the vertex

The egg-drop model $H(t) = -16t^2 + 19t + 20$ is equivalent to

$$H(t) = -16\left(t - \frac{19}{32}\right)^2 + \frac{1641}{64}$$

This version tells us at a glance that the graph of $H(t)$ involves four transformations of the generic parabola $y = x^2$:

- Right shift $\frac{19}{32}$ unit
- Reflection in the t-axis
- Vertical stretch by a factor of 16
- Upward shift $\frac{1641}{64}$ units

It also shows the axis of symmetry, $x = \frac{19}{32}$, and the vertex, $\left(\frac{19}{32}, \frac{1641}{64}\right)$.

Check Your Understanding 5.3

1. The quadratic function $h(x) = \frac{1}{3}(x + 2)^2 - 10$ is related to the function $y = x^2$ by three transformations. Identify them.

   down 10
   shift left 2
   compress

2. Find the axis of symmetry and the vertex of $h(x)$.

   x = -2

3. Find the x- and y-intercepts of the graph of $h(x)$.

4. Find the x- and y-intercepts of the graph of $y = 6x^2 - 13x - 5$.

5. Write two different quadratic functions whose zeros are -2 and 3.

In a Nutshell   The graph of every quadratic function $y = ax^2 + bx + c$ is a parabola whose vertex lies on the axis of symmetry, $x = -\frac{b}{2a}$. The y-value of the vertex is the maximum or minimum value of the function. The average rate of change on a narrow interval gives a good approximation to the instantaneous rate of change of the function at a point within the interval.

5.2 POLYNOMIAL FUNCTIONS AND MODELS

She'll be comin' round the mountain

On a straight section of highway, a driver begins to pick up speed. Noticing a curve ahead, she lifts her foot from the accelerator and allows her speed to decrease a bit. Part way into the turn, she realizes that the curve is only a small portion of a winding section of the highway, and she is forced to apply the brakes, reducing her speed more rapidly in order not to fly off the road. Coming out of the final turn, she gives the car some gas and then floors the accelerator.

Figure 5.7 could represent the example just described. The graph is a picture of how the *speed* of the car changes with time. (It does not represent the *path* of the car.) The entire episode, we will assume, lasted just over half a minute. The curve cannot be a parabola, because it changes direction twice. We'll need to look to a broader class of functions to represent this graph's shape.

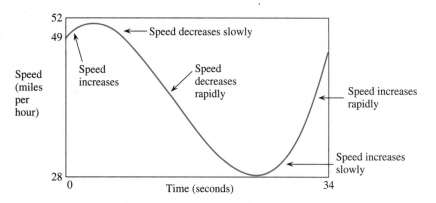

Figure 5.7 Mathematical Model of the Speed of a Car

Power Functions

The quadratic models you studied in Section 5.1 belong to a more general class, the **polynomial** functions. We begin our study of polynomial functions with single-term functions of the form ax^n, where a is a real number and n a positive integer. Because the variable x is raised to a power, functions having this form are called **power** functions.[3] Examples of power functions are $-5x^2$, $0.001x^7$, and x^{100}.

Power function: $f(x) = ax^n$

▷   To better understand how these functions behave, complete Table 5.1. Be careful counting turning points. A **turning point** is one at which the graph stops decreasing and starts increasing or vice-versa.

The "wiggle" in the graph of x^3 is a point of inflection, not a turning point; x^3 has no turning point.

---

[3]All functions of the form ax^r, with a and r *any* real numbers, are power functions. The function $\sqrt{x}$, for instance, is a power function because $\sqrt{x} = x^{1/2}$. In this section, we consider only those power functions whose exponent is a nonnegative integer.

Concavity _Possible_

	Domain	Range	Global Behavior	Turning Points
x^1	$\mathbb{R}$	$\mathbb{R}$	Down on left, up on right	0
x^2	$\mathbb{R}$	Nonnegative reals	Up on both sides	1
x^3	R	R	_down left up right_	_2_
x^4	R	_nonnegative_	_up on both sides_	_3_
x^5	R			
x^6	R			
x^7	R			

Table 5.1

▷ Use the pattern in the table to predict the domain, the range, the global behavior, and the number of turning points for x^9 and for x^{20}.

Multiplying any of the power functions in Table 5.1 by a positive constant will make its graph steeper but will not change anything that you wrote there.

▷ What will happen if you multiply the functions in Table 5.1 by a *negative* constant? What information changes? What remains the same?

reflects   _Domain + Range + Turning points_
             wont change.

Understanding the behavior of power functions with positive integer powers is the key to understanding the whole class of polynomial functions.

Polynomial Functions

A polynomial function is a sum of one or more power functions.[4]

> Every polynomial function can be expressed in the form
>
> $$f(x) = a_n x^n + a_{n-1} x^{n-1} + \cdots + a_1 x + a_0,$$
>
> where n is a nonnegative integer and $a_n, \ldots, a_0$ are constants, with $a_0 \neq 0$. The value of n, the highest power of the variable appearing in the expression, is called the **degree** of the polynomial.

Examples of polynomial functions are

$$p(x) = x^{10}$$
$$q(x) = -5 + 3x - 2x^8$$
$$S(t) = 0.005t^3 - 0.21t^2 + 1.3t + 49$$

---

[4]Power functions, that is, in which the power is a nonnegative integer.

Function p has degree 10; q has degree 8; and S has degree 3. A nonconstant linear function (such as $f(x) = 3x + 4$) has degree 1; a constant function (such as $g(x) = -7$) has degree 0. Functions q and S each contain a constant, but constant terms are really power functions of degree 0: $-5x^0$ and $49t^0$.

This business of degree is important, because it almost singlehandedly decides the global behavior of a polynomial function.

The term in which the highest power appears is called the **leading term** of the polynomial. The leading term of $p(x)$ is its only term, x^{10}. The leading term of $q(x)$ is $-2x^8$. (The leading term is not necessarily the *first* term.)

▷ What is the leading term of $S(t)$?

degree: highest power
leading term: term of highest degree
leading coefficient: the coefficient of the leading term

The coefficient (the constant multiplier) of the leading term is called the **leading coefficient.** The leading coefficient of p is 1; that of q is -2; that of S is 0.005. The *sign* of the leading coefficient and the *degree* of the polynomial, together, completely determine the global behavior of every polynomial function.

Global Behavior of Polynomial Functions

"From a distance, we all look the same"
From a Distance, by Julie Gold

The function that produced the speed-of-the-car graph in Figure 5.7 was this polynomial function of degree 3:

$$S(t) = 0.005t^3 - 0.21t^2 + 1.3t + 49$$

Two turning points

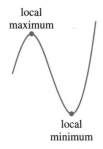

local maximum

local minimum

We can obtain a variety of pictures of its graph by changing the viewing window.

The graph of S in Figure 5.8 shows no t-intercept (we should not expect one, because the car did not come to a halt), but we observe two places at which the graph changes direction:

- At the first turning point, the graph changes from increasing to decreasing, forming a peak. The y-value at the top of the peak is a **local maximum.**

- At the second turning point, the graph changes from decreasing to increasing, forming a valley. The y-value at the bottom of the valley is a **local minimum.**

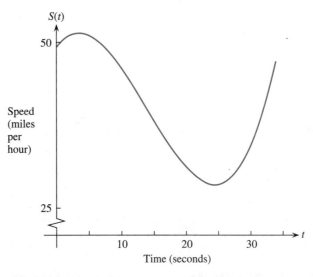

Figure 5.8 Graph of $S(t) = 0.005t^3 - 0.21t^2 + 1.3t + 49$

A local maximum or minimum isn't necessarily the most extreme value of the entire function. It is simply the most extreme value in the neighborhood of that turning point. In this model, for example, the car's speed decreased to about 28 miles per hour before it began to rise, making 28 the *local minimum* for this function.

Figure 5.9 shows $S(t)$ in successively larger viewing windows, ones that no longer make sense in our context but that allow us to see additional features of the graph.

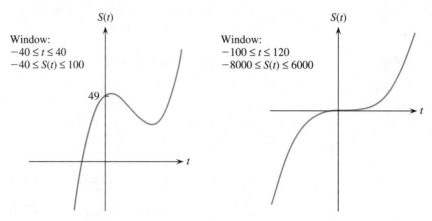

Window:
$-40 \le t \le 40$
$-40 \le S(t) \le 100$

Window:
$-100 \le t \le 120$
$-8000 \le S(t) \le 6000$

Figure 5.9   Two More Views of $S(t) = 0.005t^3 - 0.21t^2 + 1.3t + 49$

The graph on the left in Figure 5.9 shows that $S(t)$ has its y-intercept at 49 (why should we have expected that from its formula?) and one t-intercept. It also hints at the overall behavior of the graph: down on the left side, up on the right side. The graph on the right reveals the strong family resemblance between $S(t)$ and x^3. In that picture, we see a *global* view of the function S.

The graph of $S(t)$ has features common to the graphs of all polynomial functions:

- It is a smooth curve with no breaks, jumps, or corners. Going for a ride on the graph might feel like riding a rollercoaster, but there are no "surprises" in the graph of a polynomial function.

- As $|t|$ becomes very large, $|S(t)|$ becomes even larger. We see this property in Figure 5.9: The graph becomes increasingly steep as t moves far away from the origin in either direction.

A polynomial graph is a smooth curve that, globally, resembles the graph of its leading term.

- Globally, it resembles the power function given by its leading term. The graph of $S(t)$, from a distance, looks like the graph of $y = 0.005t^3$.

This final feature is the reason we had you spend time with power functions a few pages ago. If you understand power functions, you know the global behavior of every polynomial function.

Check Your Understanding 5.4

1. Identify the degree, the leading term, and the leading coefficient of the polynomial function
   $$p(r) = 4\sqrt{3} + 17r - 4.002r^3 - \pi r^4.$$

2. Globally, what power function does $p(r)$ most closely resemble?

3. Describe the global behavior of $p(r)$.

The following properties apply to every polynomial function $y = f(x)$:

• Its graph is a smooth, continuous curve.

• As x becomes large in magnitude, the magnitude of y increases without limit. Symbolically, as $|x| \longrightarrow \infty$, $|y| \longrightarrow \infty$.

• The global behavior of $f(x)$ matches the global behavior of its leading term.

Zeros, Roots, Factors, and Turning Points

"Fasten your seatbelts; it's going to be a bumpy night."
Bette Davis in *All About Eve*, script by Joseph Manckiewicz

The first graph in Figure 5.10 shows two degree-4 polynomial functions. $f(x) = x^4 - 5x^2 + 4$ and $g(x) = x^4$, in the window $-10 \le x \le 10$, $-10 \le y \le 1000$. (The high upper limit on the window lets us see at least some of the upper reaches of the graphs. Polynomial functions grow very large very fast.) As we expect, the graphs look very much alike.

The second graph in Figure 5.10 shows the same two functions in a much smaller window, $-4 \le x \le 4$, $-4 \le y \le 10$. Now their individual personalities emerge.

Wider windows show global behavior. Narrower windows reveal local features.

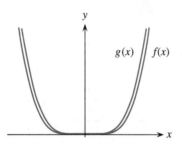
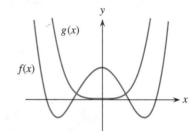

Figure 5.10   Graphs of $f(x) = x^4 - 5x^2 + 4$ and $g(x) = x^4$

Your roots are showing

The close-up view of $f(x)$ in Figure 5.10 reveals four x-intercepts that we couldn't see in the global view. A little algebra tells us exactly where they are.

▷   Factor the expression $x^4 - 5x^2 + 4$ until you obtain four linear factors.

For a review of factoring, see the Algebra Appendix, pages 470–472.

Each factor yields a root of the equation $x^4 - 5x^2 + 4 = 0$, so now we know for sure that the graph of $f(x)$ crosses the x-axis at ± 1 and at ± 2.

We were lucky; when a polynomial is expressed in factored form, finding its zeros is a piece of cake. The bitter truth, though, is that few polynomials (except those found in algebra books) can be factored easily. Mathematicians have devised clever and efficient methods for locating the intercepts of polynomials, but thanks to graphing technology, you won't have to learn them in this course. What you can do, though, is become adept at locating the intercepts whose existence you suspect.

Check Your Understanding 5.5

1. Use "grouping" to factor the polynomial $3x^3 + x^2 - 3x - 1$.

2. Give the zeros of the function $g(x) = 3x^3 + x^2 - 3x - 1$.

3. Describe the global behavior of the function g.

4. Without a grapher, use the zeros, the y-intercept, and the global behavior to sketch a graph of $g(x)$. Check your picture with a grapher.

The preceding examples and problems make use of the **Factor Theorem:**

The Factor Theorem: If r is a zero, then $(x - r)$ is a factor and vice-versa.

> If $(x - r)$ is a factor of any polynomial $p(x)$, then r is a root of the equation $p(x) = 0$ and a zero of the function $p(x)$. Similarly, if the polynomial has a zero at $x = r$, then $(x - r)$ is a factor of $p(x)$.

You're about to learn another way to pry loose some of those roots.

STOP & THINK   Explain why a polynomial of degree n can have no more than n zeros.

Divide and conquer: long division as a root-finding tool

A grapher is a terrific tool, but the values it supplies are only *approximate*. Sometimes, though, it will suggest an exact intercept, which we can verify algebraically. If we know one zero of a polynomial, we can use it to simplify the polynomial's formula, giving us the possibility of finding other exact intercepts. Here's an example.

We examine the graph of the function $f(x) = -x^3 + 4x + 3$ and see three x-intercepts, one of them at approximately $x = -1$. Is -1 really a root of $f(x) = 0$? Check it out:

▷ $f(-1) =$ _____   Yes? Yes!

Just in case long division of polynomials isn't in your current bag of tricks, please consult the Algebra Appendix, page 472.

An x-intercept at -1 means that $x + 1$ must be one of the factors of $-x^3 + 4x + 3$. All we need are the other factors, but $-x^3 + 4x + 3$ does not yield to our usual factoring techniques. It's time for the heavy artillery. We know that

$$f(x) = -x^3 + 4x + 3 = (x + 1)(\text{something})$$

and we find the "something" by long division.

▷ You perform the long division: $x + 1 \overline{)\, -x^3 \qquad\quad + 4x + 3}$

[handwritten work:]

$x = \dfrac{-b \pm \sqrt{b^2 - 4ac}}{2a}$

$-x^2 + x + 3$

$\dfrac{-x^3 \; x^2}{}$

$x = \dfrac{-1 \pm \sqrt{1 - 4(-1)(3)}}{2(-1)} = \dfrac{+1 \pm \sqrt{13}}{+2}$

$x^2 + 4x$

$x^2 + x$

$\overline{\quad 3x + 3}$

$\underline{3x + 3}$

$-x^2 + 3x + 3$

Your result shows that $p(x)$ can also be written in the following form:

$$p(x) = (x + 1)(-x^2 + x + 3)$$

Keep going. Although the quadratic expression $-x^2 + x + 3$ cannot be easily factored, it will still give us exact values for the other two x-intercepts.

▷ Use the quadratic formula to solve the equation $-x^2 + x + 3 = 0$.

▷ List the three roots of $p(x) = 0$. Do they agree with the picture your grapher shows?

Turn, turn, turn

At the start of this section, you completed Table 5.1 for power functions. You should have found that power functions of odd degree have no turning points, while all power functions of even degree have one turning point.

> Odd degree means an even number of turns. Even degree means an odd number of turns. At most $n - 1$ turns.

You've seen polynomial functions, though, with several turning points. The number of turning points is all a matter of odds and evens. A polynomial function whose degree is odd has an *even* number of turning points, while a polynomial function whose degree is even has an *odd* number of turning points. A polynomial function of degree n has no more than $n - 1$ turning points.

Let's see how this works in the following example.

We're rooting for you

Figure 5.11 shows the graph of a polynomial function. Finding its formula will be easier than you might think.

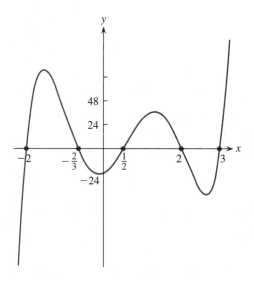

Figure 5.11 Graph of a Polynomial Function $J(x)$

- The global behavior (down on the left, up on the right) says that the degree of the polynomial is *odd*.

- Four (count 'em) turning points say that the degree is *at least 5*.

▷   Use the *x*-intercepts (roots of $J(x) = 0$) to create the factors of $J(x)$.

▷   Construct a polynomial from the factors you found. (Don't multiply it out; leave the expression in factored form.)

$$J(x) = 6(x+2)\left(x+\tfrac{2}{3}\right)\left(x-\tfrac{1}{2}\right)(x-2)(x-3) \; \text{~~~~}$$

▷   What is its *y*-intercept? Does it match the *y*-intercept of the graph in Figure 5.11? (Look carefully. The intercept probably won't match, but we can fix that.)

If your graph has a different *y*-intercept, you need to modify your formula. You don't want to add a constant to the formula, because that would change the *x*-intercepts. You could *multiply* your formula by a constant, though, because that would leave the zeros alone while changing the *y*-intercept.

▷   Fix your formula for $J(x)$, if necessary: Multiply by the constant that makes the *y*-intercept correct. Your graph should match the one in Figure 5.11.

The formula for $J(x)$ is probably a degree-5 polynomial. Remember, though, that the degree of J could be higher; we know only that the degree is *at least* 5 and that it is odd.

Check Your Understanding 5.6

1. Find the formula for a function that could represent this graph. Be sure it has the correct *y*-intercept as well as the correct zeros.

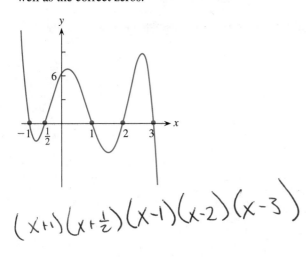

$$\left(x+1\right)\left(x+\tfrac{1}{2}\right)(x-1)(x-2)(x-3)$$

2. Find a possible formula.

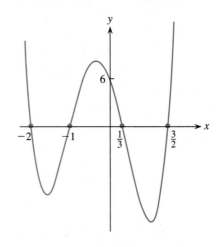

3. Could any of these graphs represent a polynomial
function? Give reasons.

(a)

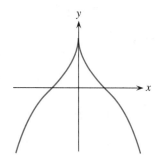

(b)

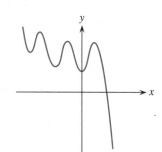

(c)

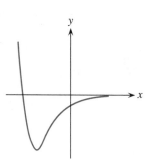

Why Polynomial Functions Are Important

Polynomial functions of degree 2 (quadratic) and degree 3 (cubic) are used frequently
as models. You see several examples in this chapter. Beyond degree 3, however, polynomial models are less common.

But polynomial functions have enormous importance in mathematics itself. Surprisingly, most of the other functions you will encounter can be approximated by polynomials. This fact is particularly important to computer scientists because evaluating
a polynomial involves only those operations that are at the heart of every computer:
addition and multiplication.

▷ To see something impressive, graph these two functions simultaneously on the interval
$[-\pi, \pi]$:

$$y = \sin(x) \qquad \text{and} \qquad y = x - \frac{x^3}{6} + \frac{x^5}{120} - \frac{x^7}{5040}$$

Chances are, you won't be able to tell them apart.

▷ Now widen the viewing window, and you will see that the polynomial approximation
works well only within a limited interval near the origin.

However (this is truly amazing!), by adding more and more carefully chosen terms,
we can make the polynomial function develop additional turning points so that it matches
the graph of $\sin(x)$ over a wider interval—as wide as we wish!

You'll have to wait until the second semester of calculus to learn about these
polynomials: why they work the way they do and how to pick the terms that each one
needs. In the meantime, though, you might ponder the fact that, when your calculator
approximates $\sin(0.123)$, it is really evaluating a polynomial.

In a Nutshell   The graph of a polynomial function is a smooth, unbroken curve whose global behavior is determined by its leading term. A polynomial function of degree n has at most n zeros and $n - 1$ turning points. A number r is a zero of a polynomial p if and only if $(x - r)$ is a factor of $p(x)$.

5.3 RATIONAL FUNCTIONS AND MODELS

If you did Project 5.4, "Galileo," you examined the average rate at which an object falls during different time intervals and saw that this rate increased with the time the object has been falling. In fact, the acceleration, or increase in an object's rate of fall, is due to the earth's gravity. Newton's Law of Universal Gravitation states that every object in the universe attracts every other. So, a spaceship traveling from the earth to the moon is affected by the gravitational attraction of the earth and also by the gravitational attraction of the moon.[5]

Newton's law is an **inverse square law.** The gravitational attraction on the spaceship is a function of the reciprocal of the square of its distance from the earth and the reciprocal of the square of its distance from the moon. Here's the algebraic formula for the gravitational acceleration experienced by the spaceship, measured in meters/sec^2:

$$g = 9.8 \left(\frac{1}{r^2} - \frac{1}{81(60 - r)^2} \right)$$

where the independent variable r is the spaceship's distance from the center of the earth, measured in earth-radii. (The distance from the earth's center to the moon's center is approximately 60 earth-radii, and the radius of the earth is approximately 4000 miles.)

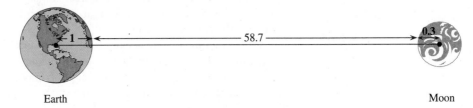

Earth   Moon

Algebra can help you understand Newton's law. The two fractional terms have different signs because, from the spaceship's point of view, the moon's gravitational attraction acts in the opposite direction from the earth's gravitational attraction (assuming that the spaceship is on a line between the earth and the moon). You're accustomed to using a minus sign to denote the opposite direction. The constant $\frac{1}{81}$ comes from the ratio $\frac{m}{M}$ of the masses of the moon and the earth.

▷ What is the value of g when $r \approx 1$, that is, when the spaceship is near the surface of the earth?

▷ What is the value of g when $r \approx 59.7$, that is, when the spaceship is near the surface of the moon?

The Algebra Appendix, page 473, has suggestions for handling algebraic fractions.

Suppose we wish to find the position of the spaceship at which the gravitational attraction of the moon exactly cancels out the gravitational attraction of the earth. How could we determine this information from g's equation? Algebraically, we want to find an r-value for which $g = 0$: an r-intercept of the graph of $g(r)$. We can use algebra to

---

[5]The earth and moon are also subject to a gravitational acceleration toward the spaceship, but this is minuscule because of the enormous differences in their masses.

combine g's two fractional terms over a single common denominator:

$$g = \frac{392(2r^2 - 243r + 7290)}{81r^2(60 - r)^2}$$

▷   Do the math. Show the steps involved in changing the algebraic formula for g on page 244 into the form you see here.

If a fraction equals 0, its numerator is 0; $\frac{a}{b} = 0$ implies that
$a = 0$.

For a fractional expression to be zero, its numerator must equal zero. Solving

$$2r^2 - 243r + 7290 = 0$$

we obtain $r = 54$ or $r = 67.5$. Only one of these values makes sense in this context.

▷   Which one of these two r-values gives the position at which the gravitational attractions of the earth and the moon are in balance? How do you know?

A rational function has the form
$\frac{p(x)}{q(x)}$, where p and q are polynomials.

We have seen two algebraically equivalent formulas for g. But, no matter how we write it, g is not a polynomial function. Like all inverse square laws, g is a quotient or *ratio* of two polynomials, and is an example of what mathematicians call a **rational function.** A rational function is any function that can be written in the form $\frac{p(x)}{q(x)}$, where p and q are polynomials. In this section, you will examine the formula for a rational function to determine important aspects of its local and global behavior.

Vertical Asymptotes

"Infinity is a fathomless gulf, into which all things vanish."
Marcus Aurelius Antoninus (121–180), Roman emperor and philosopher

We'll begin with a simple variety of rational function, one that has a first-degree polynomial on both top and bottom:

$$r(x) = \frac{2x - 1}{x + 3}$$

Before using a grapher, let's see what local features of this graph we can expect. There's a good reason for doing this: In certain viewing windows, a grapher will present misleading pictures, and we want to help develop your intuition so that you can correctly interpret the information from the grapher.

First, find the intercepts:

▷   Solve $r(x) = 0$ to get the x-intercept(s).

▷ Evaluate $r(0)$ to find the y-intercept.

You should obtain $(\frac{1}{2},0)$ and $(0,-\frac{1}{3})$. Next, observe that $r(-3)$ is undefined. But what do the function values look like when x is *near* -3?

▷ Complete Table 5.2.

x	-3.1	-3.01	-3.001	-3	-2.999	-2.99	-2.9
$r(x)$				Undefined			

Table 5.2

The actual numbers are less important than the pattern you see: *The closer x gets to -3, the closer the denominator is to zero and the farther the function value is from zero.* A mathematician would say this: As x approaches -3 from the left, $r(x)$ approaches infinity, and as x approaches -3 from the right, $r(x)$ approaches negative infinity.

In mathematical shorthand, as $x \to -3^-$, $r(x) \to \infty$; as $x \to -3^+$, $r(x) \to -\infty$.

> Infinity (∞) is not a *number*. When we say that a function approaches infinity, we mean that its values increase without limit; when we say that a function approaches negative infinity, we mean that its values decrease without limit.

Moreover, very tiny changes in x when x is near -3 result in enormous changes in $r(x)$. Not only do the function values themselves increase (or decrease) without limit; they change faster and faster as x gets closer and closer to -3.

▷ Examine the graph of $r(x)$ with your grapher. Experiment with the interval and observe that, in the vicinity of -3, the graph of the function becomes nearly vertical. The left side shoots upwards and the right side plunges downwards, so that the graph itself actually resembles the vertical line $x = -3$. Compare and contrast your graphs with the ones in Figure 5.12. What's the same? What's different?

Figure 5.12 shows three views of the graph of $r(x)$. The first resembles a hand-drawn view, the second is similar to what a graphing calculator produces, and the last comes from a computer graphing program. The y-axis does not appear in any of the three views.

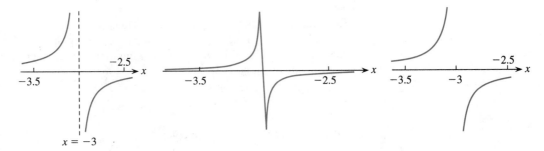

Figure 5.12   Three Views of the Graph of $r(x) = \frac{2x-1}{x+3}$ near $x = -3$

Notice that in the hand-drawn view, the vertical line $x = -3$ is shown as a dashed line. The dashes indicate that the line is not part of the graph of the function $r(x)$ but is sketched in for reference.

The view from the graphing calculator shows a solid line at $x = -3$. If you didn't know better, you might think that it is part of the graph of $r(x)$. The calculator doesn't know that this graph comes in two separate branches. Instead, it connects a pixel on the left side of -3 to a pixel on the right side of -3, resulting in what looks like a vertical piece of graph. The calculator might, for instance, join the point $(-3.01, 702)$ to the point $(-2.99, -698)$, giving a false impression of the behavior of the graph. It would be incorrect for *you* to draw such a graph!

Use "dot mode," if your grapher has that feature, to eliminate the bogus connection.

The computer graph is drawn correctly, but it does not show any vertical reference line. If you use a computer graphing package, you might have to sketch in by hand the dashed vertical asymptote unless the graphing program does it for you.

When the values of a function become increasingly extreme, so that its graph becomes nearly vertical in the vicinity of some x-value k that is not in the domain of the function, we say that the function and its graph have a **vertical asymptote** at $x = k$. For example, the function $r(x)$ in Figure 5.12 has a vertical asymptote at $x = -3$.

> If a function F has a vertical asymptote at $x = k$,
>
> - The number k is not in the domain of F.
>
> - The values of F increase (or decrease) without limit as x approaches k.
>
> - The graph of F, when x is near k, resembles the vertical line $x = k$.

A complete sketch of the graph should include the vertical asymptote as a dashed line, regardless of how the grapher deals with the asymptote.

"None of us really understands what's going on with all these numbers."
(David Stockman, budget director under U.S. President Ronald Reagan)

Let's look at the function that introduced this section, the gravity function

$$g = \frac{392(2r^2 - 243r + 7290)}{81r^2(r - 60)^2}$$

Ignoring context for a moment, we should see that there are only two numbers that are not in the domain of the abstract function g.

▷   What are those two numbers?

▷   Use a grapher to examine the graph of $g(r)$. Experiment with the interval you use for r until you can detect two vertical asymptotes. (Be patient; finding them is tricky; hunt them down one at a time.)

▷   Describe the graph of $g(r)$ as r approaches the asymptote numbers: does the graph shoot up steeply on each side, or does it plunge down steeply on each side, or does it shoot up on one side and plunge down on the other?

Check Your Understanding 5.7

1. Give the equations of the two vertical asymptotes of $g(r)$.

2. Complete the following statement for one of the asymptotes:

As r approaches _____ from the left, $g(r)$

approaches _____ .

As r approaches _____ from the right, $g(r)$

approaches _____ .

3. Write a similar statement about the other vertical asymptote.

Handwritten notes in left margin:

$r(x) = \dfrac{P(x)}{q(x)}$

1) degree of $p(x) > q(x)$

ex. $\dfrac{3x^3 - 2x^2 - x + 7}{4x^2 - 7x + 2}$

$\dfrac{x^2 + 2x + 1}{x+1} \cdot \dfrac{(x+1)(x+1)}{x+1} = x+1$

No horizontal Asymptotes

2) degree of $p(x) < q(x)$

$-\dfrac{\#}{x}$ $\dfrac{3x-1}{x^2-4}$ $\dfrac{3x-1}{(x-2)(x+2)}$

horizontal at X axis

3) degree of $p(x)$ + $q(x)$ are the same $\dfrac{x}{\#}$

$\dfrac{3x+7}{2x-5}$ Ratio of leading cofficents

Horizontal Asymptotes

Figure 5.13 shows three more views of the function $r(x) = \frac{2x-1}{x+3}$. From one picture to the next, nothing changes but the width of the viewing window. In the first view, the vertical asymptote is prominent, and we can estimate the intercepts. In the middle picture, we can still see the vertical asymptote, but we would not be able to guess the intercepts. The overall horizontal trend of the graph becomes apparent. In the last view, the vertical asymptote and the intercepts (that is, "local" features of the graph) have become a mere glitch in a picture that looks more like a horizontal line than anything else.

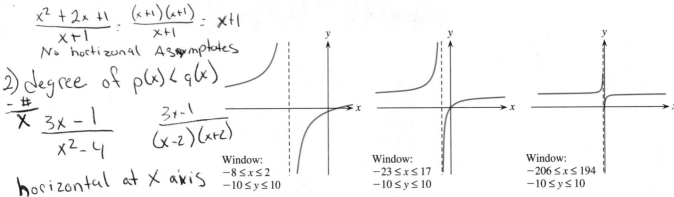

Window:
$-8 \le x \le 2$
$-10 \le y \le 10$

Window:
$-23 \le x \le 17$
$-10 \le y \le 10$

Window:
$-206 \le x \le 194$
$-10 \le y \le 10$

Figure 5.13   Three Views of One Function, $r(x) = \frac{2x-1}{x+3}$

▷ Using your grapher, obtain a picture similar to the third one in Figure 5.13. Determine the second coordinate of the points at the left and right sides of the graph. Those y-values are very close to the same integer value; what is that integer?

The graph of $r(x)$, in the long run, seems to resemble the horizontal line $y = 2$. We call that line the **horizontal asymptote** for the function r. *Globally*, the function r resembles its horizontal asymptote, $y = 2$. *Locally*, for a reasonably narrow interval containing $x = -3$, the graphs of $r(x)$ and $y = 2$ appear to have nothing in common.

In the long run, the graph of $r(x)$ resembles a horizontal line. Algebra helps us find the equation of that horizontal line.

Long division, page 472   ▷ Using long division (Last chance! Learn it now!), rewrite the formula for $r(x)$ as follows:

$$r(x) = \frac{2x-1}{x+3} = 2 - \frac{7}{x+3}$$

When the magnitude of x is very large (that is, when x is far from zero, in either the positive direction or the negative direction), the magnitude of the *denominator* of the fraction $\frac{7}{x+3}$ is also very large, so the fraction itself is very close to 0.

▷   Check it out. Complete Table 5.3.

x	1000	5000	10000	−1000	−5000	−10000
$\dfrac{7}{x+3}$						

Table 5.3

Eventually, the fraction $\frac{7}{x+3}$ becomes negligible in comparison to the 2. (Being a constant, 2 is not affected by changes in x.) So the function as a whole, looked at over a very wide interval, resembles the line $y = 2$.

Figure 5.14 shows a complete graph of the function $r(x) = \frac{2x-1}{x+3}$, including the horizontal asymptote and the vertical asymptotes as dashed lines.

In mathematical shorthand,
as $x \to +\infty, r(x) \to 2$;
as $x \to -\infty, r(x) \to 2$.

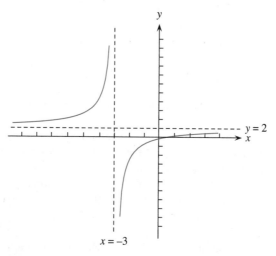

Figure 5.14   Complete Graph of $r(x) = \frac{2x-1}{x+3}$

Check Your Understanding 5.8

Writing $r(x)$ in the form $2 - \frac{7}{x+3}$ reveals it as a relative of the generic reciprocal function, $y = \frac{1}{x}$.

1. Converting $\frac{1}{x}$ into $2 - \frac{7}{x+3}$ involves four separate transformations, one for each constant and one for the negative sign. Identify each transformation.

2. Find any asymptotes, vertical and horizontal, of the function $f(x) = \frac{3x^2}{x^2-4}$.

Maybe you notice a shortcut. If the numerator and the denominator have the same degree, the quotient of just their leading terms gives the horizontal asymptote for the function. So for instance, the function $\frac{12x^4-7x^3+5x-9}{4x^4+20x^2-5}$ has the line $y = 3$ as its horizontal asymptote because $\frac{12x^4}{4x^4} = 3$.

Globally, a rational function behaves like the quotient of the leading term of its numerator and the leading term of its denominator.

> Just as the leading term determines the global behavior of a polynomial function, the *quotient* of the leading terms determines the global behavior of a rational function.

On the level

Let's look again at the gravity function g that opened this section. What is its global behavior? Does it have a horizontal asymptote?

▷   View the graph of

$$g(r) = \frac{392(2r^2 - 243r + 7290)}{81r^2(60 - r)^2}$$

on your grapher, using a wide interval, such as $-100 \leq r \leq 100$. (We're treating the function g in the abstract. We recognize that the distance r in our *model* cannot be negative.) The graph should appear to level off for large values of $|r|$.

▷   At what y-value does the graph of $g(r)$ appear to level off?

▷   What is the equation of the horizontal asymptote for $g(r)$?

We can predict the horizontal asymptote for $g(r)$ without a grapher. The degree of the numerator is 2, and the degree of the denominator is 4. This implies that the denominator grows much more rapidly, in the long run, than the numerator. By the time $|r|$ gets to a respectable size, the denominator is far more impressive than the numerator. For example, when $r = 100$, the numerator is 1,172,080, but the denominator is 1,296,000,000—more than a thousand times as large.

▷   What can you say about a fraction whose denominator is huge in comparison to its numerator? To what integer is that fraction a very close approximation?

The reasoning that you've just done should convince you that your grapher was correct in suggesting that the horizontal asymptote for $g(r)$ is the r-axis.

> In a rational function, if the degree of the denominator exceeds the degree of the numerator, the function has the x-axis as its horizontal asymptote.

It had to be U

This time you get to do most of the work. Let's determine the important local and global features of the graph of

$$q(x) = -\frac{2}{x^2 - 1}$$

▷   One time-saving clue to the graph of $q(x)$ is symmetry. What type of symmetry should we expect, and how do you know?

▷   Find the important local features of the graph: intercepts and vertical asymptotes. (Knowing that q is a symmetric function will help; if you find one vertical asymptote, there must be another one as well.)

▷   Plot the information you have so far on the axes in Figure 5.15.

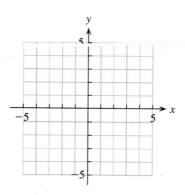

Figure 5.15   Student-Drawn Graph of the Function $q(x) = -\frac{2}{x^2-1}$

▷   Now determine the global behavior of the function q. As the independent variable becomes very large in magnitude, what happens to the value of the function itself?

Let's check: You should have found a y-intercept at 2, no x-intercepts, vertical asymptotes at -1 and 1, and a horizontal asymptote of $y = 0$ (that is, the x-axis is the horizontal asymptote). Now what? What other information do you need to draw a meaningful picture? You'll probably want to plot a few more points, but choose them wisely to save time.

▷   Plot the point that has an x-value of $\frac{1}{2}$. Now use symmetry to plot the point whose x-value is $-\frac{1}{2}$. (Notice how we can take advantage of the symmetry of a graph to obtain half the points free of charge.)

These two points, together with the y-intercept and the two vertical asymptotes, suggest a U-shaped branch in the middle of this graph.

▷   With a narrow window, use your grapher to verify that U shape.

Note: Not every U-shaped curve is a parabola! This curve, in particular, is not a parabola. The only parabolas you will see in this course are graphs of quadratic functions; they all represent polynomials of degree 2. They never represent any other type of formula, particularly one with a variable in the denominator.

STOP & THINK   How could you tell simply by looking at the graph that the middle branch cannot be a parabola?

▷   To complete the graph, plot points corresponding to x-values of 2 and 3, and then draw the right-hand branch. Be guided by those two points, the vertical asymptote $x = 1$, and the horizontal asymptote $y = 0$.

▷   Now use the symmetry of the function to sketch the left-hand branch, and verify the result with your grapher.

▷   How many separate branches does this graph have?

▷   How many did the graph of $r(x) = \frac{2x-1}{x+3}$ in Figure 5.14 have?

When there are n vertical asymptotes, expect $n + 1$ branches.

Notice the relationship between the number of branches in the graph of a rational function and the number of vertical asymptotes. The number of branches in the graph of a rational function is one more than the number of vertical asymptotes. When you view or sketch the graph of a rational function, be sure to include the branches on both sides of each vertical asymptote. That means being very careful in your choice of a viewing window so that you don't miss a branch entirely.

Check Your Understanding 5.9

1. Analyze the function $y = \frac{4x^2-1}{x^2}$.
   (a) Find the x- and y-intercepts.

   (d) What is the horizontal asymptote?

   4

   (e) Now sketch the graph.

   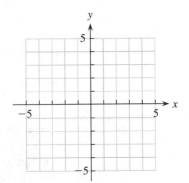

   (b) Find any vertical asymptote(s).

   (c) How many branches will the graph have?

2. Analyze the function $h(x) = 4\left(\frac{x-1}{x^2+2}\right)$.
  (a) Find the x- and y-intercepts.

(d) What is the horizontal asymptote?

(e) Sketch the graph of $h(x)$.

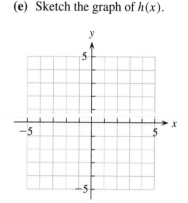

(b) Find any vertical asymptote(s).

(c) How many branches will the graph have?

Not all rational functions behave globally like horizontal lines. The function $\frac{5x^2}{x+1}$, for example, eventually resembles the diagonal line $y = 5x - 5$, while the function $\frac{5x^3}{x+1}$, in the long run, looks like the parabola $5x^2 - 5x + 5$. To learn more about non-vertical asymptotes, do Project 5.8, "From Here to Eternity."

In a Nutshell   A rational function $r(x)$ is the quotient, $\frac{p(x)}{q(x)}$, of two polynomial functions. Unless p and q have a factor in common, the roots of $p(x)$ are the x-intercepts of $r(x)$ and the roots of $q(x)$ are the vertical asymptotes of $r(x)$. Globally, the function behaves as the quotient of the leading term of $p(x)$ and the leading term of $q(x)$.

What's the Big Idea?

- The graph of a quadratic function $g(x) = ax^2 + bx + c$ is a parabola, symmetric about the line $x = -\frac{b}{2a}$. The vertex of the parabola lies on the axis of symmetry and gives the maximum or minimum value of the function.

- In the long run, the leading term of a polynomial function dominates the function and governs its global behavior. The degree of the polynomial determines the possible number of roots and turning points of the function's graph.

- A rational function is the quotient of two polynomials. Many rational functions have vertical asymptotes, dividing the graph into separate branches. Many have a horizontal asymptote, showing the function's global behavior.

Progress Check

After finishing this chapter, you should be able to do the following:

- Sketch a graph of any quadratic function. This includes locating any intercepts and finding the coordinates of the vertex. (5.1)

- Approximate the instantaneous rate of change of a function at a point. (5.1)

- Interpret the coordinates of the vertex of a parabola in the context of an optimization problem. (5.1)

- Find the zeros of a polynomial function. (5.2)

- Determine the global behavior of a polynomial function. (5.2)

- Find the vertical asymptotes of a rational function. (5.3)

- Find the horizontal asymptote of a rational function, if it has one. (5.3)

- Recognize polynomial and rational functions from their graphs. (5.2, 5.3)

- Sketch a rough graph of a polynomial or rational function, using a grapher to refine your picture. (5.2, 5.3)

Key Algebra Skills Used in Chapter 5

- Factoring (pages 470–472)

- Long division of polynomials (pages 472–473)

Answers to Check Your Understanding

5.1

1. $y = -16t^2 + 50$

2. The egg lands at $t = \sqrt{\frac{50}{16}} = \frac{5}{4}\sqrt{2} \approx 1.768$ seconds. Its impact velocity is approximately -56.56 feet per second (using $t = 1.767$ and 1.768).

3. The negative sign indicates direction; the height of the egg was decreasing.

5.2

1. $g(0) = 45$; 45 minutes

2. $g(10) = \frac{335}{3} \approx 111.7$; nearly 2 hours!

3. She should begin driving at 11 A.M.; $-\frac{-10}{2(\frac{5}{3})} = 3$

4. $g(3) = 30$; half an hour

5.3

1. Shift left 2 units; compress vertically by a factor of 3; shift down 10 units

2. Axis: $x = -2$; vertex: $(-2, -10)$

3. $x = -2 \pm \sqrt{30}$; $y = -\frac{26}{3}$

4. $(-\frac{1}{3}, 0)$, $(\frac{5}{2}, 0)$, $(0, -5)$

5. Possible functions are $y = (x + 2)(x - 3)$, $y = -100(x + 2)(x - 3)$, $y = C(x + 2)(x - 3)$, C any real-number constant.

5.4

1. Degree 4; leading term $-\pi r^4$; leading coefficient $-\pi$

2. $y = -\pi r^4$

3. As $|r| \longrightarrow \infty$, $p(r) \longrightarrow -\infty$; both sides of the graph point down.

5.5

1. $x^2(3x + 1) - 1(3x + 1) = (3x + 1)(x^2 - 1) = (3x + 1)(x + 1)(x - 1)$

2. $-1, -\frac{1}{3}, 1$

3. Graph goes down on left, up on right.

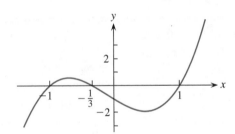

5.6

1. $y = -2(x + 1)(x + \frac{1}{2})(x - 1)(x - 2)(x - 3)$
$= -(x + 1)(2x + 1)(x - 1)(x - 2)(x - 3)$

2. $y = 6(x + 2)(x + 1)(x - \frac{1}{3})(x - \frac{3}{2})$
$= (x + 2)(x + 1)(3x - 1)(2x - 3)$

3. **(a)** The graph has a "corner," so it cannot represent a polynomial function.

   (b) This could be a polynomial function—It's a smooth, continuous curve; as $|x|$ becomes large, $|y|$ becomes even larger.

   (c) The curve appears to level off at the x-axis, so it cannot represent a polynomial function.

5.7

1. $r = 0, r = 60$

2. As r approaches 0 from the left, $g(r)$ approaches ∞. As r approaches 0 from the right, $g(r)$ approaches ∞.

3. As r approaches 60 from the left, $g(r)$ approaches $-\infty$. As r approaches 60 from the right, $g(r)$ approaches $-\infty$.

5.8

1. Shift left three units; stretch vertically by a factor of 7; reflect in the x-axis; shift up two units.

2. Vertical asymptotes are $x = \pm 2$; horizontal asymptote is $y = 3$.

5.9

1. (a) $(-\frac{1}{2}, 0)$ and $(\frac{1}{2}, 0)$; no y-intercept
   (b) $x = 0$ (the y-axis)
   (c) Two branches
   (d) $y = 4$
   (e)

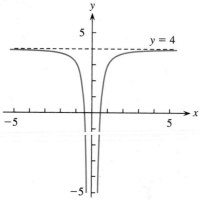

2. (a) $(1, 0)$, $(0, -2)$
   (b) No vertical asymptote ($x^2 + 2 = 0$ has no real-number solution.)
   (c) The graph consists of a single branch, because no values of x are excluded from the domain.
   (d) $y = 0$ (the x-axis)
   (e)

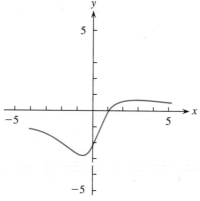

EXERCISES

stretch shifts

1. The egg-drop function $E(t) = -16t^2 + 20$ involves three distinct transformations of the graph of $y = t^2$. Identify each.

2. You have seen that there is one line through two points. In general, it takes three points to determine one specific parabola (unless one of the points is the vertex). Suppose you know that $(1, 6)$, $(0, 3)$, and $(-1, -4)$ lie on the parabola $f(x) = ax^2 + bx + c$. In the general formula for f, substitute the coordinates of each point to obtain a system of three linear equations in a, b, and c. Use the algebraic technique of solving systems of simultaneous equations to solve for a, b, and c, and so determine the formula for f.

In Exercises 3 through 10, write a formula that could represent the graph. Give your reasoning. (The Factor Theorem will help with several of these.)

3.

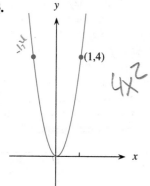

$4x^2$

4.

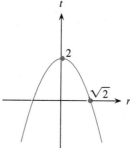

5.

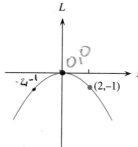

6.

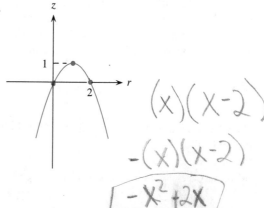

$(x)(x-2)$

$-(x)(x-2)$

$\boxed{-x^2 + 2x}$

7.

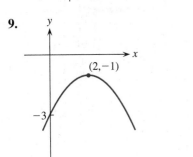

8.

9.

10.

11. Galileo (see Project 5.4) decides to leave the object where it is and hurl himself into the air.[6] Carrying a pendulum clock and a tape measure to get more accurate readings, he jumps at $t = 0$. From then until he hits the ground, his height from the ground is

$$h(t) = -9t^2 + 18t + 27$$

where t is measured in seconds and h in units known only to Galileo.

---

[6]Thanks to Dan Carter for contributing this problem.

(a) How high was Galileo when he jumped? (No reflections on his state of mind, please; just say how many units.)

(b) What was the highest point he reached, and when did he reach it?

(c) When does Galileo hit the ground?

(d) Approximately how fast was he traveling when he landed?

12. An arrow shot from a bow follows a path that is very close to parabolic. An archer, therefore, aims the arrow above, rather than straight at, the target. The flight of the illustrated arrow might be represented by the function $h(x) = -0.00033x^2 + 0.036x + 4.75$, where h is the height above ground level and x is the horizontal distance from the archer's hand. Both h and x are measured in feet.

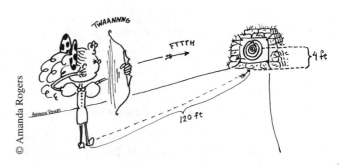

© Amanda Rogers

(a) How high was the arrow at the moment it left the bow?

(b) What is the greatest height the arrow reaches during its flight?

(c) If the center of the target is 4 feet above the ground and 120 feet from the archer and if the bull's-eye is 9.6 inches in diameter, does the archer score a bull's-eye? Does the arrow land above or below the center of the target? Above, below, or within the bull's-eye area? Give mathematical evidence for your answer. (We will assume that the arrow does not veer to the right or left as it flies. This is not entirely reasonable, of course, but the sideways motion would make a separate problem.)

13. (a) Suppose that Jack (see page 232) has only 150 feet of fencing for the rectangular enclosure. What dimensions yield the greatest area, and what is that area? Answer this by writing a quadratic function to represent the area and find that function's axis of symmetry.

(b) Suppose Jack has L feet of fencing. Write a quadratic function to represent the area, and use it to prove that the largest area is a square.

Circumference and area, page 507.

14. (a) Jill, a mathematics student, will use 150 feet of fencing to enclose a circular pen. Using the formulas for the circumference and the area of a circle, find the area of Jill's field. With the same amount of fence, is Jill's area larger or smaller than Jack's (Exercise 13)?

(b) Now let Jill have L feet of fencing. Prove mathematically that for a fixed perimeter L, a circle yields more area than a square.

15. Jack now has 300 feet of fencing to make two rectangular enclosures (one for his cow and one for his llama). How should he use the fencing if he wants to enclose the greatest area? How much area does he get?

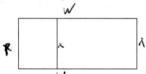

16. Jack (from Exercise 15) gets wise to the fact that he can fence in more area if he uses his barn as one side of the total enclosure. He still needs two sections and he still has 300 feet of fence. What's the largest rectangular area he can enclose with this new arrangement, and how should he use the fencing?

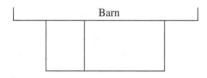

Barn

17. With a grapher, graph the parabolas
$$f(x) = x^2 - 4x + 3 \text{ and } g(x) = -2x^2 + 12x - 15.$$
(a) Use the graph to solve the inequality
$$f(x) < g(x).$$
(b) Next, view the graph of $h(x) = f(x) - g(x)$, either doing the subtraction yourself or having the grapher do it. What kind of function is $h(x)$?
(c) Use the graph of $h(x)$ to solve the inequality $h(x) < 0$.
(d) Your answers to parts (a) and (c) should be the same. Explain why.

18. Find the largest possible value for the expression

$$\frac{1}{x^2 + 6x + 11} \qquad \text{-3}$$

(You can make a positive fraction larger by making its denominator smaller, so you want to make the denominator as small as possible.)

19. Which of the following graphs could represent the function $y = a_4x^4 + a_3x^3 + a_2x^2 + a_1x + a_0$, where the a_i are nonzero (but not necessarily positive) constants? Some graphs work; some don't. Give a reason for choosing or rejecting each graph.

(a)

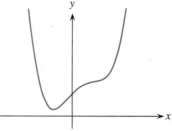

(b)

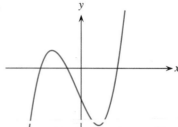

(c)

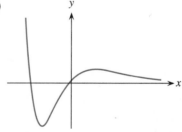

(d)

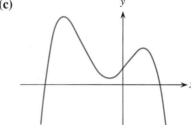

20. Which of the following graphs might represent polynomial functions? Give a reason for choosing or rejecting each one.

(a)

No, too sharp of a point

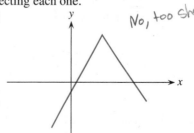

(b)

Yes, cross the x-axis and continues increase

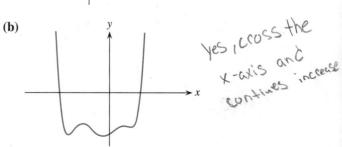

(c)

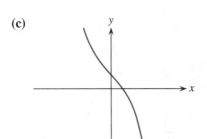

(d)

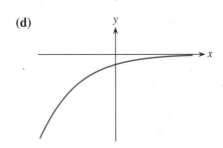

21. How can you tell from the graph of $y = |x|$ that $|x|$ is not a polynomial?

22. How can you tell from the graph of $y = \sin(x)$ that $\sin(x)$ is not a polynomial? *because its a line.*

For each function in Exercises 23 through 27, describe the global behavior of the function; use algebra to find all zeros and the y-intercept; use a grapher to help you draw an accurate graph that shows all zeros and turning points.

23. $p(x) = 2x^3 - x$

24. $y = x^4 - x^3 - x^2$ *up on both sides!*

25. $f(t) = -t(2t - 1)(2t + 3)$ $(x-1)(x-1)(x+1)(x+1)$

26. $q(r) = 4r^5 + 4r^4 - r - 1$ *down left up right*

27. Let $p(x) = 9x^3 + 5x^2 - 6x - 2$.
   (a) By computing $p(-1)$, verify that -1 is a zero of p.
   (b) Use long division to factor $p(x)$ into the product of $(x + 1)$ and a quadratic factor.
   (c) Find all the zeros of p.
   (d) Sketch a graph of $p(x)$.

28. Let $q(x) = -6x^3 + 11x^2 + 4x - 4$.
   (a) By computing $q(2)$, verify that 2 is a zero of q.
   (b) Use long division to factor $q(x)$ into the product of $(x - 2)$ and a quadratic factor.
   (c) Find all the zeros of q.
   (d) Sketch a graph of $q(x)$.

For Exercises 29 through 32, write a formula that could represent the graph. Give your reasoning.

29.

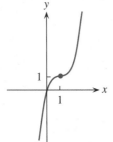

30.

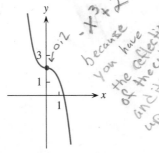

$y=0.12 - x^3 x^2$
because you have the reflection along the y-axis of the cubic but shifted and its up 2

31.

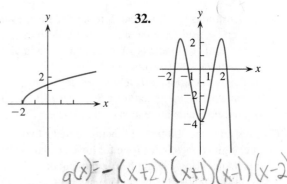

32.

$g(x) = -(x+2)(x+1)(x-1)(x-2)$

33. In a small ceramics studio, the cost of a day's production depends on how many pots are produced. When only a few are made, the per-item cost is high, and total costs rise rapidly with each additional pot produced. As production increases, efficiency reduces the cost per item, so the *rate of change* in the cost of production decreases. If the studio attempts to overproduce, however, the extra associated costs cause the rate of change in the total cost to rise again. The cost C (dollars) of operating the studio on a day on which x pots are produced is given by the function

$$C(x) = 0.01x^3 - 0.65x^2 + 14x + 20$$

   (a) Draw a graph of $C(x)$ over the interval $0 \le x \le 40$.
   (b) What is the fixed cost? (Fixed cost means the expenses incurred even if nothing is produced.)
   (c) In general, as more pots are made, the total cost increases. Is that always the case for the function $C(x)$? Investigate the graph of $C(x)$, and support your answer with specific production levels and their associated costs.
   (d) Economists use the term *marginal cost* to mean the cost of producing one additional item. The marginal cost for $x = 7$, for example, is the cost of making the eighth pot: $C(8) - C(7)$. Calculate the marginal cost for $x = 2$, $x = 10$, $x = 20$, and $x = 38$,
   (e) Describe what happens to the rate of change of C as x goes from 0 to 40. Do not describe the function itself. Words to use in your description of the rate of change include "positive," "negative," "increasing," and "decreasing."

34. *What a difference a degree makes!*   Using a grapher, overlay the graphs of any cubic polynomial and any degree-4 polynomial of your choosing. Experiment with the viewing window so that each function displays its own identity—so that it looks, in other words, like a representative of its own family. Now form a new function by adding the two functions together, and look at its graph. The cubic polynomial should appear to have been "swallowed up" by the polynomial of degree 4. Write a few sentences describing what you see and explaining why the sum of the two functions becomes completely identified with the family of the degree-4 function.[7]

35. The rate of change of $S(t) = 0.005t^3 - 0.21t^2 + 1.3t + 49$, the speed-of-a-car function from page 237, at any particular moment, tells how rapidly the speed was changing at that moment.

(a) An *average* rate of change over a narrow interval gives a good approximation to the *instantaneous* rate of change at a point within that interval. Approximate the instantaneous rate of change of S at $t = 15$ (use the interval $[14.99, 15.01]$) and at $t = 26$ (use $[25.99, 26.01]$). Include units.

(b) Explain why one result is positive and one is negative.

(c) At which time, $t = 15$ or $t = 26$, was the speed changing more rapidly? Examine the graph of $S(t)$ on page 237, and explain why your answer makes sense.

36. Repeat the steps of Exercise 35, using $t = 10$ and $t = 30$ instead.

37. After the engine of a moving motorboat is cut off, the boat's velocity decreases according to the model

$$v(t) = \frac{300}{15 + t}$$

where t is the elapsed time in seconds since the motor was cut off and v is the velocity in feet per second.

(a) Sketch a graph of the abstract function $v(t)$. In a contrasting color, highlight the portion that represents the model.

(b) Is the domain of the model a discrete set or a continuous set? Explain.

(c) How fast was the boat moving when the engine was cut off?

(d) After how many seconds did the velocity reach 10 ft/sec? Solve this problem both algebraically and graphically, and compare your answers.

(e) The rate of change of velocity is called **acceleration.** By estimating the instantaneous rate of change of v at $t = 5$, approximate the acceleration of the boat when the engine has been off for 5 seconds.

(f) Estimate the acceleration when the engine has been off for 10 seconds.

(g) Interpret your results for (e) and (f) in this context of this model.

38. You need a graph of $y = \frac{x-1}{x+2}$, so you enter the function as $x - 1/x + 2$. You have second thoughts, then, and you try $(x - 1)/(x + 2)$. With two graphs now on the screen, how can you tell which is which? Which one is the correct graph? What features in the other one tell you that it's not the graph you want?

In Exercises 39 through 42, use algebra to find all intercepts and asymptotes, and sketch a preliminary graph. Then use a grapher to help you finish the graph.

39. $r(x) = \dfrac{2x + 2}{x - 3}$

40. $q(x) = \dfrac{x - 3}{2x + 2}$

41. $v(x) = \dfrac{x + 1}{x^2 - 4}$

42. $w(x) = \dfrac{x^2 + 1}{x^2 - 4}$

In Exercises 43–48, write the formula for a function that could represent the graph. Give your reasoning.

43.

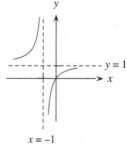

[7]Thanks to former student Amanda Rogers for observing this phenomenon and seeing it as an analogy for certain human relationships in which the more powerful person subsumes the identity of the weaker one.

44.

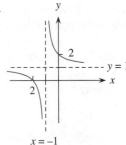

45.

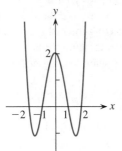

46.

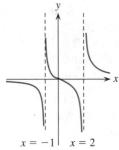

47.

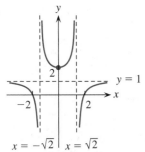

48.

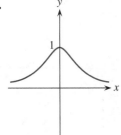

49. What is the domain of the function g on page 244 when it is used as a mathematical model of the gravitational pull on a spaceship? Use the diagram on page 244, and keep in mind what the independent variable represents.

50. The function $C(x)$ in Exercise 33 gives the total cost of a day's production, based on the number of ceramic pots produced. Sometimes, though, we are more interested in knowing the *average cost* of producing a single item. Let $A(x)$ be the average cost of producing each ceramic pot on a day on which x pots are made.

   (a) Use the formula for $C(x)$ to write a formula for $A(x)$.

   (b) Using a viewing window that shows the key features of the graph, sketch a graph of the abstract function $A(x)$. In a contrasting color, highlight the portion that represents the model, assuming that the studio will produce no more than 40 pots in a single day.

   (c) Is the domain for the model a discrete set or a continuous set? Explain.

   (d) The model has a local minimum. Find both coordinates of the turning point, correct to two decimal places.

   (e) To minimize the average cost per pot, how many should the studio make in a given day and what would be the average cost of each? Compare your answer with the coordinates of the turning point you found in part (d), and explain why the answers are slightly different.

5.1 DIAGONALS

God ever geometrizes.   (Plato)

In this project, you'll investigate a surprising property of some common geometric objects. The figures shown are polygons (from the Greek *poly*, many, and *gon*, angle). The first two polygons are **convex** (they have no dips or dents); the third is nonconvex. Polygons have their own names reflecting the number of angles they have. Thus, the first polygon is called a pentagon, since the prefix *penta* is the Greek name for five, and the second is called a heptagon, because *hepta* means seven.

1. A hexagon is a polygon with six angles. Draw two different convex hexagons.

2. A **diagonal** of a polygon is a line connecting two nonadjacent vertices or angles. Draw all possible diagonals on each of the two convex polygons above and on your two home-grown hexagons. (Did you get nine diagonals for each hexagon?)

3. In fact, the number of diagonals of a convex polygon depends only upon the number of its angles. That is, if n is the number of angles of a convex polygon and D is the number of its diagonals, then D is a function of n. (We use n instead of x for the independent variable to indicate that the only allowable values for n are integers.) Fill in just the second column of the table, giving the values of the dependent variable D corresponding to each of the values of the independent variable n. You'll use the remaining two columns later.

n	D	First Differences	Second Differences
3	0		
		2	1
4	2		
		3	1
5	5		
		4	1
6	9		
		5	1
7	14		
		6	1
8	20		
		7	—
9	27		

Now let's determine a formula for D.

4. Could D be a linear function of n? Give numerical evidence for your answer.

5. Here's a way to double-check your answer to Question 4. Notice that each n-value is separated from the next by one unit. If D were a linear function, a one-unit change

in n would always result in the same amount of change in D. In the third column of the table, enter the difference between successive D-values (subtracting each one from the one that follows it). Are these first differences constant?

6. Using the table to find a formula for D in terms of n is tricky. It would help if we knew what kind of function D is. Here's one way to find out. In your table, fill in the last column with successive second differences by subtracting each value in the first-differences column from the one following it. What do you observe?

7. The phenomenon that you should note in Question 6 tells a mathematician that D could be a *quadratic* function of n. Convince yourself of this by plotting the ordered pairs (n, D) from the table, either on graph paper or on your grapher, and noticing that they appear to lie along the right side of a parabola. But which parabola? We're about to find out.

8. One allowable value of n gives a D-value of 0. Which is it? This value of n is called a **zero,** or n-intercept, of D, so we know that n-minus-this-value must be a factor of the quadratic expression for D. In other words, if r is the zero you found, then D is the product of $(n - r)$ and some other linear factor. Experiment with the numbers in your table until you discover the other linear factor of D, and write the function equation that gives D as a function of n. Test the function to be sure it works for every entry.

9. Use the your formula for D to compute $D(25)$.

If you checked your formula against the table entries, you know that it works for $3 \leq n \leq 9$. But how can we be sure that it's valid for $n = 25$? That is, how do we know whether or not $D(25)$ actually gives the number of diagonals of a convex polygon with 25 angles? Unlike scientific theories, we can *prove* mathematical results. That's why we call our results theorems rather than theories. We can demonstrate that the quadratic function D works for all convex polygons.

10. Count the diagonals that meet at each vertex of an octagon.

11. If each vertex of a octagon has 5 diagonals, and the octagon has eight vertices, that makes $5 \cdot 8$, or 40 diagonals for the octagon. Does this result agree with the table? What's the flaw in the reasoning?

12. Did you figure out that $5 \cdot 8 = 40$ counts every diagonal twice—once at each end? How can we fix that flaw? In other words, what do we need to do to the number 40 to arrive at the correct number for the octagon?

13. Now let's perform the same computation with a polygon with n (that is, any number) of angles. How many diagonals meet at each vertex? (You should write this in terms of n.) If you have trouble generalizing from the octagon, think of it this way: a vertex is joined to the other $n - 1$ vertices; however, two of those connections are not diagonals but part of the boundary and should not be counted in the total.

14. Multiply the expression from Question 13 by n, the total number of angles, and correct the answer so that each diagonal is counted once, not twice. The result should agree with the formula you found in Question 8.

Now that you've shown that $D(n)$ is valid for all convex polygons, you can use it with confidence for figures too large for the table.

15. How many diagonals are there in a convex polygon with 40 sides?

16. Solve a quadratic equation to determine the number of angles in a convex polygon with 104 diagonals.

17. The variables n and D in this mathematical model are called **discrete,** because their values are separated by intervals. Write the domain of the model (the set of numbers that make sense as inputs), remembering that it is a discrete set.

5.2 IT'S NOT EASY BEING GREEN

Polyethylene film is often used for greenhouse coverings in Japan. Manufacturers of the film would like to make as thin a film as possible so they can produce more square feet of material for less money. The customers (greenhouse owners) want a film thick enough to resist tearing from wind and rain, but not so thick that the light that passes through is insufficient to grow healthy plants. A film more than 2 mm thick will be too opaque for use as a greenhouse covering.

Assume that the cost to the consumer can be modeled by the function

$$C_1(x) = 0.3x^2 - 1.2x + 6.2$$

and the cost to the manufacturer by the function

$$C_2(x) = 0.5x^2 + 0.4x + 4.08$$

where x represents the thickness of the film in millimeters and where the cost is measured in units of \$100,000.

1. Film thicknesses for which this problem makes sense lie in the interval from 0 to 2 mm. Graph the two cost functions, restricting the x values to the interval $(0,2]$ (that is, $0 < x \le 2$). As the thickness x increases, what happens to the cost to the consumer? What happens to the cost to the manufacturer? (Mathematicians say that $C_1(x)$ is *decreasing* on the interval $(0,2]$, while $C_2(x)$ is *increasing* on that interval.) For what thicknesses is the cost to the producer less than the cost to the consumer?

2. Japanese business philosophy (quite different from that in the United States) concerns itself with making a quality product that suits both the manufacturer and the consumer. The government sets a standard thickness for the film, selecting the thickness that will minimize the **loss**, or total cost, to society. One way to assess the loss to society is to form a new function, $T(x)$, by taking the sum of the cost to the consumers and the cost to the manufacturers. Graph $T(x)$, the function representing the loss to society, and determine the standard thickness that the government will require.

3. If the producer decides to cheat on the standard just a bit and produces a film that is 0.45 mm thick, how much will the company save? How much will the consumer lose? By how much will society be cheated? (Note that the amount by which society is cheated is not simply the value of $T(x)$ for $x = 0.45$ mm, but rather the difference between that value and the loss to society that should have been expected.) Don't round off your results. Use all the accuracy your calculator provides and express your answers in *dollars*, recalling that C_1, C_2, and T are measured in \$100,000 units.

5.3 COMPLETING THE SQUARE

Graphing a quadratic function such as $y = 3(x - 1)^2 - 2$ is straightforward once you've learned the lessons of the "Graph Trek" lab. The formula itself announces a right shift of one unit, a vertical stretch by a factor of 3, and a downward shift of two units, all applied to the basic parabola $y = x^2$.

With a quadratic function like $y = 2x^2 - 5x + 4$, though, we cannot immediately tell what transformations of x^2 are involved. Whenever the algebraic formula for a quadratic function does not explicitly indicate the transformations, we can use a technique called **completing the square** to rewrite the expression in the form that makes the transformations clear. We'll show how to do this using $y = 2x^2 - 5x + 4$.

• Factor out the coefficient of the quadratic term from all three terms:

$$y = 2\left(x^2 - \frac{5}{2}x + 2\right)$$

This might create fractions; don't worry.

- Inside the parentheses, add and subtract the *square* of half the coefficient of the linear term x. In this case, we add and subtract $\left(\frac{5}{4}\right)^2$. (These operations do not unbalance the equation.)

$$y = 2\left(x^2 - \frac{5}{2}x + \frac{25}{16} - \frac{25}{16} + 2\right)$$

- The first three terms inside the parentheses constitute the square of a binomial. Write them in the form $(x + a)^2$:

$$y = 2\left(\left(x - \frac{5}{4}\right)^2 - \frac{25}{16} + 2\right)$$

- Combine the remaining constant terms inside the outer parentheses:

$$y = 2\left(\left(x - \frac{5}{4}\right)^2 + \frac{7}{16}\right)$$

- Multiply, and remove the outer parentheses:

$$y = 2\left(\left(x - \frac{5}{4}\right)^2\right) + \frac{7}{8}$$

That's it! The final version of the formula, algebraically equivalent to the original, gives the axis of symmetry for the parabola (the line $x = \frac{5}{4}$) and the coordinates of the vertex, $\left(\frac{5}{4}, \frac{7}{8}\right)$. The coefficient of the squared factor indicates any vertical stretching, compression, or reflection. This parabola was stretched vertically by a factor of 2 before being shifted up $\frac{7}{8}$ unit.

1. Use *completing the square* to rewrite each of the following quadratic functions in the form $a(x - h)^2 + k$. With the information encoded in the rewritten formula, describe the transformations of the basic parabola x^2.
    (a) $y = 3x^2 + 4x - 6$
    (b) $y = x^2 - 3x - 2$
    (c) $y = 2x^2 + 5x$
    (d) $y = -x^2 + 3x + 1$ (*Note:* The coefficient to be factored out is -1.)

In Section 5.1, you learned that the axis of symmetry for a quadratic function $y = ax^2 + bx + c$ is $x = -\frac{b}{2a}$, the vertical line midway between the two x-intercepts of the parabola. You will now prove that fact in two different ways.

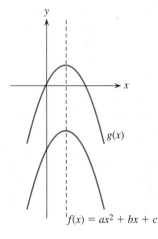

$f(x) = ax^2 + bx + c$

2. Any vertical shift of the graph of any quadratic function $y = ax^2 + bx + c$ doesn't change the axis of symmetry. In particular, the graph can be shifted up or down until it passes through the origin.
    (a) If g represents a vertically shifted version of f, and if the graph of g contains the point $(0,0)$, write a formula for $g(x)$.
    (b) Find the x-intercepts of g.
    (c) Find the axis of symmetry by averaging the x-intercepts.
    (d) Explain how your work has just provided a demonstration that the axis of symmetry of *any* quadratic function is the line $x = -\frac{b}{2a}$.

3. Apply the technique of completing the square to the generic quadratic function $y = ax^2 + bx + c$. Explain how the resulting expression demonstrates that the axis of symmetry is $x = -\frac{b}{2a}$.

5.4 GALILEO

No one will be able to read the great book of the Universe if he does not understand its language, which is that of mathematics.   (Galileo)

© 1991 by Sidney Harris

In the seventeenth century, the great Italian scientist Galileo discovered a famous function that models the physical world. Until Galileo's experiments with falling bodies, many scientists accepted Aristotle's contention that heavy objects fall faster than light ones do. Galileo studied the motion by measuring the distance an object falls in a given amount of time. Translating his data into our own units of measure, we have the following table of values for his two variables: t, which gives the time, measured in seconds, that the object has been falling, and s, which gives the total distance the object falls, measured in feet.

t	0	1	2	3	4
s	0	16	64	144	256

1. Before conducting his experiments, Galileo postulated that the distance fallen was proportional to the time elapsed. Remember that s is *proportional* to t if $s = k \cdot t$, where k is some fixed constant (called the constant of proportionality). Do you think Galileo's data confirmed his hypothesis? Test your answer using the data presented in the table.

2. Now try to reproduce Galileo's work and guess the relationship between s and t. In other words, find an algebraic expression that turns 1 into 16, 2 into 64, 3 into 144, and so on. (*Hint*: All values of s in the table—including zero!—are multiples of the same number.) Write your formula for the function $s = f(t)$.

3. Galileo's *model* for describing the fall of an object from a stationary position works well only near the surface of the earth (from the roof of a tall building, for example). Decide on a reasonable set of values for the domain and range of Galileo's function, explaining your reasons for choosing those values. (This task might seem difficult because there is not one single correct answer. You need to use the information here and your common sense and to make sure that the domain and range you give correspond to each other.)

4. Draw a graph of the function $s(t)$ that represents the data in the table. Describe the shape of your graph.

Today, we can gather data on falling objects more easily and accurately than in Galileo's time. The data in the table that follows were collected by dropping a calculus book over a motion detector, which started recording distance and time before the book was released. The variable T represents time, measured in seconds, after the device began recording data. The variable D gives the distance in feet between the motion detector and the book. A portion of the data relating to the book's fall is shown.

T	0.90	0.94	0.98	1.02	1.06	1.10	1.14	1.18	1.22
D	3.835	3.742	3.583	3.374	3.118	2.823	2.477	2.085	1.638

5. **(a)** Use data from the table to calculate the average rate of change in D:

   - From $T = 0.90$ to $T = 0.94$
   - From $T = 1.02$ to $T = 1.06$
   - From $T = 1.14$ to $T = 1.18$

   (b) What do these rates tell you about the book's motion?

6. **(a)** Use a grapher to plot the distance-versus-time data from the table. Do you think a quadratic function might describe the pattern of these points? Explain.
   (b) Fit a quadratic model to these data. Write a mathematical formula for $D(T)$, rounding each coefficient to two decimal places.[8]
   (c) Overlay a graph of the quadratic model. Does the function appear to be a reasonable model for these data?

7. The book-drop model you found is a relationship between D, the number of feet between the book and the motion detector, and T, the number of seconds since the device began recording. (Remember that the motion detector was turned on before the book started falling.) A more useful relationship, however, would be the one between h, the height of the book above the floor, and t, the time since the book was released. Your goal in this part will be to write a formula for $h(t)$; we'll guide you now in making the change of variables.

   (a) The top of the motion detector was $1\frac{1}{8}$ inches above the floor. Write the relationship between h and D. (Make the units consistent!)
   (b) Use your answer to part (a) to write h as a function of T.
   (c) Graph $h(T)$. The graph should be a parabola. Approximate to two decimal places the coordinates of the vertex. Interpret the meaning of each coordinate in the context of this project.

---

[8]Because the coefficient of the T^2 term represents the effect of gravity on the book, you might expect to obtain the same number you found for Galileo's model. This number will be different, though, because the surface area of the book produced significant air resistance.

    (d) Your answer to part (c) will enable you to determine the relationship between T, the time since the motion detector began recording data, and t, the time since the book was released. Write that relationship.

    (e) Finally, write a model $h(t)$, expressing the book's height, h, as a function of the time, t, since it was released.

8. The graph of D versus T is related to the graph of h versus t by a sequence of transformations. (Recall the shifts, stretches, and reflections introduced in the "Graph Trek" lab.) Specify precisely what transformations converted the parabola $D(T)$ into the parabola $h(t)$.

9. **(a)** Use your model from Question 7(e) to determine how long it would take for the book to hit the floor had the motion detector not been in the way.

    (b) Estimate the book's impact velocity, its instantaneous velocity when it hits the floor.

5.5 A HIGHER POWER—Repeated Zeros

When you investigated polynomial functions in Section 5.2, you might have noticed that, in order for a cubic polynomial to have exactly two zeros, the graph needed simply to *touch* the horizontal axis at a turning point. You created a **double zero** or a zero of **multiplicity two** at that point. If you were to place a microscope over the spot, the bit of graph that you'd see would resemble a ∪ or a ∩. If a function with a double zero is expressed in factored form, the factor associated with that zero appears twice (that is, raised to the second power). The "Packages" lab has two volume functions each with a double zero at the origin. If you've done that lab, you might remember that the first volume function had x^2 as a factor and the second one had r^2 as a factor.

1. Use your grapher to explore the effect of a double zero on the graph of a polynomial function. Try a whole series of graphs. Here are some suggestions:
    (a) $y = x(x - 1)(x - 2)$
    (b) $y = x(x - 1)^2(x - 2)$
    How does graph (a) differ from graph (b)? In what ways are they the same?

2. Try keeping the factors the same, but square a different factor instead. How does the graph change?

3. Now square two or more of the factors at once, and describe the resulting changes in the graph.

4. Why stop with squares? Investigate the effect of a **triple zero** on the graph of a polynomial function. You will need to raise one factor to the third power. It's difficult, sometimes, to get a good picture of this from a grapher, but a triple zero, under a microscope, resembles the "wiggle" in the graph of x^3 or $-x^3$. You can get some interesting pictures by starting with a simple polynomial such as $y = x(x - 1)(x - 2)$ and raising one factor to successively higher powers. Put a few graphs on the screen at once so that you can compare them, but keep the window narrow so that the local features aren't overwhelmed by the global stuff.

5. What do you think a zero of multiplicity four would do to the graph? Find out. (You might start by comparing x^2 to x^4. Which of these is "fatter" at the bottom?)

6. How about multiplicity five?

7. Write down any conclusions you've drawn from your investigations.

5.6 MAKING A DIFFERENCE

In this project, you'll discover an intriguing numerical property of every polynomial function. There's a lot of computation to do; make this an opportunity to work with a group.

Linear Functions

1. Choose any linear function f and write its algebraic formula. Then complete the middle column of this table by filling in function values.

x	$f(x)$	Differences
-3		
-2		
-1		
0		
1		
2		
3		

2. Now find the difference between each pair of successive function values, and complete the third column of the table. You should always obtain m, the slope of your chosen linear function. That's no surprise! Those differences are actually average rates of change of f between successive integer values of x, and we know that linear functions have a constant rate of change.

Quadratic Functions

3. To keep the calculations simple, start with the basic quadratic, $f(x) = x^2$. Fill in the second and third columns of the table, noticing that we always subtract a function value from the one that follows it: $f(n+1) - f(n)$. The order of subtraction is important.

x	x^2	First Differences	Second Differences
-3	9		
		-5	
-2	4		2
		-3	
-1	1		
0			
1			
2			
3			

4. We don't expect the first differences to be equal, because x^2 isn't linear. But now calculate the second differences; you should obtain the same result, 2, each time! (Remember the order of subtraction: $-3 - (-5) = 2$.)

5. Choose another quadratic function $g(x) = ax^2 + bx + c$, letting a be something other than 1, perhaps even a negative number. Calculate function values, first

differences, and second differences to complete the following table. Again, the second differences should be equal, though this time they will not equal 2.

x	$g(x)$	First Differences	Second Differences
−3			
−2			
−1			
0			
1			
2			
3			

6. Do you suppose that we could have predicted the second differences for this function? Look at the second-difference value and compare it to the value you used for the coefficient a. How are the two numbers related? Look back at x^2: What was a in that function? Was the second difference for x^2 related to the coefficient of x^2 in the same way that we see in this example?

Higher-Degree Polynomials

7. Will the pattern continue? Check it out for a cubic polynomial such as $2x^3$. First make a prediction. Then complete the table and see whether your theory holds up.

x	$2x^3$	First Differences	Second Differences	Third Differences
−3				
−2				
−1				
0				
1				
2				
3				

8. Do you see any relationship between the value of the third difference and the value of the coefficient of x^3? (This will be a new relationship, not the same one you saw with the quadratic functions.)

9. At this point, you might feel ready to venture a theory. For a polynomial function of degree n, how many successive differences must we take before arriving at differences that are constant? Test your theory on any polynomial p of degree 4.

x	$p(x)$	First Differences	Second Differences	Third Differences	Fourth Differences
−3					
−2					
−1					
0					
1					
2					
3					

10. Test it again with a polynomial q of degree 5.

x	$q(x)$	First Differences	Second Differences	Third Differences	Fourth Differences	Fifth Differences
−4						
−3						
−2						
−1						
0						
1						
2						
3						
4						

11. You realize, of course, that a bunch of examples does not constitute a mathematical proof. Nevertheless, what you have seen is true: for a polynomial of degree n, the nth differences are constant. What's more, you can predict the value of that nth difference from the polynomial itself. Here's how: Compute $n!$. (The symbol $n!$ is pronounced "n factorial," and it means $n \cdot (n-1) \cdot (n-2) \cdots 2 \cdot 1$, so that 4! is $4 \cdot 3 \cdot 2 \cdot 1$, or 24.) Multiply $n!$ by the coefficient of x^n in the original polynomial, and you should obtain the value of the nth difference. Check it out for the examples you already did in this investigation.

5.7 TAKING THE PLUNGE—Vertical Asymptotes and Holes

> *Il est donc possible que les plus grands corps ... de l'univers,*
> *soient ... invisibles.   (Pierre Simon, Marquis de Laplace)*
> *It is therefore possible that the largest bodies ... of the universe,*
> *are ... invisible.   (Authors' translation)*

In this exploration, you will investigate the behavior of the graph of a rational function in the vicinity of a number that isn't in the domain of the function, that is, a zero of the denominator. You might be thinking that a variable in the denominator will always produce a vertical asymptote. Don't get too comfortable with that idea, because there are exceptions, as we are about to see.

1. Using a grapher, view the graphs of several functions having the form

$$f(x) = \frac{x^2 - k}{x - 2}$$

   (You choose values for k.) Experiment with the viewing window until you see two distinct branches for each graph (or for all but one). Your best viewing window will probably be one that's only about six or eight units wide. (Reminder: Enclose the numerator and the denominator in parentheses.)

2. Notice that 2 is not in the domain of any of these functions. Arrange the viewing window so that $x = 2$ is at the center and observe the dramatic behavior of the curves as they approach $x = 2$. You should see that they become nearly vertical. Why? What is happening to the denominator of each function as the value of x approaches 2? Explain how the size of the denominator affects the value of the functions.

   The line $x = 2$ is called a **vertical asymptote** for those functions. The closer x gets to 2, the more their graphs actually resemble that vertical line.

3. You might have noticed that what was just stated is not entirely true! Was this one of your functions? If not, graph it now.

$$g(x) = \frac{x^2 - 4}{x - 2}$$

   What's going on here? Why does the graph of $g(x)$ look so different from all the others? Examine the formulas, and explain the algebraic difference between $g(x)$ and the other functions you graphed.

4. If you look closely at the screen, you might be able to find a hole in the graph of $g(x)$ at $x = 2$. Even if you don't see any hole, you know that the function isn't defined for $x = 2$, and so, despite what the grapher seems to indicate, the graph actually stops on one side of 2 and starts up again on the other side. Except for the hole, to what simpler function is $g(x)$ equivalent?

5. Sketch a graph of $g(x)$. How will you indicate that the number 2 is not in its domain? Label the coordinates of the hole.

6. Write the formula for another rational function that has a hole. Sketch its graph, labeling the coordinates of the hole.

7. Sketch the graph of $f(x) = \frac{x+2}{x^2+x-2}$, showing clearly any vertical asymptotes or holes. (*Hint*: This function has one of each; factor the denominator and examine the factors carefully.)

8. Find the formula for a function that could represent this graph.

5.8   FROM HERE TO ETERNITY—Long-Term Behavior of Rational Functions

If you did the "Doormats" lab, you studied the graphs of three rational functions. You saw that for very large values of the independent variable, the graph of each rational function resembled the graph of a simpler function—a line or a parabola. In this assignment, you will investigate several other rational functions and learn a method for predicting their nonvertical asymptotes—that is, the simpler functions they eventually resemble.

A **rational function** is defined as the quotient of two polynomial functions, where the polynomial functions can be as simple or as complicated as we wish. In the "Doormats" lab, the denominator for each of the rational functions was the very uncomplicated polynomial x, and it was relatively easy to figure out what those graphs were going to look like in the long run.

1. Define a function that has the form $\frac{ax+b}{cx+d}$, that is, a linear expression divided by another linear expression. (You choose values for a, b, c, and d.) What sort of asymptotic behavior might be expected? Investigate the graph of your function in a very wide viewing window. Does it level off? Does it resemble a curve? If you've typed the function correctly, you should see the graph leveling off at both sides of the screen and beginning to resemble a horizontal line. Use the graph to write an (approximate) equation for the nonvertical asymptote, which in this case will be a horizontal line.

Your Function	Its Asymptote

2. Now get some help from algebra: Using long division, divide the denominator into the numerator. You should get a *number* plus a remainder. In other words, you've shown that

$$\frac{ax + b}{cx + d} = n + \frac{m}{cx + d}$$

where m and n are other constants.

When x is very large, what can you say about the quantity $\frac{m}{cx+d}$? If you weren't sure about the asymptote before, you should be able to read its equation from the calculations you've just done.

3. Choose two other functions that have the same form (linear over linear). Predict their nonvertical asymptotes. Use a grapher to check. Are all of those asymptotes horizontal?

Your Functions	Their Nonvertical Asymptotes
(a)	
(b)	

4. Try a function that's a quotient of two quadratic expressions, such as

$$\frac{2x^2 + 3x - 1}{3x^2 + 2}$$

Could you have predicted its asymptote, simply from inspecting the formula?

5. Experiment with other forms of rational functions. Try one that has a linear expression on the top and a quadratic on the bottom. Do you have any expectations about the nonvertical asymptote?

Try dividing the bottom into the top. You're stuck, aren't you? Why? This has implications about the nonvertical asymptote. Graph the function and describe how it looks for large values of x. What does this tell you about the asymptote?

Your Function	Its Nonvertical Asymptote

6. Look for a pattern by trying a couple more functions in which the bottom has a higher degree than the top. They all have the same horizontal asymptote. What is it?

Your Functions	Their Nonvertical Asymptotes
(a)	
(b)	

7. Now switch tops and bottoms so that the top has a higher degree than the bottom. Try to figure out what's happening, both algebraically and graphically. For the graph, you need only look—in a very wide viewing window. For the algebra, you need to perform long division and think about what portion of the result is negligible when x becomes very large. Be sure that you're doing some examples that are other than a quadratic over a linear; you should include some expressions of higher degrees, and you should include some functions in which there's at least a two-degree difference between the top and the bottom. (Aren't you glad you don't have to draw all these graphs by hand?)

Your Functions	Their Nonvertical Asymptotes
(a)	
(b)	
(c)	

8. After enough exploration, you should be ready to draw some conclusions and write a summary of the long-term behavior of rational functions. There are really only three separate cases: (a) The degree of the numerator and the degree of the denominator are the same; (b) the degree of the numerator is less than the degree of the denominator; (c) the degree of the numerator is greater than the degree of the denominator. Tell what you learned about each case.

9. *Globally, the graph of a rational function resembles that of a polynomial.* (Recall that constant functions and linear functions are special cases of polynomials.) Tell how your work in this assignment would support that statement.

5.9 LET'S BAN WASTE!

In recent years, people concerned about the environment have campaigned to encourage moderation and conservation in our use of resources. The U.S. Public Interest Research Group (PIRG) presents Wastemaker Awards to manufacturers that it claims use unnecessary amounts of packaging for some of their products. Health and beauty aids appear to be among the worst offenders, winning all seven of the 1991 awards.

One of these dubious honors went to Bristol-Myers Products for its Ban Roll-On deodorant. One-and-a-half fluid ounces (44 cubic centimeters) of liquid come packaged in a plastic container, mainly cylindrical in shape but pinched in the middle. The cylinder is 5.1 cm high and has a diameter of 3.8 cm. It is topped by a plastic cap 4.1 cm high, having the same diameter. All of this rests in an outer container of cardboard, a rectangular box whose dimensions are 13.3 cm by 7.0 cm by 4.1 cm. There is, in addition, some inner cushioning, also made of cardboard.

1. Calculate the approximate surface area of the packaging materials used for a 1.5 ounce container of Ban Roll-On. How much is cardboard? (You may ignore the inner cushioning as well as any overlapping of flaps that would be found in a box.) How much is plastic? (Again, you may ignore the region where the cap overlaps the main container.) The plastic would be tricky, because the shape isn't a perfect cylinder, so you'll have to pretend that it is one. Do you think your answer is an overestimate or an underestimate of the actual amount of plastic used? Explain.

2. Now you will figure out just how efficient Bristol-Myers might have been with its plastic container. Suppose you wanted to package 44 cubic centimeters of liquid in a cylindrical container. (We're going to ignore details such as providing a way to open the package.) Write an equation expressing the surface area S as a function of r and h. Remember to include the top and the bottom.

3. Express the volume V as a function of r and h.

4. Since you have determined that V is to be 44 cubic centimeters, use 44 for V. Now you can solve for one of the variables in terms of the other. Substitute that expression into your surface-area function. The surface-area function should now be a function of a single variable.

5. Your goal is to minimize the surface area, and thus the amount of plastic. Use your grapher to graph the function. Choose a sensible interval for the independent variable and look for a minimum in that interval. (You should see a curve that dips down and then rises. If your graph doesn't have the right shape, check to see whether you've enclosed the denominator in parentheses.) Zoom in on the turning point (the local minimum) until you can report its coordinates accurate to three decimal places. Use the coordinates to determine the best value of the independent variable. Make a sketch of the relevant portion of the graph, labeling the coordinates of the local minimum. Put scales on the axes and label them.

6. Give the dimensions (r and h) of the most efficient cylinder and tell how much plastic it uses (surface area)

7. What is the ratio of h to r?

8. Note your answer to Question 7. That ratio is not any old random result but an example of the elegant patterns to be found in mathematics. In fact, for a cylindrical container of *any* size, we can minimize its surface area by making the height twice as large as the radius—or, to put it another way, by making the height equal to the diameter. If you were to hold such a container upright in front of a light, what would be the shape of the shadow it would cast on an opposite wall? Does this remind you of Jack's field from Section 5.1?

9. Conserving materials doesn't appear to be one of the major concerns of this manufacturer. In designing the package, what might be some other considerations? Do you think your optimal cylinder would serve its purpose well? Do you think the cardboard is necessary or important? Give reasons.

5.10 DRUG TESTING

How reliable is a drug test?

The concern over increasing drug use among teens has prompted some schools to consider mandatory drug testing of its students. In addition, worried parents now can administer relatively inexpensive tests to their own children. But caution is in order. Drug tests, while generally reliable, are not perfect. Sometimes a person who does use drugs gets a negative result (a **false negative**). Sometimes a person who does not use drugs tests positive (a **false positive**). Companies that produce tests for drugs usually provide information on two characteristics of their test: p, the probability that their test correctly identifies a drug user, and r, the probability that it correctly reports the absence of drugs. The numbers p and r are decimal values between 0 and 1, with 1 representing a probability of 100%.

A large high school in your community is considering a mandatory drug testing program. The local school board has asked your mathematics instructor to quantify the benefits and the dangers of instituting such a program. The board is particularly concerned that the percentage of positive tests from nonusers might be too high. Applying principles of conditional probability, your instructor derived this relationship between E, the percentage of positive test from users, and U, the percentage of drug users in the group being tested:

$$E = \frac{100\,pU}{(p + r - 1)U + 100(1 - r)}$$

You, the members of her precalculus class, are to investigate the implications of this function and prepare a summary for the school board.

1. The school is considering a test that correctly identifies 95% of the users ($p = 0.95$) and correctly reports the absence of drugs in 90% of the nonusers ($r = 0.90$). Write the function describing the relationship between E and U for this drug test.

2. Graph that function, using a window that shows the key features of the *abstract* function. Sketch the graph, identifying the horizontal and the vertical asymptotes.

3. Highlight the portion of the graph that represents the drug-test model, recalling that E and U are *percentages*. Does either of the asymptotes you identified have meaning in the context of drug testing? Explain.

4. Suppose that 25% of the students are drug users ($U = 25$). What percentage of the positive tests will be from users?

5. Suppose instead that only 2% of the students use drugs ($U = 2$). What percentage of the positive tests will be from users? What does this imply about the rest of the positive tests?

6. The answers to Questions 4 and 5 show that there would be a high percentage of false positives—positive test results for students who do not use drugs. The committee, therefore, is considering a more expensive but more reliable test whose characteristics are $p = 0.99$ and $r = 0.95$. Analyze the more expensive test if 25% of the students are users and again if 2% are users.

7. Compare and contrast the results, and summarize your findings in a presentation you could make to the school board.

8. The function your instructor derived applies equally well to medical tests, such as the one for the HIV virus that causes AIDS. Tracking people infected with the HIV virus might be helpful in controlling the spread of the disease. One particular test correctly detects the presence of HIV antibodies 99.7% of the time and correctly

reports the absence of HIV antibodies 98.5% of the time. Do you think the government should institute a program of mandatory testing for HIV? Explain why or why not, and support your answer mathematically.

To use the function, you will have to estimate U, the percentage of the population that is HIV-positive. Be sure to state your assumptions.

You might have found some of the results in this project to be startling or disturbing. The function E shows that, even with a highly reliable test, the percentage of true positives will be relatively low and the percentage of false positives alarmingly high whenever the condition you're testing for is one that affects only a small percentage of the population. For this reason, among many, the issue of mandatory drug testing is controversial.

5.11 JUST ALGEBRA

1. Use the quadratic formula to determine exact solutions. Then use your calculator to find decimal approximations.
   (a) $2x^2 + 5x - 6 = 0$
   (b) $5w + 5 = 4w^2 + 2$
   (c) $3z^2 + 17 = 6z$

2. Solve the following equations by factoring.
   (a) $2p^3 + 4p^2 - 6p = 0$
   (b) $3x^5 - 4x^4 = 16 - 12x$
   (c) $t^4 - 3t^2 - 4 = 0$

3. Factor the numerator and denominator of each of the following rational functions. Use the factored form to determine the domain of the function and its x-intercepts.
   (a) $f(x) = \dfrac{x^2 - x - 2}{x^2 + x - 42}$

   (b) $g(x) = \dfrac{2x^3 + 7x^2 - 4x}{4x^2 - 4x + 1}$

   (c) $h(x) = \dfrac{x^3 - 3x^2 + x - 3}{x^4 - 10x^2 + 9}$

4. Below are four rational functions. Using long division, divide the denominator into the numerator. You should get a quotient $q(x)$ and a remainder $r(x)$. Then express the rational function in the form $q(x) + \frac{r(x)}{d(x)}$, where $q(x)$, $r(x)$, and $d(x)$ are all polynomials and the degree of $r(x)$ is less than the degree of $d(x)$. Here is an example:

$$\frac{x^3 - 2x^2 + x - 2}{x^2 + x - 2} = x - 3 + \frac{6x - 8}{x^2 + x - 2}$$

   (a) $\dfrac{2x^3 - 4x^2 - 10x + 12}{x + 1}$

   (b) $\dfrac{2x^3 - 4x^2 - 10x + 12}{x^2 + 1}$

   (c) $\dfrac{2x^3 - 4x^2 - 10x + 12}{x^3 + 1}$

As the degree of the denominator increases in parts (a), (b), and (c), what happens to the degree of the quotient?

   (d) $\dfrac{5x^4 + 1}{x^2 - x + 1}$

Signed, sealed, and delivered

Preparation

Before meeting with your group, read through the lab sheets to get an understanding of the problem as a whole. Do the volume calculations for the test-case packages (two rectangular boxes and one cylinder). Work on the volume formulas and try to come up with functions $V(X)$ for the rectangular box and $V(r)$ for the cylinder.

At the beginning of the lab period, you should compare your numbers and formulas with those of your partners and iron out any discrepancies.

$$V(X) = \underline{\hspace{6cm}}$$

$$V(r) = \underline{\hspace{6cm}}$$

The following note appeared in the June 27, 1991 *Postal Bulletin*, a publication of the U.S. Postal Service—one more example of the primal urge to standardize!

LENGTH OF UNIFORM GARMENTS/CARRIER AND MOTOR VEHICLE SOCKS AND HOSE

The following clarifications respond to field inquiries about certain provisions of the Uniform and Work Clothes Program. A future revision of the *Employee and Labor Relations Manual* will include these revisions.

Length of Shorts, Culottes, Skirts, Jumpers

Garments should not be more than 3 inches above mid-knee. Employees should not alter the length of their garments, and vendors are not authorized to make alterations that have hems falling more than 3 inches above mid-knee.

However, since everyone is not the perfect ratio of height to girth, some alterations may be necessary. Common sense must prevail in some situations. For example: If an individual stands 6 foot, 4 inches tall with a 34-inch waist, it is likely that the hem of his/her garment will fall more than 3 inches above mid-knee even when unaltered. If left as originally manufactured, the garment should still have a reasonable appearance.

Carrier/Motor Vehicle Socks and Hose

Currently, only black knee-length socks are authorized with walking shorts. However, specific requirements for skirts and culottes do not exist. Although these apparel items are an option for female employees, the National Uniform Committee intended that employees wear either the black knee-length hose, neutral-colored hose, or a coordinated, colored sock. Bright, fluorescent hose and socks are not permitted.

New socks have been designed to complement the carrier and motor vehicle uniforms and will be available for purchase soon. The *Postal Bulletin* will announce their availability.

—Labor Relations Dept., 6-27-91

The Packages Lab

The U.S. Postal Service will accept as fourth-class domestic parcels objects of a variety of shapes and sizes, provided that the weight does not exceed 70 pounds and that the length added to the girth of the parcel does not exceed 108 inches. The length is defined to be the measurement of whatever is the longest side of the parcel. The girth is the distance around the parcel at its thickest remaining part. When in doubt, postal employees refer to pictures such as the following in their manual.

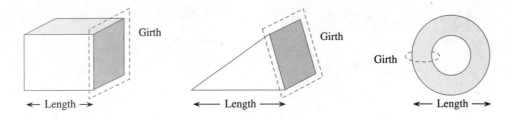

Let us suppose that we have a substance to mail, that it can be packaged in any shape whatsoever, and that we would like to be able to ship as much of it as possible in a single package. (We will assume that the substance is sufficiently lightweight that we don't have to worry about the 70-pound restriction.)

The Rectangular Box

First, we will consider a rectangular box, and we'll suppose that the cross section of the box is square. For a given perimeter, a square yields more area than any other rectangle. It seems reasonable, therefore, that we would want to make the cross section of this box a square.

Just to be sure, though, let's check out this assumption by computing the volume for a couple of test cases. First, consider a rectangular box whose length is 44 inches, and let the cross section be a 12-inch $\times$ 20-inch rectangle. Verify that length plus girth equals 108 inches. What is the volume? Now, keeping the length at 44, make the cross section a 16-inch square. The volume should be greater. This, of course, doesn't prove that a square is best, but it does at least support our intuition that a "flattened" box wouldn't hold as much as a "boxy" one. (If you laid a sealed half-gallon carton of milk on its side and jumped on it, you would have a mess, because the flattened carton can't hold as much, even though its girth hasn't changed.)

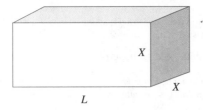

Now let both L and X vary, and write a formula for the volume of the box in terms of L and X. Then use the fact that the length plus the girth equals 108 inches (assume that we'll use up the allowable 108 inches) to express the relationship between L and X.

To optimize V (that is, to find the greatest volume), we need to express V as a function of X alone. Use the relationship that you just wrote to assist you in writing a formula for V that has no variable other than X. As soon as your group agrees on a formula for $V(X)$, you're ready to roll.

What type of function is $V(X)$? Use your grapher to draw its graph. Be sure to choose a viewing window large enough to see its important features: any X-intercepts, any turning points, and how it behaves beyond the intercepts. You might need to experiment with viewing windows and changes of scale before you get a meaningful picture. Is there a maximum or minimum value for $V(X)$?

You should observe that there is no maximum value for $V(X)$ because the function increases without bound as X moves away from the origin to the left. Similarly, there's no minimum value for $V(X)$. Why not? However, the graph has two turning points, one a **local minimum** and the other a **local maximum.**

Now consider only the portion of the graph that makes sense for a package. What is the smallest value of X that you would consider for the model? Remember that X represents a dimension of the package. Some values, even positive ones, are too small to be reasonable. How large an X-value would you consider? Give reasons for your answers.

Look at the portion of the graph of $V(X)$ on the X-interval you just chose. You should see a local maximum. With your grapher, locate the highest point. What is the value of X at that point? (Zoom in enough to obtain X correct to three decimal places. It should be a "nice" number.) What is V? These numbers give you not only the volume of the package, but also its shape. How long should it be? What is its girth, and how did you find it?

We need to distinguish between the shape of the box and the shape of the function representing the volume of the box. The curve $V(X)$ on your screen certainly does not suggest the shape of the mailing container, yet it expresses information about the volume of the container. When X is a small positive number, what can you say about the volume of the box? As X increases, what happens to the volume? As X continues to increase beyond its optimal value, what happens to the volume? To visualize how the capacity and the dimensions of the box vary with the value of X, think of an object that you might ship in a box for which $x = 3$ and in one for which $X = 21$. These objects will not be at all alike!

The Cylinder

In two dimensions, a circle is the shape that gets the most area out of its perimeter. (If you have a fixed length of string and want to enclose the greatest area, you should make a circle rather than a square or some other shape.) Therefore, it seems reasonable that a cylindrical container satisfying postal regulations might hold more than a rectangular box.

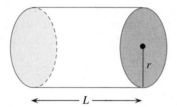

As a test, try $L = 44$ inches and see what the volume would be. (Notice that here, the girth is the circumference, so you would have 64 inches for the circumference. Also, to find the volume, you need to know the radius. How can you find the radius if you know the circumference?) Compare the volume with the volumes you already calculated for the two rectangular boxes of the same length. It should be larger.

Let's try to find the best cylindrical container. As before, you will need to write a formula for the volume and then rewrite it in terms of a single variable. First write V in terms of L and r. Then use the fact that length plus girth will equal 108 inches to rewrite V as a function of r alone. (You'll need to write the girth in terms of r.)

What type of function is $V(r)$? Does it have a maximum or minimum value? Consider only the portion of the graph that makes sense in this context: On what interval can r represent the radius of the cylinder? Look at the graph on that interval; you should see a local maximum. Do some serious zooming to find the value of r that produces that maximum. (This value will not be a "nice" number, as the optimal X-value was, and you'll need three-decimal-place accuracy to make good comparisons.) Explain what the coordinates mean in terms of the package. How long should the package be? What will be its girth? Its radius?

Compare the dimensions (length and girth) of the largest rectangular box to the dimensions of the largest cylinder. Does the result surprise you, or is this what you expected?

Compare the volumes of the optimal cylindrical parcel and the optimal rectangular box. Is a circular cross section a more efficient use of girth than a square cross section? All other things being equal, which shape would you choose? Why?

The Sphere

Because the circle is such an efficient shape, perhaps it would be advantageous to make the package round in every direction. The Postal Service permits spherical packages if they meet the weight and dimension restrictions. (Wrapping the package is, of course, a separate problem!) Perform the same analysis for a sphere and determine the maximum volume you would be permitted to mail in a single package. What makes this problem fundamentally different from, and simpler than, either the box or the cylinder?

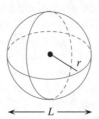

The Lab Report

What is the best shape for a package? Your report should show how you used polynomial models of container volumes to help answer this question. Distinguish between the entire polynomial function and the portion that represents the model. Tell how you used the graph to determine the dimensions that yield the maximum volume for each shape. Give the maximum volume and the best dimensions (length and girth) for each case, noting the ideal proportions (girth-to-length ratio). The sphere presents a situation that is different mathematically from either the box or the cylinder. Explain what's unique about the sphere. Discuss the relative merits of different shapes under the particular restrictions of the Postal Service, and explain in everyday terms why you think one shape yields more volume than either of the other two. Illustrate your report with graphs of $V(X)$ and $V(r)$.

"I myself have never been able to find out precisely what feminism is: I only know that people call me a feminist whenever I express sentiments that differentiate me from a doormat." (Rebecca West, 1913)

Preparation

Prepare for this project by reading the lab pages and writing formulas for the three average-cost functions F, G, and H. Bring the formulas with you to the lab, and begin by comparing your functions with those of your partners.

$F(x) = $ _____

$G(x) = $ _____

$H(x) = $ _____

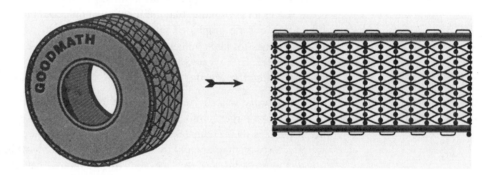

The Doormats Lab

A western Massachusetts couple runs a small business from their home. They recycle discarded automobile tires by slicing them up into strips and making doormats, somewhere between 500 and 2000 doormats each year. Their somewhat imprecise method of cutting the tires results in a product of irregular thickness, perfect for scraping off mud and snow from shoes. Worn-out tires are a disposal problem nationwide, so they have no difficulty acquiring their raw materials. Additional materials are few and inexpensive—some colored plastic beads for spacing and some thick aluminum wire for holding everything together.

The couple has figured out that, counting the cost of the raw materials and the value of their labor, it costs them $5 to produce each doormat. In addition, they estimate that their business has fixed annual expenses of $200. (This is a very low-tech operation.) In the language of economics, the capital cost is $200, and the variable cost is $5 per unit. They want to determine the average cost of each mat they make this year.

Keeping It Simple

First write a formula for the total cost of producing x doormats this year. Now write a function that expresses the average cost of a doormat. Using a grapher, explore the behavior of this function, which we'll call F. Is the average cost a constant? If so, what constant? If not, why not?

What's the average cost if they produce 200 doormats this year? 500? What happens to the average cost if they produce very few?

As the number of doormats produced increases, what happens to the average cost? (Economists call this effect "spreading the overhead.") Is it reasonable to suppose that as more and more doormats are made, the average cost would continue to drop? If you study the graph for large values of x, you should see that the curve appears to level off. Is there a lowest possible average cost? Verify that although the values of the function get smaller and smaller, there's a limit below which the average cost cannot go. From the graph, determine this limit. Does the function ever attain that value? In other words, if we call the limit L, can you find a value of x such that $F(x) = L$? Give a reason for your answer.

Enlarge the x-interval of the viewing window so that it's at least $10,000$ units wide. You may include negative values of x if you wish. The graph of $F(x)$, except near the y-axis, should resemble a horizontal line. Give the equation of that horizontal line. Because the graph of $F(x)$, in the long run, resembles the line, we call that line the **horizontal asymptote** for the function F.

The function F as you've written it probably has the form $F(x) = \frac{a+bx}{x}$. Rewrite it now (by dividing) in the form $F(x) = \frac{a}{x} + b$.

If the magnitude of x is very large, what can you say about the value of the term $\frac{a}{x}$? Explain how your answer to that question shows that for large values of x, the graph of $F(x)$ resembles the line $y = b$. We say that the graph of $F(x)$ has the horizontal line $y = b$ as an asymptote, that the function approaches the value b asymptotically.

View the graphs of $F(x)$ and $y = b$ together so that you can appreciate their similarities for large values of x.

The More, The Merrier

The doormat makers have discovered that they become more efficient economically if they make larger numbers of mats. Perhaps they can get a better price on the beads, or perhaps their enthusiasm propels them to work more quickly. You can express this efficiency mathematically by including a quadratic term in the *total cost* function. (Be sure to use *total cost*, not *average cost*.) Suppose that term has been determined to be $0.0003x^2$.

Rewrite the total cost function, including the quadratic term (you will need to decide whether to add or subtract this term). Now write a new average-cost function; call it G.

Use a grapher to compare the behaviors of $F(x)$ and $G(x)$ over several different intervals, in a variety of viewing windows. Find intervals over which they look nearly the same. Find out approximately where they begin to diverge. At approximately what number of mats does the x^2-term appear to "kick in" and become significant? (This is a matter of opinion; your answer will depend on how you define *significant*.) Look at the formulas for the two functions and explain why the graphs are very close for certain values of x.

If you use sufficiently large values of x, you should observe that the graph of $G(x)$ begins to look like a straight line rather than a curve. This line, though, should appear slanted rather than horizontal. It is called an **oblique asymptote** or **slant asymptote** for the function G. Here's a quick way to find an equation for that line.

As you did with the first model $F(x)$, rewrite the formula for $G(x)$ by dividing, so that it now has the form $\frac{a}{x} + b + cx$. Think about what happens to each term as x grows very large. The expression has three terms, but one of them is so close to zero when x is very large that its impact on the function G is negligible. Compare the results with those you obtained with the first method; they should be similar, though not identical, because the first method gives only an *approximation* to the slant asymptote. Do you see now how you can use the formula for the function to determine its asymptote? This division trick probably took much less time than the first method; use it from now on when you want to determine the long-range behavior of a rational function such as F or G.

What does this have to do with making doormats? Go back to some realistic values for x, and interpret the behavior of the graph in the context of this situation.

Too Much of a Good Thing

The doormat makers have also observed that if they spend all of their energies making doormats, their productivity declines, and therefore the cost to them begins to rise again. You can express this mathematically by including a cubic term in the *total cost* function. Decide whether the term ought to be added or subtracted.

Suppose the cubic term has been determined to be $0.0000001x^3$. Rewrite the total cost function to include this term. (Hang onto the quadratic term as well.) Find a third average-cost function, $H(x)$, that includes the cubic term. Then use a grapher to examine all three functions over an interval that makes sense in context.

The graph of $H(x)$ suggests an optimal number of doormats for the business to produce. According to the graph, if production is kept within a certain interval, the cost per mat is as low as possible. Approximately what is that interval?

Now investigate the long-run behavior of $H(x)$. Enlarge the viewing window so that the dramatic behavior near the y-axis is minimized. You'll probably need an x-interval of $40,000$ units or so. In this large viewing window, what is the apparent shape of the graph of $H(x)$?

Check this out algebraically: Rewrite the formula for $H(x)$ by dividing, as you did in the last two sections. How many terms are there? Which one is of negligible size whenever x is a huge number? Explain why, for enormous values of x, the graph of $H(x)$ looks the way it does.

The function H has an asymptote that is neither horizontal nor oblique; its asymptote is another curve. Give the formula for that curve. (Remember that you can obtain it from the formula for $H(x)$.) Overlay the graph of the curved asymptote, try several different windows, and observe that the graph of $H(x)$, from a distance, resembles the graph of its asymptote.

So what about doormats? Interpret the asymptotic behavior of the function H in the context of this cottage industry.

The Lab Report

You are serving as mathematics consultants to the doormat makers. They've given you the basic information about their business—$200 fixed annual cost, $5 per-mat variable cost—and have told you that they currently manufacture between 500 and 2000 doormats per year.

Your firm has prepared three different average-cost models, one of which they will select as most representative of their operations. Recommending one model over another is not your job; your role is to explain each model objectively. Tell them what each term in the function represents and why the term is positive or negative. Compare and contrast the functions for their ability to model a concrete situation.

Describe the long-term behavior of each of the functions. Include graphs. Show how the algebraic formula for each function can be used to predict its asymptote.

In light of these long-term behaviors, conclude with three separate recommendations for the doormat makers. What might they want to adopt as production goals if they were to use model F? What advice would you give them if they decided that G were a more accurate model for their business? Finally, how many mats would you recommend that they produce if model H appeared to be the best fit for their operation?

Periodic Functions

6

© Yvette Cordozo/Words and Pictures/PictureQuest

All my life's a circle
Sunrise and sundown
Moon rolls through the nighttime
Till daybreak comes around
All my life's a circle
Still I wonder why
Seasons spinning 'round again
Years keep rolling by.

Harry Chapin, "Circles" (1971, 1974 American
Broadcasting Music by ASCAP)

Every January, many newspapers publish a graph similar to the one in Figure 6.1, showing the day-by-day fluctuations of the temperature over the course of the previous year. The graph, despite its jagged peaks and valleys, nonetheless traces an overall pattern of low temperatures during the winter months climbing to high temperatures during the summer.

If, instead of plotting the actual high and low daily temperatures for a particular year, we were to plot the average, over several decades, of the high and low temperatures for each date, the jaggedness would be smoothed out and a graph similar to the one in Figure 6.2 would appear.

Figure 6.1 Daily High and Low Temperatures

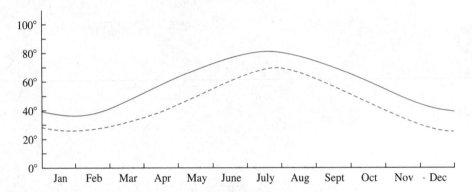

Figure 6.2 Average Daily High and Low Temperatures in New York City

To appreciate the cyclical nature of this temperature function, we extend the horizontal axis for more than a single year—four years, to be precise. Our output variable will still be the average high temperature for the given day, and the result would look like Figure 6.3, a wavelike pattern that repeats itself every 365 days. (Figure 6.3 shows only the high temperatures; a graph of the low temperatures would have a similar shape but would be shifted downward.)

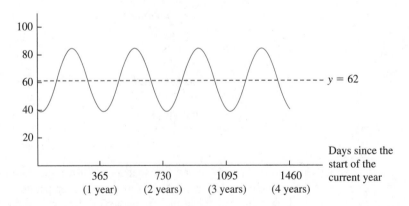

Figure 6.3 Four-Year Pattern of Average Daily High Temperatures (New York City)

Daily high temperatures over an entire year average out to about 62°F. Figure 6.3 shows a dashed horizontal line at that value. The entire temperature curve oscillates about that horizontal line, swinging above it during the warm months and below it during the cold ones.

Striking a new note

Musicians often use a tuning fork to help them tune their instruments to the proper pitch. Tuning forks that will produce any desired note can be manufactured (the larger the fork, the lower the tone), but today's standard fork is designed to produce the A above middle C. When it is struck, the tuning fork vibrates at 440[1] cycles per second, causing the air around it to vibrate as well and transmitting the sound to our ears.

If we were to attach a pen or stylus to the tip of the tuning fork and pull a strip of paper at uniform speed beneath it as it vibrated, we would see a pattern similar to the one in Figure 6.4. The dashed line about which the curve oscillates represents the line that the pen would have traced if the fork had not been struck.

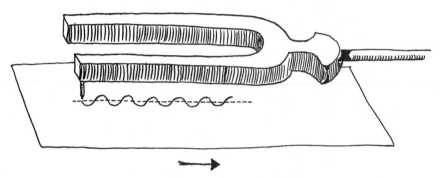

Figure 6.4 A Tuning Fork Vibrating at 440 Cycles per Second

Musical notes, temperature patterns—what's the connection? Although these examples come from unrelated fields, they are linked mathematically by their periodic nature, that is, by the fact that they exhibit a fundamental pattern that is repeated at regular intervals. Moreover, the fundamental pattern of the musical tone has the same shape as that of the daily temperatures. There is a whole class of natural behaviors whose most striking characteristic is their cyclical nature. Tides, phases of the moon, sleep-wake cycles, and cicada invasions, to name just a few examples, follow repeating patterns that are more or less predictable. A course in elementary functions, such as this one, ought to develop mathematical tools for modeling such periodic behavior, so that's exactly what we're about to do in this chapter.

You have already encountered a function whose graph has the shape seen in Figures 6.3 and 6.4: the **sine**[2] function. In this chapter, we will study the sine function as a mathematical model of periodic behavior. We'll begin by studying the way in which the function is defined, to understand why the graph has its particular shape, and then look at ways to modify the sine function and its relatives to represent many real-world phenomena.

> A periodic function repeats its pattern over and over.

> You met the graph of the sine function on page 102.

---

[1]That particular frequency is a relatively recent international compromise. Right up through the nineteenth century, a variety of different pitches were used for A, pitches that varied from country to country and even from one church organ to another. The music of Mozart is sung today at a pitch more than half a tone higher than the composer intended. A soprano singing *The Magic Flute* today has an even greater challenge than she would have had back in Mozart's day.

[2]The word "sine" comes from the Latin word *sinus*, meaning "a bending or turning" or "a curve." Another meaning of *sinus* is "a fold of a toga." You might imagine the hem of a Roman toga rippling in a pattern similar to a sine curve.

6.1 THE SINE AND COSINE FUNCTIONS

Let's go 'round one more time.   (Harry Chapin, *"Circles"*)

The circle in Figure 6.5, with a radius of one unit and center at the origin, can be represented by the equation $p^2 + q^2 = 1$. Note that this is our old familiar coordinate plane but that we've renamed the variables p (on the horizontal axis) and q (on the vertical axis) instead of the usual x and y. Among mathematicians, this figure is affectionately known as the **unit circle,** because its radius has a length of one unit. We do not specify the unit; we might choose to think of it as an inch, or a meter, or a mile—so a particular unit circle may be of any size.

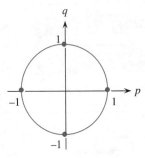

Figure 6.5 The Unit Circle, $p^2 + q^2 = 1$

▷ How many units are there in the circumference of the unit circle?

Points on the Unit Circle

A point on the unit circle can be specified in either of two ways:

- By referring to its coordinates in the rectangular p-q system: p measures position *left or right* of the origin; q measures position *up or down* from the origin.

- By referring to it in terms of distance on the circle itself: measuring *how far around the circumference* the point is located.

In this section, we will show how the two ways are related.

▷ Suppose that a point P with coordinates $(0.6, q)$ is on the unit circle. Use Figure 6.6 to show approximately where P would be located. There are two possible locations for P; show both.

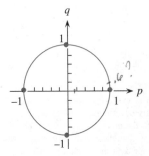

Figure 6.6 Unit Circle, Working Model

▷   Use algebra to find the values of q. (Remember that the equation $p^2 + q^2 = 1$ holds true for every point on the circle.) Be sure that the two values you find agree with the picture.

$$(0.6)^2 + q^2 = 1$$
$$0.36 + q^2 = 1 - .36$$
$$\sqrt{q^2} = .64$$
$$q = \pm .8$$

Arcs on the Unit Circle

Seems like I've been here before
Can't remember when . . .
(Harry Chapin, *"Circles"*)

Imagine yourself walking around the unit circle. You need a place to start, and you need a direction in which to walk. We'll define the unit circle point (1,0) to be your starting point and the counterclockwise direction to be the positive direction, as shown in Figure 6.7.[3] If you walk once around the entire circle, you have traveled 2π units of arc and are back at your starting point.

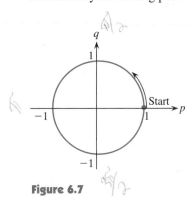

Figure 6.7

▷   Where does the number 2π come from?
half of a circle is π distance from start
⟿ full circle is 2π distance from start

▷   If you walk halfway around the circle, how far have you gone (how many units of arc)? Give the coordinates of the point where you end up.
⟿ π units
⟿ $(-1,0)$

The counterclockwise direction is considered, arbitrarily, to be the positive direction. The *opposite* direction is considered to be the negative direction, so that an arc that extends clockwise from its origin is said to have a negative **directed distance.** (Recall that the *opposite* of a number is indicated in mathematics by a negative sign.)

For example, an arc of $-\frac{3\pi}{2}$ units is one that begins at the point (1,0) and extends in the clockwise direction three quarters of the way around the circle. Notice in Figure 6.8 that it has the same terminal point as an arc of $\frac{\pi}{2}$ units. Two *different* arcs can share the same terminal point.

The positive direction is counterclockwise.

▷   To help yourself understand the relationship between directed arcs and their terminal points, complete Table 6.1.

---

[3]You could, of course, choose to start your trip at any point on the circle, but why buck mathematical tradition?

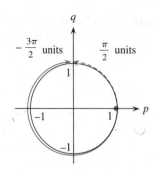

Figure 6.8 Two Arcs That Share a Terminal Point

Directed Arc	-2π	$-\frac{3\pi}{2}$	$-\pi$	$-\frac{\pi}{2}$	0	$\frac{\pi}{2}$	π	$\frac{3\pi}{2}$	2π
Terminal Point	$(0,1)$	$(0,1)$	$(-1,0)$	$(0,-1)$	0	$(0,1)$	$(-1,0)$	$(0,-1)$	$(1,0)$

Table 6.1 Certain Arcs and Their Terminal Points

A Variable to Represent Arc Length

It is important to distinguish between the distance and direction traveled (signed arc length) and the terminal point (two coordinates in the p-q system). To help sort out these separate measurements, we introduce another variable, t, to stand for arc length. Keep in mind that a directed arc (t-value, one of those trips on the unit circle) can have any real-number value—positive, negative, or zero.

Check Your Understanding 6.1

CYU answers begin on page 324.

1. Make a copy of the unit circle and draw on it, using separate colors, labeled arcs with the following t-values:

   (a) $\frac{3\pi}{2}$ (b) $-\pi$ (c) 4π

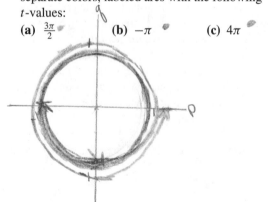

2. Suppose the terminal point of an arc is $(0,-1)$. Give four possible values for t. Make at least one of them negative.

$$\frac{3\pi}{2}, \; -\frac{\pi}{2}, \; \frac{7\pi}{2}, \; -\frac{5\pi}{2}$$

3. Give the unit-circle coordinates (that is, the coordinates in the p-q system) of the terminal point of the arcs with the following t-values:

   (a) 3π (b) $-\frac{5\pi}{2}$ (c) $\frac{5\pi}{2}$

    $(-1,0)$ $(0,-1)$ $(0,1)$

   (d) -13π (e) 2004π

    $(-1,0)$ $(1,0)$

You might wonder why we used only multiples of π as values for t in CYU 6.1. You might even complain, as one of our students did, "Aren't we ever going to use 'regular' numbers?" The sole reason for using π in all these examples is that the t-values we measure represent arcs of a circle, and the circumference of a circle can't be divorced from the number π. We use multiples of π for our own convenience; they help us locate our exact stopping place on the circle. Otherwise, you would not have been able to locate those terminal points precisely with your "bare hands."

Note: $2\pi \approx 2(3.14) = 6.28$

But the number π is simply that: a number. Thus, going completely around the unit circle means covering a distance of just over six units, and going halfway gives a distance of approximately 3.14 units. An arc (a "trip" on the unit circle), however, can be of any directed length—positive, negative, or zero. In other words, *the variable t has as its domain all real numbers.*

Hint: $\frac{\pi}{2} \approx 1.57$

▷   To remind yourself that the t-values really are "regular numbers," use the unit circle in Figure 6.9 to mark the approximate endpoints of arcs that measure 1, 2, 3, 4, 5, and 6 units; that is, $t = 1, t = 2$, and so on. (Just *mark* the endpoints on the circle; don't worry about their p-q coordinates.)

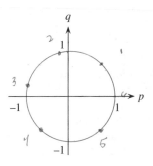

Figure 6.9 Working Model of the Unit Circle

Check Your Understanding 6.2

The diagram shows the conventional numbering for the four quadrants of the Cartesian plane. Notice that we number the quadrants in the order in which we would encounter them while traversing the unit circle in the positive direction, starting at (1,0).

q

II	I

→p

III	IV

1. Here is some information about six different arcs. Determine the *quadrant* in which each arc terminates. (You won't be able to find the exact coordinates of any terminal points, but you can estimate them closely enough to find the quadrant in which they land.)

   ✓(a) $t = 4$ ✓(b) $t = 7$?(c) $t = -3$
   III   I   II III

(d) $2 < t < 3$ (e) $-1 < t < 0$ (f) $5 < t < 6$
   II   IV   IV

2. Use the diagram you created in Figure 6.9 to find a rough approximation for the measurement of an arc that would terminate:
   (a) At the point $(-0.6, 0.8)$

   (b) At the point $(0.6, -0.8)$

3. Now find a second answer to each question in (2); that is, find a *different* arc that would terminate, approximately, at the point $(-0.6, 0.8)$ and one that would terminate, approximately, at the point $(0.6, -0.8)$.

Because we know the radius of the unit circle, we're able to determine, or at least estimate, the length of any arc on it. When we say that an arc has length 3, we mean that its length is equivalent to the total length of three radii placed end to end. We can also say that the arc has a length of 3 **radians,** emphasizing that we're measuring how many times the radius will fit into the arc. This allows us to generalize our arc-length idea to larger circles: thus, an arc of 3 radians (three radii) will extend nearly halfway around any circle, regardless of the circle's size.

Calvin and Hobbes   by Bill Watterson

We have shown that every real number t is associated with a point on the unit circle by letting t represent the length of an arc that begins at the unit-circle point $(1,0)$ and ends at the point with coordinates (p,q). In short, any real number t determines one specific point (p,q).

That last sentence should sound suspiciously like the definition of a function. Now we'll see precisely how it becomes the definition of the sine and the cosine functions.

Defining the Sine and Cosine Functions

Any real value t, used as input, determines two distinct values, the p and q coordinates of the terminal point of the arc of length t, as output. The first coordinate p is called the **cosine** of t; the second coordinate q is called the **sine** of t. The single independent variable t yields two functions, linked to t and to each other by means of the unit circle.

Note: The *first* definition uses the *second* coordinate.

> The sine function, $\sin(t)$, is the second coordinate, q, of the unit-circle point (p,q) associated with the number t.
>
> The cosine function, $\cos(t)$, is the first coordinate, p, of the unit-circle point (p,q) associated with the number t.

The number t represented by the arc in Figure 6.10, for instance, has a cosine value (the p-coordinate) of approximately -0.9 and a sine value (the q-coordinate) of roughly 0.4.

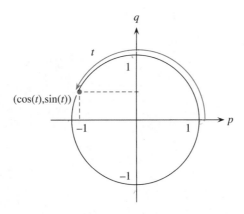

Figure 6.10 Defining the Sine and Cosine Functions

STOP & THINK

The definitions of the sine and cosine functions are more elaborate than the definitions of any of the other functions we have studied. We cannot clarify for an outsider what is meant by "sine" or "cosine" without leading the person through the entire unit-circle explanation. These functions, unlike all the functions that preceded them in this text, aren't defined by simple mathematical operations such as multiplications and exponentiation. Nevertheless, they qualify as functions according to our definition. How would you convince a skeptic that $\sin(t)$ and $\cos(t)$ are functions?

Check Your Understanding 6.3

1. Give the exact value of each of the following:

   (a) $\cos(7\pi)$ **(b)** $\sin\left(\frac{5\pi}{2}\right)$

   (c) $\cos\left(-\frac{2001\pi}{2}\right)$ **(d)** $\cos(4444\pi)$

2. Decide whether each of the following is positive or negative:

   (a) $\sin(7)$ **(b)** $\cos(-7)$ **(c)** $\sin(4)$

3. If $\cos(t) = 1$, find three possible values for t. Now give *all* possible values. (Either write a mathematical expression or state the rule for obtaining them.)

4. What is the domain of the functions $\sin(t)$ and $\cos(t)$? What is their range? Give reasons.

© Anne Kaufman

A sine of the times?

In a Nutshell   The sine and cosine functions are defined as the coordinates of points on the unit circle $p^2 + q^2 = 1$. Every real number t corresponds to a unit-circle point (p,q). The p-coordinate is the cosine of t; the q-coordinate is the sine of t. The sine and cosine functions have all real numbers as their domain and the interval $[-1, 1]$ as their range.

6.2   CIRCULAR FUNCTIONS AND THEIR GRAPHS

Sine on the dotted line

In this section, we will see how the unit circle generates the sine and cosine waves.

The Graph of sin(t)

> *No straight lines make up my life*
> *All my roads have bends*
> *No clear-cut beginnings*
> *So far, no dead ends.*
> (Harry Chapin, *"Circles"*)

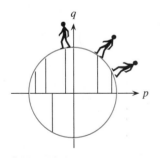

Think of vertical displacement as *height*. A negative vertical displacement puts you below the p-axis.

Imagine yourself walking around the unit circle again, starting at $(1,0)$. The t-value will represent the length of your trip so far, and sin(t) will be the q-coordinate of your position as you move. The q-coordinate measures your *vertical displacement* from the horizontal axis. (If q is negative, what does that say about your location?)

▷   Complete the description:

• As the t-value increases from 0 to $\frac{\pi}{2}$, $\sin(t)$ increases from 0 to 1 (rapidly at first but slowing down).

• As the t-value increases from $\frac{\pi}{2}$ to π, $\sin(t)$ decreases from __1__ to __0__ (slowly at first but speeding up).

• As the t-value increases from π to $\frac{3\pi}{2}$, $\sin(t)$ decreases from __0__ to __-1__ (rapidly at first but slowing down).

• As the t-value increases from $\frac{3\pi}{2}$ to 2π, $\sin(t)$ ....   −1 to 0

Notice that *you* are back to your starting point, and your vertical displacement is back to zero, but that you have covered a distance of 2π radians. The value of t is 2π, not zero.

▷   Suppose the trip continues around the circle in the same direction. What happens, then, to the value of $\sin(t)$, as measured by the changing q-coordinate?

it follows the same pattern

▷   Suppose the trip had started out in the clockwise (negative) direction:

• Describe what happens to $\sin(t)$ as the t-value decreases from 0 to $-\frac{\pi}{2}$.

decreases from 0 to −1

• Describe what happens to $\sin(t)$ as the t-value continues to decrease, from $-\frac{\pi}{2}$ to $-\pi$, then from $-\pi$ to $-\frac{3\pi}{2}$, and finally from $-\frac{3\pi}{2}$ to -2π.

→ decreases from −1 to −1
→ increases from −1 to 1
→ decreases from 1 to 0

The verbal description you just completed provides instructions for drawing the graph of the function $f(t) = \sin(t)$ shown in Figure 6.11. Observe the changing slopes of the curve and the locations of the t-intercepts.

Domain: $\mathbb{R}$
Range: $-1 \leq \sin(t) \leq 1$
t-intercepts: $n\pi$

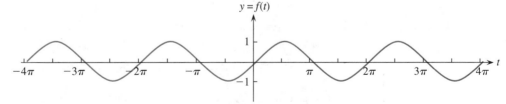

Figure 6.11 Graph of the Function $f(t) = \sin(t)$

▷ Using a contrasting color, emphasize the portion of the sine curve between $t = 0$ and $t = 2\pi$.

Fundamental sine wave

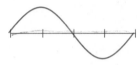

We will refer to the shape of the emphasized portion as a **fundamental sine wave**. A fundamental sine wave consists of one upward loop and one downward loop. Figure 6.11 shows a total of four fundamental waves.

Intercepts   The graph of $\sin(t)$ intersects the horizontal axis at the origin and again at every integer multiple of π in either direction. In short, the intercepts of $\sin(t)$ occur at $t = n\pi$, where n can be any integer (positive, negative, or zero). Shorthand notation such as $t = n\pi$ allows us to specify an infinite number of values.

amplitude $= \frac{max - min}{2}$

Amplitude   The graph of $\sin(t)$ goes up to 1 in the vertical direction and down to -1, covering all the values in between. We say that the function $\sin(t)$ has an **amplitude** of 1, meaning that its never strays from its horizontal axis by more than one unit.

▷ Find a number t for which the function $\sin(t)$ achieves its maximum value.

▷ What is that maximum value? (Remember that "maximum value" always refers to the *output* of the function, not to the independent variable.)

Period   The maximum value of the sine function is 1, occurring (on the graph in Figure 6.11) at $t = -\frac{7\pi}{2}, -\frac{3\pi}{2}, \frac{\pi}{2}$, and $\frac{5\pi}{2}$. Each of those t-values is separated from the next by 2π units, because the function $\sin(t)$ is periodic with a **period** of 2π. To include *all* the values of t for which the function $\sin(t)$ produces its maximum output, we can write $t = \frac{\pi}{2} + 2n\pi$.

period: the width of one fundamental wave

The Graph of cos(t)

To draw the graph of the cosine function, we retrace our mental trip around the unit circle, focusing now on the p-coordinate (*horizontal displacement*) of our position:

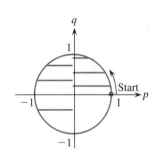

- As the t-value increases from 0 to $\frac{\pi}{2}$, $\cos(t)$ decreases from 1 to 0 (slowly at first but gathering speed).

- As the t-value increases from $\frac{\pi}{2}$ to π, $\cos(t)$ decreases from 0 to -1, (rapidly at first but slowing down).

- As the t-value increases from π to $\frac{3\pi}{2}$, $\cos(t)$ increases from -1 to 0 (slowly at first, then speeding up).

- As the t-value increases from $\frac{3\pi}{2}$ to 2π, $\cos(t)$ increases from 0 to 1 (rapidly at first, then slowly).

- As the t-value continues to increase, the pattern repeats.

The preceding narrative describes the fundamental pattern of a cosine wave and will enable you to draw its graph. If you have any trouble with the portion to the left of the y-axis, just imagine walking around the circle counterclockwise.

▷ On the axes of Figure 6.12, sketch a graph of the function $g(t) = \cos(t)$. Your graph should show the precise location of all the intercepts.

▷ Using a contrasting color, emphasize the portion of the cosine curve between $t = 0$ and $t = 2\pi$.

Domain: $\mathbb{R}$
Range: $-1 \leq \cos(t) \leq 1$
t-intercepts: $\frac{\pi}{2} + n\pi$
Amplitude: 1
Period: 2π

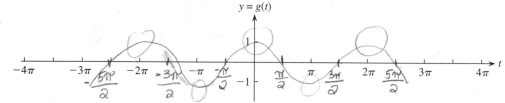

Figure 6.12 Student-Drawn Graph of the Function $g(t) = \cos(t)$

Fundamental cosine wave

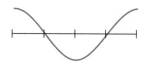

We will refer to the shape of the emphasized portion as a **fundamental cosine wave.** A fundamental cosine wave starts at its maximum value, crosses the horizontal axis one quarter of the way and again three quarters of the way through its pattern, dips to its minimum in the middle, and returns to its maximum at the end. Figure 6.12 should show four fundamental waves.

The graph of $\cos(t)$ intersects the horizontal axis at $\pm\frac{\pi}{2}, \pm\frac{3\pi}{2}, \pm\frac{5\pi}{2}, \ldots$ ("every π on the half-π"). A compact form of writing this is $\frac{\pi}{2} + n\pi$, where once again we are using n to represent any integer, positive, negative, or zero.

Like the sine function, the cosine function is periodic with period 2π.

Check Your Understanding 6.4

1. For what values of t does the graph of $\cos(t)$ have its maximum value? Express your answer in compact form, so that you include all possible t-values.

   ✓

2. What is the minimum value of the function $g(t) = \cos(t)$?

✓

-1

3. For what values of t does the graph of $\cos(t)$ have its minimum value? Give your answer in compact form.

$-\pi n$ $\times$

4. For what values of t does the graph of $\sin(t)$ have its minimum value? Give your answer in compact form.

$-\frac{\pi}{2}n$ $\times$ $\frac{3\pi}{2} + 2\pi n$

Side by side

Mathematicians call a curve exhibiting the wave pattern characteristic of the sine and cosine functions a **sinusoid.** Surely you have noticed that the sine and cosine graphs have exactly the same shape.

▷   Does Figure 6.13 represent a sine graph or a cosine graph? What additional information would you need to answer that question?

Figure 6.13 Graph of a Sinusoid

The connection between the sine function and the cosine function is straightforward. Because $\sin(t)$ is the second coordinate (the q-value) of the unit-circle point and $\cos(t)$ is the first coordinate (the p-value), and because the relationship $p^2 + q^2 = 1$ holds true for every point on the circle, the equation

The fundamental identity:
$\sin^2(t) + \cos^2(t) = 1$

$$[\sin(t)]^2 + [\cos(t)]^2 = 1$$

is valid for every value of t. The equation is usually written in the following form:

$$\sin^2(t) + \cos^2(t) = 1$$

The expressions $\cos^2(t)$ and $(\cos(t))^2$ mean the same thing to a mathematician. Don't try $\cos^2(t)$ on a calculator, though; it won't understand. Use $(\cos(t))^2$.

The two equations mean exactly the same thing; don't be misled by the notation in the second version. This relationship between the sine and the cosine functions is called an **identity,** an equation that's true for every possible value of the variable.

Let's look at an example. If we know that $\cos(t) = 0.3$ and that $-\frac{\pi}{2} \leq t \leq 0$, we can find $\sin(t)$. A picture will help; see Figure 6.14. The statement $-\frac{\pi}{2} \leq t \leq 0$ tells us about the arc; it tells us, specifically, that the arc goes in the negative direction and stops before reaching the bottom of the circle. It also lets us know that, though the cosine value is positive, the sine value will be negative. Now we use the relationship $\sin^2(t) + \cos^2(t) = 1$ to find $\cos(t)$:

$$\sin^2(t) + \cos^2(t) = 1$$
$$\sin^2(t) + (0.3)^2 = 1$$
$$\sin^2(t) = 1 - 0.09 = 0.91$$
$$\sin(t) = \pm\sqrt{0.91} \approx \pm 0.95$$

The sine value is negative because q-coordinates of all points in Quadrant IV are negative.

We know to choose the negative *sign* because we expect a negative *sine!* Thus, $\sin(t) \approx -0.95$.

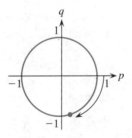

Figure 6.14

Check Your Understanding 6.5

1. If $\sin(t) = 0.25$ and an arc of length t terminates in Quadrant II, find $\cos(t)$.

2. If $\cos(t) = \frac{2}{3}$ and if $\frac{3\pi}{2} < t < 2\pi$, find $\sin(t)$.

3. If $\cos(t) = \frac{2}{3}$ and if $-2\pi < t < -\frac{3\pi}{2}$, find $\sin(t)$.

4. Find a much simpler way to write the function $L(x) = \sin^2(x) + \cos^2(x)$. Describe the graph of $L(x)$.

Could $\cos(N)$ equal $\sin(N)$ for some number N? What would be the value of N? Is there more than one answer to that question?

Instantaneous Rates of Change

An important feature of all sine and cosine graphs is the way their slopes change as we move along the curve from left to right. Look at your graph of $\sin(t)$ in Figure 6.11. You should observe that the curve is steepest wherever it crosses the t-axis. The greatest positive slopes occur at $t = -4\pi, -2\pi, 0, 2\pi$, and 4π (even multiples of π). The most negative slopes occur at $t = -3\pi, -\pi, \pi$, and 3π (odd multiples of π).

Let's find out just how steep the curve is at one of those points, $(0,0)$. First, we'll compute the average rate of change of $\sin(t)$ over an interval near the origin. The average rate of change of $\sin(t)$ from $t = -\frac{\pi}{2}$ to $t = \frac{\pi}{2}$ is given by the following quotient:

$$\frac{\sin\left(\frac{\pi}{2}\right) - \sin\left(-\frac{\pi}{2}\right)}{\frac{\pi}{2} - \left(-\frac{\pi}{2}\right)},$$

which simplifies to

$$\frac{1 - (-1)}{\pi} = \frac{2}{\pi} \approx 0.64$$

The dashed line in Figure 6.15 joining the points $\left(-\frac{\pi}{2}, -1\right)$ and $\left(\frac{\pi}{2}, 1\right)$ has a slope of $\frac{2}{\pi}$.

As Figure 6.15 shows, however, the sine curve itself is steeper than the dashed line as they pass through the origin. Just how steep is it? If we zoom in on the graph, the curve will appear straight. Since it looks linear, let's treat it as a line and find the slope between two points that lie close to the point of interest (see Figure 6.16):

Use radian mode.

$$\frac{\sin(0.1) - \sin(-0.1)}{0.2}$$

or approximately 0.998334. Hmmm ... awfully near 1, isn't it? Let's get even closer to the origin. Table 6.2 will help you to see what happens when we systematically narrow the window.

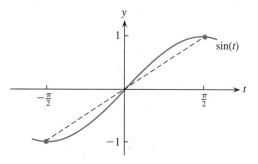

Figure 6.15

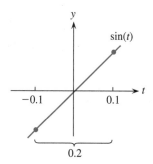

Figure 6.16

Δt	Slope
0.2	0.9983341665
0.02	
0.002	
0.0002	

Table 6.2

For $\Delta t = 0.02$, for instance, the slope is
$$\frac{\sin(0.01) - \sin(-0.01)}{0.02}.$$

$\triangleright$ With a single sequence of keystrokes for each calculation, complete Table 6.2. (Don't write down any intermediate results, or you'll lose accuracy.)

Convince yourself that the slope becomes closer and closer to 1 as the window narrows. Your calculator, eventually, will simply report the result as 1. The slope of the graph of $\sin(t)$ at $t = 0$ is exactly 1.

If you've already studied Chapter 4 or 5, you are familiar with this technique, which we present here for those who leapt directly into this chapter from Chapter 3.

By zooming in on a graph and using two points very close to our target to calculate a slope, we are estimating the slope of the curve at the target point. Remember, this gives the **instantaneous rate of change** of the function at that point.

Now let's approximate the slope of the sine graph at another point, $t = 2.4$. Again, we choose two points close to our target and calculate an average rate of change; the points $(2.39, \sin(2.39))$ and $(2.41, \sin(2.41))$, straddling the target point, will do nicely (see Figure 6.17).

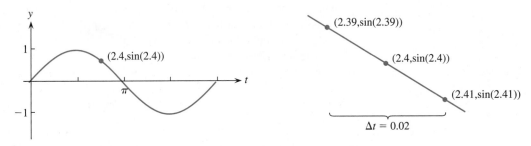

Figure 6.17   Approximating the Slope of $\sin(t)$ at $t = 2.4$

$\triangleright$ Using those two points, calculate a slope. Round the result to three decimal places.

The instantaneous rate of change of $\sin(t)$ at $t = 2.4$ is approximately -0.737. We expect a negative result because at $t = 2.4$, the function is decreasing.

Check Your Understanding 6.6

1. Using the points $(1.01, \cos(1.01))$ and $(0.99, \cos(0.99))$, estimate the instantaneous rate of change of $\cos(t)$ at $t = 1$.

2. Refer to the graph of $\cos(t)$ to explain why your answer to Question 1 is reasonable.

3. Find the instantaneous rate of change of $\cos(t)$ at $t = \pi$. (No calculator!)

Other Circular Functions

The sine and the cosine are called **circular functions** because they are defined in terms of the unit circle. They are the most familiar of all **periodic functions**—functions whose output values repeat with unending regularity. They are also known as **trigonometric functions,** for reasons that will be given in Chapter 7.

There are four additional circular functions: the tangent, the cotangent, the secant, and the cosecant, each defined in terms of the original two. Here they are:

$$\tan(t) = \frac{\sin(t)}{\cos(t)}, \quad \cot(t) = \frac{\cos(t)}{\sin(t)}, \quad \sec(t) = \frac{1}{\cos(t)}, \quad \csc(t) = \frac{1}{\sin(t)}$$

Notice that all four are defined as *quotients*. Thus, in determining the domain of these functions, we must pay attention to the possibility that a denominator could become zero.

We'll use the secant function to illustrate what happens when one of these denominators approaches zero. Figure 6.18 is a graph of $\cos(t)$ and of its reciprocal, $\sec(t)$. The graph of $\sec(t)$ consists of an infinite number of separate U-shaped sections.

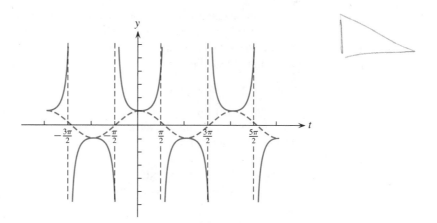

Figure 6.18 Graphs of $\sec(t)$ (Solid) and $\cos(t)$ (Dashed)

• The positive portions of the cosine wave (above the t-axis) account for the upward-opening U's of the secant, because the reciprocal of any positive value is also positive.

- Likewise, when $\cos(t)$ is negative, so is $\frac{1}{\cos(t)}$, creating the upside-down U's.

▷ Explain why the two functions share a point at $t = 0$, $\pm\pi$, and $\pm 2\pi$.

Recall that the cosine function equals zero at $t = \pm\frac{\pi}{2}$, $\pm\frac{3\pi}{2}$, ... We need to investigate the behavior of its reciprocal, $\sec(t)$, as t approaches any of those values, because $\frac{1}{0}$ isn't a real number.

▷ Complete Table 6.3, observing the behavior of $\sec(t)$ as t approaches $\frac{\pi}{2} \approx 1.5708$.

t	1.30	1.40	1.50	1.55	1.57	1.58	1.60	1.70
$\cos(t)$	0.2675	0.1700						
$\sec(t)$	3.7383	5.8835						

Table 6.3

As $t \longrightarrow \frac{\pi}{2}^-$, $\sec(t) \longrightarrow \infty$.

In the first five entries, the secant values increase dramatically as t approaches $\frac{\pi}{2}$ from the left (through values less than $\frac{\pi}{2}$). The curve climbs and becomes nearly vertical as the function values increase without limit.

As $t \longrightarrow \frac{\pi}{2}^+$, $\sec(t) \longrightarrow -\infty$.

Between $t = 1.57$ and $t = 1.58$, we witness an abrupt change. Suddenly, the secant values are negative, and the closer we come to $\frac{\pi}{2}$ on the right side, the more negative are those function values. Now we're on an upside-down U. As t approaches $\frac{\pi}{2}$ from the right (through values greater than $\frac{\pi}{2}$), the secant function decreases without limit.

On either side of $t = \frac{\pi}{2}$, the curve approaches a vertical line. That line is called a **vertical asymptote.** On the graph of the function, we show a vertical asymptote as a dashed line, an indication that the function is not defined for that value and that, near the value, the behavior of the function is extreme.

In these four functions—the tangent, the cotangent, the secant, and the cosecant—we anticipate a vertical asymptote wherever a denominator approaches zero. Because each function has $\sin(t)$ or $\cos(t)$ on the bottom, its denominator has an infinite number of opportunities to become zero, so its graph has an infinite number of vertical asymptotes.

Of the four functions, the one most frequently used in mathematics is the tangent. Let's use what you already know about the sine and the cosine to sketch the graph of $\tan(t)$ without a grapher.

▷ Decide where the vertical asymptotes should go. (For which t-values does $\cos(t) = 0$?) Draw them as dashed lines in Figure 6.19.

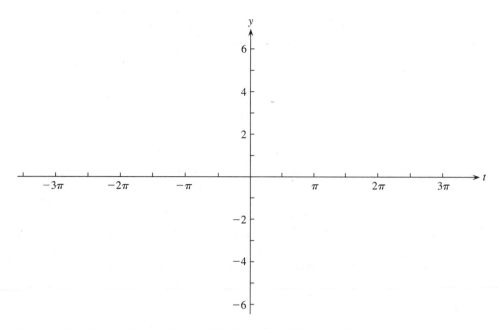

Figure 6.19 Student-Drawn Graph of the Function $h(t) = \tan(t)$

▷   Plot the t-intercepts—all the t-values for which the numerator, $\sin(t)$, equals 0.

With a few more well-chosen values, you can sketch an excellent graph. As Figure 6.20 illustrates, $\sin(t)$ and $\cos(t)$ are equal at two points on the unit circle, where $t = \frac{\pi}{4}$ and where $t = -\frac{3\pi}{4}$. If $\sin(t) = \cos(t)$, the value of the tangent is 1. Sine and cosine values are equal in magnitude but opposite in sign at two other points, where $t = \frac{3\pi}{4}$ and where $t = -\frac{\pi}{4}$. If $\sin(t) = -\cos(t)$, the value of the tangent is -1.

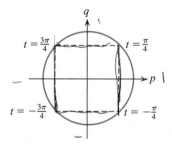

Figure 6.20 $\sin(t) = \pm \cos(t)$

▷   Plot the points $\left(\frac{\pi}{4}, 1\right)$ and $\left(-\frac{3\pi}{4}, 1\right)$ in Figure 6.20. Then plot four other points with a y-coordinate of 1.

You'll find graphs of the secant, the cosecant, and the cotangent functions in the Trigonometry Appendix, Section T.2.

▷   Plot $\left(\frac{3\pi}{4}, -1\right)$ and $\left(-\frac{\pi}{4}, -1\right)$ and four additional points whose y-coordinate is -1.

▷   Draw smooth curves through the plotted points, making the graph extremely steep wherever it approaches a vertical asymptote. You have completed a graph of the tangent function. Check your artwork by comparing it to the graph on page 512.

Check Your Understanding 6.7

1. Give the value, if it exists, of each of the following:

    (a) $\tan(-4\pi)$ **(b)** $\csc\left(\frac{3\pi}{2}\right)$ **(c)** $\tan\left(\frac{29\pi}{2}\right)$

2. Give the domain and the range of the function $\tan(t)$.

3. Surprise: What is the period of the function $\tan(t)$?

In a Nutshell   The graphs of the sine and cosine functions are periodic waves that repeat their fundamental pattern every 2π units. To approximate the instantaneous rate of change of a function at a point, we zoom in until the graph appears straight, then calculate the slope of that "line." In addition to the sine and the cosine, the unit circle generates four more functions, which are defined as reciprocals or quotients of the sine and the cosine. Of these, the tangent function is most frequently used. The tangent function is defined for all real numbers except odd multiples of $\frac{\pi}{2}$; its range is $\mathbb{R}$; it repeats itself every π units.

6.3 SINUSOIDAL MODELS

To use sine and cosine functions for modeling, we have to be able to stretch them up, squash them down, pull them out, squeeze them together, and shift them horizontally and vertically. Our goal in this section is to enable you to find a formula for any sinusoidal graph and to draw the graph of any sine or cosine function whose formula you have. You'll be applying all of the techniques you learned in Chapter 3 for stretching, compressing, shifting, and reflecting graphs, so this is an opportunity to firm up your skills and understanding of these techniques.

The Tuning Fork—A Speeded-up Sinusoid

> *"Take but degree away, untune that string, and hark! what discord*
> *follows . . ."*   (Shakespeare, *Troilus and Cressida*)

Let's examine the pattern produced by the vibrating tuning fork on page 287 and compare it to our basic sine wave. A major difference between the two is the frequency of the oscillations. The graph of $\sin t$ completes one cycle in the course of 2π units[4] along the t-axis, while the wave generated by the tuning fork needs $\frac{1}{440}$ of a second to complete one cycle (see Figure 6.21). We know this because the fork vibrates at 440 cycles per second, or it would be producing some note other than A above middle C. If we were to use the same scale (that is, one unit equals one second) on each horizontal axis, we would see the second graph waving madly up and down while the first one swung along in leisurely fashion. Now we will modify the basic sine function so that it represents the tuning-fork function.

    We need to compress the basic sine wave in the horizontal direction (squeezing the accordion) to create a formula for the tuning-fork vibrations.

For a horizontal compression, multiply the input variable by a constant greater than 1.

    Let's see: We *have* a period of 2π; we *want* a period of $\frac{1}{440}$. We need to give the basic sine graph a hard squeeze. From Chapter 3, you know that a serious horizontal compression demands a large multiplier for the independent variable.

---

[4]In our basic sine wave, we have not assigned any specific unit, such as "seconds" or "inches," to the variable t, which is known as a "dimensionless variable." The function $\sin(t)$ and its relatives have units associated with the variables only when they're used as mathematical models.

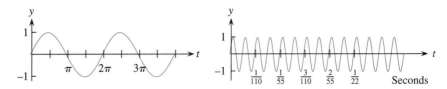

Figure 6.21 Graph of $\sin(t)$ and of the Tuning-Fork Wave

Hang on; in a few pages, we'll show you a general procedure for finding that number. For now, trust us on this: The factor by which we need to compress the generic sine wave to represent the tuning-fork wave is

$$\frac{\text{the basic sine period}}{\text{the period we want}} \quad \text{or} \quad \frac{2\pi \text{ units}}{1/440 \text{ second}}.$$

▷ Simplify that double-decker fraction to obtain the correct compression factor.

A formula to represent the tuning-fork function, therefore, is $F(t) = \sin(880\pi t)$. Notice how the dimensions work out:

$$\sin\left(880\pi \frac{\text{units}}{\text{second}} \cdot t \text{ seconds}\right) = \sin(880\pi t \text{ units}).$$

We haven't said anything about how high or low the tuning-fork wave goes. That depends upon the force with which the fork is struck: A louder sound means a taller wave. (The oscillations will still occur at a frequency of 440 cycles per second, but the wave will rise higher and dip lower.) For now, let's assume that the wave height is one unit up and one unit down, making our formula OK as it stands.[5]

The High-Temperature Graph—Four Adjustments Needed

The Fantastic Four

The wave representing average high temperature in New York City (see Figure 6.3, page 286) is a little more complicated, but you can still use shifting and stretching techniques to write its formula. We will have to make four modifications to the no-frills sine graph to produce a formula for the graph in Figure 6.3.

The Period The temperature wave shows an annual pattern, so its period ought to be 365 days. (We're ignoring leap years.)

▷ With the same strategy that we used for the tuning fork (start with the period we have; divide by the period we want), show that a sine function with a period of 365 days will have the approximate form $\sin(0.0172t)$. That is, justify the 0.0172.

---

[5]The number associated with wave height is called the *amplitude* of the sinusoid. With good reason, the device used to crank up the volume of an electric guitar is called an *amplifier*.

The Vertical Shift   The average high temperature for the whole year is approximately 62 degrees Fahrenheit. That means that the horizontal axis about which the temperature wave oscillates is the line $y = 62$. We need to shift our wave up by 62 units: $\sin(0.0172t) + 62$.

The Amplitude   The function we have so far has an amplitude of only one unit; the curve will rise to 63 degrees and drop to 61 degrees. That doesn't sound much like the annual temperature pattern in New York City. What we need, instead, is a wave that will rise to the warmest temperature (about 86 degrees, the average high temperature for a specific date in midsummer) and drop to the coolest (about 38 degrees, the average high temperature for a particular date in the depth of winter). We need to *stretch* the sine wave vertically.

> Show that an amplitude of 24 degrees will accomplish that goal, so that the function $24\sin(0.0172t) + 62$ comes close to matching the graph in Figure 6.22.

If these temperatures don't sound sufficiently extreme to you, recall that they represent averages *over many years. Truly impressive cold snaps and heat waves get swallowed up in an average.*

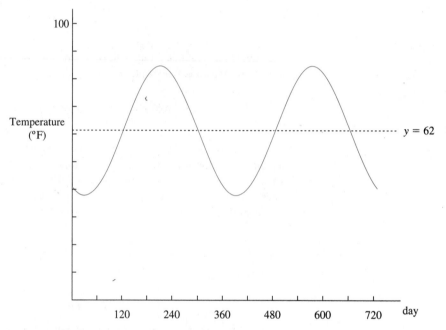

Figure 6.22 Average Daily High Temperatures in New York City

The Horizontal Shift   One big problem with the graph in Figure 6.22 is that, even though it is definitely a relative of the sine function, it does not "begin" in the right place. An ordinary, unshifted, sine graph reaches its highest point one fourth of the way through its cycle; it reaches its low point three fourths of the way through. If the temperature graph did that, it would give the warmest temperatures of the year to March and April, and the coolest to September and October. We need to determine how much of a horizontal shift (how many days, that is) to give our working model so that it lines up with the one in Figure 6.22. Reasonable people will disagree with one another on this, but it would not be far off to suggest that the actual graph appears to be a bit less than 4 months to the right of an unshifted sine graph with the same period. Let's call it 120 days. So we want to introduce into our formula a 120-unit shift to the right. Remember, that means *subtracting* 120 from the independent variable:

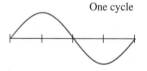

One cycle

$$24\sin[0.0172(t - 120)] + 62$$

> Perform the multiplication by 0.0172 to rewrite the given function in an equivalent form:

$$24\sin[0.0172(t - 120)] + 62 = 24\sin(0.0172t - \underline{\hspace{2cm}}) + 62$$

The expression you just wrote hides the fact that the shift is 120 days; yet, because it is equivalent to the other one, it gives the correct graph.

It's tempting to try to induce a 120-day shift by writing $24 \sin(0.0172t - 120)$. *Be careful!* That formula subtracts 120 days from $0.0172t$ rather than from t itself, so it produces a shift very different from the one we want.[6]

Here's the final formula for our temperature graph model:

$$HOT(t) = 24 \sin[0.0172(t - 120)] + 62$$

▷ Use a grapher to graph $HOT(t)$. It should be an excellent match for the graph in Figure 6.22, which you can now label with its mathematical formula.

The General Form of a Sinusoid

Learning your A, B, C's (and D's)

Now let's apply the same techniques we used for the temperature graph and the tuning-fork wave to any function having the form

$$A \sin[B(t - C)] + D \quad \text{or} \quad A \cos[B(t - C)] + D$$

Our goal is to learn to write a formula for any sinusoid by determining its four parameters, A, B, C and D, and to sketch any sinusoidal graph, given its formula.

The Amplitude   Figure 6.23 shows the graphs of four functions: $y = \sin(t)$, $y = \frac{1}{2}\sin(t)$, $y = 4\sin(t)$, and $y = -2\sin(t)$.

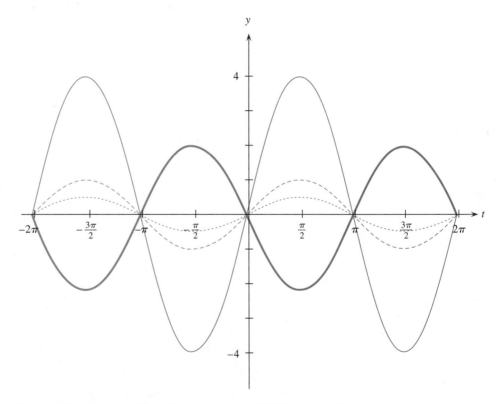

Figure 6.23 Graphs of Four Sine Functions of Differing Amplitudes

---

[6]The incorrect formula $24 \sin(0.0172t - 120)$ yields a shift of $120/0.0172$ days—approximately 6977 days, or a bit more than 19 years—not at all what we have in mind.

▷   Identify each graph by labeling it with its formula.

The graphs in Figure 6.23 present examples of vertical stretching and compression. When $\sin(t)$ is multiplied by a positive constant, each y-value is multiplied by that constant. If the constant is greater than 1, the graph is "taller" than a standard sine wave; if the constant is between 0 and 1, the graph is "flatter" than a standard sine wave.

▷   Describe what happens when the constant multiplier is negative. Consider constants less than -1 and constants between -1 and 0.

> The absolute value of the coefficient A is called the **amplitude** of the sine or cosine function.

▷   Give the amplitude of each of the four functions in Figure 6.23.

STOP & THINK   Notice that changing the amplitude of a sine function $A\sin(t)$ does not affect any of the intercepts. Explain why the intercepts do not change.

The Period   Figure 6.24 gives the graphs of $y = \sin(t)$, $y = \sin(2t)$, and $y = \sin(\frac{1}{2}t)$.

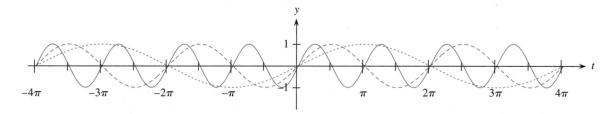

Figure 6.24 Graphs of Three Sine Functions of Differing Periods

▷   Label each graph with its formula.

The graphs in Figure 6.24 give an example of horizontal compression and stretching. We can think of $\sin(2t)$ as a "speeded-up" sine wave, one that completes its pattern twice as fast. Similarly, we can think of $\sin(\frac{1}{2}t)$ as a "slowed-down" sine wave, one that needs twice as much time (or space) to complete its pattern.

▷   Use a grapher to help you complete Table 6.4, paying attention to any numerical patterns that emerge.

Function	Period
$\sin(t)$	2π
$\sin(2t)$	
$\sin(3t)$	
$\sin(4t)$	
$\sin(2\pi t)$	
$\sin(\pi t)$	
	4π
	6π
	2

Table 6.4

▷ For any positive constant B, what is the period of the function $\sin(Bt)$?

You should have just written a formula for the period of any sine function:

The period of $\sin(Bt)$ is $\dfrac{2\pi}{|B|}$.

Use what you know. If you know B, period $= \frac{2\pi}{|B|}$. If you know the period, $B = \frac{2\pi}{\text{the period}}$.

From this relationship, it follows that if you already know the period and need to find B, you use

$$B = \frac{2\pi}{\text{the period}}$$

The Horizontal Shift Figure 6.25 shows two graphs. One represents $\sin(t)$; the other is a shifted version of $\sin(t)$.

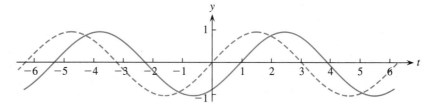

Figure 6.25 Graphs of a Sine Wave, with and without a Horizontal Shift

▷ Write a formula for the shifted sine wave in Figure 6.25. Use a grapher to check your formula, making sure that your graph has the correct intercepts.

▷   With a grapher, view the graph of $\sin(\pi t)$. Then write the formula for a function that shifts $\sin(\pi t)$ one unit to the right. Be sure that the new graph is one full unit to the right. If it's less than that, check your parentheses.

> In any function of the form $A\sin[B(t-C)]$ or $A\cos[B(t-C)]$, the value of C gives the amount of horizontal shift.

If you know how many units to shift and in which direction, the horizontal shift of any sinusoid is easy to handle: Add or subtract the appropriate number from the independent variable. Likewise, if you see a formula such as

$$10\sin\left[120\pi\left(t-\frac{1}{240}\right)\right]$$

you know that its graph will be the same as the graph of $10\sin[120\pi t]$, shifted $\frac{1}{240}$ unit to the right. Confusion arises, however, when the function is presented in the following form, which is the way it's most likely to show up in scientific applications:

$$10\sin\left(120\pi t-\frac{\pi}{2}\right)$$

Looking quickly, you might think that the horizontal shift is really $\frac{\pi}{2}$ units. But watch what happens when we factor out 120π:

$$10\sin\left(120\pi t-\frac{\pi}{2}\right)=10\sin\left[120\pi\left(t-\frac{\pi}{2\cdot120\pi}\right)\right]=10\sin\left[120\pi\left(t-\frac{1}{240}\right)\right]$$

The two formulas are algebraically equivalent. The horizontal shift is the same: $\frac{1}{240}$ units to the right.

The Vertical Shift   Figure 6.26 shows the graphs of $y=-2\cos(2\pi t)$, $y=-2\cos(2\pi t)-2$, and $y=-2\cos(2\pi t)+3$. All three graphs have exactly the same

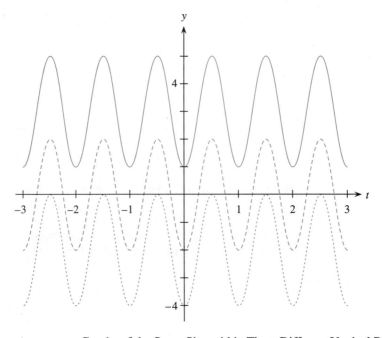

Figure 6.26 Graphs of the Same Sinusoid in Three Different Vertical Positions

size and shape; their amplitudes are the same. Notice, though, that they differ in their t- and y-intercepts.

▷   Match the formulas to the graphs of Figure 6.26.

Check Your Understanding 6.8

1. What is the period of $\cos(0.25t)$?

2. If the period of $\sin(Bt)$ is one unit, what is the value of B?

3. If the period of $\cos(Bt)$ is N units, what is the value of B?

4. Find a function that shifts $\sin(\pi t)$ one unit to the left.

5. What is the horizontal shift of $\cos\left(3t - \frac{\pi}{4}\right)$?

6. Find the amplitude, period, and horizontal shift of $y = -3\sin[0.5(t+1)]$.

7. Find the amplitude, period, and horizontal shift of $y = -3\sin(0.5t + 1)$.

8. Find a formula for this graph. (There are infinitely many possible answers.)

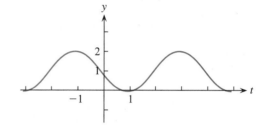

STOP & THINK   Students sometimes confuse the effect of a vertical shift with the effect of a change in amplitude. The graphs of $2\cos(t)$ and $\cos(t) + 1$ both have the same y-intercept, but they are not the same graph. What are some of the ways in which these two graphs differ?

Sketching a Sinusoidal Graph

Sine and cosine waves are easy to draw if you choose a convenient horizontal scale, and we're about to give you some pointers. Before attempting to draw anything, determine the period of the function, because that will tell you how to subdivide the horizontal axis. What you definitely do *not* want to do is to mark off the t-axis $1, 2, 3, \ldots$ before you have found out what numbers would actually be convenient.

Here's an example involving electricity. An alternating electric current can be represented by a function like this one:

$$I(t) = 15\cos(120\pi t)$$

where t is measured in seconds and $I(t)$ in amperes. The standard cosine wave has a period of 2π units; this one is compressed horizontally by a factor of 120π, so its period is

$$\frac{2\pi}{120\pi}, \quad \text{or} \quad \frac{1}{60}$$

This is a model of our ordinary household electricity, 60 cycles per second.

The period suggests a convenient numbering scheme for the horizontal axis. Our fundamental unit is $\frac{1}{60}$ of a second, which we'll subdivide into four equal sections, each representing $\frac{1}{240}$ of a second. After that, it's pretty straightforward to draw the wave, using the horizontal marks as guideposts and remembering the amplitude of 15.

If the wave has a horizontal shift as well, draw the *unshifted* version first, then slide it right or left by the appropriate amount. So, for instance,

$$J(t) = 15\cos\left[120\pi\left(t + \frac{1}{480}\right)\right]$$

is $I(t)$ shifted $\frac{1}{480}$ of a second to the left. See the dashed graph in Figure 6.27.

In scaling the t-axis, $\frac{\text{period}}{4}$ is the basic unit. Place tick marks, equally spaced, at integer multiples of $\frac{\text{period}}{4}$.

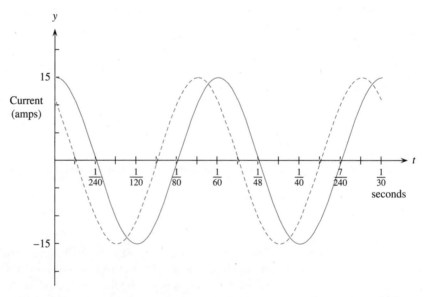

Figure 6.27 Graph of $15\cos(120\pi t)$ (Solid) and of $15\cos\left[120\pi\left(t + \frac{1}{480}\right)\right]$ (Dashed)

▷ Do the algebra to show that the formula for $J(t)$ can also be written

$$15 \cos\left[120\pi t + \frac{\pi}{4}\right],$$

which is the form in which it is more likely to appear in a physics book.[7]

Cosine waves work mathematically in the same way as sine waves. The terms *amplitude* and *period* have the same meaning, and they affect the formula in the same way. The only difference is that the fundamental cosine wave starts and ends with its highest value. Sometimes, it is more convenient to use a cosine formula rather than a sine formula, but either one can always be written in terms of the other by means of an appropriate shift or reflection.

Check Your Understanding 6.9

1. Put a shift into the formula $\cos(t)$ so that its graph ends up looking like the graph of $\sin(t)$.

2. Find three different formulas that could represent this graph. You might try a sine, a cosine, and a reflected cosine or sine.

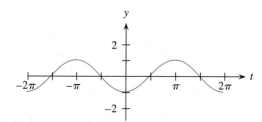

3. Sketch a graph of the function $\cos[0.5(t - \pi)]$. Then give another formula to represent the same graph.

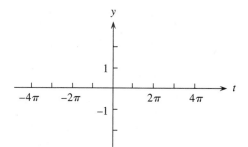

In a Nutshell Any sinusoidal function $A \sin[B(t - C)] + D$ or $A \cos[B(t - C)] + D$ is characterized by its amplitude $|A|$ (a vertical stretch or squash), its period $\frac{2\pi}{B}$ (the horizontal length of one fundamental wave), its horizontal shift C, and its vertical shift D. The parameter B is its horizontal compression factor (or stretch, if $0 < B < 1$).

---

[7] In physics and engineering, the quantity $\frac{\pi}{4}$ is called the **phase shift** of the wave.

6.4 INVERSE CIRCULAR (TRIGONOMETRIC) FUNCTIONS

Solving $\sin(x) = 1 - 2x$

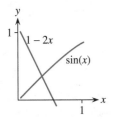

You're adept at using graphs to solve equations that can't be solved by algebraic methods. For example, although the equation $\sin(x) = 1 - 2x$ cannot be solved for x by any straightforward algebraic techniques, you could approximate the solution to any desired degree of accuracy by finding the intersection of $y = \sin(x)$ and $y = 1 - 2x$. In this section, you'll continue to use graphs to solve equations, but you will also learn to get support from your calculator and, yes, even algebra.

Visualizing an Algebraic Solution

> *Personally, I like short words and vulgar fractions.*   (Winston Churchill)

Suppose we want to solve the equation $\sin(t) = \frac{1}{3}$. As the graph of $\sin(t)$ in Figure 6.28 indicates, there are an infinite number of solutions: every place that the horizontal line $y = \frac{1}{3}$, intersects the sine curve gives another t-value. A calculator and the unit circle will help us find all those t-values.

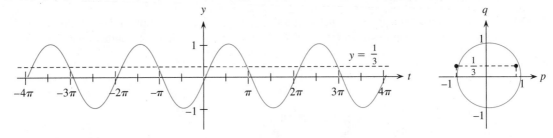

Figure 6.28 Solving the Equation $\sin(t) = \frac{1}{3}$

Use radian mode throughout this chapter.

First, be sure that your calculator is in **radian mode** and not in **degree mode.** Degree mode is used for operations involving angles measured in degrees (you'll use it in Chapter 7) but will not give us meaningful results here. The "radian mode" option signals to the calculator that the t-values count *radii* of the unit circle, measured along its circumference.

You need to ask your calculator the question, "What real number has $\frac{1}{3}$ as its sine value?" That is, you want to *undo* the sine function by using its inverse. The appropriate function key will be labeled $\sin^{-1}$ or 2nd sin or *INV* sin. (See the instructions for your particular calculator.)

The result should be approximately 0.34. What does that signify? We know that there are an infinite number of real numbers whose sine value is $\frac{1}{3}$. The calculator had to choose *one* of them. But the single one it produced will allow you to obtain all the others.

Figure 6.28 shows the two unit-circle points at which the sine value is $\frac{1}{3}$.

▷ Which one corresponds to an arc of 0.34? How can you be sure? (*Hint*: An arc of $\frac{\pi}{2}$ reaches one fourth of the way around the circle. How does the number 0.34 compare to the number $\frac{\pi}{2}$?)

▷   What about that other unit-circle point? Use the symmetry of the circle to find an arc to which it corresponds. (There are, of course, an unlimited number of such arcs. You need only one, and you'll have to do a little subtraction to get it.)

An approximate value for
$\pi - 0.34$ is 2.80

OK. Now you should have two t-values. We hope they're 0.34 and 2.80.[8] All the other solutions to the equation $\sin(t) = \frac{1}{3}$ come directly from these two because the sine function repeats itself every 2π units. Here are all of the solutions:

$$t = \begin{cases} 0.34 + 2n\pi \\ 2.80 + 2n\pi \end{cases}$$

where n means any integer—positive, negative, or zero.

Check Your Understanding 6.10

1. Find all solutions to the equation $\sin(t) = -\frac{1}{3}$. (Don't reinvent the wheel! Make sketches and adapt the preceding example.)

2. Use the inverse sine function to find the coordinates of points A, B, C, and D.

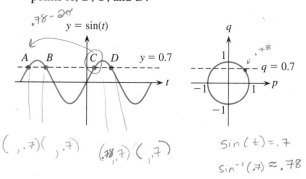

Keeping cool

The preceding problems involved simple equations. Here's a more complicated one. We'll use the high-temperature function $HOT(t)$ from Section 6.3 to find the days of the year that have a certain average high temperature, say, 50°F. The equation to solve is

$$24 \sin[0.0172(t - 120)] + 62 = 50$$

The easiest way to do this is to look at the graph of $HOT(t)$, as illustrated in Figure 6.29, and see where the function values equal 50, but you can solve the problem algebraically as well. We present this example to illustrate some useful strategies for solving equations that involve a trigonometric function.

---

[8]If you went clockwise around the circle instead, your second t-value would be $-\pi - 0.34 \approx -3.48$, an equally valid solution.

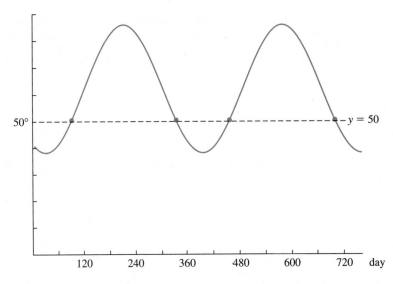

Figure 6.29 An Average High Temperature of 50° Fahrenheit

Our aim will be to get the equation into this form:
$\sin(\text{some mess}) = \text{a number}.$

The inverse sine is a really great tool, but it won't work until the equation is in the proper form. Here we go; hang onto your hat.

▷ Next to each step, write a brief algebraic justification.

$$24\sin[0.0172(t - 120)] + 62 = 50 \qquad \text{We can't use the inverse sine yet.}$$
$$24\sin[0.0172(t - 120)] = -12$$
$$\sin[0.0172(t - 120)] = -0.5$$

Calculate $\sin^{-1}(-0.5)$

Now comes the crucial step. The sine of some quantity, $0.0172(t - 120)$, is -0.5. What could the quantity itself be?[9] It's time to pull out our tool, the inverse sine function. Ask your calculator what number has a sine value of -0.5. You should obtain approximately -0.5236. So the next step in our algebra is

$$0.0172(t - 120) \approx -0.5236$$

You know, though, that -0.5236 is only one of the numbers whose sine is -0.5. The unit circle gives us another one: -2.6180, or 3.6652, depending on which way you go around the circle (see Figure 6.30).

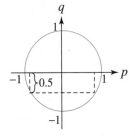

Figure 6.30 A Sine Value of -0.5

---

[9]From previous exposure to trigonometry, you might realize that the number in question is one of the "special values" on the unit circle that can be found without a calculator. There's more about special values in the Trigonometry Appendix, Section T.1. Armed with calculators, however, we needn't restrict ourselves to problems that use special values.

▷ Show where those two numbers, -2.6180 and 3.6652, came from.

▷ We're not finished, though, because we don't know t. Keep going, remembering that whatever you do to one side of the equation you must do to the other. Beside each step, write a brief algebraic justification.

$$0.0172(t - 120) \approx \begin{cases} -0.5236 + 2n\pi \\ 3.6652 + 2n\pi \end{cases}$$

$$t - 120 \approx \begin{cases} -30 + 365n \\ 213 + 365n \end{cases}$$

$$t \approx \begin{cases} 90 + 365n \\ 333 + 365n \end{cases}$$

Yes, that was a lot of algebraic manipulation just to find out that day 90 (March 31) and day 333 (November 29) in New York City have an average high temperature of 50°F, but we wanted to convince you that it can be done (and that *you* can do it!).

▷ Check the numerical results with a grapher to be sure the graph of $HOT(t)$ gives a temperature of 50°F for days 90 and 333. Then, graphically, find two more t-values that give an average temperature of 50°F.

STOP & THINK

Were you surprised to see the term $365n$ appear in the result? What's its significance? Show algebraically where the term came from, and explain why it makes real-world sense.

The Inverse Sine and Cosine Functions

Sines and wonders

In the preceding examples, we came up with a sine value and asked ourselves, "What numbers have this value as their sine?" In other words, we had the *output* of a sine function, and we wondered what the *input* could have been. From your work with inverses in Chapter 3, you're prepared for what's to come.

When you pressed the $\sin^{-1}$ key on your calculator, you used an inverse function. You gave the calculator a sine value; you asked it to return a number having that value as its sine. The only mystery is how it knew which one of the unlimited possibilities to choose.

Refresh your memory: inverse functions begin on page 116. If you've studied Chapter 4, you remember asking the same question of the exponential function. That's how we got logarithms.

▷ In the graph of $y = \sin(t)$ in Figure 6.31, interchange the input and output values and sketch the resulting graph on the same axes. You should see a wave snaking its way up the y-axis. The graph you just drew isn't the graph of a *function*. Why not?

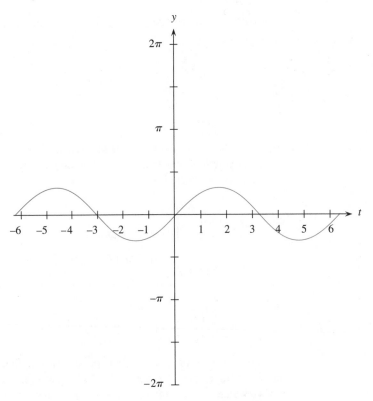

Figure 6.31 Graphs of $y = \sin(t)$ and of Its Reflection

Only a one-to-one function has an inverse that is also a function.

To have an inverse function, a function must have exactly one input associated with any given output. Not every function, therefore, has an inverse function. In its unadulterated form, the sine function does not have an inverse function. To define an inverse for the sine function, we must restrict its domain to a one-to-one portion of the graph. The most straightforward portion to use is the piece right in the middle, in which the t-values are between $-\pi/2$ and $\pi/2$.

▷   In Figure 6.31, color the portion of the graph of $y = \sin(t)$ on the interval $\left[-\frac{\pi}{2}, \frac{\pi}{2}\right]$. Use the horizontal line test to show that the function $y = \sin(t)$ is one-to-one on that interval.

On the interval $\left[-\frac{\pi}{2}, \frac{\pi}{2}\right]$, $\sin(t)$ *does* have an inverse function.

$$y = \sin^{-1}(x) \qquad \text{or} \qquad y = \arcsin(x)$$

is the number y, $-\frac{\pi}{2} \leq y \leq \frac{\pi}{2}$, whose sine value is x.

The output of $\sin^{-1}(x)$ is a number whose sine value is x.

The statement $y = \sin^{-1}(x)$ says, "y is the number whose sine is x." The statement $y = \arcsin(x)$ says, "y is the arc whose sine is x." Both statements have the same meaning. In each case, y is a value between $-\frac{\pi}{2}$ and $\frac{\pi}{2}$.

Check Your Understanding 6.11

1. Without a calculator, evaluate

    (a) $\sin^{-1}(1)$ **(b)** $\arcsin(0)$

    (c) $\sin^{-1}(\cancel{1})$ **(d)** $\sin(\sin^{-1}(0.6))$

       $-\dfrac{\pi}{2}$

2. Give the domain and the range of the inverse sine function.

▷ Now use your calculator (radian mode!) to determine the inverse sine values of several numbers: 0.5678, −0.5678, 0.9876, 0.0002, −0.9999, and so on. (Remember, you're asking for the arc that has the given number as its sine value.) Notice that all results lie between −1.57 and 1.57. Why is this so?

▷ What does your calculator say if you ask it for the inverse sine of a value that's greater than 1? Why?

The cosine function also has an inverse function, which is defined similarly but with one important difference. With the sine function, we used the piece from $-\frac{\pi}{2}$ to $\frac{\pi}{2}$, because it is one-to-one. Do you see what would go wrong if we chose the same portion from the cosine function? The function $\cos(t)$ is not one-to-one on that interval. We restrict the cosine to the interval $[0, \pi]$, because there it *is* one-to-one.

$$y = \cos^{-1}(x) \qquad \text{or} \qquad y = \arccos(x)$$

is the number y, $0 \le y \le \pi$, whose cosine value is x.

Table 6.5 summarizes the domains and ranges of $\sin(t)$, $\cos(t)$, and their inverses.

	Restricted Sine	Inverse Sine	Restricted Cosine	Inverse Cosine
Domain	$[-\frac{\pi}{2}, \frac{\pi}{2}]$	$[-1, 1]$	$[0, \pi]$	$[-1, 1]$
Range	$[-1, 1]$	$[-\frac{\pi}{2}, \frac{\pi}{2}]$	$[-1, 1]$	$[0, \pi]$

Table 6.5

The Inverse Tangent Function

Going off on a tangent

We can restrict the domains of all six of the circular functions so that their inverses are functions. The only other one, though, with much importance in mathematics and science is the inverse tangent function, $y = \arctan(x)$ or $y = \tan^{-1}(x)$.

The vertical asymptotes of the graph of $\tan(t)$ in Figure 6.32 separate the graph into an unlimited number of branches that reach downward and upward without bound. Even though the function $y = \tan(t)$ is not one-to-one (many different inputs can produce the same output), any individual branch is one-to-one. To define the inverse tangent function, we'll choose one branch, the obvious choice being the one right in the middle.

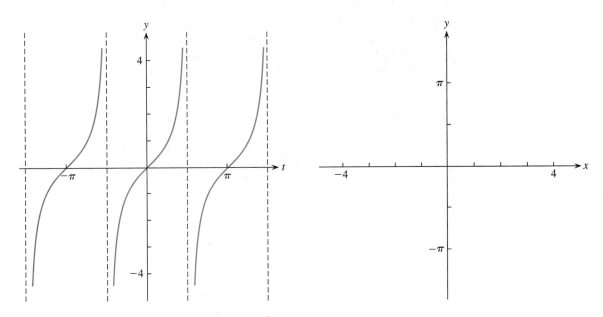

Figure 6.32 Graphs of $y = \tan(t)$ and $y = \tan^{-1}(x)$

$$y = \tan^{-1}(x) \qquad \text{or} \qquad y = \arctan(x)$$

is the number y, $-\frac{\pi}{2} < y < \frac{\pi}{2}$, whose tangent value is x.

As $t \longrightarrow \frac{\pi}{2}$ from the left, $\tan(t) \longrightarrow \infty$. As $x \longrightarrow \infty$, $\tan^{-1}(x) \longrightarrow \frac{\pi}{2}$.

▷ Sketch a graph of $y = \tan^{-1}(x)$ on the second set of axes in Figure 6.32. Notice that the tangent's vertical asymptotes at $t = \pm\frac{\pi}{2}$ become the arctangent's horizontal asymptotes, at $y = \pm\frac{\pi}{2}$. Check your graph by looking in Section T.2 of the Trigonometry Appendix.

Check Your Understanding 6.12

1. Without a calculator, evaluate
   (a) $\arctan(-1)$ (b) $\tan^{-1}(0)$

   (c) $\arccos(-1)$ (d) $\cos^{-1}(\cos(3\pi))$

2. Give the domain and the range of the inverse tangent function.

3. Find all solutions to the equation $\tan(t) = 3$.

4. Complete these statements:
   - As $t \longrightarrow -\frac{\pi}{2}$ from the right, $\tan(t) \longrightarrow$ _____.
   - As $x \longrightarrow -\infty$, $\tan^{-1}(x) \longrightarrow$ _____.

Just as geologists use the logarithmic function for rescaling earthquake data, animal biologists sometimes use the arctangent function for rescaling population information. A biologist wishing to compare populations of animals in several locations by using the ratio of one year's count to a second year's count gets into trouble if all the animals in one location disappear by the second year. Suddenly, there's a zero in the denominator, and the ratio itself becomes infinitely large. With arctangent scaling, however, the difficulty disappears: An infinitely large input (look at the graph you drew in Figure 6.32) corresponds to an output of $\frac{\pi}{2}$. Converting raw ratios by means of the arctangent function takes numbers ranging from 0 to $+\infty$ and returns numbers ranging from 0 to $\frac{\pi}{2}$. If the application were one in which negative values made sense, arctangent scaling could accept any real number, always returning a value between $-\frac{\pi}{2}$ and $\frac{\pi}{2}$.

▷   Use your calculator to find the arctangent, or inverse tangent, of several numbers of widely differing magnitudes: 8000, 0.002, −49, −50000, for instance. Observe that all of the outputs are between $-\frac{\pi}{2}$ and $\frac{\pi}{2}$.

Using Algebra to Analyze a Graph

t for 2

▷   Draw the graphs of $\sin(t)$ and $\sin(2t)$ together on the axes in Figure 6.33, using the entire interval. You'll see that they intersect each other in nine places. We'll use algebra to find the coordinates of each point of intersection.

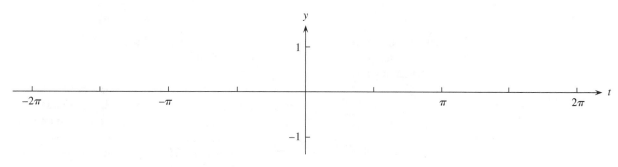

Figure 6.33 Student-Drawn Graphs of $\sin(t)$ and $\sin(2t)$

First, we set up an equation:

$$\sin(2t) = \sin(t)$$

Use the identity $\sin(2t) = 2\sin(t)\cos(t)$ to replace $\sin(2t)$ with its equivalent. The Trigonometry Appendix, Section T.3, gives additional identities, and Project 6.2, "Aliases," helps you investigate them.

Algebra can't help us yet because we cannot simply divide through by the "2" or yank it out through the parentheses. We can, though, use an identity to make a substitution for the term $\sin(2t)$. Because $\sin(2t)$ is equivalent to $2\sin(t)\cos(t)$ for every number t, the equation becomes

$$2\sin(t)\cos(t) = \sin(t)$$

Now we can proceed with algebra. Gather everything on one side of the equation:

$$2\sin(t)\cos(t) - \sin(t) = 0$$

Factor out $\sin(t)$:

$$\sin(t)\,[2\cos(t) - 1] = 0$$

Thanks to the *zero product principle* from algebra, we can set each factor separately equal to zero. Therefore,

$$\sin(t) = 0 \qquad \text{or} \qquad 2\cos(t) - 1 = 0$$

If the sine of a number is zero, we know that the number is an integer multiple of π. Do you see the two curves intersecting at each multiple of π in your graph in Figure 6.33?

▷ Give both coordinates of those five points of intersection.

There are still four more intersections; we get them from the second factor:

$$2\cos(t) - 1 = 0$$
$$\cos(t) = \frac{1}{2}$$

Whose cosine is equal to $\frac{1}{2}$? If you've studied the "special values," you might know these numbers in terms of π, but they're equally easy to approximate with a calculator.

▷ Choose radian mode, and ask your calculator for the inverse cosine (which may be labeled $\cos^{-1}$ or 2nd cos) of 0.5; the result should be approximately 1.047.

▷ Use the unit circle to help you determine other t values whose cosine is $\frac{1}{2}$. (See Figure 6.34, and take advantage of the symmetry of the cosine function.) Find the t values of the remaining three points of intersection.

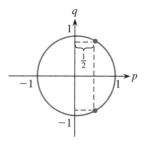

Figure 6.34 $\cos(t) = \frac{1}{2}$

▷   We're nearly done. Each point of intersection needs two coordinates. You have 1.047 and three other t-values; use them to find the second coordinate of each intersection point.

Here's the final score:[10] In order, from left to right, the points are $(-2\pi,0)$, $(-5.236,0.866)$, $(-\pi,0)$, $(-1.047,-0.866)$, $(0,0)$, $(1.047,0.866)$, $(\pi,0)$, $(5.236,-0.866)$, and $(2\pi,0)$. How did you do?

You might have found the preceding example a little heavy. It included many different concepts and skills: sines and cosines, graphing, symmetry, the unit circle, the calculator, algebra, and who knows what else. For that reason, it's a problem worth spending time on and lends itself especially well to group effort.

STOP & THINK

We warned you that dividing both sides of the equation $\sin(2t) = \sin(t)$ by 2 would not help solve the equation. Show, by means of a numerical example, that dividing $\sin(2t)$ by 2 doesn't give you $\sin(t)$.

KEEP THINKING!

In the equation $2\sin(t)\cos(t) - \sin(t) = 0$, we factored out the expression $\sin(t)$. Would we have lost anything if we had simply divided through by $\sin(t)$ instead? Try it. Which solutions does that reduced equation yield? Which ones are lost? Why does dividing through by $\sin(t)$ cause the loss of solutions?

Some students, particularly those who have studied trigonometry[11] in high school, think of the subject as being extraordinarily detailed, full of identities to prove and numbers to memorize. In this chapter, we have presented only the bare bones of the topic, peeling away all the information that isn't essential to precalculus mathematics and focusing entirely on the unit-circle approach to the sine and cosine functions, considered

---

[10]Those who like to use the "special values" might be happier with $\left(\frac{-5\pi}{3}, \frac{\sqrt{3}}{2}\right)$ and so forth.

[11]The name **trigonometry** comes from two Greek words that mean "triangle measurement," and refers to the original uses for the sine and cosine functions. Although what we're doing in this chapter has nothing to do with triangles, the name has stuck. See Chapter 7 for an overview of right-triangle trigonometry.

as models of periodic behavior. The Trigonometry Appendix serves as a reference for the graphs of all the circular and inverse circular functions and for any identities you might need.

In a Nutshell   If we restrict the domain of the sine, the cosine, or the tangent to an interval on which it is one-to-one, then each of those circular functions has an inverse that is also a function. The inverse sine, inverse cosine, and inverse tangent help to solve equations involving the sine or cosine or tangent of an unknown value.

What's the Big Idea?

- The sine and cosine functions are defined, respectively, as vertical and horizontal coordinates on the unit circle.

- Because their values repeat themselves in a regular manner, the sine and cosine functions serve as mathematical models of periodic behavior.

- The four other circular, or trigonometric, functions are defined in terms of the sine and cosine functions.

Progress Check

After finishing this chapter, you should be able to do the following:

- Use the unit circle to define the sine and the cosine functions. (6.1)

- Use compact notation to represent an infinite number of values. (6.2)

- Approximate the instantaneous rate of change of a sine or cosine function. (6.2)

- Given a graph or a description of periodic behavior, write a model. (6.3)

- Given the formula for a sine or cosine function, draw its graph. (6.3)

- Solve an equation involving sines or cosines graphically and, if appropriate, algebraically. (6.4)

- Use a calculator correctly to evaluate sines and cosines and to find inverse sines and cosines. (6.4)

Trigonometry Appendix Topics

- T.1 Special values on the unit circle

- T.2 Graphs of trigonometric and inverse trigonometric functions

- T.3 Handy identities

Answers to Check Your Understanding

6.1

1. (a)

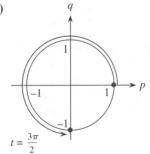

(b)

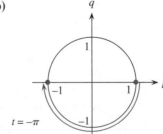

(c)

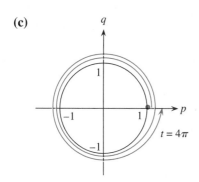

2. $t = \ldots, -\frac{5\pi}{2}, -\frac{\pi}{2}, \frac{3\pi}{2}, \frac{7\pi}{2}, \frac{11\pi}{2}, \ldots$

3. **(a)** $(-1,0)$ **(b)** $(0,-1)$ **(c)** $(0,1)$

    (d) $(-1,0)$ **(e)** $(1,0)$

6.2

1. **(a)** III **(b)** I **(c)** III

    (d) II **(e)** IV **(f)** IV

2. **(a)** $t \approx 2$ (or $t \approx -4$)

    (b) $t \approx -1$ (or $t \approx 5$) (These are very rough approximations.)

3. You could use the alternative answers for 2(a) and 2(b). If you choose arcs that wrap more than once around the circle, there are many other possibilities.

6.3

1. **(a)** -1 **(b)** 1

    (c) 0 **(d)** 1

2. **(a)** Positive **(b)** Positive

    (c) Negative

3. Some possible values are 0, 2π, -6π, and 2000π (even multiples of π). The expression $2n\pi$, where n can represent any integer, gives all possible values.

4. The functions $\sin(t)$ and $\cos(t)$ have all real numbers as their domain, because there are no restrictions on the directed arc-length t. The range of each function is the interval $[-1, 1]$—all the numbers between -1 and 1, inclusive—because the radius of the unit circle is one unit.

6.4

1. The graph of $\cos(t)$ has its maximum value when t is an even multiple of π, that is, when $t = 2n\pi$.

2. -1

3. $t = \pi + 2n\pi = (2n+1)\pi$

4. $t = \frac{3\pi}{2} + 2n\pi$

6.5

1. Approximately -0.97

2. $-\frac{\sqrt{5}}{3}$ **3.** $\frac{\sqrt{5}}{3}$

4. $L(x) = 1$. $L(x)$ is a constant function, because $\sin^2(x) + \cos^2(x)$ is constant for all values of x. It always equals 1. The graph of $L(x)$ is a horizontal line.

6.6

1. -0.841457

2. We expect a negative rate because, at $t = 1$, $\cos(t)$ is decreasing.

3. 0. (At $t = \pi$, the graph of $\cos(t)$ is horizontal.)

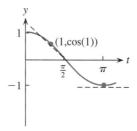

6.7

1. **(a)** 0

    (b) -1 (the reciprocal of $\sin(\frac{3\pi}{2})$)

    (c) Undefined

2. Domain: the set of all real numbers except odd multiples of $\frac{\pi}{2}$; $t \neq \frac{\pi}{2} + n\pi$. Range: $\mathbb{R}$

3. The tangent pattern repeats itself every π units; each wave has a width of π units; the period of $\tan(t)$ is π.

6.8

1. $\frac{2\pi}{0.25}$, or 8π **2.** $B = 2\pi$

3. $B = \frac{2\pi}{N}$

4. $\sin[\pi(t+1)]$ or $\sin(\pi t + \pi)$

5. $\frac{\pi}{12}$ units (to the right)

6. Amplitude 3, period 4π, shifted 1 unit to the left

7. Amplitude 3, period 4π, shifted 2 units to the left

8. The period of the function is 4 units; $1 - \sin(\frac{\pi}{2} t)$ is one possibility.

6.9

1. The graph of $\cos(t)$, shifted $\frac{\pi}{2}$ units to the right, looks like $\sin(t)$: $\cos(t - \frac{\pi}{2})$. Other answers are possible.

2. Possible formulas are $\sin(t - \frac{\pi}{2})$, $\cos(t - \pi)$, and $-\cos(t)$.

3. Period 4π

shift right π units

a.k.a. $y = \sin(0.5t)$

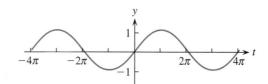

6.10

1. $t \approx \begin{cases} -0.34 + 2n\pi \\ -2.80 + 2n\pi \end{cases}$

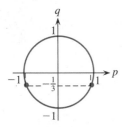

2. A $(-5.51, 0.7)$; B $(-3.92, 0.7)$;
C $(0.78, 0.7)$; D $(2.36, 0.7)$
(t-values rounded to two places)

6.11

1. (a) $\frac{\pi}{2}$ (b) 0   (c) $\frac{\pi}{2}$

2. Domain: $[-1, 1]$. Range: $[-\frac{\pi}{2}, \frac{\pi}{2}]$

6.12

1. (a) $-\frac{\pi}{4}$ (b) 0   (c) π
    (d) π (Remember the range of the inverse cosine.)

2. Domain: $\mathbb{R}$. Range: $(-\frac{\pi}{2}, \frac{\pi}{2})$

3. $t \approx 1.25 + n\pi$

4. As $t \longrightarrow -\frac{\pi}{2}$ from the right, $\tan(t) \longrightarrow -\infty$.
As $x \longrightarrow -\infty$, $\tan^{-1}(x) \longrightarrow -\frac{\pi}{2}$.

EXERCISES

1. Without a calculator, find the exact value of $\cos(-2468\pi)$, $\sin(-2468\pi)$, $\cos\left(\frac{5\pi}{2}\right)$, and $\sin\left(\frac{5\pi}{2}\right)$.

2. Without a calculator, find the exact value of $\cos(13579\pi)$, $\sin(13579\pi)$, $\cos\left(-\frac{5\pi}{2}\right)$, and $\sin\left(-\frac{5\pi}{2}\right)$.

3. For the arc t shown, sketch the following arcs.
  (a) $\pi - t$ (b) $2\pi + t$
  (c) $-2t$

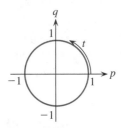

4. For the arc t shown, sketch the following arcs.
  (a) $\pi + t$ (b) $\pi - t$
  (c) $2t - \pi$

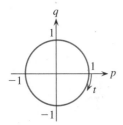

5. Arcs on the unit circle are sometimes measured in degrees, rather than in real numbers, with 360° representing the entire circumference. With your calculator in *degree mode*, evaluate $\sin(90°)$ and $\cos(90°)$. Then, in *radian mode*, evaluate $\sin(90)$ and $\cos(90)$. Explain what the four results mean and why the two pairs of results are so different from one another.

6. Without a calculator, rank these numbers from least to greatest: $\sin(1)$, $\sin(2)$, $\sin(3)$, $\sin(4)$, $\sin(5)$, $\sin(6)$, $\sin(7)$. Make a unit-circle diagram to demonstrate that the order is correct.

Finding the sine and cosine of most numbers requires a calculator. Some values, however, we can calculate on our own with a little geometry. In Exercises 7–9, you will find the sine and the cosine of three of these "special values."

7. The sketches below show an arc of $\frac{\pi}{4}$, one eighth of the unit circle.
  (a) In the right triangle pictured, how long is the hypotenuse? How do you know?
  (b) How many degrees are there in angle α? In angle β? Give reasons.
  (c) Let the base of the triangle have length x. Use geometry to show that the height of the triangle also has length x.
  (d) Apply the Pythagorean theorem to the triangle. Solve that equation to find the exact value of x, which involves a square root.
  (e) Explain how the result of (d) gives you $\sin\left(\frac{\pi}{4}\right)$ and $\cos\left(\frac{\pi}{4}\right)$ exactly. What are those values?

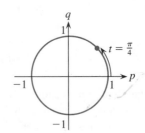
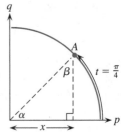

8. The sketch below shows an arc of $\frac{\pi}{6}$, one twelfth of the unit circle. Triangle OBC is the mirror image of triangle OAC in the p-axis.
   (a) Find lengths OA and OB. How do you know?
   (b) How many degrees are there in $\angle AOC$? $\angle BOC$? $\angle AOB$? Give reasons.
   (c) Explain carefully why $\triangle AOB$ is equilateral.
   (d) Explain why $AC = \frac{1}{2}$.
   (e) Apply the Pythagorean theorem to $\triangle OAC$ and find OC exactly. (The exact length contains a square root.)
   (f) Tell how the results of parts (d) and (e) give you exact values for $\sin(\frac{\pi}{6})$ and $\cos(\frac{\pi}{6})$, respectively. State those values.

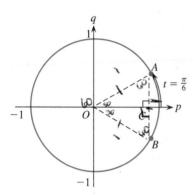

9. Adapt the ideas of the preceding exercise to calculate the exact values of $\sin(\frac{\pi}{3})$ and $\cos(\frac{\pi}{3})$. Justify each step mathematically.

10. Use the unit circle to demonstrate that for any number t, $\cos(-t) = \cos(t)$. Similarly, show that $\sin(-t) = -\sin(t)$. (Do not substitute any specific number for t; make your diagram and your argument generic, valid for any value of t.)

A function f is said to be **even** if $f(-x) = f(x)$ and **odd** if $f(-x) = -f(x)$ for every x in the domain of f. The graph of an even function is symmetric about the y-axis, and the graph of an odd function is symmetric about the origin. Exercise 10 says that $\cos(t)$ is an *even* function and that $\sin(t)$ is an *odd* function.

11. Show, algebraically, that $\tan(-t) = -\tan(t)$ and that, therefore, $\tan(t)$ must be an odd function.

12. **(a)** Determine, algebraically, the symmetry of these two functions (one at a time):

$$f(x) = \frac{\sin(x)}{x} \quad \text{and} \quad g(x) = \frac{\cos(x) - 1}{x}$$

   (b) What is the domain of f and g? Examine their graphs; even though neither function is defined for $x = 0$, each function will *appear* to have a value there. There is actually a "hole" in each graph at $x = 0$. Write the coordinates of the "hole" in the graph of $f(x)$; write the coordinates of the "hole" in the graph of $g(x)$.

13. If $\cos(t) = -1$, find all possible values for t.

14. If $\sin(t) = 1$, find all possible values for t.

15. If $\tan(t) = 1$, find all possible values for t.

$n\pi$

16. Find all solutions to the equation $\sin(t) + \cos(t) = 0$.

17. If an arc of length t terminates in the second quadrant and if $\sin(t) = \frac{3}{4}$, find $\cos(t)$ and $\tan(t)$. (Pay attention to signs!)

18. If an arc of length t terminates in the third quadrant and if $\cos(t) = -\frac{3}{4}$, find $\sin(t)$ and $\tan(t)$.

19. If $-\frac{\pi}{2} < t < 0$ and if $\sin(t) = -\frac{2}{3}$, find $\cos(t)$, $\tan(t)$, $\sec(t)$, $\csc(t)$, and $\cot(t)$.

20. If $\frac{\pi}{2} < t < \pi$ and if $\sin(t) = \frac{5}{7}$, find $\cos(t)$, $\tan(t)$, $\sec(t)$, $\csc(t)$, and $\cot(t)$.

21. Give the intervals on which the illustrated function is
   (a) positive   **(b)** increasing
   (c) concave up

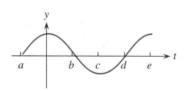

22. Find all values of t for which the graph of $\cos(t)$ has its greatest positive slope.

23. **(a)** Form the composition $g \circ f(x)$, if $f(x) = \cos(x)$ and $g(x) = x^2$.
   (b) Write the function $C(x) = \sin^2(3^x)$ as a composition of three simpler functions.

24. **(a)** Form the composition $p(q(x))$, if $p(x) = \sin(x)$ and $q(x) = \cos(x)$.
   (b) Write the function $D(x) = 3^{\sin^2(x)}$ as a composition of three simpler functions.

25. The cosecant function is the reciprocal of the sine function: $\csc(t) = \frac{1}{\sin(t)}$.

   (a) Sketch a graph of $y = \sin(t)$ on the interval $-4\pi \leq t \leq 4\pi$.

   (b) On the same axes, using a contrasting color, sketch the vertical asymptotes of $y = \csc(t)$. Write a sentence saying why the graph has those vertical asymptotes.

   (c) If a number is between 0 and 1, its reciprocal is greater than 1. Locate all the portions of the graph of $y = \sin(t)$ where the y-values are positive. At the same t-values, use your second color to sketch an approximate graph of the reciprocals of those y-values. You have drawn the positive portion of the graph of $y = \csc(t)$.

   (d) If a number is between -1 and 0, its reciprocal is less than -1. Repeat the work of part (c) for the negative y-values of $y = \sin(t)$ to draw the negative portion of $y = \csc(t)$ on your graph.

   (e) Explain why the graph of $\csc(t)$ has no t-intercepts.

   (f) What kind of symmetry does the graph display? Is $\csc(t)$ odd or even?

   (g) Give the domain and the range of $\csc(t)$.

26. The cotangent function is the reciprocal of the tangent function: $\cot(t) = \frac{1}{\tan(t)} = \frac{\cos(t)}{\sin(t)}$.

   (a) Sketch a graph of $y = \tan(t)$ on the interval $-3\pi \leq t \leq 3\pi$. Prepare another set of axes using the same scale and the same interval.

   (b) The cotangent function has vertical asymptotes at all the t-values for which $\tan(t)$ or $\sin(t)$ is zero. What are those values? On your second set of axes, draw the vertical asymptotes of $y = \cot(t)$.

   (c) To find the t-intercepts of a function given by a quotient, we look at the numerator. The cotangent function is zero at all the t-values for which $\cos(t)$ is zero. What are those values? On your second set of axes, mark the t-intercepts of $y = \cot(t)$.

   (d) Because $\cot(t)$ is the reciprocal of $\tan(t)$, $\cot(t)$ will be close to zero whenever $\tan(t)$ is far from zero, and vice-versa. On your second set of axes, sketch a graph of $y = \cot(t)$ that shows it to be the reciprocal of your graph of $y = \tan(t)$.

   (e) What kind of symmetry does the graph of $y = \cot(t)$ have? Is the cotangent an odd or an even function?

   (f) Give the domain and the range of $\cot(t)$.

27. Sketch $y = \sin(t)$ and $y = \cos(t)$ on the interval $[0, \frac{\pi}{2}]$.

   (a) Give the exact t-value of their point of intersection.

   (b) Approximate the instantaneous rate of change of $\sin(t)$ at the point of intersection. Use the

technique you learned in Section 6.2. (*Hint*: Don't use your own decimal approximation for π; use your calculator's π key.)

   (c) Approximate the slope of the graph of $\cos(t)$ at that same point.

   (d) Use your graph to explain why the results of parts (b) and (c) are reasonable.

28. The **linear approximation** to a function f at a point P is the line containing P whose slope is the same as the slope of the function's graph at P. Find the following linear approximations:

   (a) To $y = \sin(t)$ at $t = 0$

   (b) To $y = \sin(t)$ at $t = 2\pi$

   (c) To $y = \cos(t)$ at $t = 0$

   (d) To $y = \cos(t)$ at $t = \frac{\pi}{2}$

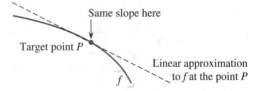

29. Explain why the two functions $\cos^2(x^3 - 4) + \sin^2(x^3 - 4)$ and $\sin^2(\sqrt[3]{x}) + \cos^2(\sqrt[3]{x})$ produce the same graph. What is a much simpler formula for that graph?

30. What is the slope of the graph of $y = \sin^2(10^t + 1) + \cos^2(10^t + 1)$? Explain.

For Exercises 31–34, write a formula for a sine function with the given characteristics. Then give its y-intercept, correct to two decimal places.

31. Amplitude 10, period $\frac{\pi}{4}$, vertical shift five units downward

32. Amplitude $\frac{1}{2}$, period 6π, horizontal shift two units to the right

33. Amplitude 1, period 6, horizontal shift two units to the left

34. Amplitude 3, period 3π, horizontal shift $\frac{\pi}{4}$ units to the right

35. Here are the graphs of $\sin(2x)$, $\sin(\frac{1}{2}x)$, and $2\sin(x)$. Which is which? Give the exact values of a, b, and c.

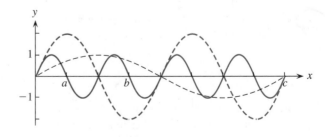

36. In Section 6.3, we presented a sine function representing the average high temperatures, day by day, in New York City. A graph of the average daily *low* temperatures in New York City also looks like a sinusoid (see Figure 6.2, page 286). The minimum average low temperature is 25°F, occurring on or about January 30 (day 30); the maximum average low temperature is 69°F, occurring on or about July 31 (day 212).

    (a) Sketch a graph of the average daily low temperature function over a year's time.

    (b) Write a formula for the function, with time measured in days.

    (c) Write another formula for the same function but with time measured in months. You may assume that the low point occurs after one month and the high point after seven months.

    (d) If you used a sine function in part (c), write an equivalent formula involving the cosine. (If you did part (c) with a cosine, then use a sine here.)

    (e) At what two times of the year does the average temperature change most rapidly?

37. You're driving at night behind a cyclist, both of you maintaining a steady speed. Your headlights illuminate the reflectors on the pedals, and you notice that the height above the road of each reflector varies sinusoidally with time. The cyclist is pedaling at 60 revolutions per minute. The height above the road of each pedal varies from 4 to 18 inches.

    (a) Sketch a graph showing a height of the reflector as a function of time. Choose the time unit carefully.

    (b) Write a formula for the function.

38. In Shanghai, the ferris wheel that will be the world's tallest (200 meters) is slated for completion in 2005. It's so large that a full revolution of the wheel will take 45 minutes. In this problem, we will assume that the diameter of the wheel is 198 meters and that the bottom car clears the ground by 2 meters. We'll also assume that new riders board at the low point of the cycle and that you are the last one to climb aboard.

    (a) Sketch a graph of your elevation above the ground as a function of time during one complete revolution.

    (b) Write a formula for the function, using a cosine.

    (c) Write a formula for the function, using a sine.

    (d) At the top of the wheel, you feel as though you are floating. What is the instantaneous rate of change of your height, in meters per minute, at the moment you reach the top?

    (e) Where are you when your height is changing most rapidly? What is that instantaneous rate of change?

A pendulum clock keeps time because the back-and-forth motion of its pendulum is periodic. The length of time, T, needed to complete one back-and-forth cycle is called the *period* of the pendulum. The period depends on the length of the pendulum: $T = 2\pi\sqrt{\frac{l}{g}}$, where l is the length of the pendulum in inches and g is the acceleration due to gravity in inches per second per second. For Exercises 39 and 40, use the value 384 $\frac{in.}{sec^2}$ for g.

39. An antique clock displayed at Old Sturbridge Village (a recreated colonial village in Massachusetts) has a pendulum 39.375 inches long. The pendulum is centered in a case 12 inches wide.

    (a) Determine the period of the pendulum, showing algebraically that the units for T are *seconds*. (Use the preceding formula for T.)

    (b) As the pendulum swings, it oscillates symetrically between a point 3.75 inches from the left side of its case and a point 3.75 inches from the right side. Write a sinusoidal model representing the distance of the pendulum from the left side as a function of time.

    (c) At approximately what point in the swing is the pendulum moving most rapidly away from the left side of the case? How fast is it going then?

40. An antique clock tower at Sturbridge Village has a pendulum 350.25 inches long. (Information on pendulums precedes Exercise 39.)

    (a) Determine the period of the pendulum, justifying the units.

    (b) As the pendulum swings, it oscillates between a point 11 inches from the left side and a point 36 inches from that same side. (This pendulum doesn't hang symetrically; the tower also contains a bell.) Write a sinusoidal model representing the distance of the pendulum from the left side of the tower as a function of time.

    (c) At approximately what point in its swing is the pendulum moving most rapidly toward the left side of the tower? How rapidly is it moving?

For Exercises 41–46, write a formula of the form
$A \sin[B(t - C)] + D$ or $A \cos[B(t - C)] + D$ that could
represent the given graph.

41.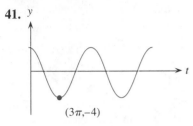
(3π,−4)

42.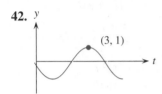
(3, 1)

43.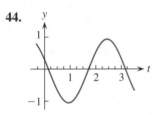

44.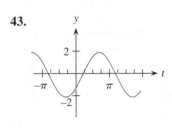

45.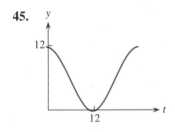
12
12

46.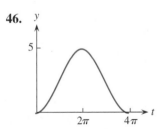
5
2π   4π

47. Find the instantaneous rate of change of $\tan(t)$ at
$t = 0$. (Keep narrowing the window until you feel
confident of the exact slope value.) Compare your
result with the slope of $\sin(t)$ at $t = 0$. (See page
300.)

48. Find the instantaneous rate of change of $\sin(2t)$ at
$t = 0$. (Keep narrowing the window until you feel
confident of the exact slope value.) Compare your
result with the slope of $\sin(t)$ at $t = 0$. (See page
300.) Explain, using the graphs of $\sin(t)$ and $\sin(2t)$,
why your result is reasonable.

49. Sketch a graph of one fundamental wave of the
function $3 \cos(3t + \frac{\pi}{4})$.

50. Sketch a graph of one fundamental wave of the
function $10 \sin(120\pi t - \frac{\pi}{4})$.

51. Find a function equivalent to $\sin(t)$ that uses the
cosine function instead.

52. Find a function equivalent to $\sin(2t)$ that uses the
cosine function instead.

In Exercises 53–62, find all solutions or state that none
exist.

53. $\cos(t) = 0.44$

54. $\cos(t) = -0.25$

55. $3 \cos(4t) = 1$

56. $\cos(2t - 1) = 0.25$

57. $\tan(2t) = 2$

58. $\sin(2t) = 2$

59. $\cos^2(3t) + \sin^2(3t) = 1$

60. $\cos^2(t) - \sin^2(t) = 0$

61. $\sin(t) = 5 \cos(t)$

62. $\sin(2t) + \cos(t) = 0$ (*Hint*: Replace $\sin(2t)$ with
an equivalent expression whose argument is simply t.
Then perform some factoring.)

63. Find all the solutions to the equation
$1 - \sin(\frac{\pi}{2} t) = 1.25$ on the interval $-4 \leq t \leq 4$.
(See CYU 6.8, page 311, for a graph of the related
function.)

64. Sketch graphs of $\cos(t)$ and $\sin(2t)$ on the same set
of axes. Use the entire interval $-2\pi \leq t \leq 2\pi$. The
graphs should intersect in eight places. Use algebra
to find the coordinates of the eight points. Verify
your results with a grapher.

65. The graph below represents a cosine wave. Find the
values of a and b both graphically and algebraically.
(You'll need a formula for f first.)

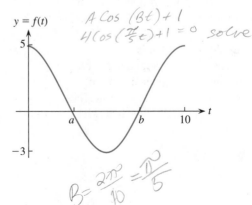
y = f(t)

66. The graph below represents a sine wave. Find the values of a and b both graphically and algebraically. (You'll need a formula for g first.)

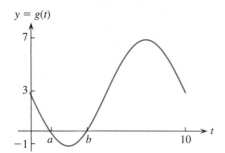

$y = g(t)$

67. A clock is mounted on the wall, its bottom 80 inches above the floor. The diameter of the clock is 12 inches, and the length of the hour hand is 4 inches.
   (a) Write a sinusoidal model that gives the position $H(t)$ (that is, its height above the floor) of the tip of the hour hand t hours after midnight.
   (b) At 4:00 A.M., how far from the floor is the tip of the hour hand?
   (c) What time is it when the tip of the hour hand is 86 inches from the floor. (This question has two answers.)
   (d) What time is it when the tip of the hour hand is 87 inches from the floor (again, two answers)?

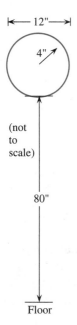

68. A weight suspended from a spring is pulled downward 10 centimeters from its equilibrium position. At $t = 0$, the weight is released. It bobs up and down, making one complete (up and down) oscillation every 2 seconds.

(a) Write a sinusoidal model $H(t)$ to represent the height of the weight, relative to its equilibrium position, t seconds after release. (We will ignore the effects of friction, which would cause the amplitude of the oscillations to diminish with time.)
(b) At what time, to the nearest 0.01 of a second, does the weight reach the height of 6 cm above equilibrium for the first time? Show an algebraic solution.
(c) At what time, to the nearest 0.01 of a second, does the weight pass the 6-cm mark for the second time? Show calculations.

69. Does the equation $\cos(t) = t$ have a solution? If so, is there more than one solution? Show how you decided. Give a two-place decimal approximation for each solution you find.

70. Does the equation $\tan(t) = t$ have a solution? If so, is there more than one solution? Explain. (For an enlightening view, get the equation into a form in which one side involves $\sin(t)$ while the other involves $\cos(t)$. Then graph each side of the revised equation.)

6.1 COPYCATS—Exploring Sine and Cosine Functions

In this activity, you will investigate the effects of various modifications to the basic sine and cosine graphs, and will see them as examples of transformations that you have already learned—shifts, stretches, and so on.

1. On the interval $[0, 2\pi]$, draw a graph of $\sin(x)$ and of $\cos(x)$. You have drawn what we call a **fundamental wave** of each of those functions. Notice that the fundamental sine wave starts and ends on its center line, while the fundamental cosine wave starts and ends at its highest value.

   (a) $y = \sin(x)$

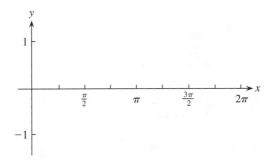

   (b) $y = \cos(x)$

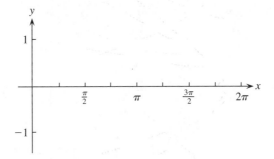

2. Experiment with the sine and cosine graphs by multiplying the functions by various constant values: $A \sin(x)$ and $A \cos(x)$. Use both positive and negative values for A, including some numbers between -1 and 1.

   (a) Write a summary of your observations, interpreting the effects you see as the result of one of the transformations you have studied.

   (b) The absolute value of the coefficient A is called the **amplitude** of the sine or cosine function. Does this seem to you to be an appropriate name for $|A|$?

3. The **period** of a sine or cosine function means the width of the smallest interval that contains one complete fundamental wave. What is the period of $\sin(x)$? Of $\cos(x)$?

   (a) Speed up the wave: Fit two sine waves into the interval $[0, 2\pi]$. What formula did you use to create the graph?

   (b) Now fit three waves, four waves, five waves into that interval. Write the formula for each function you used.

   (c) How would you slow the wave down, causing it to use twice as much horizontal space to complete its fundamental pattern? Write the function.

Transformations: shifts, reflections, stretches, compressions

(d) Can you find a sine wave with a period of one unit? Give it a try. If you find yourself frustrated, move on and come back to this.

(e) Interpret the effects you see as the results of another transformation you have learned.

The constant B controls the period, but the period is not B.

4. To complete Question 3, you needed to write functions of the form $\sin(Bx)$. The multiplier B does not have a mathematical name, but it is important to each of these functions because it determines the period. There is a connection between the value of B and the period of a sine or cosine function; now you'll discover their relationship.

(a) As B grows, does the period lengthen or shorten?

(b) As B shrinks toward 0, what happens to the period?

(c) If your answers to parts (a) and (b) are correct, they suggest that the value of B and the length of the period of $\sin(Bx)$ or $\cos(Bx)$ are *inversely proportional*— that is, that

$$\text{period} = \frac{k}{B}$$

where k is a constant that we will determine. Use the fact that the period is 2π when $B = 1$ to find k.

(d) Write a formula to find the period of $\sin(Bx)$ or $\cos(Bx)$ if you know B.

(e) Write another formula for finding B if you know the period.

5. Sine and cosine waves often come with a horizontal shift that causes them to begin their fundamental pattern somewhere other than in the accustomed place. You already know how to shift a graph laterally (that is, horizontally). Now you will apply that technique to sine and cosine graphs.

(a) Shift the graph of $\cos(x)$ one unit to the left. What formula did you use?

(b) Shift the graph of $\sin(2x)$ one unit to the left. Write the formula you used. Be sure that the graph shifted exactly one unit. If it didn't go far enough, check parentheses. Remember that to cause a horizontal shift, we add a constant directly to the independent variable (before doing anything else to that variable).

(c) Shift the graph of $\sin(x)$ to coincide with $\cos(x)$. What formula did you use?

(d) Shift the graph of $\cos(x)$ so that it coincides with itself. Give the formula.

(e) Shift the graph of $\sin(2x)$ so that it coincides with $\cos(2x)$.

6. Not all waves oscillate about the x-axis. Some are higher; others are below the axis.

(a) Find a sine function that oscillates about the line $y = 12$.

(b) Find a sine function with an amplitude of 3 whose top bumps the x-axis but does not cross it. (Mathematicians would say that the graph is *tangent* to the x-axis at its peaks.)

7. Now that you have investigated four different ways in which a sine or a cosine wave can be transformed, here are some practice problems. Predict the appearance of each wave in terms of its amplitude, its period, its horizontal shift, and its vertical shift. Then check your prediction by viewing the graph.

(a) $-10\sin(2x) + 10$

(b) $3\cos(0.25(x - 1))$

(c) $1 - \cos(\pi x)$

(d) $\sin(2\pi(x + 0.1))$

(e) $\cos(3x + 1)$ (Be careful with the horizontal shift!)

6.2 ALIASES—Investigating Trig Identities

In mathematics, the equals sign has three different meanings:

- To indicate a definition, as in $f(x) = \cos(x)$. Here, the equals sign says, "The function f is defined by this mathematical expression."

- To denote a conditional equation, as in $\cos(x) = \sin(x)$. This statement is not *always* true, but it is true for certain values of x.

- To specify an **identity,** as in $\cos(-x) = \cos(x)$. This statement is true for *any* value of x.

Traditional trigonometry texts present pages of identities for students to memorize or to derive. Identities deserve mathematical proof before we accept them, but graphs can at least provide persuasive arguments in favor of an identity and can show that a particular relationship is *not* an identity.

Remember that if two expressions are mathematically equivalent, they have the same graph. In this investigation, you will examine the graphs of several expressions and decide which ones are equivalent.

1. The central identity of trigonometry is $\sin^2(x) + \cos^2(x) = 1$. Explain why that statement is always true. A sketch of the unit circle would help.

2. What sort of graph should the function $y = \sin^2(x) + \cos^2(x)$ have? Make your prediction; then view the graph. You will probably have to enter the function as $(\sin x)^2 + (\cos x)^2$.

3. Examine the graphs of $\sin(-x)$, $\cos(-x)$, and $\tan(-x)$ and rewrite each function in a form that uses x, rather than $-x$, as its argument.

4. Every expression on the left can be matched with an expression on the right to form an identity. Use your grapher to assist you in writing the ten identities.

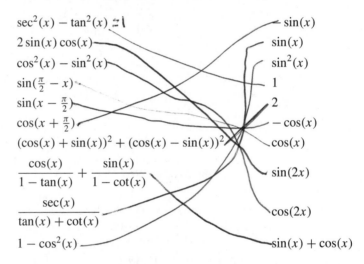

5. Four of the following functions have the same graph. Compare the graphs to discover the four mathematically equivalent functions.

$$\cos(2t) \qquad 2\cos(t) \qquad \sin^2(t) - \cos^2(t) \qquad 1 - 2\sin^2(t)$$

$$\cos(t/2) \qquad 2\cos^2(t) - 1 \qquad \cos^2(t) - \sin^2(t) \qquad \sin^2(t) + \cos^2(t)$$

6.3 DOWN THE DRAIN

Engineers concerned with the public water supply need to know the daily water usage patterns for their community. A municipality typically has water storage tanks to provide a reserve for periods of high demand. When large quantities of water are being used, the amount of water stored goes down. During periods of light demand, the tanks refill from the municipal wells or reservoirs. Engineers monitor the water level in the storage tanks and can control the flow by opening or closing valves. They try to smooth out the flow so that a reserve is maintained.

The circular graph shows one week's readings from a device that records the water level in a town's storage tank. A pen traces a path far from the center when the water level in the tank is high; the pen moves toward the center when the water level is low. Although there is some variation and some asymmetry, we can see a pattern: The level is at its highest every morning and at its lowest every evening.

The circular graph was created when a disk of graph paper revolved slowly under a moving pen. If we were to imagine the graph "unrolled," we could see time (in hours) on the horizontal axis and water level (in feet) on the vertical axis. The resulting curve would resemble a rough sine wave.

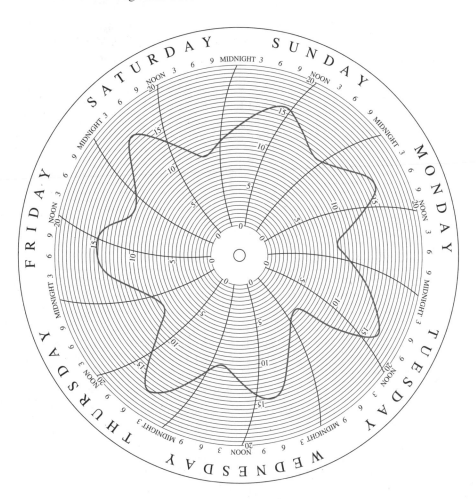

A mathematical model simplifies reality, capturing the key points of the situation in a function that can be analyzed mathematically. What you learn from the mathematical analysis sheds light on the real-world situation.

Your task is to find a function that would approximate that sine wave. You will need to take some liberties with the data. The amplitude is not constant from day to day; decide upon a representative amplitude. The high and low points don't occur at the same time each day. To write a sine function, you need them to be equally spaced, so feel free to assign them to 8:00 or 9:00 or whatever seems right to you. Once you've made those leaps into imprecision, you'll be able to write a sine function. (We should note that such leaps have to be taken if we're to get anywhere. If we refuse to use anything but the exact numbers, the model that we get to represent our reality will be just as large and unwieldy as the reality itself.) Let the independent variable represent time in hours since midnight.

1. Write the formula that you came up with. Explain how you decided the value for each constant.

2. Draw a graph of your water-level model, putting scales and labels on the axes.

3. The water-level model (the idealized sine function you just wrote) also tells us something about water usage patterns in the town. Whenever more water is coming in from wells or reservoirs than people are using, the tank fills up. If people are using more water than is coming in, the tank empties.

(a) At what time of day is the greatest amount of water being used? (This is a tricky question. You need to think not about the level of water in the tank, but about the rate at which water is being drawn out of the tank.)

(b) At what time is the least amount of water being used?

Explain carefully how you determined your answers to the last two questions.

6.4 FUELS RUSH IN

In a home that uses natural gas for heating and cooking, the owners saved all their gas bills for 1994–95 (except the bill for March 1995, which accidentally went out with the recycling). Each bill reports the amount of gas used in therms (a *therm* is 100 cubic feet) and the average temperature for the month in degrees Fahrenheit. The data from 1994 and 1995 appear in the table. The months have been coded so that 1 represents January 1994 and 13 represents January 1995.

1994	1	2	3	4	5	6	7	8	9	10	11	12
Therms	81	75	57	37	17	13	11	10	11	17	22	48
Temp.	18	15	28	41	54	61	62	65	62	55	52	36

1995	13	14	15	16	17	18	19	20	21	22	23	24
Therms	72	68	?	40	30	14	10	10	11	12	27	54
Temp.	32	23	?	40	53	60	63	65	62	59	46	29

1. Make two plots on the same set of axes: gas usage versus time and temperature versus time. Use different symbols for gas and for temperature so that you can tell them apart. Each plot should look roughly sinusoidal. What is similar about the two plots? What is different? What, if anything, do these plots tell you about the relationship between temperature and gas consumption?

2. Assume that the gas-usage data can be modeled by a function of the form $G(t) = A \sin[B(t - C)] + D$. The following questions will help you to determine a sinusoidal model for gas usage versus time.

   (a) What should be the period of the model?

   (b) The average of the 1994 maximum and the 1995 maximum is a good estimate for the maximum value of the model G. Similarly, the average of the two minimum values will work as the minimum value of G. Using these estimates, determine the amplitude of the model and its axis of oscillation.

   (c) Determine appropriate values for A, B, and D based on your answers to parts (a) and (b). Write a preliminary formula for G, letting $C = 0$ for the moment.

   (d) The maximum gas usage first occurs when $t = 1$. For what positive value of t does your preliminary version of G first reach its maximum value? Use that information to find a good value for C, and finish the formula for $G(t)$.

   (e) Graph your model and the gas-usage data on the same set of axes. For what months does the model appear to fit well? For what months is it not such a good fit? In those months, does the model appear to overestimate or underestimate the actual gas usage?

   If your grapher is capable of fitting a sinusoidal function, complete part (f).

   (f) Use your calculator's regression capabilities to fit a sinusoidal model to the gas-usage-versus-time data. Does this model appear to do a better job of describing the data than the model that you determined on your own? Do you see any problems with the calculator model? Explain.

3. Now consider the relationship between temperature and time.

    (a) Using any method, determine a sinusoidal model for temperature versus time.

    (b) Superimpose a graph of your model on a plot of the data. Does the model appear to represent the data reasonably well? Do you see any problems with the model? Explain.

4. Use your models to estimate gas usage and temperature for the missing month, March 1995.

6.5 SAD AND LATITUDE

This project is a follow-up to the "Daylight" lab.

The incidence and severity of seasonal affective disorder seem to depend upon latitude. *Scientific American* (January 1989) reports that 24% of the population of Tromsø, Norway (latitude 69°N.), may suffer from midwinter insomnia, another manifestation of SAD. Tromsø is so far north (more than 200 miles above the Arctic Circle) that the people there do not even see the sun between November 20 and January 20. (They do, however, enjoy 24 hours a day of sun during the summer.)

1. Sketch a possible graph for Tromsø's length-of-day function. Put a scale on each axis. Does your sketch show a sine wave? Explain.

2. You might not be able to write a precise mathematical formula for the function whose graph you have drawn. It is a piecewise function, some segments of which are constant. Which segments are constant? What are their constant values?

3. In the southern hemisphere, where the pattern of daylight is reversed, SAD reaches its peak during June and July. Put scales on the axes and sketch a graph of the length-of-day function for Wellington, New Zealand, or Puerto Montt, Chile, both of which have an approximate latitude of 42°S. (They are the same distance from the equator as Boston but in the opposite direction.)

4. Show how to modify one of the mathematical models you derived in the SAD lab to obtain a formula for the southern hemisphere graph.

6.6 DAY BY DAY—An Alternative Variety of SAD

This project is a follow-up to the "Daylight" lab.

Some people with seasonal depression are more prone to feel symptoms in spring and fall than in winter. Their depression appears to be triggered not by a shortage of light, but rather by changing amounts of it. Psychologists wishing to help such patients might want to know the rate at which the amount of daylight is changing at various times of the year.

1. Using the length-of-day function you wrote for Boston, examine the graph and find the season of the year during which the days are lengthening most rapidly. (How does the graph show which time of year that is?) Estimate the instantaneous rate of change of the function at that point. With that information, determine how rapidly, in *hours per day*, the days are lengthening at that time of year. Now convert your calculation into *minutes per day*, to the nearest whole minute. How many minutes per week is this? Would a person be likely to notice the difference from one week to the next?

2. Do the same steps for the part of the year when the days are shortening most rapidly. Compare the results with your previous ones.

3. Repeat the two sets of rate-of-change calculations, but this time use the length-of-day function for Reykjavik. Compare your results with those for Boston. The number of daylight hours changes from day to day much more rapidly in Reykjavik than it does in Boston; approximately how many times as rapidly does it change?

4. Examine the region of the graph (either graph) where the greatest number of daylight hours occurs. What is the approximate rate of change, in minutes per day, for that region? Explain how you can read this information from the graph without having to do any calculations.

5. A psychologist studying this particular form of seasonal depression asks what you're working on. Explain, in a brief paragraph, the correlation between the two sine functions and the variety of SAD described at the beginning of this project.

6.7   JUST ALGEBRA

t	$\sin(t)$	$\cos(t)$	$\tan(t)$
0	0	1	
$\frac{\pi}{6}$	$\frac{1}{2}$	$\frac{\sqrt{3}}{2}$	
$\frac{\pi}{4}$	$\frac{1}{\sqrt{2}}$	$\frac{1}{\sqrt{2}}$	
$\frac{\pi}{3}$	$\frac{\sqrt{3}}{2}$	$\frac{1}{2}$	
$\frac{\pi}{2}$	1	0	

Complete the table by filling in the values for $\tan(t)$. Then use the table to assist you in finding exact solutions for the following conditional equations. (Find all solutions in the interval $0 \le t \le 2\pi$.) Factoring will help with many of these.

1. $2\sin^2(t) = 1$
2. $2\cos^2(3t) = 1$ (*Hint*: Let $x = 3t$; solve for x and then for t.)
3. $4\cos(t)\sin(t) = 2\sin(t)$
4. $\tan^2(4t) = 3$
5. $8\sin^2(t) - 8\sin(t) + 2 = 0$
6. $5 + 3\sin(t) = 2\sin^2(t)$
7. $\tan^2(t) - \tan(t) = \sqrt{3}(\tan(t) - 1)$
8. $2\cos^4(t) + \cos^2(t) - 1 = 0$
9. $4\cos^3(2t) = 3\cos(2t)$

The identity $\sin^2(t) + \cos^2(t) = 1$ establishes the fundamental relationship between the sine and cosine functions.

10. Divide both sides of $\sin^2(t) + \cos^2(t) = 1$ by $\cos^2(t)$. Simplify the resulting equation to obtain an identity involving only the tangent and secant functions.

11. Divide both sides of $\sin^2(t) + \cos^2(t) = 1$ by $\sin^2(t)$. Simplify the resulting equation to obtain an identity involving only the cotangent and cosecant functions.

Daylight and SAD

You Are My Sunshine

Preparation

Before coming to the lab, you should make yourself familiar with the graph of the sine function. You should know the meaning of the terms **amplitude** and **period.** Sketch the following graphs. Show scales on the axes.

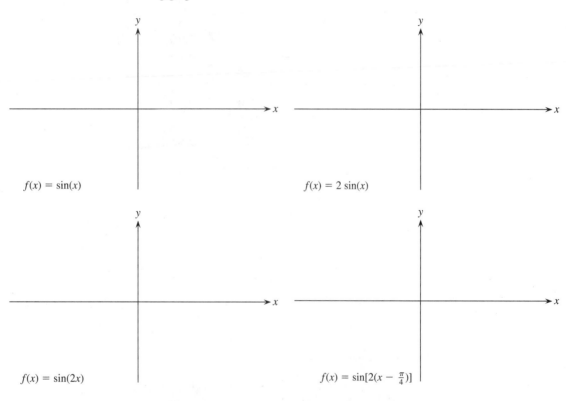

$f(x) = \sin(x)$

$f(x) = 2 \sin(x)$

$f(x) = \sin(2x)$

$f(x) = \sin[2(x - \frac{\pi}{4})]$

Now prepare the data for this lab. To see the annual pattern more clearly, use two years' worth of information for Boston and for Reykjavik. (For year 2, add 365 to each *day* value; don't change *hours of daylight.*) If you're working with your own graphing calculator, enter both data sets as lists before you come to the lab. If you don't have a grapher at home, then write out the two-year tables on paper so that you can plot the points quickly when you begin the lab.

The Daylight Lab

Over the course of a year, the length of the day—that is, the number of hours of daylight, calculated by subtracting the time of sunrise from the time of sunset—changes every day. Here is a table giving the length of day, rounded to the nearest tenth of an hour, for Boston, latitude 42°N.

Date	Day	Hours of Daylight	Date	Day	Hours of Daylight
1/2	2	9.2	7/5	186	15.2
1/10	10	9.3	7/13	194	15.0
1/18	18	9.6	7/21	202	14.8
1/26	26	9.8	7/29	210	14.5
2/3	34	10.1	8/6	218	14.3
2/11	42	10.4	8/14	226	13.9
2/19	50	10.8	8/22	234	13.6
2/27	58	11.1	8/30	242	13.3
3/7	66	11.5	9/7	250	12.9
3/15	74	11.9	9/15	258	12.5
3/23	82	12.3	9/23	266	12.1
3/31	90	12.7	10/1	274	11.8
4/8	98	13.0	10/9	282	11.4
4/16	106	13.4	10/17	290	11.0
4/24	114	13.8	10/25	298	10.7
5/2	122	14.1	11/2	306	10.3
5/10	130	14.4	11/10	314	10.0
5/18	138	14.8	11/18	322	9.6
5/26	146	14.9	11/26	330	9.5
6/3	154	15.1	12/4	338	9.3
6/11	162	15.2	12/12	346	9.2
6/19	170	15.3	12/20	354	9.1
6/27	178	15.2	12/28	362	9.2

Plot these points (hours of daylight versus day of year) on your grapher, using two years' worth of data so that you can see the pattern repeating itself. The shape you see should look like a rough approximation of a sine wave. In fact, the graph can be approximated by a function of the form

$$f(x) = A \, \sin[B(x - C)] + D$$

Your challenge in this lab will be to determine values for the four parameters A, B, C, and D.

Although many graphers can fit a sine function to a data set, we won't use that capability in this lab. Here's why: We want a model that not only fits the data well, but also has certain theoretical properties, such as a particular period that we already know from our experience of the natural world.

Let's first consider A, the **amplitude,** which determines the height of the wave. The constant A tells how far the wave will stray from its center line. How tall does the daylight wave need to be? Use the table to give you some ideas. Because we don't have data for every day of the year and because the values we do have are rounded off, we don't know exactly what the highest and lowest values are or on which days they occur. Find reasonable numbers to use for the highest and lowest values, given the data that we have. Decide a value for A, and check to see that it gives your sine wave the proper height.

Next, let's find B, which provides information on the **period** of the function. A standard sine function repeats its pattern every 2π units; that is, it has a period of 2π. What is the period of the daylight function? (After how many days does the length-of-day pattern repeat itself?) We will need to stretch out the standard sine wave if we want it to match the daylight pattern. Use the desired period, along with the 2π, to determine the necessary stretching factor, B. Keep in mind that the B-value itself is not the period of the function, but the horizontal stretching factor that produces the correct period.

How are you doing? Graph what you have so far, $A \sin(Bx)$, using your values for A and B, and compare its graph with the daylight graph. You should see a curve that has the correct size and shape but is in the wrong place.

Skip over C for the moment and consider D, which gives the average value of the sine wave. You probably see that your graph is too low, oscillating about the x-axis instead of up where the data points are. How much higher should it be? How did you decide? Use D to give your graph the boost it needs.

Once you determine D, you know the horizontal axis of the sine wave (overlaying the line $y = D$ would be helpful), and you can work on finding C, which gives the horizontal shift of the function.

Start by determining visually how much of a horizontal shift (how many days) your graph needs and in which direction. Remind yourselves how to write a horizontal shift into a function, and don't forget that you'll need to modify the independent variable *directly* before you do anything else (such as multiplication) to it. In other words, pay attention to parentheses.

Don't forget to relate this to the natural world: The start of spring is a significant day in the annual daylight cycle.

Try the completed model, $f(x) = A \sin[B(x - C)] + D$, using your values for A, B, C, and D. Its graph should hug the data points closely. If it does not, decide which constant needs to be adjusted. No matter how hard you try, though, you'll never achieve a *perfect* fit.

Remember that the data are approximations and so are the plotted points. When the grapher plotted the points, you might have noticed that few of the points went exactly where you tried to place them, because the screen has a limited number of pixels and the grapher does the best it can. Much of what you did in choosing the constants involved intelligent estimation. You found a function that models the data fairly well. There is more than one such function, and no function will be a perfect match.

Another difficulty inherent in modeling daylight is that the given pattern is *not* a perfect sine wave, because the apparent path of the sun is slightly elliptical rather than perfectly circular. That's a problem we are not going to attempt to fix in this lab!

The seasons affect everyone's moods to some degree, but some people are so strongly affected that they experience severe depression during the part of the year with the least amount of daylight. Many articles have been written about this condition, which has been termed *seasonal affective disorder*, or SAD. A typical person with SAD feels depressed for two or three months, sometime between the end of October and late February. She (women are affected more often than men) may experience a lack of energy and a craving for carbohydrates, and she may respond by oversleeping, overeating, and withdrawing from society. An estimated 6% to 8% of the population of New England suffers from full-blown SAD.

Unlike the traditional treatments for other forms of depression, an effective therapy for SAD has the patient sit in front of full-spectrum lights every morning. If we assume that 1 hour of light therapy is equivalent to an hour of natural daylight, approximately how many hours of light therapy might a person with SAD require on a day in early January if she wanted to make up for the "missing" hours of natural daylight (compared to March 21, the first day of spring)?

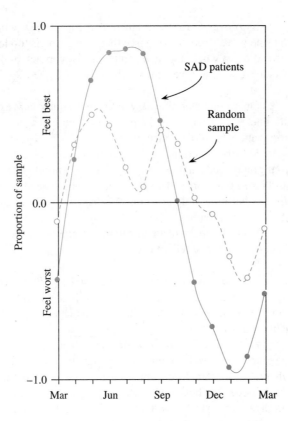

The graph represents results from a study* of SAD patients and a group selected at random from the New York City telephone book, in which they were asked to specify the months in which they felt best or worst. Each point shows the proportion of people feeling at their best or worst in a particular month. "Feeling worst" is counted as a negative value.

If the seasonal mood fluctuations of SAD patients could be approximated by a sine function, would the function be in phase (that is, in synch) with the length-of-day function or out of phase? Explain.

Notice that the effect is reversed during spring and summer: SAD patients may, in fact, feel better than the average person. They are full of energy and usually lose the extra weight they put on during the winter.

SAD appears to be even more prevalent farther north. The following table shows data for Reykjavik, Iceland, latitude 64°N.

When you plot these points, you will see that they do not seem to lie nearly so close to a smooth sine curve as Boston's points did. The Reykjavik data are less precise, and therefore, you must not take any single point too seriously. Nevertheless, it's possible to find a sine function that fits the data, taken as a whole, fairly well.

Writing such a function, $A \sin[B(x - C)] + D$, won't involve much more work than you've already done. Think about what features of the graph remain substantially the same. Which letters (A, B, C, or D) control those features? What features of the Reykjavik graph makes it different from the Boston graph? Which constants control those features? Calculate new values for those constants, and try out your new mathematical model.

*Michael Terman (1988) On the Question of Mechanism in Phototherapy for Seasonal Affective Disorder: Considerations of Clinical Efficacy and Epidemiology, in *Journal of Biological Rhythms*, Vol. 3, No. 2, pages 155–172.

Date	Day	Hours of Daylight	Date	Day	Hours of Daylight
1/2	2	4.5	7/5	186	20.5
1/10	10	5.0	7/13	194	19.9
1/18	18	5.7	7/21	202	19.1
1/26	26	6.5	7/29	210	18.3
2/3	34	7.3	8/6	218	17.4
2/11	42	8.2	8/14	226	16.5
2/19	50	9.1	8/22	234	15.7
2/27	58	9.9	8/30	242	14.8
3/7	66	10.8	9/7	250	13.9
3/15	74	11.7	9/15	258	13.1
3/23	82	12.5	9/23	266	12.2
3/31	90	13.4	10/1	274	11.4
4/8	98	14.3	10/9	282	10.5
4/16	106	15.1	10/17	290	9.7
4/24	114	16.0	10/25	298	8.8
5/2	122	16.9	11/2	306	8.0
5/10	130	17.8	11/10	314	7.1
5/18	138	18.6	11/18	322	6.3
5/26	146	19.5	11/26	330	5.5
6/3	154	20.2	12/4	338	4.9
6/11	162	20.7	12/12	346	4.3
6/19	170	21.0	12/20	354	4.2
6/27	178	20.9	12/28	362	4.3

The Lab Report

Your lab report should explain how you decided on the value of each parameter in the Boston length-of-day function and how you modified that function to write one for Reykjavik. Mention any difficulties you encountered in attempting to fit a sine curve to the data. Explain what information these models might provide about seasonal affective disorder, including (among other things) the effect of latitude. Include any observations you made about the graph of seasonal mood fluctuations and its relationship to the daylight models. Illustrate your report with appropriate graphs.

On the light side . . .

From 1948 until 1951 Japan practiced Daylight Savings Time. From April to September there was a summer time schedule of an additional hour of sunlight. This system was abolished because of the following reasons:

1. The sun set too late. An additional meal was required because the day was so long.

2. Longer hours for laborers.

3. Lack of sleep.

More information is available on request.

—Japanese Embassy Information and Culture Center publication, reprinted in The New York Times.

Follow-up Activities for The Daylight Lab

Projects 6.5 and 6.6 extend the ideas of Lab 6A. If the changing hours of daylight at various times of the year and in various parts of the globe interest you, check them out.

Once in a blue moon

Preparation

If you look at the moon on a series of clear nights, you'll notice that the size of the bright portion changes from night to night. The portion of the moon that we see from the earth appears as a circular disk, illuminated to some degree by direct sunlight. In reality, the moon's sphere is always half illuminated by the sun, but as the moon orbits the earth, we see an ever-changing fraction of the illuminated half. We give names to four phases: new moon, first quarter (when half the disk is illuminated), full moon, and last quarter (again, half is illuminated).

The photos display the moon's appearance on five nights in January 1999.

Courtesy: NASA

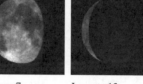

| January 1 | January 7 | January 13 | January 19 | January 25 |

The illuminated fraction of the apparent disk is the ratio of the illuminated area to the total area. On January 1, the moon was nearly full; hence, the illuminated fraction was close to 1. A new moon occurred on January 19. The moon's disk appeared totally dark and thus the illuminated fraction was 0.

Use the information from the five photos to sketch a graph of the illuminated fraction versus time over the three-month period from January 1 to March 31. This graph will be only an estimate, because you don't have enough information yet for an accurate graph. Define the variables; add appropriate scales to the axes.

Use your graph to predict how many full moons occurred between January 1 and March 31. Then predict the number of new moons in that time interval.

On the basis of your graph, do you think that the illuminated fraction changes at a constant rate from day to day? Explain.

The Moonlight Lab

In preparation for this lab, each of you drew a graph of the illuminated fraction of the moon's disk versus time. Compare graphs. How are they alike? How are they different? Did you agree on the number of full moons and new moons?

Creating a Model

As the photos show, the illuminated portion of the moon's circular disk changes from night to night. However, the five pictures do not provide enough information to create an accurate model. Table 1 provides data on the fraction illuminated (expressed as a decimal) every six days from January 1999 through March 1999.

X   y

30 Days {

Date	Day	Illuminated Fraction
1/1	1	0.99
1/7	7	0.73
1/13	13	0.19
1/19	19	0.03
1/25	25	0.55
1/31	31	1.00
2/6	37	0.71
2/12	43	0.17
2/18	49	0.05
2/24	55	0.62
3/2	61	1.00
3/8	67	0.70
3/14	73	0.15
3/20	79	0.08
3/26	85	0.69

Table 1 Fraction of the Moon Illuminated at Midnight* (E.S.T.) by Date in 1999

Use a grapher to plot the Table 1 data. (What is your independent variable?) Could a sine or cosine wave be used to describe the pattern that you see?

In fact, that's your next task: to approximate the pattern of the plotted data with a sinusoidal model by finding a function of the form $A\cos(B(t - C)) + D$. We'll provide suggestions to help you determine reasonable values for the four parameters, that is, the constants, A, B, C, and D.

First, consider the amplitude A (which determines the height of the wave) and the axis of oscillation $y = D$ (which determines the center of the wave). The moon is never fuller than full nor darker than new. Hence, the illuminated fraction oscillates between 0 and 1. Use this fact to determine A and D.

Next focus on the value of B. Recall that the standard cosine function $y = \cos(t)$ repeats its pattern every 2π units; the period of $\cos(t)$ is 2π. What is the period of the illumination function? That is, approximately how long is it between full moons or new moons?

The value of B in the standard cosine function is 1, creating a period of 2π units. Since the period of the illumination function is greater than 2π (≈ 6.28) days, you need to stretch out the standard cosine wave to make it fit the pattern of the moon data. Determine the stretching factor B that yields a cosine function with the desired period. (Remember that B is not itself the period of the function. It is, instead, inversely related to the period: smaller B's make longer periods.)

Now check your progress. Graph what you have so far, $A\cos(Bt) + D$, using your values for A, B, and D. The curve should have the correct size and shape, but it needs to be shifted horizontally in order to fit the moon data.

Finally, work on C: Estimate visually how much of a horizontal shift your graph needs and in which direction. (The moon, that year, was full on January 2.)

---

*A calendar day *begins* at midnight, 12:00 A.M.

Using this value of C, test the completed model, $f(t) = A\cos(B(t - C)) + D$. The graph should hug the data points closely. If it doesn't, adjust your model until it fits the data well. (Change the size or the sign of C if the graph seems to need a different horizontal shift. Adjust the value of B if your function appears to repeat its pattern too quickly or too slowly in comparison to the data.)

We had you use a cosine function rather than a sine function for this lab because the given data already resembled a cosine wave. With only a bit of additional work, though, you can now transform your cosine model into an equivalent sine model. Which of the parameters A, B, C, and D needs to change? Discuss the role of each parameter, and determine which ones can remain the same as you write a moonlight model of the form $A\sin(B(t - C)) + D$.

Now use your model (either version) to determine the number of full moons and the number of new moons that occurred from January 1 to March 31, 1999.

You've probably heard the expression "once in a blue moon." A *blue moon* is the second full moon to appear in a single month. Blue moons are relatively rare, occurring on average once every 2.7 years. January 1999 had a blue moon: The moon was full on January 2 and again on January 31. Was there another blue moon in 1999? Use your model and Table 2 to decide whether or not a second blue moon occurred in 1999 and, if so, in which month. If it did, then 1999 was a most unusual year, moonwise.

Month	Jan	Feb	Mar	Apr	May	Jun	Jul	Aug	Sep	Oct	Nov	Dec
Day	1	32	60	91	121	152	182	213	244	274	305	335

Table 2 Day Number of the First Day of Each Month

Rates of Change

In preparation for this lab, you were asked whether or not the size of the illuminated portion was changing at a constant rate. Explain how the data in Table 1 provide evidence that the rate of change is variable, not constant.

If you pay attention to the night sky, you'll notice that the size of the illuminated portion of the moon seems to change more rapidly during certain parts of its cycle than it does at other times. Let's figure how fast it was changing on a particular date, January 20, 1999, a date on which the moon was neither full nor new but somewhere in between. Your mathematical model will help; here's how:

• Set up a narrow interval centered on $t = 20$, perhaps 0.1 day on either side.

• Use your model to calculate the average rate of change in the illuminated fraction over that time interval.

Because the time interval was brief, your result is a good approximation to the instantaneous rate of change on January 20. What are the units for that rate?

You might have noticed that when the moon is about half illuminated, its appearance changes rapidly from night to night. At what rate is that fraction changing? Follow the above procedure, although this time you must first determine a t-value that corresponds to a half-illuminated disk. Compare this result with the rate you obtained for January 20. Does the result jibe with our experience that the moon's appearance changes rapidly when it's half full?

The preceding calculation isn't actually as simple as it might have appeared, because two different answers to the rate question are possible. If you obtained a single answer, go back and think about how you could obtain a second answer to the question "What is the rate of change of the illumination function when the moon is half full?" Now how are these two different rates related?

You might also have observed that it's difficult to tell, just by sky-gazing, the exact date of the full moon. Notice, for instance, that the illuminated fraction on January 1 was 0.99, and the full moon occurred the following night. Our eyes are unreliable judges; we're hard put to distinguish between 99% and 100%. Thus, our experience suggests that, around the time of the full moon, the moon's appearance must be changing very slowly from night to night. What is the instantaneous rate of change of the illumination function at the time of the full moon?

What is the instantaneous rate of change of the illumination function at the time of the new moon (when the lighted portion is zero)?

Now summarize your rate findings. At what time(s) in the moon's cycle is its appearance changing most rapidly? At what time(s) is it changing most slowly? Under what conditions is the rate of change negative? Positive? Zero?

The Lab Report

What do sines and cosines have to do with the moon? Your report should give both versions of your illumination function, explaining how you determined the value of each parameter. Tell how you used your model to decide how many blue moons occurred in 1999. Discuss the rate of change in the illuminated fraction with respect to time. In particular, how does this rate vary with the four phases of the moon—new, first quarter, full, and last quarter. Finally, why do you suppose that a half-illuminated disk is called a "quarter" instead of a "half"? Illustrate the report with an annotated graph of the illumination function.

Follow-Up to the Moonlight Lab

A Calculator Model

Use your calculator's regression capabilities to fit a sinusoidal model to the data in Table 1. In the same viewing window, graph the data from Table 1, your hand-fit model from the Moon lab, and the calculator's model. Which of the two models appears to be a better fit for the data?

What are the period and the horizontal shift of the calculator-fit model? (You might have to re-express it in the form $A \sin(B(t - C)) + D$ to answer this.) Compare these to the period and the horizontal shift of your own model.

Do you think that any of the calculator-generated parameters should be adjusted to make the model more consistent with reality? Explain.

Triangle Trigonometry

© Hiroyuki Matsumoto/Black Star Publishing/PictureQuest

Good engineering practice recommends that sewer pipes be laid with a $\frac{1}{4}$-inch vertical drop for each foot of pipe laid.[1] What should be the angle between the sewer pipe and the horizontal?

How can we measure the height of a mountain without leaving the plain?

A home-built picnic table requires an angled brace to stabilize the legs. At what angle should the amateur carpenter cut the piece of wood that forms the brace?

The local public library plans a wheelchair ramp for its main entrance. To create an angle of 15 degrees, how long a ramp is needed?

All of these problems can be solved by using a branch of mathematics known as **right-triangle trigonometry.**

The Greek astronomers Ptolemy and Hipparchus laid the foundations of trigonometry nineteen and twenty-one centuries ago, respectively. They developed mathematics to predict the paths and positions of heavenly bodies and to aid navigation, geography, and calendar-reckoning. These early astronomers were concerned with the measurement

---

[1]Too much slant and the liquid flows too rapidly, leaving solid waste behind. Too little slant and the flow hasn't enough force to carry the solid waste.

of angles (or arcs) and the relationships among the angles and the sides of a triangle. We still use the Greek name *trigonometry*, which means "triangle measurement," for the branch of mathematics they founded.

In Chapter 6, we defined the sine and the cosine functions in terms of the unit circle and used them as models for periodic phenomena. We then defined four more circular functions in terms of the sine and the cosine. The same six functions can be defined instead as ratios of the lengths of the sides of a right triangle. They are used in this way in engineering and many sciences, including geology, astronomy, and physics.

In this chapter, we define all six trigonometric functions in terms of right triangles. Then we show how to use them for any triangle, whether or not it has a right angle. Finally, we establish the connection between the two separate definitions of the sine, the cosine, and the other four functions.

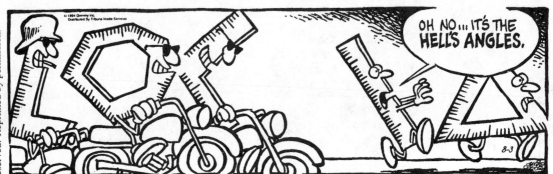

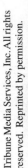

7.1   RIGHT TRIANGLES AND TRIGONOMETRIC FUNCTIONS

Similarity

You probably remember learning about similar triangles in high school. Two geometric figures are **similar** if they have the same shape, regardless of their size. A 1:200 scale model of a Jaguar—the car, not the beast—is *similar* to the real thing. The model's length, wheel diameter, and every other linear dimension is $\frac{1}{200}$ of the corresponding dimension of the actual car. The angle formed where the windshield meets the hood is the same in the model as it is in the car.

For triangles, it's simple: Two triangles are similar if each angle of one equals the corresponding angle of the other or if the ratio of any two sides of one equals the ratio of the corresponding two sides of the other. The right triangles in Figure 7.1 are similar because their respective angles are equal, giving both figures the same proportions. As a consequence of their similarity, we have the following equalities:

> Triangles are similar if corresponding sides are proportional or if corresponding angles are congruent.

$$\frac{a}{c} = \frac{d}{f}, \qquad \frac{b}{c} = \frac{e}{f}, \qquad \frac{a}{b} = \frac{d}{e}$$

If, in Figure 7.1, we were to change the size of an acute angle in one right triangle but not in the other, the sides would no longer be proportional. In other words, the size of angle θ, rather than the size of the triangle, completely determines the ratios $\frac{a}{c}$, $\frac{b}{c}$, and $\frac{a}{b}$. The converse is also true: In a right triangle, the ratio of any two sides determines θ. This relationship between angle and ratio allows us to define six functions—the sine,

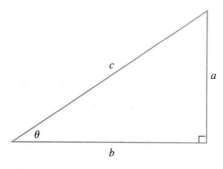
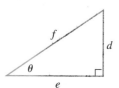

Figure 7.1 Similar Right Triangles

the cosine, the tangent, the cosecant, the secant and the cotangent—as follows:

$$\sin(\theta) = \frac{a}{c}, \qquad \cos(\theta) = \frac{b}{c}, \qquad \tan(\theta) = \frac{a}{b}$$

$$\csc(\theta) = \frac{c}{a}, \qquad \sec(\theta) = \frac{c}{b}, \qquad \cot(\theta) = \frac{b}{a}$$

You'll notice that these functions have exactly the same names as the ones in the preceding chapter. This is no coincidence. To understand that both sets of functions are compatible, consider the unit-circle representation for the sine and the cosine of an arc that terminates in Quadrant I, with a right triangle inscribed as in Figure 7.2.

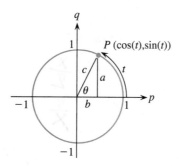

Figure 7.2 Unit-Circle Definition of Sine and Cosine

Using the unit-circle, we defined $\cos(t)$ to be the first coordinate of arc t's terminal point P. But you can see from Figure 7.2 that this is the same as b, the horizontal distance from P to the vertical axis. In the same way, we defined $\sin(t)$ as the second coordinate of P. But P's second coordinate is also given by a, its distance up from the horizontal axis. Moreover, c is 1, because this is the *unit* circle.

Let's compare the two sets of definitions:

Right-Triangle Definition	Unit-Circle Definition
$\sin(\theta) = \frac{a}{c} = \frac{a}{1} = a$ (in this case, $c = 1$)	$\sin(t) = a$ (P's second coordinate)
$\cos(\theta) = \frac{b}{c} = \frac{b}{1} = b$ (in this case, $c = 1$)	$\cos(t) = b$ (P's first coordinate)
$\tan(\theta) = \frac{a}{b}$	$\tan(t) = \frac{\sin(t)}{\cos(t)} = \frac{a}{b}$

Because angle θ determines arc t and vice-versa, the right-triangle definitions are completely compatible with the unit-circle definitions, and the other three functions are simply the reciprocals of these three. Angle θ is measured in degrees; arc t is measured in radians. This is why we switch modes on the calculator.

SOHCAHTOA

Maybe this isn't your first experience with right-triangle trigonometry. If that's the case, you might have heard the mnemonic device *SOHCAHTOA*. If, instead of using the letters a, b, and c for the side lengths, we think of those lengths in terms of their relationship to the angle θ—the side *opposite* θ, the side *adjacent* to θ, and the *hypotenuse*—we can relabel our triangle as in Figure 7.3.

$$\sin(\theta) = \frac{\text{opp}}{\text{hyp}} \qquad \csc(\theta) = \frac{\text{hyp}}{\text{opp}}$$

$$\cos(\theta) = \frac{\text{adj}}{\text{hyp}} \qquad \sec(\theta) = \frac{\text{hyp}}{\text{adj}}$$

$$\tan(\theta) = \frac{\text{opp}}{\text{adj}} \qquad \cot(\theta) = \frac{\text{adj}}{\text{opp}}$$

Figure 7.3 The Trigonometric Functions

▷   Explain what *SOHCAHTOA* helps you remember.

Check Your Understanding 7.1

CYU answers begin on page 368.

1. In triangle ABC, which side is opposite angle θ? Which side is opposite angle ϕ? Which side is the hypotenuse, and which angle is it opposite?

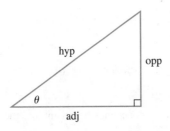

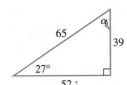

opposite $\overline{CD}$
hyp = $\overline{AB}$
adj = $\overline{AC}$

2. Referring to triangle ABC again, write an expression for the following trigonometric functions in terms of side lengths x, y, and z, remembering that "opposite" and "adjacent" sides depend on the angle you're considering.

$$\cos(\theta) = \frac{x}{z} \qquad \sin(\phi) = \frac{x}{z}$$

$$\tan(\theta) = \frac{y}{x} \qquad \tan(\phi) = \frac{x}{y}$$

3. How large is angle α?

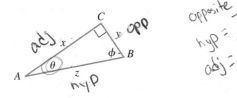

$90 - 27 = 63$

4. Find $\sin(\alpha)$.

$\sin(\alpha) = \frac{\text{opp}}{\text{hyp}}$

$\frac{52}{65} = \frac{65}{52}$

Solving a Right Triangle

Armed with trigonometric functions, we can find unknown angles and side lengths in any right triangle as long as we know two pieces of information, including the length of one side. Finding the remaining side lengths and angles is called **solving** the triangle. First, we'll show two purely numerical examples. Then we'll answer the sewer-pipe question and determine the elevation of a mountain.

Reminder: Degree mode.

A Side Length and an Angle

The first triangle in Figure 7.4 shows the problem. We know one side length and one angle. The second triangle shows its solution. The calculations that support the solution follow.

$$\sin(61°) = \frac{x}{47} \qquad\qquad \cos(61°) = \frac{y}{47}$$

$$x = 47\sin(61°) \approx 41.1 \text{ units} \qquad y = 47\cos(61°) \approx 22.8 \text{ units}$$

$$\text{angle } \gamma = 90° - 61° = 29°$$

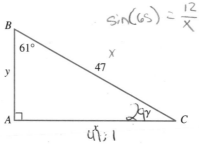

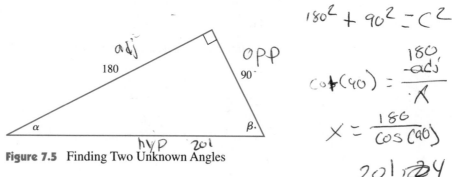

Figure 7.4 Finding Two Unknown Sides

Knowing the length of each side and the measurement of each angle, we have solved triangle ABC.

The Pythagorean theorem

$$a^2 + b^2 = c^2$$

▷ The cosine function isn't our only tool for finding y. Use the Pythagorean theorem to find y, knowing that $x = 41.1$ and that the hypotenuse measures 47 units.

Two Side Lengths

Next, we'll solve a right triangle for which we know two side lengths. In this example, we will use an *inverse* trig function, and you'll get to do some of the work.

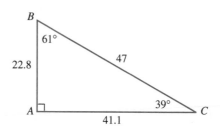

Figure 7.5 Finding Two Unknown Angles

▷ With respect to angle α, which sides (opposite, adjacent, hypotenuse) do we know?

Your answer should suggest the tangent function:

$$\tan(\alpha) = \frac{90}{180} = \frac{1}{2}$$

The symbol $\tan^{-1}\left(\frac{1}{2}\right)$ means "the angle whose tangent is $\frac{1}{2}$." It does not mean the reciprocal of a tangent.

Therefore, α is the angle whose tangent value is $\frac{1}{2}$. We don't know α, but we have its tangent, which is just as good. Such a situation calls for an inverse function.

$$\alpha = \tan^{-1}\left(\frac{1}{2}\right)$$

The $\tan^{-1}$ button on your calculator[2] (using degree mode) will tell you the value of α, approximately 26.6 degrees.

▷   Use a similar procedure to find β. Check your result by verifying that $\alpha + \beta = 90$.

▷   Complete the triangle by finding the unknown third side. You can choose among many methods; why not try the Pythagorean theorem first and then check the answer by using your favorite trig function?

$$\sin\left(26.6°\right) = \frac{90}{x}$$

$$x = \frac{90}{\sin(26.6)}$$

Flushed with success

Now we'll solve the sewer-pipe problem from page 349. We don't know the length of the pipe, but we do know the recommended ratio: $\frac{1}{4}$ inch of vertical drop for every foot of pipe.

▷   Which function of angle α do those specifications determine? Refer to Figure 7.6.

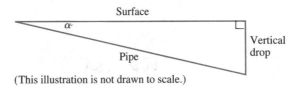

(This illustration is not drawn to scale.)

Figure 7.6   Maintaining a Fixed Angle of Depression for a Sewer Pipe

With respect to angle α, the two known sides of the sewer-pipe triangle are the *opposite* and the *hypotenuse*. Therefore, we'll use the sine function:

$$\sin(\alpha) = \frac{\frac{1}{4}\text{ inch}}{1\text{ foot}}$$

---

[2]The inverse tangent function is often called the *arctangent*. The symbol $\tan^{-1}$ fits better on a calculator key, but $\arctan\left(\frac{1}{2}\right)$ and $\tan^{-1}\left(\frac{1}{2}\right)$ have the same meaning.

▷ Finish the problem. You'll need the inverse sine function, $\sin^{-1}$ on your calculator. Remember to convert feet to inches first.

$$\tan(\alpha) = \frac{opp}{adj}$$

$$\tan = \frac{98}{302} = \left(\frac{49}{151}\right)$$

$$\alpha = \tan^{-1}\left(\frac{49}{151}\right)$$

$$= 18°$$

You should have found α to be a very small angle, only about 1.2 degrees. Do you understand why we did not draw the diagram to scale?

Check Your Understanding 7.2

1. In triangle ABC, angle C is a right angle. Side AC has length 14. Angle A has 39 degrees.
   (a) Draw a labeled diagram

2. In triangle DEF, angle F is a right angle. Side DE has length 100; side EF has length 47. Draw and solve triangle DEF.

   (b) Find angle B and sides AB and BC.

Right-triangle trigonometry is even more powerful than it appears at first glance. In the following example, we work with two right triangles. Neither, by itself, gives enough information, but together they allow us to solve the problem.

Angles we have seen on high

Every September, runners from all over New England meet at Mitch's Marina in South Hadley, Massachusetts, for the annual Summit Run. The 5-km race begins on a flat stretch of roadway along the Connecticut River and then turns up the steep road to the old Summit House atop Holyoke Mountain. The winner typically covers the course in under 20 minutes. How much vertical distance does each runner travel from start to finish? We'll use Figure 7.7 to find out.

From a point A on the plain near the base of the mountain, we used a surveyor's tool and measured the angle of elevation α as 53.5°. From a position B, 1000 feet farther from the mountain, we found angle β to be 28°. The task is to approximate height h, the distance from C to D. To set up ratios of side lengths of right triangles ACD and BCD, we introduce another variable, x, the distance from A to C. (Note that we can't measure

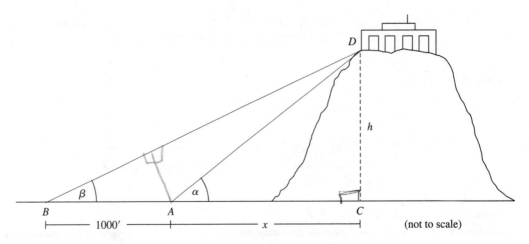

Figure 7.7   Finding the Height of a Mountain by Triangulation

the distance from A to C directly, because the mountain stands in the way.) Then we use the tangent function to set up two ratios involving x and h:

$$\tan(53.5°) = \frac{h}{x}$$

$$\tan(28°) = \frac{h}{1000 + x}$$

Use degree mode.

▷ Approximate $\tan(53.5°)$ and $\tan(28°)$ with your calculator. Substitute those values in the following equations:

$$\frac{1.35}{} = \frac{h}{x}$$

$$\frac{.53}{} = \frac{h}{1000 + x}$$

Trouble with simultaneous equations? See the Algebra Appendix, page 489.

▷ Solve the two equations for h. (*Hint:* Start by solving each one separately for x; then equate the two expressions that represent x.)

$$h = 1.35x \quad ; \quad h = 530 + .53x$$

$$1.35x = 530 + .53x$$

$$.82x = 530$$

$$x =$$

If your algebra is correct, you found Holyoke Mountain to be a small mountain, only about 877 feet from base to summit. Nevertheless, if you were one of the runners, you would feel that your feet had accomplished a monumental feat.

Check Your Understanding 7.3

From the top of a 150-foot-tall building, an observer watches a car moving toward the building. During the period of observation, the angle of depression (see diagram) of the car changes from 20° to 45°.

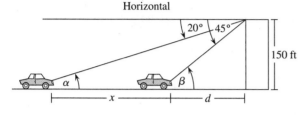

Horizontal

150 ft

1. Determine the angles of elevation α and β.

2. Write two equations involving $\tan(\alpha)$, $\tan(\beta)$, and the variables x and d.

3. How far has the car traveled in the time during which it was observed? (*Hint*: Find x by solving the system of equations in Question 2.)

In a Nutshell   The six trigonometric functions of an acute angle are ratios of the side lengths of a right triangle that contains the angle.

$$\sin(\theta) = \frac{a}{c}, \qquad \cos(\theta) = \frac{b}{c}, \qquad \tan(\theta) = \frac{a}{b}$$

$$\csc(\theta) = \frac{c}{a}, \qquad \sec(\theta) = \frac{c}{b}, \qquad \cot(\theta) = \frac{b}{a}$$

To solve a triangle means to determine any unknown angles and side lengths.

7.2   THE TRIGONOMETRY OF NON-RIGHT TRIANGLES

We were able to solve the problems in Section 7.1 because each triangle had a right angle. But trig functions are equally useful for oblique triangles—triangles in which no angle measures 90°. Because none of the angles is a right angle, we'll need *three* pieces of information, at least one of which must be the length of a side.

Distorting a square without changing any side lengths:

Initially, our strategy will be to divide the triangle into two right triangles. Then we'll derive two handy shortcuts, the Law of Sines and the Law of Cosines.

Of all polygons, only the triangle is rigid. This means that a triangle with three given sides cannot have a shape other than the one it has. This is not true of other polygons: A square, for instance, can be squashed into a rhombus. Two sides and

an angle, or two angles and a side,[3] also determine a unique triangle, though not a unique quadrilateral or other polygon. Because triangles are not distorted under pressure, engineers and architects use them as structural support elements. And because a relatively small amount of information completely determines a triangle, mathematicians like you can find out all there is to know about any triangle.

The Law of Sines

Triangle ABC has no right angle. We know two sides and one angle, as shown in Figure 7.8.

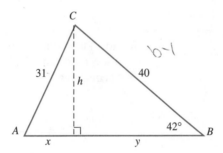

Figure 7.8

Let's begin by drawing the altitude from point C and labeling its length h. This subdivides triangle ABC into two right triangles with a common side. Now we'll use the sine of each base angle, because both of those sines involve a known side length and because they have h in common.

$$\sin(A) = \frac{h}{31} \qquad \text{and} \qquad \sin(B) = \frac{h}{40}$$

$$h = 31\sin(A) \qquad\qquad h = 40\sin(B)$$

Equate the expressions for h:

$$31\sin(A) = 40\sin(B)$$

$$31\sin(A) = 40\sin(42°)$$

$$\sin(A) = \frac{40}{31}\sin(42°) \approx 0.8634$$

To find the acute angle whose sine is 0.8634, use $\sin^{-1}(0.8634)$.

We have the sine of A. Now use an inverse sine:

$$A = \sin^{-1}(0.8634) \approx 59.7°$$

---

[3]There's one exception, called the **ambiguous case,** in which the same two sides and one angle can belong to two different triangles. If you're curious, read about it on pages 371–372 and do Exercises 19 and 20.

Check Your Understanding 7.4

In this CYU, you will finish solving the triangle in Figure 7.8.

1. Find angle *C*. (*Hint*: The sum of the angles in any triangle is constant.)

2. Using an appropriate trig function, find *x*.

3. Similarly, find *y*.

4. Using your answers to Questions 2 and 3, find *AB*.

The procedure you just completed can be adapted to any triangle. From the preceding work, however, we can gain insights to speed up the process. In finding angle *A*, we developed the following relationship:

$$31 \sin(A) = 40 \sin(B)$$

If we reprint the triangle as a generic figure without specified side lengths (Figure 7.9), we can say that

$$b \sin(A) = a \sin(B)$$

or that

$$\frac{\sin(A)}{a} = \frac{\sin(B)}{b}$$

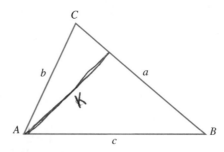

Figure 7.9

What about angle *C*, and how does it relate to the other two angles? The answer is an elegant theorem known as the Law of Sines. We'll help you derive the rest of it.

▷   In Figure 7.9, draw the perpendicular from point *A* to side $\overline{BC}$. Call its length *k*.

▷   Write an expression for sin(C) and an expression for sin(B).

▷   Solve each expression for k, then equate the two. Work with this equation until you obtain the result $\frac{\sin(B)}{b} = \frac{\sin(C)}{c}$.

Congratulations! You have just derived the following:

> ### Law of Sines
>
> $$\frac{\sin(A)}{a} = \frac{\sin(B)}{b} = \frac{\sin(C)}{c}$$

In words, we can say that the ratio of the sine of any angle to the length of its opposite side is *constant* for a given triangle.

Check Your Understanding 7.5

1. Points A and C are on opposite sides of a swamp. A surveyor measures angle A and distances AB and BC, in meters, as shown. Your task is to find AC, the width of the swamp.

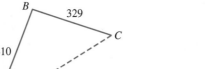

   (a) Use the Law of Sines to find angle C.

   (b) Find angle B, and don't jump to the conclusion that it's a right angle.

   (c) Use the Law of Sines again to find AC.

2. Sketch and solve triangle DEF, where angle D measures $34°$, angle E measures $48°$, and $DE = 7.4$.

The Law of Cosines

The Pythagorean theorem, $c^2 = a^2 + b^2$, applies only to right triangles. A similar relationship, however, works for *every* triangle:

> **Law of Cosines**
>
> $$c^2 = a^2 + b^2 - 2ab\cos(C)$$

▷ Show that if C is a right angle, the Law of Cosines gives the Pythagorean theorem.

Thus, the Pythagorean theorem is actually a special case of a more general theorem, the Law of Cosines. Proving that law isn't difficult; let's do it.

Triangle ABC in Figure 7.10 represents any triangle. From point A, we draw the altitude, which has length h and divides the opposite side of the triangle into segments of length x and $a - x$. By the Pythagorean theorem (using the right-hand triangle), we have

$$(a - x)^2 + h^2 = c^2$$
$$a^2 - 2ax + \underbrace{x^2 + h^2}_{b^2} = c^2$$

The Pythagorean theorem also tells us (from the left-hand triangle) that $x^2 + h^2 = b^2$. Substituting and rearranging, we have

$$a^2 + b^2 - 2ax = c^2$$

This is almost what we're aiming for. Look at Figure 7.10 again, and see that $\cos(C) = \frac{x}{b}$, making $x = b\cos(C)$. Replace x with $b\cos(C)$, and we're done:

$$a^2 + b^2 - 2ab\cos(C) = c^2$$

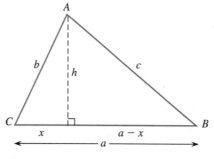

Figure 7.10

Because there was nothing special about triangle ABC (other than the implicit assumption that all its angles were acute angles), what we proved here is valid for any two sides of the triangle and the angle between them. In the exercises, you can prove that the Law of Cosines holds even when angle C is an obtuse angle.

Check Your Understanding 7.6

1. Two cyclists depart from the same place at the same time on straight roads that form an angle of 20°. One maintains a speed of 22 miles per hour, and the other rides at 18 miles per hour.

(a) After 1 hour, what is the distance between the cyclists?

2. In triangle XYZ, find the measure of angle X.

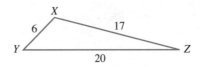

3. Sketch and solve triangle DEF, where angle D measures 125°, $DE = 9.7$, and $DF = 11.8$. Use the Law of Cosines and/or the Law of Sines, as appropriate.

(b) How far apart were they after the first 6 minutes?

In a Nutshell   Trigonometry, a branch of mathematics that was developed from right triangles, also provides two useful tools for solving oblique triangles:

Law of Sines: $\dfrac{\sin(A)}{a} = \dfrac{\sin(B)}{b} = \dfrac{\sin(C)}{c}$

Law of Cosines: $c^2 = a^2 + b^2 - 2ab\cos(C)$

The Law of Cosines is useful when two sides and their included angle (SAS) are known or when all three sides (SSS) are known. If you know one side and two angles (ASA or SAA) or you know two sides and an angle opposite one of them (SSA), try the Law of Sines.

7.3   ANGLES, ARC LENGTHS, AND RADIANS

In Chapters 6 and 7, we have defined the sine and cosine functions in two apparently unrelated ways:

- As coordinates associated with a directed arc on the unit circle

- As ratios of lengths associated with an angle in a right triangle

We already gave a hint of the compatibility of these definitions in Section 7.1. In this section, we establish the remaining connections between trigonometric functions of directed arcs and trigonometric functions of angles.

From Arc Lengths to Radians

Raiders of the lost arc

In Chapter 6, we used t as the variable representing a directed arc length on the unit circle. Recall that a t-value of 2π means traversing the unit circle once in the counterclockwise direction (the circle's circumference is 2π units), and a t-value of $\frac{\pi}{2}$ means traversing one fourth of the circumference. For a circle with a radius different from 1, we can still use directed arc length to measure the amount of rotation. In a circle of radius 5, for example, a directed arc of length $5 \cdot 2\pi$ or 10π units would wrap once around the circle in the counterclockwise direction. A directed arc of length $\frac{5\pi}{2}$ (see Figure 7.11) would wrap one fourth of the way around that circle.

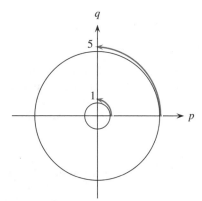

Figure 7.11   Two Arcs Corresponding to a Quarter Turn or $\frac{\pi}{2}$ Radians

Hence, for circles of radius one and five units, a quarter-turn rotation corresponds to directed arcs of length $\frac{\pi}{2}$ and $\frac{5\pi}{2}$, respectively. One way to eliminate the dependence on the radius of the circle is to measure arc length in terms of the number of radii. Thus, in the circle of radius 5, $\frac{\pi}{2}$ *radii* (of five units each) would cover an arc of length $\frac{5\pi}{2}$ units.

▷   For the circle of radius 5, how many radii are equivalent to one complete revolution? That is, how many radii fit around the circumference of that circle?

You should conclude that, independent of the size of the circle, exactly 2π radii fit around the circumference. That's another way of stating the familiar formula $C = 2\pi r$.

Directed arcs measured in terms of the number of radii are said to be measured in **radians.** Both arcs shown in Figure 7.11 have the same radian measure: $\frac{\pi}{2}$.

▷ Draw two arcs of $-\frac{3\pi}{4}$ radians in Figure 7.11. Although they differ in length, they have the same radian measure.

From Radians to Angles

A clean sweep

Another way of traversing a circle is to sweep out a central angle θ, measured in degrees. We start with a ray whose initial position is the positive p-axis and rotate the ray θ degrees in the counterclockwise direction, as shown in Figure 7.12. In the process, the ray sweeps through the shaded sector of the circle and determines an arc on the circumference. This sets up a correspondence between the degree measurement of the central angle θ and the radian measure t of the arc.

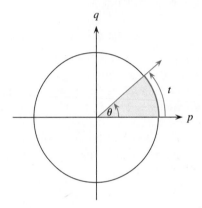

Figure 7.12 Central Angle and Corresponding Arc

A 360° rotation of this ray will sweep out one entire circumference of the circle, or an arc whose radian measure is 2π. Thus, 360° is equivalent to 2π radians. The angle measures the arc, and the arc measures the angle. This means that we can speak of any angle by giving its radian measure. So, for instance, a right angle is an angle of $\frac{\pi}{2}$.

> **Degree-Radian Equivalence**
>
> $$360° = 2\pi \text{ radians}$$

From this correspondence, we obtain two useful conversion equations:

$$1 \text{ degree} = \frac{\pi}{180} \text{ radians}$$

$$1 \text{ radian} = \frac{180}{\pi} \text{ degrees}$$

Thus, a central angle of 135° will sweep out an arc of $135 \cdot \frac{\pi}{180}$, or $\frac{3\pi}{4}$ radians.

Check Your Understanding 7.7

1. How many degrees are there in an angle of $\frac{5\pi}{6}$ radians?   **2.** What's the radian measure of an angle of 12°?

STOP & THINK   Write a formula that converts any angle θ, measured in degrees, into radian measure. Then write a formula to convert any angle of t radians into degrees.

Extending the Trigonometric Functions

When we drew a right triangle inside the unit circle in Section 7.1, we gave a hint of the close relationship between central angle and arc. Now we can extend those ideas to negative angles and to angles larger than 180°. An angle of 207° cannot be part of a triangle, nor can a negative angle. Nevertheless, the unit circle gives meaning to expressions such as $\cos(207°)$ and $\sin(-56°)$. As Figure 7.13 suggests, we can use **reference right triangles** for both 207° and −56°.

The 207° central angle in Figure 7.13 is formed by rotating the ray made by the positive p-axis through 207° counterclockwise. Let C be the point where the ray in its terminal position cuts through the unit circle. A vertical line drawn from C to the horizontal axis forms a right triangle with the p-axis. That triangle is the reference triangle for an angle of 207°. The **reference angle** is the angle in the reference triangle between the horizontal leg and the hypotenuse. For an angle of 207°, the reference angle is the difference between 207° and 180°, or 27°.

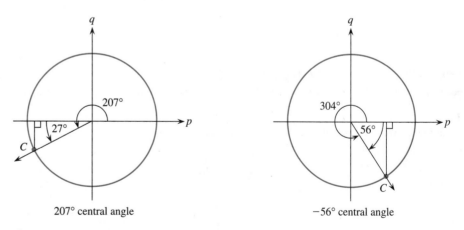

207° central angle   −56° central angle

Figure 7.13 Reference Triangles for Nonacute Angles

For the angle of −56°, rotate the ray in the *clockwise* direction from its initial position on the p-axis. Notice that the ray has the same terminal position as it would have after a counterclockwise rotation of 304°. Again, form a triangle with the ray in its terminal position, the horizontal axis, and the vertical line drawn from C. The reference angle is 56°.

The amount of rotation of the ray can be more than a single revolution. An angle of 405°, made by rotating the ray $1\frac{1}{8}$ turns, has a reference angle of 45°, or one eighth of a complete revolution.

▷   What is the reference angle for an angle of −405°?

To extend the definitions of the sine and cosine functions to nonacute angles, we determine the *magnitude* of the sine or cosine value from the reference angle and the *sign* of the sine or cosine value from the quadrant in which the reference triangle lies (see Figure 7.14). Here's how.

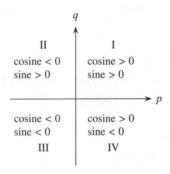

Figure 7.14

The reference triangle (see Figure 7.13) for 207° lies in the third quadrant, where both sine and cosine are negative. The reference angle is 27°; $\cos(27°) \approx 0.8910$ and $\sin(27°) \approx 0.4540$. Therefore, $\cos(207°) \approx -0.8910$ and $\sin(207°) \approx -0.4540$.

By a similar process, we see that $\cos(-56°) = \cos(56°)$ and $\sin(-56°) = -\sin(56°)$.

▷   Use your calculator (degree mode) to find $\cos(207°)$ and $\cos(27°)$. Verify that $\cos(207°) = -\cos(27°)$.

▷   Verify with a calculator that $\sin(-56°) = -\sin(56°)$. What is the approximate value of $\sin(-56°)$?

Check Your Understanding 7.8

1. Sketch a central angle of 135° and find its reference angle α. With a calculator, approximate $\cos(\alpha)$, $\cos(135°)$, $\sin(\alpha)$, and $\sin(135°)$. How are $\cos(\alpha)$ and $\cos(135°)$ related? How are $\sin(\alpha)$ and $\sin(135°)$ related?

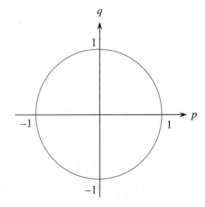

2. Sketch a central angle of $-420°$ and find its reference angle α. With a calculator, approximate $\cos(\alpha)$, $\cos(-420°)$, $\sin(\alpha)$, and $\sin(-420°)$. How are $\cos(\alpha)$ and $\cos(-420°)$ related? How are $\sin(\alpha)$ and $\sin(-420°)$ related?

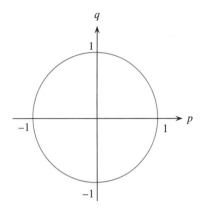

We have come full circle, as it were. Initially, triangle trigonometry appeared quite distinct from unit-circle trigonometry, but we have seen strong connections between the two approaches. Understanding either one adds to your understanding of the other. Because we can convert from degrees to radians and vice-versa, we see the sine and the cosine as functions having all real numbers (radians), or all possible amounts and directions of rotation (degrees), as their domain.

In a Nutshell   The six trigonometric functions, first defined for directed arcs on the unit circle, are also defined for angles measured in any number of degrees, positive, negative, or zero. We convert between degree and radian measure of angles by using the equivalence 360 degrees = 2π radians. To represent an arc in radians, we use the radius of the circle as our unit of measure. To interpret the trigonometric functions for a negative angle or an angle greater than 90°, we use a reference right triangle.

What's the Big Idea?

- The six trigonometric functions are ratios of side lengths of right triangles.

- These definitions can be reconciled with the unit-circle definitions of the trigonometric functions by relating angle measurements to the length of an arc on the unit circle.

Progress Check

After finishing this chapter, you should be able to do the following:

- Use trigonometric functions to find the missing sides or angles of a right triangle. (7.1)

- Work with ratios and proportions. (7.1, 7.2)

- Use the Law of Sines and the Law of Cosines in solving oblique triangles. (7.2)

- Use trigonometric functions to solve problems from the physical world. (7.1, 7.2)

- Use a calculator correctly to evaluate sines, cosines, and tangents and to find inverse sines, cosines, and tangents. (7.1, 7.2)

- Convert the measure of an angle from degrees to radians or vice-versa. (7.3)

- Use reference angles to find sines and cosines for nonacute angles. (7.3)

Trigonometry Appendix Topics

- Definitions and Handy Identities

Answers to Check Your Understanding

7.1

1. side $\overline{BC}$; side $\overline{AC}$; side $\overline{AB}$, opposite the right angle

2. $\cos(\theta) = \frac{x}{z}$; $\sin(\phi) = \frac{x}{z}$ (same as $\cos(\theta)$); $\tan(\theta) = \frac{y}{x}$;
$\tan(\phi) = \frac{x}{y}$ (reciprocal of $\tan(\theta)$)

3. $\alpha = 90° - 27° = 63°$

4. $\frac{52}{65} = \frac{4}{5}$

7.2

1.

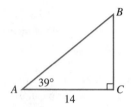

Angle $B = 51°$

$$AB = \frac{14}{\cos(39°)} \approx 18.0$$

$$BC = 14\tan(39°) \approx 11.3$$

2.

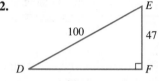

$$\text{Angle } D = \sin^{-1}\left(\frac{47}{100}\right) \approx 28°$$

$$\text{Angle } E = 90° - 28° = 62°$$

$$DF = \sqrt{100^2 - 47^2} \approx 88.3$$

7.3

1. $\alpha = 20°$, $B = 45°$

2. $\tan(20°) = \frac{150}{x+d}$; $\tan(45°) = \frac{150}{d}$

3. $x \approx 262$ feet

7.4

1. Angle $C = 180° - 42° - 59.7° = 78.3°$

2. $x = 31\cos(59.7°) \approx 15.6$

3. $y = 40\cos(42°) \approx 29.7$

4. $AB = x + y \approx 45.3$

7.5

1. (a) $\dfrac{\sin(38.2°)}{329} = \dfrac{\sin(C)}{410}$

$$\sin(C) = \frac{410\sin(38.2°)}{329} \approx 0.7707$$

angle $C = \sin^{-1}(0.7707) \approx 50.4$ degrees[4]

(b) angle $B = 180° - 38.2° - 50.4° = 91.4°$

(c) $AC = \dfrac{329\sin(91.4°)}{\sin(38.2)} \approx 532$ meters

2. Angle $F = 180° - 34° - 48° = 98°$

$$DF = \frac{7.4\sin(48°)}{\sin(98°)} \approx 5.6$$

$$EF = \frac{7.4\sin(34°)}{\sin(98°)} \approx 4.2$$

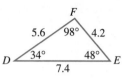

7.6

1. (a)

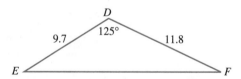

$$d^2 = 22^2 + 18^2 - 2(22)(18)\cos(20°)$$

$$d \approx 8 \text{ miles}$$

(b) In the first 6 minutes ($\frac{1}{10}$ mile), the riders travel 2.2 and 1.8 miles, respectively. The entire triangle is scaled down by a factor of 10, so their distance apart is 0.8 mile.

2. $6^2 + 17^2 - 2(6)(17)\cos(X) = 20^2$
$\cos(X) \approx -0.3676$
$X = \cos^{-1}(-0.3676) \approx 111.6°$

3. $EF \approx 19.1$ by the Law of Cosines.
Angle $E \approx 30.4°$, angle $F \approx 24.6°$, by either Law of Sines of Law of Cosines.
Check: $125° + 30.4° + 24.6° = 180°$

7.7

1. $\dfrac{5\pi}{6}$ radians $\cdot \dfrac{180 \text{ degrees}}{\pi \text{ radian}} = 150°$

2. 12 degrees $\cdot \dfrac{\pi \text{ radians}}{180 \text{ degree}} = \dfrac{\pi}{15}$ radians

---

[4]In theory, angle C could be the obtuse angle $180° - 50.4° = 129.6°$, whose sine value is also 0.7707. In practice, a surveyor would have at least a rough idea of the size of angle C.

7.8

1.

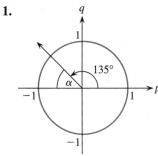

$\alpha = 45°$
$\cos(\alpha) \approx 0.7071$, and $\cos(135°) \approx -0.7071$,
its opposite
$\sin(\alpha) = \sin(135°) \approx 0.7071$

2.

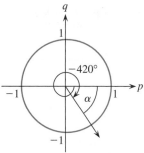

$\alpha = 60°$
$\cos(\alpha) = \cos(-420°) = \frac{1}{2}$ (This is exact.)
$\sin(\alpha) \approx 0.8660$; $\sin(-420°) \approx -0.8660$, its opposite

EXERCISES

1. Solve right triangle ABC if angle $A = 22°$, $\overline{AC}$ is the hypotenuse, $AB = 14.1$.

2. Solve right triangle DEF if angle $E = 90°$, $DE = 98$, $EF = 302$.

3. If possible, find the missing sides and angles in each triangle. If a triangle cannot be solved, tell what additional information you would need.

(a)   **(b)**

(c)

4. (a) Find *two* pairs of similar triangles among the six triangles shown.
(b) Fill in the missing angles and side lengths for each triangle.

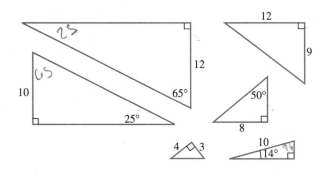

5. To comply with federal regulations, the owners of a building plan to construct an access ramp from the ground to their doorway 6 feet above the ground. Only 16 feet separates the building from a brick wall.

(a) What is the longest ramp that could fit between the doorway and the wall?
(b) For the ramp to be usable, they must leave room for people in wheelchairs to enter and exit. Recommend how much space they should allow, and recalculate the length of the ramp.
(c) Find the rise-to-run ratio of the ramp in part (b), and determine its angle of elevation from the ground.
(d) The resulting angle will make the ramp rather steep. Suggest a way to reconfigure the ramp to provide a gentler slope.

6. The public library plans a wheelchair ramp for its main entrance, 28 inches above ground level. The ramp is to have an angle of 10° with the horizontal.
   (a) How long a ramp does the library need?
   (b) How far out from the building will the ramp extend?

7. Here's a problem one of the authors encountered. The wood stove in her living room has an 8-inch-diameter vent pipe that passes through the pitched ceiling. A metal ceiling plate with a hole 8 inches in diameter came with the stove—perfect if the ceiling is horizontal (perpendicular to the pipe) but inadequate for the actual slanted ceiling.

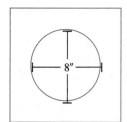

The hole in a plate for a pitched ceiling that will fit around the pipe will have the shape of an ellipse. The shorter dimension, called the **minor axis** of the ellipse, is still 8 inches, but the longer dimension a, or **major axis,** will be more than 8 inches. How much more? You'll figure this out in two different ways.

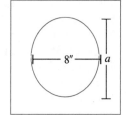

Plate for flat ceiling   Plate for pitched ceiling

The carpenter determined the pitch of the roof, or what he called the rise-to-run ratio, by measuring along a horizontal beam for 12 inches and then vertically from that point to the ceiling. That vertical distance was 6 inches.

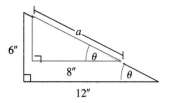

(a) Use trigonometry to find a.
(b) Recalculate a using similar triangles and the Pythagorean theorem.

8. If the sewer pipe in Figure 7.6, page 00, is 1000 feet long and maintains a vertical drop of $\frac{1}{4}$ inch for every foot of pipe, what is the horizontal distance from one end of the pipe to the other?

9. In Edgar Allen Poe's story "The Gold Bug," a coded inscription on a treasure map reads as follows:

> . . . main branch, seventh limb east side—shoot from the left eye of the death's head—a bee line from the tree through the shot fifty feet out.

The tree is climbed, and a weighted line is dropped mistakenly through the *right* eye of the skull. This error causes the treasure hunters' mark to be off by about 3 inches.

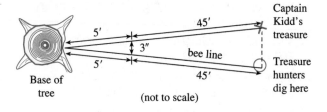

(not to scale)

(a) The treasure hunters dig a hole at the end of their 50-foot line. Needless to say, they don't find the treasure. Use similar triangles and proportionality to find the distance from the center of their hole to the spot where the treasure was buried.
(b) Suppose the instructions had read "a bee line from the tree through the shot one hundred feet out." How far from the treasure would they then have been?
(c) Did the treasure hunters ever find the treasure? You'll have to read the story to find out.

10. Knowing two sides or one side and an acute angle, we can solve any right triangle. Explain why we can't solve the triangle if we know only the two acute angles.

11. Conditions for proving quadrilaterals similar are more stringent than they are for triangles.
    (a) Draw a pair of similar, but not congruent, rectangles.
    (b) Draw a pair of rectangles that are not similar.

(c) Each angle in the first quadrilateral pictured below is equal to the corresponding angle in the second. However, the quadrilaterals are not similar, because their corresponding sides are not proportional. Draw two quadrilaterals whose corresponding sides *are* proportional but that nevertheless are not similar.

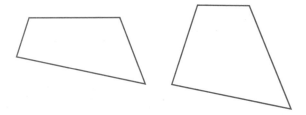

12. Here is the carpentry problem faced by one of the authors when she needed to cut a piece of wood at each end, at the correct angle, so that the resulting brace would snug up against the horizontal surface of the picnic table at one end and against the vertical at the other end. To the nearest degree, find angles θ and ϕ.

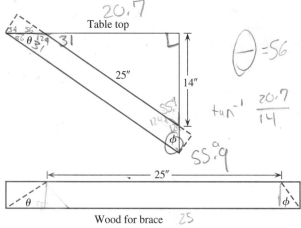

13. Runners who compete in the grueling Mount Washington road race in New Hampshire cover a distance along the road of 7.6 miles. Use the information in the illustration to find h, the vertical distance each runner travels from the lowest section of the course to the highest.

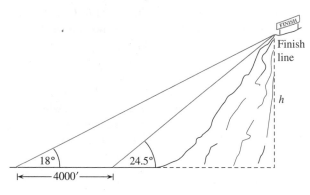

14. Given the information in the illustration, find the height of the tree.

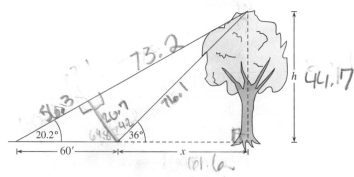

15. Solve triangle ABC if angle $A = 68°$, angle $C = 40°$, and $AC = 175.5$.

16. Solve triangle DEF if angle $D = 115°$, angle $F = 25°$, and $DE = 12.3$.

17. Solve triangle GHI if angle $G = 145°$, $GH = 9.6$, and $GI = 13.3$.

18. Solve triangle JKL if angle $J = 65°$, $JK = 505$, and $JL = 510$.

The Ambiguous Case

Under certain circumstances, having three facts about a triangle isn't enough. We've already seen that the same three angles can exist in many different-sized triangles of the same shape. Now we'll look at a situation in which the same two sides and one angle can produce two different triangles. Here's an example.

In triangle MNO, $MN = 10$, $NO = 12$, and angle $O = 50°$. Because angle O isn't *between* the two known side lengths, we'll apply the Law of Sines:

$$\frac{\sin(50°)}{10} = \frac{\sin(M)}{12}$$

$$\sin(M) = \frac{12}{10}\sin(50°) \approx 0.919$$

$$M = \sin^{-1}(0.919) \approx 66.8°$$

But another angle, $180° - 66.8° = 113.2°$, also has a sine value of 0.919. So angle M has two possibilities: It might be $66.8°$, or it might be $113.2°$. Let's look at the picture.

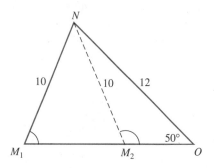

Note that both triangles M_1NO and M_2NO fit the description given: $M_1N = M_2N = 10$, $NO = 12$, angle $O = 50°$. But angle M_1 is acute, while angle M_2 is obtuse.

19. Finish solving triangle MNO. Treat each case separately, obtaining two different possible triangles.

20. Solve triangle PQR if angle $P = 43°$, $QR = 236$, $PR = 306$. Like the triangle in the preceding exercise, this has two solutions. Find both.

21. Find the angle measurements of triangle STU.

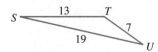

22. In proving the Law of Cosines in Section 7.2, we implicitly assumed that the triangle was acute. Now consider triangle ABC, in which angle C is obtuse, and prove that the identity $a^2 + b^2 - 2ab\cos(C) = c^2$ continues to hold. Adapt your proof from the work on page 361, and use the fact that $\cos(C) = -\cos(180° - C)$.

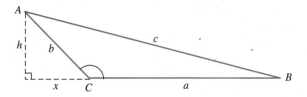

23. Recalculate the height of Holyoke Mountain (see Figure 7.7 on page 356) by another method: find angles DAB and BDA first; then apply the Law of Sines to triangle BDA to find BD; finally, use an appropriate trigonometric function to find h. (The result, of course, should agree with the height we found earlier, about 877 feet.)

24. Recalculate the height of the tree in Exercise 12 via the Law of Sines.

25. A microwave tower stands atop a rise in an open field. At a point P downhill, 220 feet from the base of the tower, a surveyor measures the angle between the ground and the top of the tower as 40°. But the ground itself slopes down from the base of the tower, forming an angle of 95°. Find the height, h, of the tower.

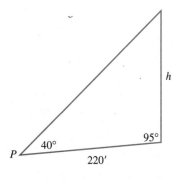

26. In CYU 7.5 on page 360, you assisted a surveyor in computing the width of a swamp. Suppose that the surveyor has measured angle B instead of angle A. (This would be a more natural measurement to take.) Find the width of the swamp again, using this revised information, and verify that it agrees with the answer you got in CYU 7.5.

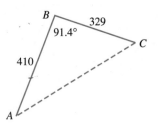

27. Find the angle in degrees that corresponds to each radian measurement.
  (a) $\frac{2\pi}{3}$ **(b)** $-\frac{5\pi}{4}$ **(c)** 5.9

28. Find the angle in degrees that corresponds to each radian measurement.
  (a) $\frac{3\pi}{4}$ **(b)** $-\frac{7\pi}{6}$ **(c)** 6.3

29. Find the radian measure of each angle.
  (a) 30°   **(b)** 344°   **(c)** $-12°$

30. Find the radian measure of each angle.
  (a) 60°   **(b)** 215°   **(c)** $-71°$

31. Consider the three concentric circles shown.

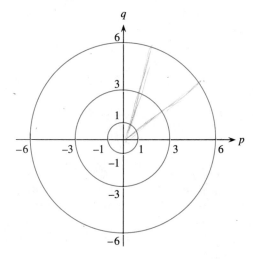

(a) Suppose you start from the point (6,0) and move counterclockwise on the outermost circle. What is the directed arc length associated with one eighth of a complete revolution? What is the equivalent measure in radians?

(b) If you start at the point (3,0) and move 3π units counterclockwise on the middle circle, where do you stop? What fraction of a complete revolution have you made? What is the equivalent measure in radians?

(c) What fraction of a complete revolution is a directed arc length of $\frac{2\pi}{3}$ units on the unit circle? What is its radian measure?

32. A model train travels once around an 8-foot-diameter circular track in 60 seconds. What is the train's velocity in feet per second? What is its angular velocity in radians per second? (Angular velocity is the rate of change of the angle of rotation.)

33. Explain the *Calvin and Hobbes* phonograph-record cartoon on page 292.

34. The sine or the cosine of an angle is positive or negative according to the quadrant in which its reference triangle lies. Explain Figure 7.14 on page 366. That is, for each quadrant, tell why the sine and cosine functions have the given signs.

35. Find the reference angle for each central angle. In which quadrant does the corresponding reference triangle lie?
(a) 275°   **(b)** 740°   **(c)** −148°
For which of these angles is the cotangent negative?

36. Find the reference angle for each central angle. In which quadrant does the corresponding reference triangle lie?
(a) 382°   **(b)** −75°   **(c)** 462°

For which of these angles is the secant negative?

In Exercises 37–40, find two different angles θ for which the given sine and cosine values are good approximations. Of each pair of angles, only one should be between 0° and 360°.

37. $\sin(\theta) = -0.6428, \cos(\theta) = 0.7660$
38. $\sin(\theta) = -0.5878, \cos(\theta) = -0.8090$
39. $\sin(\theta) = 0.2588, \cos(\theta) = -0.9659$
40. $\sin(\theta) = 0.9063, \cos(\theta) = 0.4226$

2

1.

760

7.1 SPECIAL ANGLES—Some Exact Trig Values

In this project, you will use two special right triangles to determine exact trigonometric function values for angles of 30, 45, and 60 degrees.

The 30°-60°-90° Triangle

To form this triangle, begin with an equilateral triangle of side length one unit. Each of its angles contains 60 degrees. Then divide the triangle into two equal halves, as shown.

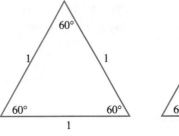 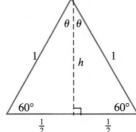

1. What is the degree measurement of angle θ?
2. Find the exact value of the altitude h. (The value will contain a radical.)
3. Use the values of θ and h to determine the exact value of each of the following:
   (a) $\sin(30°)$ (b) $\cos(30°)$
   (c) $\tan(30°)$ (d) $\sin(60°)$
   (e) $\cos(60°)$ (f) $\tan(60°)$

The 45°-45°-90° Triangle

To form this triangle, begin with a square of side one unit, and divide it diagonally into equal halves.

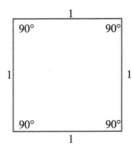 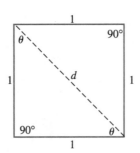

4. What is the degree measurement of angle θ?
5. Find the exact length of the diagonal d.
6. Use your values for θ and d to find the following:
   (a) $\sin(45°)$
   (b) $\cos(45°)$
   (c) $\tan(45°)$

Sums and Differences of Angles

The special angles will let you find the exact trigonometric values for a few more angles as well. Note, for example, that $75° = 45° + 30°$, and so we might be tempted to write $\sin(75°)$ as the sum of $\sin(45°)$ and $\sin(30°)$. The trigonometric functions, however, aren't distributive.

7. Without a calculator, find the exact value of $\sin(45°) + \sin(30°)$, and tell how you know that this number could not be the sine of any angle.

There are, however, formulas giving the sine and the cosine of the sum of any two angles. For any two angles α and β,

$$\sin(\alpha + \beta) = \sin(\alpha)\cos(\beta) + \cos(\alpha)\sin(\beta)$$
$$\cos(\alpha + \beta) = \cos(\alpha)\cos(\beta) - \sin(\alpha)\sin(\beta)$$

8. Use the preceding formulas to find the exact values of $\sin(75°)$, $\cos(75°)$, and $\tan(75°)$.

9. Using the identities $\cos(-\beta) = \cos(\beta)$ and $\sin(-\beta) = -\sin(\beta)$, write formulas for $\sin(\alpha - \beta)$ and $\cos(\alpha - \beta)$.

10. Use your formulas from Question 9 to find the exact values of $\sin(15°)$, $\cos(15°)$, and $\tan(15°)$.

7.2 HOW HIGH IS THE MOON?

In his treatise *On the Sizes and Distances of the Sun and the Moon*, written around 260 B.C., Aristarchus states this observation: "When the moon is just half full, the angle between the line of sight to the sun and the line of sight to the moon is one-thirtieth of a quadrant less than a right angle." The following diagram is a geometric representation of what Aristarchus wrote.

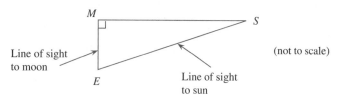

The sun's rays illuminate an entire hemisphere of the moon. If an observer on the earth can see exactly half of the illuminated portion, then a line from sun to moon must be perpendicular to a line from moon to earth. Thus, Aristarchus' statement implies that angle EMS is a right angle. (During a different phase of the moon, angle EMS would be either greater or less than 90 degrees.)

1. According to Aristarchus, what is the degree measurement of angle MES, the angle made by the lines of sight to the sun and to the moon? (A quadrant measures 90 degrees.)

2. Use an appropriate trigonometric function to determine the ratio of ES to EM. That ratio allows us to compare the earth-to-sun distance to the earth-to-moon distance; if the ratio is r, then the sun is r times as far from us as the moon is. Had Aristarchus been correct, what would that ratio be?

3. In reality, angle MES is larger than Aristarchus thought, approximately 89.8°. Using the modern approximation of angle MES, recalculate the ratio $\frac{ES}{EM}$.

4. Compare and comment on the results of Questions 2 and 3.

7.3 DON'T LEAN ON ME

The Leaning Tower of Pisa, pictured on page 349, is noteworthy both for its beauty and for its incline. A freestanding cylinder, it consists of eight tiers of round-arched arcades. During construction (1174–1350), an uneven settling of the ground caused the building, whose foundation was too shallow to support its weight, to lean. From the top of this famous tower, Galileo is said to have conducted experiments on gravity, the basis for the Sidney Harris cartoon in Project 5.4, *Galileo*, on page 265.

If Galileo were to visit the Leaning Tower today, he would find a structure that leans considerably more than it did in his time. As part of ongoing efforts to preserve the fragile tower, Italian engineers monitor the angle of inclination of the structure. Measuring the tower's tilt requires some trigonometric calculation. This project asks you to use information similar to what might be supplied to the engineers to determine the angle of the tower's slant, the vertical distance from the ground to the top, and the length of its shadow.

At a distance of 150 feet from the center of the tower's base, the angle of elevation to the top of the tower is 54.2° The original structure would have stood approximately 184.5 feet tall if it had not started leaning prior to its completion. The situation pictured differs, as we'll see, from a standard right-triangle problem.

1. Triangle ABC isn't a right triangle:
   (a) If $\angle A$ were a right angle, would we call this the Leaning Tower? Explain.
   (b) If $\angle C$ were a right angle, side AB would be the hypotenuse. Explain how the known side lengths make that impossible.

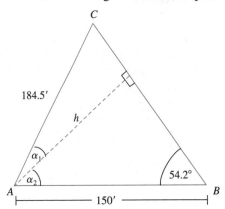

2. Now let's calculate α, the angle of inclination of the tower. One method of attack is to divide triangle ABC into two right triangles, as shown. Then use trigonometry to find h, and use that h-value to approximate α as the sum of α_1 and α_2.

3. Next, draw the line through C that makes a right angle with the ground. The length of that line is the height of the top of the tower above the ground. What is that height?

4. If the sun were shining directly overhead, how long a shadow would the tower cast?

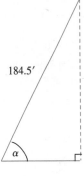

Shadow

5. Pisa is actually too far north for the sun ever to shine directly overhead. Suppose, instead, that the rays of the sun come in at an angle of 60° with the ground. How long is the shadow cast by the tower? (*Hint*: Use your results from Questions 3 and 4.)

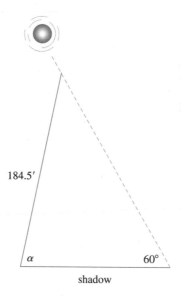

184.5′

α 60°

shadow

7.4 FORESHORTENING

Imagine yourself an artist. Canvas before you, you sketch a person who is standing several feet away. Then you have the person lie down, keeping his feet in roughly the same spot, with his head positioned away from you. You make another sketch. If you have followed the principles of perspective drawing, the sketch of the standing figure is taller than the sketch of the reclining figure. The principle of reducing the height of a figure that is tilted away from the plane of the canvas is called *foreshortening*, a famous example of which is Andrea Mantegna's *Dead Christ*, reproduced here. Observe that Christ's body fills less vertical space on the canvas than it would occupy if the artist had painted Christ standing. This foreshortening creates the illusion of depth.

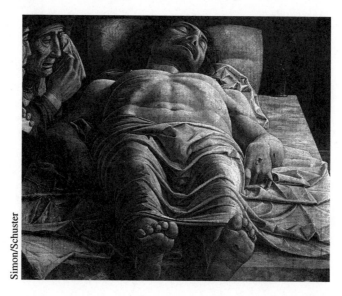

Simon/Schuster

The basic question is this: By how much should you shorten the figure of an object tilted away from the picture plane? The answer, as you may have guessed, is based on mathematics.

Let's start simply. Assume that you want to draw a 9-inch rod that is held vertically against the picture plane. (See Figure 1.) If, on the picture plane, you draw what you see, you would draw a 9-inch rod.

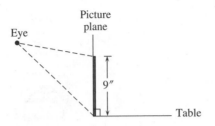

Figure 1 Rod Held against Picture Plane

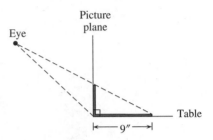

Figure 2 Rod Placed on Table

Now imagine placing the rod on the table so that one end touches the picture plane, as in Figure 2. The rod's apparent length in the picture plane is less than in Figure 1. How much shorter is it? We'll use geometry to find out.

1. Suppose your eye is 2 feet above the table and the picture plane is 2 feet away from you. Figure 3 represents the situation geometrically, with measurements given in feet.

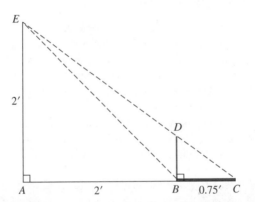

Figure 3

   (a) Use similar triangles to find BD, the apparent length of the 9-inch rod in the picture plane.
   (b) Now let's generalize: Suppose the rod is r feet long. Write an expression giving its apparent length as a function of its actual length r.

2. (a) In Question 1, you assumed that one end of the rod was touching the picture plane. What if, instead, you were to slide the 9-inch rod 6 inches away from the picture plane. (Figure 4 shows this modification.) What is the rod's apparent length, given by DF, now? (*Hint*: First find BD.)
   (b) Suppose you were to slide the rod x feet away from the picture plane. Write its apparent length as a function of x.

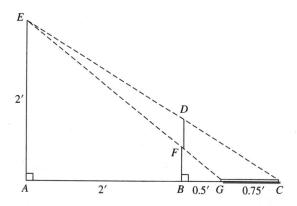

Figure 4

3. Next, consider a slightly different situation. Bring your eye down to the level of the tabletop. Start with the rod held vertically against the picture plane, and rotate it away from that plane. (See Figure 5, in which segment AC represents the rod.) Assume that the rod, as before, is 9 inches (0.75 foot). Keep EA, the distance from your eye to the near end of the rod, at 2 feet. Tilt the rod 30° from the picture plane. ($\theta = 30°$.)

   (a) Determine angles CAB and ACB.
   (b) Find lengths x and y.
   (c) Determine AD, the apparent length of the rod.

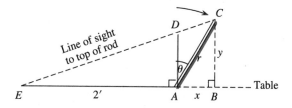

Figure 5 Rotating Rod away from Picture Plane

4. **(a)** You would hate to redo your work from Question 3 every time you change the angle. So generalize the results! Write a formula for the apparent length of the 9-inch rod when it is tilted at an angle θ from the picture plane.
   (b) The mathematical expression you derived in part (a) gives the apparent length as a function of θ. Graph that function, and use the graph to approximate the angle θ that gives the rod an apparent height of 6 inches.

5. **(a)** Rework Question 4(a), assuming a rod of length r.
   (b) Use that mathematical expression to determine the apparent length of an 18-inch rod tilted 50° away from the picture plane.

6. You must be getting terribly tired of stooping with your eye at table level. Unbend a bit, positioning your eye one foot above the table surface. (See Figure 6.) Assuming that the rod is 9 inches long and makes an angle of 60° with the picture plane, find BF, the apparent height of the rod. (There are various ways to find the answer, and each method involves several steps. As you proceed, retain all the decimal digits your calculator gives, rounding the result only at the end.)

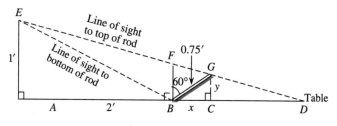

Figure 6

7.5 JUST ALGEBRA

1. Solve each equation for the indicated variable.
   (a) $a^2 + b^2 = c^2$ for c
   (b) $\dfrac{\sin(A)}{a} = \dfrac{\sin(B)}{b}$ for b
   (c) $\dfrac{\sin(A)}{a} = \dfrac{\sin(B)}{b}$ for B
   (d) $\sin^2(A) + \cos^2(A) = 1$ for $\cos(A)$ and then for A (Assume that $\cos(A) \geq 0$.)
   (e) $c^2 = a^2 + b^2 - 2ab\cos(C)$ for b (Assume that $\cos(B) \geq 0$.)

2. (a) Find an approximate solution to the equation $1764 = 900 + b^2 - 60b\cos(62°)$.
   (b) Solve $c^2 = a^2 + b^2 - 2ab\cos(C)$ for b.

3. Show that $\dfrac{a}{\sin(A)} = \dfrac{b}{\sin(B)} = \dfrac{c}{\sin(C)}$ is equivalent to the Law of Sines stated on page 360.

4. We are given the following system of equations: $\dfrac{h}{x} = a$ and $\dfrac{h}{d+x} = b$.
   (a) Solve for x in terms of a and h.
   (b) Use that expression for x to solve for h in terms of a, b, and d.

Look on the bright side

Preparation

Five planets—Mercury, Venus, Mars, Jupiter, and Saturn—are visible from the earth with the naked eye. Ancient Greeks called them *The Wanderers*, because they appear to roam among the constellations. To us, these planets look like bright stars in the night sky. This lab focuses on Mars, Jupiter, and Saturn, the three "naked-eye" planets whose orbits lie beyond that of the earth.

Courtesy: NASA

The portion of each planet that we see from the earth appears as a circular disk that, like our moon, is illuminated to some degree by direct sunlight. The illuminated portion of the moon's disk oscillates between total darkness (new moon) and total illumination (full moon). If you completed the "Moonlight" Lab (Lab 6B), then you determined a mathematical model for predicting the illuminated fraction of the moon's disk over time. The question for this lab is whether or not Mars, Jupiter, and Saturn have phases similar to those of our moon.* (Is there such a phenomenon, for instance, as "new Mars"?) In other words, you will investigate how much of the illuminated half of the planet is visible from the earth and whether that portion remains constant or changes as time passes.

---

*The article "How 'Full' Are the Outer Planets?" by Jim Shaw, published in *The Physics Teacher* (Vol. 37, December 1999, pp. 528–529) provided the central idea for this lab.

Suppose you could look down on the solar system from the North Star. The drawing in Figure 1 is a geometric model of your view of the earth and a planet (Mars, Jupiter, or Saturn) orbiting the sun. Although the actual orbits of the planets are slightly elliptical, we have drawn circular orbits in Figure 1 to simplify the mathematics.

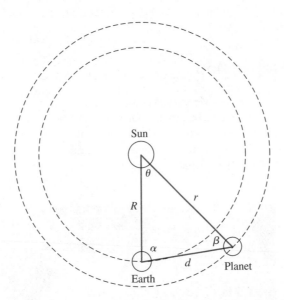

Figure 1 Earth and Planet Orbiting Sun, View from North Star

In Figure 2, a diameter is drawn through the planet perpendicular to the line connecting the earth to the planet. This line divides the planet into two hemispheres: the one that faces the earth and the one that faces away. From a vantage point on the earth, the hemisphere facing the earth appears as a circular disk.

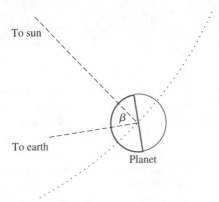

Figure 2 Magnified View of Planet

Similarly, a diameter can be drawn through the planet perpendicular to the line connecting the planet to the sun. Sketch this diameter in Figure 2. This diameter, also, divides the planet into two hemispheres: the one that is illuminated by the sun and the one that remains dark.

The two diameters divide this picture of the planet into four wedges:

- Visible from the earth and illuminated by the sun (wedge 1)

- Visible from the earth but not illuminated by the sun (wedge 2)

- Not visible from the earth but illuminated by the sun (wedge 3)

- Not visible from the earth and not illuminated by the sun (wedge 4)

In Figure 2, label these four wedges.

Now rotate your completed Figure 2 diagram so that the earth is positioned at the bottom (nearest you). Your rotated diagram should be similar to Figure 3. The angle labeled with a question mark represents the portion of the planet visible from the earth but not illuminated by the sun. Use geometry to explain why this angle has the same measure as β, the earth-planet-sun angle. How are the angles of the other wedges in the diagram related to β? (For instance, if β measures 10°, how large are the angles of the other wedges?)

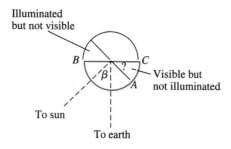

Figure 3 Magnified View of the Planet

The Planets Lab

Begin by comparing your explanations of why the earth-planet-sun angle, β, is the same size as the angle of wedge 2 (visible from the earth but not illuminated).

In the preparation, you identified two diameters of the planet:

- The one perpendicular to the line from the earth's center to the planet's center

- The one perpendicular to the line from the sun's center to the planet's center

For ease in referral, we'll call these diameters the planet's earth-diameter and its sun-diameter, respectively. Your task in this lab will be to determine the maximum and minimum fractions of the planet's earth-diameter that will ever be illuminated by the sun.

The idea is this: If the entire earth-diameter is illuminated, then the entire planet, viewed from the earth, appears bright; if a smaller amount of the earth-diameter is illuminated, then only a portion of the planet's surface appears bright to us. If we were being exact, we would compare areas—illuminated portion of the disc to the entire disc. But those areas are much more difficult to calculate, so astronomers use this method as a valid, though rough, approximation.

As you'll discover, the answer to the minimum-diameter question depends on the radius, r, of the planet's orbit and on angle β, the planet's position relative to the earth and the sun. One half of the planet is always lit by the sun, but we can't always see all of that half. Angle β is a measure of the earthbound view of the planet's illumination: As β increases, the visible portion of the illuminated half decreases.

As a reminder, Figure 4 shows a top-down view of the situation, along with a table of planetary information.

Let's start with Mars. Use the Law of Sines to set up a relationship between the angles α and β and the radii of the orbit of the earth and the orbit of Mars. Using an inverse sine function, solve for β in terms of α.

In solving for β, you wrote β as a function of α. Let's think about the domain of this function. Since the angles in any triangle sum to 180°, the triangle formed by connecting the centers of the earth, the sun, and Mars collapses when $\alpha = 0°$ or when

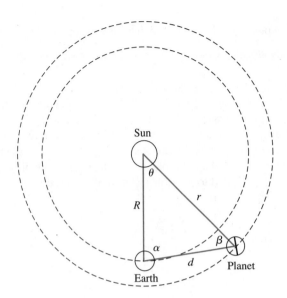

Figure 4

Planet	Average Distance to Sun (millions of miles)	Diameter (miles)
Earth	93	7930
Mars	142	4220
Jupiter	484	88,800
Saturn	891	74,900

Table 1

$\alpha = 180°$. Draw a geometric representation of the earth, the sun, and Mars when $\alpha = 0°$; draw another for $\alpha = 180°$.

When $\alpha = 0°$ or $\alpha = 180°$, Mars' earth-diameter and sun-diameter coincide. In this situation, $\beta = 0°$ and Mars' earth-diameter is totally illuminated by the sun. There-fore, $0°$ is the smallest possible value for β. (We are ignoring the possibility of an eclipse.) Graph your formula for β over the interval $0° \leq \alpha \leq 180°$, and find the maximum value for β. What value of α is associated with that maximum?

Next, you will determine what fraction of Mars' earth-diameter is illuminated when β attains its maximum value. (Remember, when β reaches its maximum, the illuminated portion of the Mars' earth-diameter is a minimum.) To do this, you'll need to figure out the relationship between angle β and the illuminated segment of Mars' earth-diameter. Here goes.

On a sketch similar to Figure 3 (but larger), draw the line that passes through point A on Mars' sun-diameter and is perpendicular to Mars' earth-diameter. Your line intersects Mars' earth-diameter in a point; label that point P. Determine the length of BP. (Information in Table 1 along with a trigonometric ratio—sine, cosine, or tangent—may prove helpful here.) What fraction of BC (the length of Mars' earth-diameter) is BP?

How do the "phases" of Mars compare to the phases of the earth's moon (new moon to full moon)?

Now adapt the process that you used for Mars to solve the problem for Jupiter and Saturn. Find the maximum and minimum illuminated fractions of the earth-diameter of each of those planets.

It's time to generalize your findings. As the planet's distance from the sun increases, does the maximum value of β increase, decrease, or remain the same? What about the corresponding α-value? The approximate distance of Uranus from the sun is 1790 million miles. Without doing any calculations, what can you say about the minimum illuminated fraction of the earth-diameter of Uranus?

The Lab Report

Do planets have phases similar to those of the moon? Your report should demonstrate your use of geometry and trigonometry to investigate this question. Justify that the earth-planet-sun angle, β, is the same as the angle of the wedge that is visible from the earth but not illuminated. For each of the three planets, show how you determined β as a function of α. Sketch the graphs of those three functions on the same set of axes, identifying each with its respective planet. Describe the process that is used to find the minimum illuminated fraction of the earth-diameter of Mars, Jupiter, and Saturn. Summarize any patterns that you observed as the planet's distance from the sun increased.

Multiple Inputs/
Multiple Outputs

8

© J. Lawrence/Image State–Pictor/PictureQuest

In the previous seven chapters, you used your grapher to help you visualize a functional relationship between *two* variables. From the function's graph, you extracted information about its domain, range, rate of change, and asymptotic behavior. At times, you were even able to work backward and determine a formula for a function from its graph.

In this chapter, we will study relationships among *three* or more variables. In the process, we'll examine ways of visualizing information that may be new for you.

The photograph in Figure 8.1, for example, depicts the three-dimensional wire model of a DNA molecule constructed by James Watson and Francis Crick in 1953 to help them understand the helical structure of DNA. The diagram of the double helix in Figure 8.1 is taken from a booklet prepared in 1988 for members of the U.S. Congress who were conducting hearings on the social and economic consequences of genetic engineering. Like Congress, we are interested in ways of compressing three (or more)-dimensional information onto the two-dimensional worlds of a sheet of paper or the computer or calculator screen.

In Section 8.1, we'll examine relationships in which the output is a function of two or more independent variables. In Sections 8.2 and 8.3, we'll investigate sets of functions, called *parametric equations*, in which two or more dependent variables are functions of a single input, the *parameter*. We'll continue to use graphs to help us understand relationships between pairs of variables, but we'll have to find ways of presenting several dimensions worth of information in a two-dimensional graph.

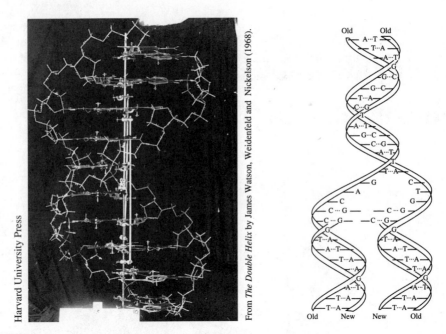

Harvard University Press

From *The Double Helix* by James Watson, Weidenfeld and Nickelson (1968).

Figure 8.1 Two Representations of the DNA Molecule

8.1 FUNCTIONS WITH MORE THAN ONE INPUT VARIABLE

Many real-world situations can be modeled by functions whose outputs depend on more than one input. The baking time for a cake is a function of the oven temperature and the elevation above sea level. The length of time it takes to pay off a car loan is a function of the amount of money borrowed, the size of the monthly payment, and the interest rate. Your final precalculus grade might be a function of your exam, quiz, homework, project, and lab grades.

Function Notation

Throughout this text, we have used function notation to express relationships between a dependent variable and an independent variable. In Chapter 1, we expressed a mathematical model for a women's shoe size w based on her foot length x, and in Chapter 2, we represented world population G as a function of the time in years since 1990:

$$w(x) = 3x - 21$$
$$G(t) = 5230(1.0174)^t$$

The symbols $w(x)$ and $G(t)$ emphasize the dependence of w and G on the single independent variables x and t, respectively.

Similarly, we can express the volume of a rectangular container using function notation:

$$V(l, w, h) = l \cdot w \cdot h$$

The value of the single output variable, *volume*, depends on the values of three input variables, *length*, *width*, and *height*. The notation $V(l, w, h)$ tells us explicitly that we must supply those three inputs—l, w, and h—to determine the value of V. Thus, V is the dependent variable, and l, w, and h are the independent variables.

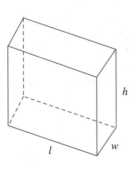

A rectangular box

▷   Determine the value of $V(7, 8, 3)$. Assume that all measurements are in inches.

To determine the domain and the range of a multivariable model, consider both the laws of mathematics and the function's context.

In Chapter 1, you learned that the *domain* of a function is the set of all inputs that make sense for the function and the *range* is the set of possible outcomes. We can apply the same criteria for determining the domain and range of a function of more than one variable. For example, the ordered triple (3,5,6) belongs to the domain of V because feeding (3,5,6) into V makes sense mathematically, producing an output of 90. As a possible output, then, 90 belongs to the range of V. The domain of the *abstract function* $V(l,w,h)$ is the set of all ordered triples (l,w,h) of real numbers, because the product of any three real numbers is also a real number. The range is the set of all possible outputs for V, that is, the set of all real numbers.

However, if we use $V(l,w,h)$ as a model to compute the volumes of rectangular packages, we must restrict the domain so that it makes sense in context. (There can't be a package whose side length is negative or zero.) Thus, the domain for the model is the set of ordered triples (l,w,h) of *positive* real numbers and the range is the set of *positive* real numbers. We would place an upper limit as well, depending upon the context, on each of the variables l, w, and h, resulting in a corresponding upper limit to the range.

Check Your Understanding 8.1

CYU answers begin on page 426.

Suppose you deposit S dollars into a savings account that earns interest at an annual rate of 5%, compounded continuously. Suppose, also, that you make no additional deposits or withdrawals. The amount M in the account t years after your original deposit is given by the expression $Se^{0.05t}$. The output M depends upon two inputs, S and t.

1. Express the amount of money in the account using function notation.

2. Evaluate and interpret $M(100,10)$ and $M(10,100)$.

3. Give the domain and the range of the function M in the abstract, that is, as a purely mathematical relationship.

4. Give a reasonable domain and range for M when it is used to model the amount of money in your bank account.

By now, you have plenty of experience graphing functions such as $w(x)$ and $G(t)$ from page 388 and using your graphs to interpret the relationship between the dependent and independent variables. But how would you graph the relationship between the volume of a box and its three independent variables—length, width, and height—for the function $V(l,w,h) = l \cdot w \cdot h$? The basic difficulty is that we have four dimensions worth of information, but the screen of your grapher is only two-dimensional.[1]

We will adopt two different strategies for compressing three or more dimensions worth of information onto a two-dimensional graph. Our first method will treat all but one input as (temporarily) constant and consider the relationship between the output and the one remaining input. Our second method will assign specific output values and focus

---

[1] In Lab 5A "Packages," we were able to sidestep the difficulty by eliminating all but one of the independent variables. The decision to give the package a square cross section allowed us to omit height as a separate variable. The U.S. Postal Service constraint—that length plus girth not exceed 108 inches—meant that the length was not independent of the width.

on the relationship between two independent variables. Both strategies require that we choose a fixed value for all but two of the variables, allowing us to use two-dimensional graphs to visualize higher-dimensional relationships.

Fixed Values for All but One Input Variable

I left my heart in San Francisco

A map is a familiar way to present information. The map in Figure 8.2 shows some of the streets in central San Francisco. Imagine that you want to determine the distance from Lafayette Square Park to Coit Tower. There are many different ways of measuring this distance: in miles on the odometer of your car, in minutes it takes to walk or drive, or in miles measured in a straight line "as the crow flies." You can see from the map that no street leads directly from the park to the tower, but if you walked up Gough Street and made a right onto Filbert to the tower, you could determine your distance from the park.

Figure 8.2 Map of Central San Francisco

▷ With the Pythagorean theorem, find the straight-line distance in miles from Lafayette Square Park to Coit Tower if the distance is 0.5 mile on Gough Street and 1.3 miles on Filbert.

With an arbitrary location as the origin, we can measure distances x and y from the origin along two perpendicular axes. We'll choose the entrance to Lafayette Square Park (at the corner of Gough and Clay) as the origin, let the x-axis run along Clay Street and the y-axis along Gough, and measure distances in miles. Then we can identify any other location on the map by specifying an ordered pair (x, y). Figure 8.3 indicates how the location of Coit Tower becomes the point $(1.3, 0.5)$.

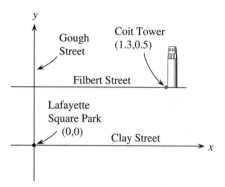

Figure 8.3 A Coordinate System for the Map of San Francisco

Furthermore, using the Pythagorean theorem, we can determine the diagonal distance from any point (x, y) on the map to the entrance of Lafayette Square Park, as shown in Figure 8.4. This distance is given by the formula $d(x, y) = \sqrt{x^2 + y^2}$.

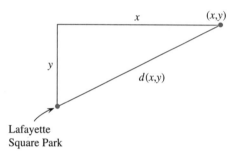

Figure 8.4 Distance d from Any Point (x, y) to Lafayette Square Park

▷   Locate, approximately, the points $(0.4, 0.3)$ and $(-0.2, 0.5)$ on the map. Then calculate $d(0.4, 0.3)$ and $d(-0.2, 0.5)$.

When we use d as a model, the ordered pairs that make sense as inputs should locate a point in downtown San Francisco. Although, for example, we can evaluate the abstract function d for the input pair $(2000, -1000)$, this input has no meaning as a location in San Francisco and so does not belong to the domain of the model.

Here, y designates an *independent*, or input, variable. Remember that y is just a symbol, and its classification as an independent or dependent variable is determined by its use.

The formula that is used to compute the distance is familiar to you. What might be new is thinking of distance as a function $d(x, y)$ requiring *two* inputs. In the formula $d(x, y) = \sqrt{x^2 + y^2}$, x and y are both independent variables; both are required as inputs before we can determine the value of the function d. The output is a single number: the distance from the point (x, y) to the origin.

In previous chapters graphs have helped us to visualize relationships between variables. Here, too, we would like to use graphs to picture the relationship between the distance to Lafayette Square Park and the x- and y-distances along Clay and Gough. The Cartesian plane allows for one input on one axis and one output on the other. To create a graph that accommodates *two* inputs and one output, we need to condense some information so that it fits onto two axes. One way to do this corresponds to the physical situation of staying on a single street. On Filbert, for example, the y-coordinate is fixed at 0.5 mile, and only the x-value, the distance along Filbert from its intersection with Gough, varies, as shown in Figure 8.5.

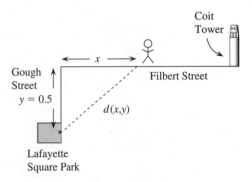

Figure 8.5   Distance from Position on Filbert Street to Lafayette Square Park

▷   In Figure 8.6, draw the straight-line path connecting your position on Filbert to the entrance of Lafayette Square Park as you arrive first at A, then B, C, and D.

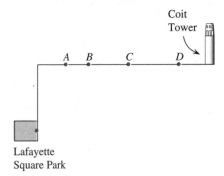

Figure 8.6   Straight-Line Paths from Park to Positions on Filbert Street

▷   As you progress east on Filbert from A to B to C to D, the x-value of your position increases while the y-value remains constant. What happens to the value of d?

At any point on Filbert Street, the diagonal distance from Lafayette Park is

$$d(x,0.5) = \sqrt{x^2 + 0.25}$$

where x is the distance along Filbert to the corner of Gough and y is fixed at 0.5. The distance d is now a function of the single input x and so can be graphed in the two-dimensional Cartesian plane. (See Figure 8.7.)

▷   Now suppose that after leaving the park, you walk along Clay Street on your way to the Embarcadero. What is the constant value of y in this case?

▷   In Figure 8.7, draw the graph of d with y fixed at this new constant value. Label this graph[2] with its fixed y-value.

---

[2]If your grapher shows negative outputs, be suspicious. The function d has only nonnegative numbers in its range.

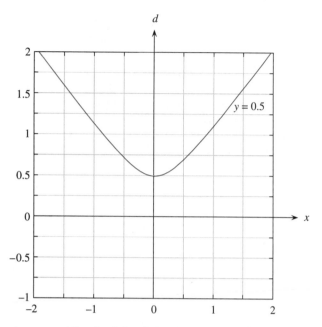

Figure 8.7 Graph of $d(x,0.5)$ with Student-Drawn Graphs of $d(x,0)$ and $d(x,0.8)$

Finally, imagine yourself returning from Fisherman's Wharf to the park along Bay Street, which is parallel to Filbert and Clay, 0.8 mile north of Clay. As before, we can express the diagonal distance from Lafayette Square Park by

$$d(x,0.8) = \sqrt{x^2 + 0.64}$$

where x measures distance along Bay Street from its intersection with Gough.

▷   Add a graph of $d(x,0.8)$ to Figure 8.7, and label it $y = 0.8$ to distinguish it from the other two.

Approximating an instantaneous rate of change

Each of the individual curves in Figure 8.7 shows how d varies with x for a particular fixed y. For any specified value of y, as the value of x increases from 0, the value of d also increases. By selecting one of the fixed y-values, say $y = 0.5$ for Filbert Street, we can analyze the rate at which the distance from Lafayette Square Park changes as x increases. Again, imagine walking up Filbert Street from Gough toward Coit Tower. When you reach the intersection of Filbert and Van Ness, your x-value is approximately 0.2 mile. Let's compute the rate at which your distance from the park is changing at that moment.

The technique is the one we used back in Chapters 4, 5, and 6; nothing is different but the notation. We use a narrow interval centered on $x = 0.2$ and calculate a slope:

Here, $d(x,y) = d(x,0.5)$
$= \sqrt{x^2 + 0.25}$

$$\frac{\Delta d}{\Delta x} = \frac{d(0.21,0.5) - d(0.19,0.5)}{0.21 - 0.19} = \frac{\sqrt{(0.21)^2 + 0.25} - \sqrt{(0.19)^2 + 0.25}}{0.02} \approx 0.37 \frac{\text{miles}}{\text{mile}}$$

That is, when you reach the corner of Filbert and Van Ness, your distance from the park entrance is increasing at a rate of 0.37 miles per mile. The "miles per mile" units mean *miles of total distance* per *mile of distance in the x-direction.*

Now use $d(0.71,0.5)$ and $d(0.69,0.5)$ to calculate the rate.

▷   Approximate the rate of change of $d(x,0.5)$ when $x = 0.7$.

Your calculations should indicate that the distance is changing more rapidly when $x = 0.7$ than when $x = 0.2$. This increase in the rate of change helps explain the upward concavity of the graph of d for $y = 0.5$.

In addition to providing information about how d varies with x, the collection of curves in Figure 8.8 *taken as a whole* can be used to visualize how d varies with y. Notice what happens, for instance, when x is held equal to 1, while y varies. As y increases from 0 to 0.5 to 0.8, we move vertically from the lower curve to the middle to the upper, as shown by the three dots in Figure 8.8. The corresponding values of d increases as well. (Remember that d is plotted on the vertical axis of this graph.)

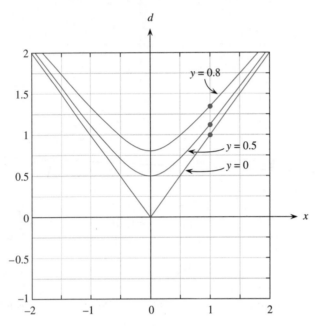

Note the axis labels: d is the output, x the input. The other input, y, is represented by the tag on each individual curve.

Figure 8.8 Graphs of $d(x,0)$, $d(x,0.5)$, and $d(x,0.8)$

▷   Hold x fixed at some other value such as -0.5 and let y vary from 0 to 0.5 to 0.8. Use Figure 8.8 to illustrate what happens to the value of d as y increases.

STOP & THINK

Look again at the formula for the distance function d, $d(x,y) = \sqrt{x^2 + y^2}$, and explain how you can determine from the algebraic formula alone that an increase in magnitude in either x or y results in an increase in d.

Check Your Understanding 8.2

1. Sketch graphs of $M(S,t) = Se^{0.05t}$ (from CYU 8.1) for S fixed at three distinct values: \$100, \$200, and \$400. Label each curve with the corresponding S-value.

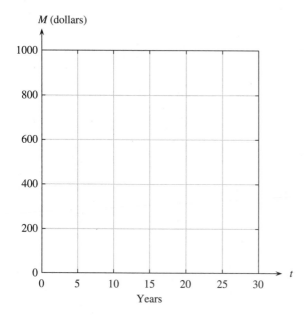

2. Each of the individual curves in Question 1 shows how M varies with t for a particular fixed S. In each case, what happens to M as t increases?

3. For the curve corresponding to $S = 200$, approximate the rate of change in M at $t = 10$ years and again at $t = 20$ years. When is the investment increasing more rapidly: at 10 years or at 20 years?

4. The collection of curves *taken as a whole* can be used to visualize how M varies with S. For any fixed value of t (say, 15 years), what happens to the value of M as S increases from 100 to 200 to 400?

The distance and savings-account functions are examples of **multivariable functions,** functions with more than one independent variable. Lab 8A and Projects 8.1 and 8.2 have models with as many as three independent variables. The strategy that we used here—holding constant all but one input and examining the relationship between the output and a single input that's allowed to vary—works for those as well. Now we'll look at one other two-input function with the same domain as the distance function. This function, however, cannot be specified by a mathematical formula, and therefore we will need a different strategy.

Fixed Values for the Output—Level Curves

I've been down so long it looks like up to me

Here's an example of another two-input function with which you might be familiar. Each location in downtown San Francisco can be assigned a number representing its height, measured in feet above sea level. If we again let the ordered pair (x,y) designate a location in San Francisco relative to the entrance to Lafayette Square Park, we can use function notation to define the altitude function a:

$$a(x,y) = \text{altitude in feet above sea level of the point } (x,y)$$

Certainly, *a* is a function: Each location in the city has its specific height above sea level. However, unlike the distance function *d*, the function *a* cannot be expressed by an algebraic formula. For an altitude function, information is most commonly conveyed visually by a topographical map like the one in Figure 8.9.

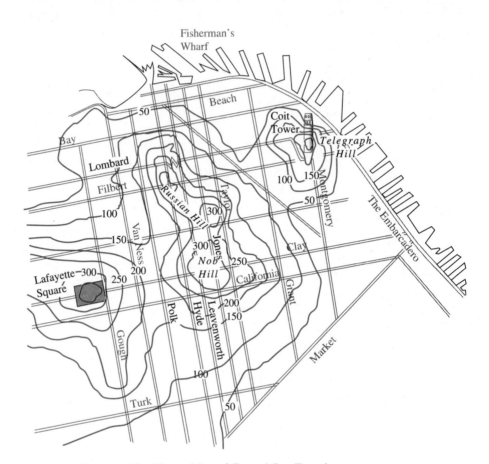

Figure 8.9 Topographical Street Map of Central San Francisco

Each curve that you sketched in Figure 8.7 was obtained by holding the independent variable *y* fixed at a specific value: 0, 0.5, or 0.8. For a topographical map, it is the *dependent variable* (here, *a*) that is set to a fixed value. The curves on a topographical map, called **contours** or **level curves,** consist of points having the same altitude. Walking along a contour, you remain at a constant altitude—you neither climb nor descend. The curves labeled 200 in Figure 8.10, for example, contain all points (x, y) on the map for which $a(x, y) = 200$, that is, all points 200 feet above sea level.

▷   Use Figure 8.9 to decide the approximate altitude of the entrance to Lafayette Square Park on Gough Street. Write the information in function notation:

$$a(0,0) =$$

Now look at the concentric curves around Russian Hill in Figure 8.10. Moving from one curve to another indicates a change in altitude. If we imagine walking east (to the right) on Vallejo Street from its intersection with Gough, we begin at an altitude of about 125 feet and climb to 150 feet when we cross the first contour curve. We then climb to 200 feet by the next curve and 250 feet by the next, finally reaching the summit, more than 300 feet above sea level. Imagine continuing along Vallejo on the other side of Russian Hill. Now you cross level curves in the reverse order because you are descending. There, the contours are closer together on the eastern side; this means

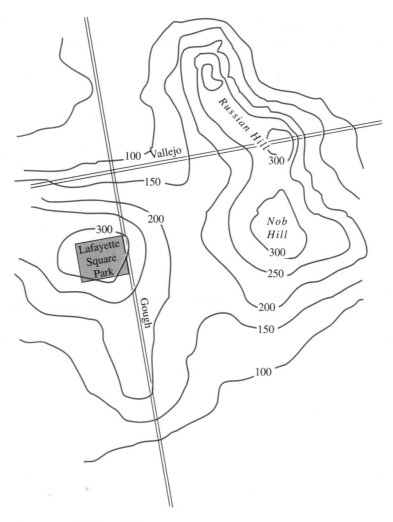

Figure 8.10 Enlarged Section of San Francisco Topographical Map

On a steep hill, level curves are close together.

that the hill is steeper. The spacing of the level curves gives us information about the rate of change of the altitude function: The more closely spaced the contours, the more rapidly the altitude changes.

▷ In Figure 8.9, locate Coit Tower. To approach the tower, you must climb Telegraph Hill, as indicated by the concentric level curves around the tower. Which side of Telegraph Hill is the steepest? How do you know?

In mathematics, a **contour plot** is an abstraction of the ideas represented in a topographical map to any function with two input variables. Just as a topographical map shows the set of locations at a specific altitude, a labeled contour of a function $f(x,y)$ provides visual representation of the set of points (x,y) for which the function has one specific output value. Try your hand at a contour plot in CYU 8.3.

Check Your Understanding 8.3

1. Use this topographical map of Mount Rainier in Washington State to answer the following.
   (a) The altitude near Sunset Ridge is 12,000 feet. Mark a location near Nisqually Glacier at the same altitude.
   (b) Which is steeper, the terrain near Sunset Ridge or near Nisqually Glacier?
   (c) Which side of Mount Rainier do you think would have an easier hiking trail: the east side or the west?

(c) Imagine two ants on the surface described by the function $f(x,y) = xy$. The first ant starts at the point $(4,0)$ follows the vertical line $x = 4$ to $(4,2)$. The second ant walks along the line $x = 2$ from $(2,0)$ to $(2,2)$. Which ant has taken the steeper path? For which path is the rate of change of f larger? How can you tell this from your contour plot?

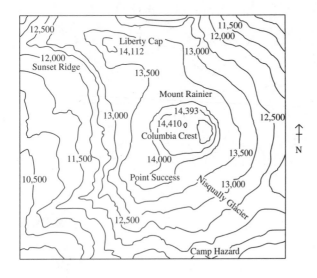

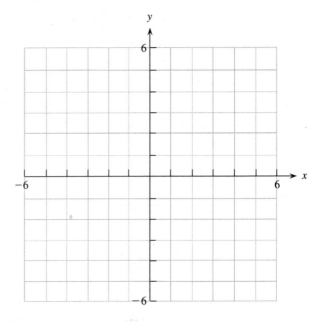

2. These questions concern the two-input function $f(x,y) = xy$.
   (a) Draw the contour for $f(x,y) = 1$. (First, express y as a function of x by solving $xy = 1$ for y; then graph y as a function of x.) Label the curve 1.
   (b) On the same axes, draw level curves for $f(x,y) = 2$ and $f(x,y) = 3$, labeling them, respectively, 2 and 3.

Note the axis labels: Both x and y are *inputs*. The output, $f(x,y)$, is shown by the tag that you put on each curve.

In this section, we found ways to visualize functions with two inputs. The dependent variable, in the case of both d and a, was a function of two independent variables, x and y. We used two different strategies to represent three variables worth of information on a two-dimensional graph. In Section 8.2, we will investigate a different technique for handling three variables at once.

Calvin and Hobbes   by Bill Watterson

In a Nutshell   A multivariable function F is one whose output depends upon more than one independent variable. To represent F on a flat surface, we can graph F with respect to one input while holding all other inputs constant. Or with exactly two input variables, we can use one axis for each input and draw contours for predetermined values of F. Either way, we usually show several different curves for F on the same axes.

8.2   PARAMETRIC EQUATIONS—MOTION ALONG A LINE

In this section, we develop ways of picturing mathematical relationships in which two dependent variables are functions of a single input, called the **parameter.** The algebraic relationship of each output variable to the parameter gives rise to a set of functions called **parametric equations.** With parametric equations, you can do the following:

• Model the motion of an object along a path

• Represent x-y curves in which y isn't a function of x

We continue to use the x-y plane. But instead of y being a function of x, both x and y are functions of the parameter t.

It's about time

Imagine yourself in San Francisco again, driving west on Filbert Street away from Coit Tower. (See Figure 8.11.) The path of your trip is a straight line. The coordinate system on this map has its origin in the lower left-hand corner, and the x- and y-axes coincide with the east-west and north-south directions, respectively. Note that no street runs exactly east-west or north-south.

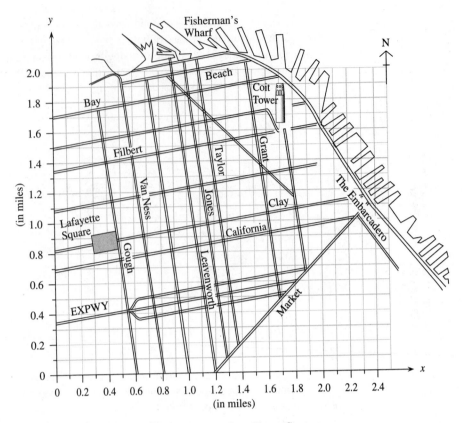

Figure 8.11 Map of San Francisco on a Coordinate System

▷   Using Figure 8.11, write an approximate equation for Filbert Street.

Equation for Filbert Street: $y =$ _____

The equation you wrote gives y as a function of x. It keeps track of your location, relative to the coordinate system, as it varies during the course of your drive. It doesn't, however, provide information about where you are at any given time.

Suppose, now, that a limousine is traveling along a path given by the equation

$$y = 4.2 - 7x \qquad \text{for } 0.35 \leq x \leq 0.54$$

where x and y are the limo's east-west and north-south positions, respectively, in miles relative to the origin.

▷   On what street is the limo? Use the x-intercept of $y = 4.2 - 7x$ to find out.

Need help solving simultaneous equations? The Algebra Appendix is ready and waiting. See page 488.

▷   Determine algebraically the point of intersection of the two lines (the one for the limo and the one for Filbert Street). Is your answer consistent with Figure 8.11?

Your path down Filbert Street intersects with the limosine's path up Gough. Is a crash imminent? To find out, you need more information than the fact that the two vehicles cross paths; you also need the time at which each car arrives at the intersection.

Think about this: As you travel down Filbert, the x-coordinate (east-west position) and y-coordinate (north-south position) of your location change over time. The equation that you wrote to represent Filbert Street relates your north-south position to your east-west position. To answer the question about an impending crash, what is really needed is the relationship between each vehicles's north-south and east-west positions and *time*. In essence, that's what this section is about: sets of equations that relate two dependent variables, x and y, to a third variable, the parameter t.

Linear Motion at a Constant Rate

Suppose we want to simulate graphically a car and a limousine driving on the streets of San Francisco. We represent the two vehicles with two dots and the vehicles' motion by the movement of the dots along their respective paths on the map. The location of each dot will be governed by a set of *parametric equations*,

$$x = x(t) \qquad \text{and} \qquad y = y(t)$$

that give the x- and y-coordinates of the vehicle at any given time t. The input variable t is called the *parameter*. (Notice that we are using x both as an east-west coordinate and as the name of a function of t. Similarly, y is both a north-south coordinate and the name of another function of t.) Suppose, for example, that the limousine's motion along its path is described by the parametric equations

$$x(t) = 0.54 - 0.05t$$
$$y(t) = 0.42 + 0.35t$$

where t is measured in minutes from some reference time and x and y are distances (in miles) east and north of the origin (the bottom left corner of the map in Figure 8.11). According to these equations, the limo's starting position, when $t = 0$, is the point (0.54, 0.42).

▷   On the map in Figure 8.11, place a dot at the limo's approximate starting position. (It should be *on a street*.) Next to the dot, write the label "$t = 0$."

After 1 minute of travel, we have

$$x(1) = 0.54 - 0.05(1) = 0.49$$
$$y(1) = 0.42 + 0.35(1) = 0.77$$

so the limo's position is (0.49, 0.77).

▷   Place another dot on the map in this new position. Label it "$t = 1$."

▷   Find the position after 2 minutes of travel. Mark it on the map; label it "$t = 2$."

▷   Does the limousine cover more, less, or the same distance from $t = 1$ to $t = 2$ seconds as it did from $t = 0$ to $t = 1$?

To learn whether the car and the limo collide, we'll need a second set of parametric equations that describe the car's motion. But first we take a brief detour.

Oh where, oh where has my little dot gone?

For the detour, we'll use a time-honored mathematical custom: Simplify the problem. The frames in Figure 8.12 trace the location of a dot moving vertically in the x-y plane. In Frame 0, the dot is located at the position (5, 1). We set the dot in motion by increasing its vertical position by two units each second. Subsequent frames show the position of the dot at 1-second intervals after the motion begins.

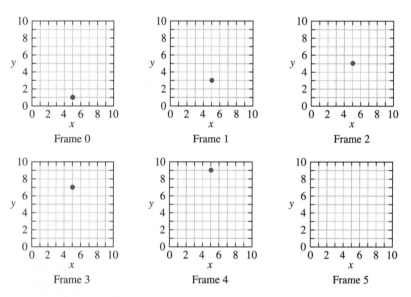

Figure 8.12 Positions of a Dot at 1-Second Intervals

Imagine putting the six frames in a stack and flipping through them (like the child's toy that simulates an animated cartoon). The resulting image is that of a dot moving upward. By combining the frames of Figure 8.12 and labeling each dot with its corresponding time, we can condense the information from the six frames into the single graph of Figure 8.13. In this way, we're able to present three variables worth of information (horizontal location, vertical location, and time) in a two-dimensional plot.

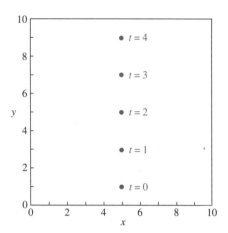

Figure 8.13   Six Frames from Figure 8.12 Condensed into a Single Graph

▷   For each value of t in Table 8.1, record the x- and y-coordinates of the dot as it moves along its path in Figure 8.13. (How will you decide the dot's position at $t = 5$?)

t	x	y
0		
1		
2		
3		
4		
5		

Table 8.1   Coordinates of Moving Dot

Study the number patterns of the x- and y-coordinates in Table 8.1: The x-coordinate is always 5, but the y-coordinate changes with t so that its value is always "one more than twice the number of seconds elapsed." We can express these relationships algebraically with a set of parametric equations:

$$x = 5 = 5 + 0t$$
$$y = 1 + 2t$$

Watch, now, what happens if we change the coefficient of t in the y-equation.

▷ Complete Figure 8.14 by drawing the six positions of the dot corresponding to $t = 0$, 0.5, 1, 1.5, 2, and 2.5 for each given set of parametric equations.

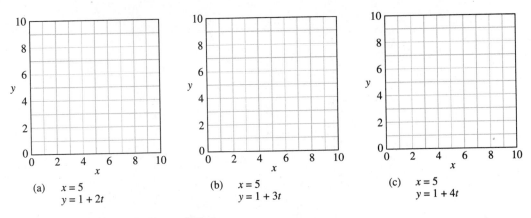

(a) $x = 5$ (b) $x = 5$ (c) $x = 5$
       $y = 1 + 2t$ $y = 1 + 3t$ $y = 1 + 4t$

Figure 8.14 Changing the Rate of Motion

Because each dot in Figure 8.14 moves at a constant rate, the points on any one graph should be equally spaced. If yours aren't, check your calculations. Notice that increasing the coefficient of t in the parametric equation for y had the effect of increasing the vertical distance between plotted points—in effect, speeding up the dot. Furthermore, the points in graph (c) should be exactly twice as far apart as those in graph (a), since the dot in (c) travels twice as fast, covering twice as much distance per second.

Check Your Understanding 8.4

Your task is to move (mathematically) a dot across a screen. The dot starts at (4,2) and moves horizontally to the right at a rate of 1.5 units per second.

3. When will the dot disappear off the edge of the screen?

1. Show the dot's position at 1-second intervals as it moves across the screen. Label each position with its corresponding time.

4. Adjust the equations so that the dot moves three times as fast horizontally.

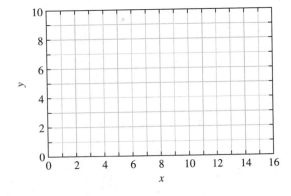

5. Using the adjusted equations from Question 4, show the position of the dot at 1-second intervals. How did increasing the rate affect the distance between consecutive points?

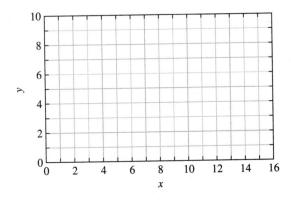

2. Write a set of parametric equations to produce the graph in Question 1.

$x = $ _____

$y = $ _____

Thus far, we've worked with equations to move a dot vertically or horizontally. In each of these situations, either x or y remained fixed while the other varied. Now consider the path of the car moving down Filbert Street. The east-west (x) and north-south (y) positions of the car were changing simultaneously. With this in mind, we'll try moving a dot simultaneously in the horizontal and vertical directions. For example, suppose that the dot starts at $(4,2)$ and that its horizontal position increases at two units per second while its vertical position increases at three units per second. In the first second, the dot moves from $(4,2)$ to $(6,5)$.

▷ Complete Table 8.2 with the x- and y-coordinates of the dot at half-second intervals, observing the numerical patterns that develop.

t	x	y
0.0	4	2
0.5		
1.0	6	5
1.5		
2.0		
2.5		
3.0		

Table 8.2 Coordinates of a Dot Moving along a Diagonal Path

▷ Express in words the relationship between x and t. Then translate your verbal description into an algebraic formula giving x as a function of t.

▷ Similarly, state the relationship between y and t and write a formula expressing y as a function of t.

Your $x(t)$ and $y(t)$ should be equivalent to this set of parametric equations:

$$x = 4 + 2t$$
$$y = 2 + 3t$$

This pair of equations incorporates the original information: that the dot originates at the point $(4,2)$ and moves horizontally at two units per second and vertically at three units per second. Furthermore, these equations can be entered into any grapher capable of handling parametric equations.[3] With the given equations, a graphing calculator produces a graph similar to Figure 8.15. (Times and scales have been added to the display.)

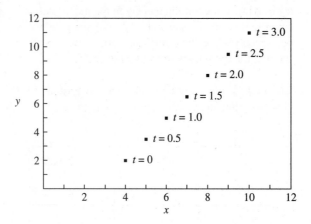

Figure 8.15 Dot Moving According to the Parametric Equations $x = 4 + 2t$, $y = 2 + 3t$, for $0 \leq t \leq 3$

Notice that the seven positions shown in Figure 8.15 are evenly spaced. This means that the dot travels along its straight-line path at a constant rate, because it covers equal distances in equal time increments.

Because of the limited time interval, this path has a definite beginning and end.

▷   To indicate the path of the moving dot during the interval $0 \leq t \leq 3$, draw a line segment in Figure 8.15 beginning at $(4,2)$ and ending at $(10,11)$.

▷   Using the Pythagorean theorem, determine the exact distance the dot travels in 1 second. (Remember that the horizontal position changes at a rate of two units per second and the vertical at three units per second.) To two-decimal-place accuracy, approximate the rate at which the dot moves along its path.

Now focus on the path of the dot—the line you drew in Figure 8.15.

▷   What is the slope of the linear path? How is the slope related to the coefficients of t in the parametric equations for x and y?

▷   Find an equation for the linear path that involves x and y but not t. (You can use Chapter 2 techniques here since you know the slope and a point on the line.)

---

[3]With a graphing calculator, choose *parametric* mode. On a computer, consult the instructions for the graphing software.

Check Your Understanding 8.5

Suppose an animated dot moves according to the parametric equations

$$x = 1 + 0.5t$$
$$y = 2 + 1.5t$$

where t is time measured in seconds.

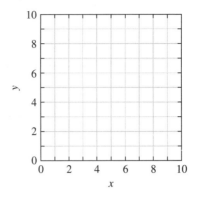

1. In the table, fill in the horizontal and vertical positions of the dot for the times indicated. Then plot the points.

t	x	y
0		
1		
2		
3		
4		
5		

2. By how much does the horizontal position of the dot change each second? By how much does the vertical position change each second?

3. To indicate the path of the dot, draw a line connecting the points on your graph in Question 1. Use the Pythagorean theorem or the distance formula to determine how far the dot travels along its path each second.

4. Write an equation for the path that expresses the vertical position y as a function of the horizontal position x. This equation should not contain the parameter t.

STOP & THINK   Suppose that the motion of a dot is governed by the parametric equations in CYU 8.5. Perform this experiment: Double the coefficient of t in the x-equation alone, in the y-equation alone, and finally in both equations. In each case, how is the path affected? How is the speed of motion affected? Are your results peculiar to this set of parametric equations, or would similar results hold if you used a different set?

We've got to stop meeting like this!

At the start of this section, we left your car and a limousine careening down the streets of San Francisco on a collision course. Now let's look at sets of parametric equations that describe the motion of both vehicles. As before, t is time in minutes and x and y are distances (in miles) east and north, respectively, of the origin.

Car's Motion	Limo's Motion
$x = 1.53 - 0.43t$	$x = 0.54 - 0.05t$
$y = 1.56 - 0.06t$	$y = 0.42 + 0.35t$

To avoid a crash at the intersection of Filbert and Gough, where their paths cross, the two vehicles need to reach the intersection, located at approximately $(0.4, 1.4)$, at different times. We can determine the time when the limo reaches the intersection by finding the value of t corresponding to an x-coordinate of 0.4 (or, equivalently, a y-coordinate of 1.4).

$$0.54 - 0.05t = 0.4$$

$$t = 2.8 \text{ minutes}$$

▷ Verify that you get the same t-value if you set the limo's y-coordinate equal to 1.4.

▷ Determine when the car reaches the intersection. Do the vehicles collide?

Luck is on your side, and there is no crash. The limousine arrives at the intersection approximately 10 seconds after your car does.

Figure 8.16 shows the paths of the car and the limo with their positions marked at one-half-minute intervals. Notice that for each vehicle, the points are evenly spaced, indicating that each is driving at a constant speed. Is this realistic? In other words, can anyone expect to drive on city streets at an unvarying speed? Figure 8.17 presents a more realistic graph of the position of a car as it travels from one red light to the next: The points are closer together when the car is going slowly and farther apart when it picks up speed.

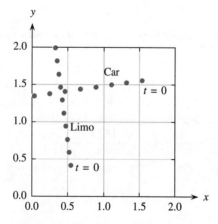

Figure 8.16   Positions of Two Vehicles on Intersecting Paths

Until now, we've worked with parametric equations in which x and y are both linear functions of t. Unlike the motion depicted in Figure 8.17, such equations simulated

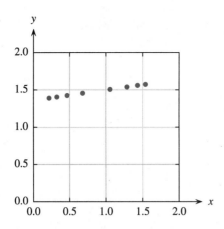

Figure 8.17   Positions of Car Whose Speed is Changing

motion along a line *at a constant rate* only. Next, we'll see how to adjust the parametric equations to simulate motion at a nonconstant rate.

Linear Motion at a Changing Rate

Marching to a different drummer

Let's start with a set of parametric equations for motion along a linear path.

$$\text{Dot 1:} \qquad x = 10 - 3t, \qquad y = 1 + 4t$$

where t is measured in seconds. Since the equations express x and y as linear functions of t, changes in x and y are proportional to changes in t:

$$\Delta x = -3\Delta t$$
$$\Delta y = 4\Delta t$$

Furthermore, for any given Δt, the corresponding ratio $\frac{\Delta y}{\Delta x}$ is $-\frac{4}{3}$. In other words, the path of dot 1 has a constant slope of $-\frac{4}{3}$ and must therefore be a line. In addition, dot 1 travels along its linear path at a constant rate of five units per second.

▷   Use the Pythagorean theorem to verify that dot 1 moves at five units per second.

To create models in which the speed varies, we replace the linear functions of t in the equations for dot 1 by nonlinear functions of t. Let's experiment with the two functions t^2 and $\sqrt{t}$. Suppose that two other dots move according to the equations resulting from these replacements:

$$\text{Dot 2:} \qquad x = 10 - 3t^2, \qquad y = 1 + 4t^2$$
$$\text{Dot 3:} \qquad x = 10 - 3\sqrt{t}, \qquad y = 1 + 4\sqrt{t}$$

How do the paths and the rates of travel differ for the three dots? We'll focus first on the paths. For each dot, we need an equation that doesn't involve t. Dot 1 starts at the point

(10,1) and follows a linear path whose slope is $-\frac{4}{3}$. Therefore, it moves along the line

$$y = 1 - \frac{4}{3}(x - 10)$$

$$y = \frac{43}{3} - \frac{4}{3}x$$

How to eliminate the parameter and obtain y as a function of x

We'll use the following strategy to find a single equation for the path of dot 2:

- Solve dot 2's x-equation for t

- Substitute the resulting expression for t into dot 2's y-equation

Solving dot 2's x-equation for t yields

$$x = 10 - 3t^2$$
$$3t^2 = 10 - x$$
$$t^2 = \frac{10 - x}{3}$$
$$t = \pm\sqrt{\frac{10 - x}{3}}$$

That looks pretty messy, but hang in there. Now substitute the expression for t into dot 2's y-equation:

$$y = 1 + 4\left[\pm\sqrt{\frac{10 - x}{3}}\right]^2$$

▷   Simplify the preceding equation to show that dot 2 travels the same path as dot 1.

Now look back at the equations for dot 2. Both the x and the y equations are quadratic functions of t, yet dot 2 follows the same linear path as dot 1. How can this be? Let's see.

In dot 2's x-equation, Δx is proportional to $(\Delta t)^2$: $\Delta x = -3\,(\Delta t)^2$. Similarly, using the y-equation, $\Delta y = 4\,(\Delta t)^2$.

▷   Compute the value of $\frac{\Delta y}{\Delta x}$ for dot 2. You should obtain $-\frac{4}{3}$, the same constant slope as before, showing that dot 2 does indeed travel along a line.

▷   It's your turn. Using the two-step procedure outlined above for dot 2, find a single equation in x and y describing the path of the third dot.

All three dots travel along the same path, namely,

$$y = \frac{43}{3} - \frac{4}{3}x$$

What, then, is different about the motion of the three dots? To answer this, we'll observe some details of their motion.

▷   Using the parametric equations governing dots 1, 2, and 3, complete Table 8.3.

Dot 1

t	x	y
0		
1		
2		
3		
4		
5		

Dot 2

t	x	y
0		
1		
2		
3		
4		
5		

Dot 3

t	x	y
0		
1		
2		
3		
4		
5		

Table 8.3 Successive Positions of Three Dots Moving along the Same Path

▷   Use the values from Table 8.3 to plot the positions of each dot at 1-second intervals in Figure 8.18.

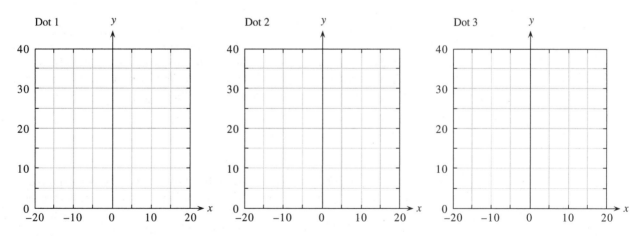

Figure 8.18 Three Dots Moving along the Same Linear Path

Notice that, unlike the first dot, dots 2 and 3 are not moving at constant rates. Since consecutively labeled points in the plot for dot 2 are spreading apart as time elapses, dot 2's speed is increasing. For dot 3, the distances between consecutive points get smaller as time passes, signifying that dot 3 is slowing down as it moves along its path.

STOP & THINK

Why does replacing t by t^2 in the equations of motion cause the speed to increase over time without changing the path of motion? Why does replacing t by $\sqrt{t}$ cause the speed to decrease as time passes but not change the path? If we had started with a different set of parametric equations and replaced t by t^2 or $\sqrt{t}$, do you think the same effects would have occurred?

For every set of parametric equations in this section, changes in the y-coordinate were directly proportional to changes in the x-coordinate. In other words, $\frac{\Delta y}{\Delta x}$ was constant and the motion was linear. In the next section, we'll see how to write parametric equations to model motion along a curve.

In a Nutshell   A set of parametric equations for x and y uses the familiar x-y plane but treats each coordinate separately as a function of a third variable, the parameter t, usually thought of as time. Thus, each moment in time becomes associated with a particular coordinate pair. Parametric equations give the (x, y) position of an object at any time t. If $\frac{\Delta y}{\Delta x}$ is constant, the path of motion of the object is a line.

8.3   PARAMETRIC EQUATIONS—MOTION ALONG A CURVE

Few things in life, however, travel in a straight line. An arrow heading toward a target, the moon orbiting our planet, and an airplane circling an airport are all examples of travel along a curved path. We will see that for a curved path, the change in the y-coordinate will not be directly proportional to the change in the x-coordinate, as it was for a linear path.

A Parabolic Arc

Look out below!

The egg-drop contest in Section 5.1 (pages 226–227) involved objects that traveled straight down or—in the case of one egg that was initially tossed upward—straight up and down. Imagine now that an engineering student throws another packaged egg, this time hurling it upward and away from the building. The following parametric equations could describe the motion of the package along its path:

$$x = 10t$$
$$y = -16t^2 + 19t + 20$$

where x gives the horizontal distance from the building, y gives the vertical distance above the ground, and t gives the elapsed time, measuring x and y in feet, t in seconds.

▷   Find the duration of the flight of the egg. (Solve $y = 0$ for t.)

▷   How far away from the building does it land? (Use that t-value in the x-equation.)

A plot of the parametric equations, with the egg's position labeled at 0.3-second intervals, appears in Figure 8.19. The graph in Figure 8.19, at first glance, looks very much like an earlier egg-drop graph, that of $H(t)$ in Figure 5.3, reproduced here for comparison as Figure 8.20. However, there is an important difference. Whereas the vertical axes of both graphs measure height above the ground, the horizontal axes represent distinct quantities. In Figure 8.19, the horizontal axis measures distance from the building, while in Figure 8.20, it represents time. Figure 8.19 shows the actual path of the egg, *its vertical position with respect to its horizontal position.* Figure 8.20 shows *vertical position with respect to time* of an egg whose path was entirely vertical. It does not display the path of the egg.

Pay attention to the horizontal axes in these two graphs; x-axis gives position; t-axis gives time.

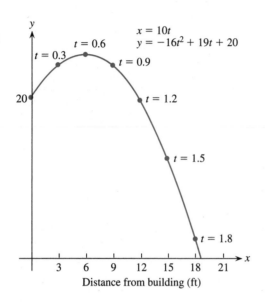

Figure 8.19 Path of Egg Thrown Up and Out

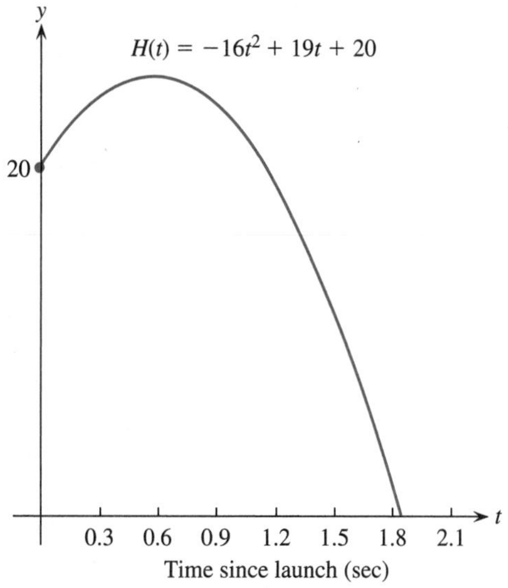

Figure 8.20 Height of Egg Thrown Vertically

Figure 8.19 is like a sequence of time-lapse photos of the egg, one shot every 0.3 second. The curve of the graph shows the actual curved path. The unequal spacing of the points shows that the speed of the egg is not constant. The increasing spacing between points beyond the vertex indicates that the egg picks up speed as it descends. By way of contrast, a sequence of time-lapse photos of the vertically thrown egg would look like Figure 8.21. (Some of the positions nearly coincide.)

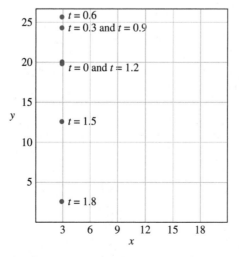

Figure 8.21 Positions of Vertically Thrown Egg

The information in Table 8.4 sheds light on why the parametric equations $x(t) = 10t$, $y(t) = -16t^2 + 19t + 20$ produce a curved path rather than a straight-line path. (Results are rounded to two decimal places.)

t	x	Δx	y	Δy	$\Delta y / \Delta x$
0.0	0.0	—	20.00	—	—
0.3	0.9	0.9	24.26	4.26	4.73
0.6	1.8	0.9	25.64	1.38	
0.9	2.7	0.9	24.14	-1.50	
1.2	3.6	0.9	19.76	-4.38	
1.5	4.5	0.9	12.50	-7.26	
1.8	5.4	0.9	2.36	-10.14	

Table 8.4   Horizontal and Vertical Coordinates of Egg

Focus on the Δx and Δy columns. For Δt equal to 0.3 second, we have a constant value of 0.9 for Δx but varying values for Δy. Thus, the ratio $\frac{\Delta y}{\Delta x}$ will not be constant.

▷   Complete the last column of Table 8.4. Explain, on the basis of the entries in that column, how you can tell that the path of motion is not linear.

The graph in Figure 8.19 certainly looks like a parabola. We can check to see whether it really is parabolic by finding the mathematical relationship between x and y.

▷   By algebraically eliminating t, find a single formula for the parametric equations $x(t) = 10t$, $y(t) = -16t^2 + 19t + 20$ that gives y as a function of x.

You should have found that y, the egg's height above the ground, is a quadratic function of x, its horizontal distance from the building. Therefore, the path of the egg in space can be represented mathematically by a parabola. The equation in x and y that you determined produces the same graph as the parametric equations from which it was derived. However, once t is eliminated, all information about the position of the egg at various points in time is lost.

The preceding example illustrates one of the advantages of parametric equations. The graph of a function $y = f(x)$ is static: it just sits there on the plane. The graph of a set of parametric equations $x = g(t)$, $y = h(t)$ is dynamic: As t increases, the curve is traced in a specific direction at a particular speed.

Parametric equations tell when *as well as* where.

Check Your Understanding 8.6

An investigative precalculus student lobs a well wrapped egg from atop a classroom building. The flight of the egg is modeled by this set of parametric equations:

$$x = 4.62t$$
$$y = 20 + 1.95t - 4.9t^2$$

where x is the horizontal distance in meters from the build-ing, y is the vertical distance in meters above the ground, and t is the elapsed time in seconds.

1. How high above the ground is the egg when it leaves the student's hand?

2. How much time elapses before the egg hits the ground?

3. How far from the building is the egg when it lands?

4. Draw a graph of the path of the egg, marking its position at 0.3-second intervals.

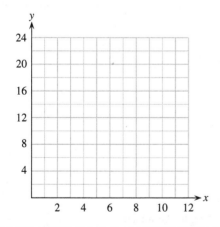

5. Find a single formula for the path of the egg that gives *y* as a function of *x*, and draw its graph. Is it the same?

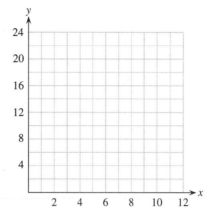

A Circular Path

Will the circle be unbroken?

In Chapter 1 (page 22), you met a man wandering in a trackless waste. His circular path, with radius *r* inches, was the result of the discrepancy *d* in his left and right stride lengths. According to the formula $r = \frac{480}{d}$, the radius of his path is inversely proportional to his stride-length difference. If this discrepancy *d* were $\frac{1}{16}$ inches, for example, then (without any landmarks to guide him) he would wander in a circle of radius 7680 inches, or 640 feet. Our purpose in reminding you of this poor man's plight is to introduce parametric simulations of circular motion. We will create a mathematical model that represents the man's journey. First, though, we'll start with a simpler problem: motion around the unit circle.

The relationship between x and y in the unit circle's equation, $x^2 + y^2 = 1$, is not a function. To plot that circle with a function grapher, we have to break it into two halves, each corresponding to a function:

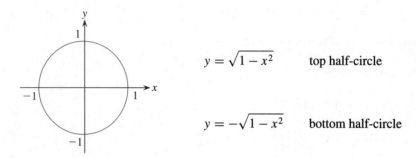

$$y = \sqrt{1 - x^2} \qquad \text{top half-circle}$$

$$y = -\sqrt{1 - x^2} \qquad \text{bottom half-circle}$$

To obtain the entire circle, we plot both functions. Here's an opportunity for parametric equations to show their versatility: they can describe paths that don't necessarily correspond to graphs of functions.

For parametric equations that produce the unit circle, we recall how the sine and cosine functions are defined. (See Figure 8.22.) The cosine function is the first coordinate of a point P that is s units around the unit circle from the point $(1,0)$. The sine function is the second coordinate of P. With s as the parameter, the variables x and y are both functions of s, and the set of parametric equations

$$x = \cos(s), \qquad y = \sin(s)$$

describes a set of points on the unit circle. Notice that this parameter, s, represents distance rather than time.[4] With these parametric equations, we will simulate the motion of a dot traveling around the unit circle.

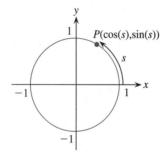

Figure 8.22   Points on Unit Circle Corresponding to Parameter s

▷   With a calculator (set in radian mode), complete Table 8.5 to three-decimal-place accuracy. Then mark the positions of the moving dot on the unit circle in Figure 8.23, starting from $s = 0$ and ending at $s = 2\pi$. Label the corresponding s-value beside each plotted point. The first two are done for you.

[4]In Chapter 6, we used t for the measure of an arc. To avoid confusion in this chapter, we use s for the radian measure of an arc or an angle, reserving t for time.

s (radians)	x = cos(s)	y = sin(s)
0	1	0
$\frac{\pi}{4}$	0.707	0.707
$\frac{\pi}{2}$		
$\frac{3\pi}{4}$		
π		
$\frac{5\pi}{4}$		
$\frac{3\pi}{2}$		
$\frac{7\pi}{4}$		
2π		

Table 8.5   Unit-Circle Coordinates

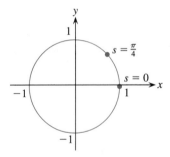

Figure 8.23   Dot Moving on Unit Circle

Now we're ready for the man going around in circles. To construct a mathematical model of his position at different times, we'll need to do two things:

- Account for the 640-foot radius of his circular path
- Change the parameter from s (distance) to t, the elapsed time

Changing the radius of the circle is easier; we need only multiply the x- and y-equations by 640.

▷   Verify that the parametric equations

$$x = 640 \cos(s), \qquad y = 640 \sin(s)$$

produce a circular path with radius 640. Use a grapher or work with the entries in Table 8.5. In either case, use square scaling.

Now (this is a bit more work) we need to change the parameter to *time*. Let's suppose that the man walks at a constant rate of 3 miles per hour.

▷ Compute the distance around a circle of radius 640 feet. Then determine how many minutes it will take the man to complete a single trip around this circle. (Pay close attention to units!)

(If you got an answer different from 15.2 minutes, go back and check the units.)

In completing one circuit, the man traverses an arc of 2π radians in approximately 15.2 minutes. Because his walking speed is constant, arc length s is proportional to time t. That is, $s = k \cdot t$ for some constant k.

▷ Solve for the proportionality constant k, using the known pair of s- and t-values.

We can now switch the parameter to time by replacing s in the parametric equations by $\frac{2\pi}{15.2}t$:

$$x = 640\cos\left(\frac{2\pi}{15.2}t\right) \quad \text{and} \quad y = 640\sin\left(\frac{2\pi}{15.2}t\right)$$

Figure 8.24 is a plot of these equations, with x and y measured in feet, showing positions at 1-minute intervals.

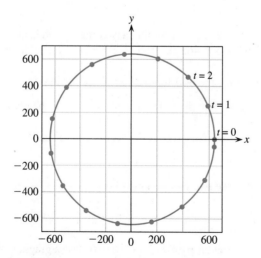

Figure 8.24 Simulation of Man Wandering in a Circle

Check Your Understanding 8.7

1. Another man's stride-length difference is $\frac{1}{8}$ inch. What would be the radius of his circular path (in feet)?

2. Adjust the parametric equations $x = \cos(s)$, $y = \sin(s)$ to produce a circular path with the radius you determined in Question 1.

3. If he walks at 2 miles per hour, how long will it take this man to complete one circuit?

4. Write another set of parametric equations, using time as the parameter, to model this man's circular travels.

STOP & THINK

How would you modify the parametric equations $x = 640\cos\left(\frac{2\pi}{15.2}t\right)$, $y = 640\sin\left(\frac{2\pi}{15.2}t\right)$ so that the man walks in the clockwise direction rather than counterclockwise?

KEEP THINKING!

All the points in Figure 8.24 are equally spaced except for the two on the far right, which are much closer together. Can you explain why?

Modeling the man's circular travels brings *us* full circle (pun intended) in this section. Although the problem was suggested in fun, there are many real-world situations involving circular motion that can be modeled with parametric equations. Some examples include planes circling in level holding patterns above a busy airport, cars rounding a rotary, a stone stuck in the tread of an automobile tire, a person on a Ferris wheel, and a button on a shirt during the spin cycle of a wash. We round out this chapter on a cosmic scale by using parametric equations to represent the motion of the planets.

Planetary Motion

Brother, can you paradigm?

Ancient models of the solar system provide another context for parametric equations of uniform circular motion. Early astronomers proposed models that described the paths of the planets with reasonable accuracy and, at the same time, conformed to strongly held contemporary metaphysical beliefs. Philosophers declared the circle to be the perfect plane figure and the sphere the perfect solid shape. Thus, the ancient astronomers proposed spherical planets whose positions in the sky were governed by uniform circular motion. Two of the best-known models of the solar system were those of Ptolemy (circa A.D. 140) and Copernicus (circa A.D. 1530). Until it was deposed by the heliocentric (sun-centered) theory of Copernicus, Ptolemy's geocentric (earth-centered) theory of the solar system stood, with some modifications, as the definitive model or paradigm.

In the late sixteenth century, Tycho Brahe compiled what were then the most complete and accurate astronomical data. His observations included thousands of position measurements for the sun, the moon, and the planets; he spent 20 years analyzing these data.[5] Fifty years after Copernicus published his heliocentric model for the solar system, Tycho proposed a compromise model in which the earth was stationary, the sun revolved around the earth once a year, and the other planets revolved around the sun. Figure 8.25 shows a schematic drawing of Tycho's model, from his book on the comet of 1577.

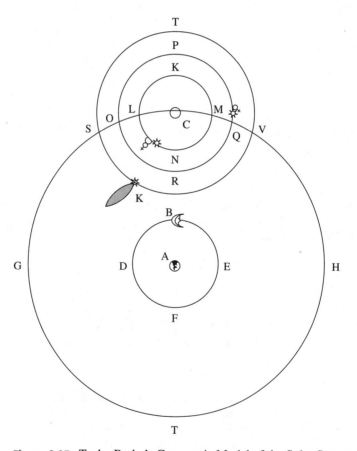

Figure 8.25 Tycho Brahe's Geocentric Model of the Solar System (Sixteenth Century)

---

[5]A complicating factor in any analysis of the path of a planet is that its apparent wanderings in the firmament result from the combination of its own motion and the earth's orbital revolution around the sun. This causes the planet, from our perspective, to appear to change direction and backtrack a bit before resuming its forward progress.

Let's use parametric methods to derive equations for the motions of the planet Mercury. Mercury, according to Tycho's theory, revolved about the sun, which in turn revolved about the earth, as illustrated in Figure 8.26. The only additional information we need is the distance from the sun of our planet and of Mercury and the period of revolution of each.

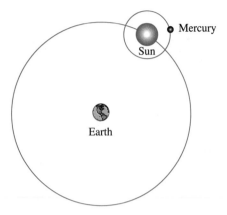

Figure 8.26   Mercury, Sun, and Earth, According to Tycho's Model

▷   Write a set of parametric equations representing a circular path for the sun about the earth (in accord with Tycho's model). Assume that the earth is at the origin, that x and y measure distances from the earth in millions of miles, and that the parameter s represents radian measurement from the positive x-axis. The distance from the earth to the sun is approximately 93 million miles.

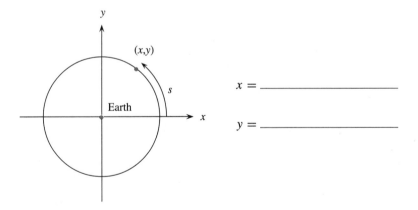

To write equations for Mercury's orbit around the sun, we create a new coordinate system whose origin is the sun. Let x_h and y_h represent distances from the sun in millions of miles, and let s_h be the radian measurement from the positive x_h-axis. (We're using h for *heliocentric*—a reminder that the sun is now the origin.)

▷ Write equations for Mercury's orbit, given that Mercury is approximately 36 million miles from the sun.

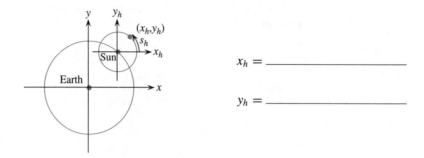

$$x_h = \underline{\hspace{4cm}}$$

$$y_h = \underline{\hspace{4cm}}$$

Now let's go back and change the parameter for the first set of equations (those for the sun revolving around the earth) from radians, s, to time t, measured in years. This won't be difficult; in Tycho's model, the sun would make one full orbit of the earth in 365 days, or one earth-year, which is exactly the time unit we're using.

▷ Given that $s = 2\pi$ radians when $t = 1$ year, and that s is proportional to t ($s = kt$) because Tycho assumed a uniform speed, find the proportionality constant k.

Now we can rewrite those equations using *time* as the parameter:

$$x = 93\cos(2\pi t)$$
$$y = 93\sin(2\pi t)$$

Mercury, however, requires only 88 days to complete its orbit around the sun. That means Mercury makes $\frac{365}{88}$, or approximately 4.15, orbits of the sun in one earth-year. With reasoning similar to what we used for the x- and y-equations, we can change the parameter in the x_h- and y_h-equations from s_h to t. (Note the beauty of this strategy: We start with two separate spatial parameters, s and s_h, but we end up with a single time parameter, t.)

Because s_h is $\frac{365}{88} \cdot 2\pi$ radians when t is one year, the x- and y-equations become

$$x_h = 36\cos\left(\frac{365}{88} \cdot 2\pi t\right)$$

$$y_h = 36\sin\left(\frac{365}{88} \cdot 2\pi t\right)$$

Table 8.6 summarizes what we've accomplished so far.

	With Parameter s (distance in radians)	With Parameter t (time in earth-years)
Sun's equations (relative to the earth)	$x = 93\cos(s)$ $y = 93\sin(s)$	$x = 93\cos(2\pi t)$ $y = 93\sin(2\pi t)$
Mercury's equations (relative to the sun)	$x_h = 36\cos(s_h)$ $y_h = 36\sin(s_h)$	$x_h = 36\cos\left(\frac{365}{88} \cdot 2\pi t\right)$ $y_h = 36\sin\left(\frac{365}{88} \cdot 2\pi t\right)$

Table 8.6 Parametric Equations for Tycho's Planetary Model

The model is nearly complete, and Figure 8.27 will help us see the final step. The x_h- and y_h-equations give Mercury's horizontal and vertical distances from the sun. Similarly, the x- and y-equations give the sun's horizontal and vertical distances from the earth. For a mathematical model of the path of Mercury as viewed from the earth, we need the horizontal and vertical distances of Mercury from the earth.

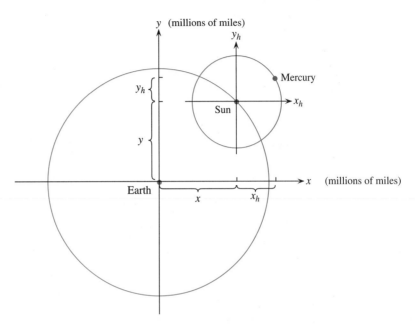

Figure 8.27 Horizontal and Vertical Distances from the Earth to Mercury

For the x-value of Mercury relative to the earth, go from the earth to the sun (that's x) and then from the sun to Mercury (that's x_h).

Figure 8.27 shows that those distances are the sums of the separate distances: the value of x for Mercury relative to the earth is the sum of our original x-value and x_h (either of which might be negative). The y-values work the same way. Therefore, the parametric model for the positions of Mercury with respect to the earth is

$$x = 93\cos(2\pi t) + 36\cos\left(\frac{365}{88} \cdot 2\pi t\right)$$

$$y = 93\sin(2\pi t) + 36\sin\left(\frac{365}{88} \cdot 2\pi t\right)$$

▷   Use a grapher, in parametric mode, to graph these equations from $t = 0$ to $t = 1$. The picture should resemble Figure 8.28, where we've marked Mercury's position at 0.1-year intervals. (Distances x and y are measured in millions of miles.)

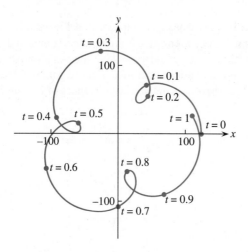

Figure 8.28 Path of Mercury Relative to the Earth, According to Tycho's Model

Imagine that you are observing Mercury from the earth, that is, from the origin. As Mercury moves along its orbital path from its position at $t = 0$, it initially appears to move counterclockwise. After a time, however, it seems to reverse direction and move briefly clockwise, after which it resumes its counterclockwise path. This, in fact, is precisely what we see from the earth but not for the reasons that Tycho gave.

Check Your Understanding 8.8

Venus is approximately 67 million miles from the sun and completes one revolution of the sun every 224 days.

1. In one earth-year, approximately how many revolutions of the sun does Venus make?

2. Write a set of parametric equations x_h and y_h for the orbit of Venus about the sun. The final parameter should be t, time in earth-years.

3. Following Tycho's model, write a set of parametric equations giving the motion of Venus relative to the earth.

4. Sketch the path produced by the equations from Question 3, for $0 \le t \le 1$.

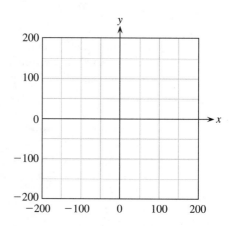

Tycho Brahe died before all the details of his model for planetary motion had been adjusted to match his meticulous astronomical observations. Near the end of his life, he collaborated with Johannes Kepler. Kepler began his own study with the planet Mars, which has the most eccentric (noncircular) orbit of the planets then known. He spent nearly a decade trying to fit various combinations of circular motions to the observed path of Mars, but without success. Finally, he was inspired to represent the orbit of Mars by an oval and realized that an elliptical orbit was an extremely good fit. Because of Kepler's work, the ancient models for the solar system based upon uniform circular motion gave way to today's models based upon elliptical orbits about the sun.

This year's model

In this section, we have discussed several paradigms, or theoretical frameworks, for the observed movements of the sun and the planets. Ptolemy's model was the authoritative world view until Copernicus proposed his "revolutionary" theory.[6] One model may be supplanted by another, particularly when new data don't fit the prevailing theory. Science has moved from Ptolemy's and Tycho's models to Kepler's, Newton's, and Einstein's. The *models* may be replaced, but the mathematics used in the models remains.

In this book, we have used the mathematics of functions to model real-world situations. Some models are crude, others more sophisticated, but each is essentially temporary—the current "best approximation" to reality. Mathematics itself, however, is neither approximate nor temporary. The equations defining the orbit of Mercury exist as pure mathematics apart from their use in expressing a now-discredited view of the solar system. Mathematical relationships don't change with the times. Ptolemy's astronomy is out-of-date, but Euclid's geometry is timeless.

In a Nutshell   Parametric equations can represent motion of an object along a curved path in the x-y plane. They have two advantages over ordinary functions of the form $y = f(x)$:

* They give not only the (x, y) position but also the time t at which the object arrives there.

* They can describe a curve that could not be the graph of a function.

What's the Big Idea?

* The concept of a function of a single independent variable can be extended to include functions with two or more independent variables. Each ordered set of inputs for the independent variable results in a single output for the dependent variable.

* Motion of an object along a path can be modeled using parametric equations. These equations not only produce the path of the motion but also provide the position of the object on the path for any given value of the parameter.

* If the speed, or rate of motion, is constant, positions corresponding to equal increments of time will be separated by equal lengths of path.

* If, for every pair of t-values, Δt gives a constant ratio $\frac{\Delta y}{\Delta x}$, a set $x(t)$, $y(t)$ of parametric equations produces a linear path. If the ratio $\frac{\Delta y}{\Delta x}$ is not constant, the path is curved.

---

[6]Copernicus presented his model in the book *On the Revolutions of the Celestial Spheres*. The title of his book led to our use of the word "revolution" to denote social upheaval as well as motion in a circular path.

Progress Check

After finishing this chapter, you should be able to do the following:

- Use function notation for functions with more than one independent variable. (8.1)

- Graph the relationship between the dependent variable and one of the independent variables by holding all the other independent variables fixed. (8.1)

- Interpret a family of graphs for a succession of fixed values of an independent variable. (8.1)

- Graph a contour, or level curve, for a fixed value of the dependent variable. (8.1)

- Interpret a contour plot for a two-input function. (8.1)

- Draw the graph of a set of parametric equations. (8.2)

- Find a set of parametric equations for linear motion at a constant rate, given either a point on the path and its corresponding time and the rate of motion, or two points on the path and their corresponding times. (8.2)

- Determine whether an object's speed is increasing, decreasing, or constant by looking at x-y positions of the object at equal time increments. (8.2)

- Find a single equation for the path of a set of parametric equations by eliminating the parameter. (8.3)

- Write parametric equations to model uniform circular motion. (8.3)

Key Algebra Skills Used in Chapter 8

- Finding the distance between two points (page 501)

- Solving systems of simultaneous equations (page 489)

- Finding an equation for a circle (page 504)

Answers to Check Your Understanding

8.1

1. $M(S,t) = Se^{0.05t}$
2. $M(100,10) \approx 164.87$ ($100 invested for 10 years yields $164.87.)
   $M(10,100) \approx 1484.13$ ($10 invested for 100 years grows to $1484.13.)
3. Domain: the set of ordered pairs of real numbers; range $\mathbb{R}$
4. Possible domain: the set of ordered pairs (S,t), where $0 \le S \le 500$ and $0 \le t \le 20$; corresponding range: $0 \le M \le 500e$ (This answer assumes that $500 is the largest amount of money you could scrape together for the deposit, and that you could leave the money untouched for 20 years.)

8.2

1.

M (dollars)

(graph showing curves labeled $S = 400$, $S = 200$, $S = 100$ with vertical axis marked 200, 400, 600, 800, 1000 and horizontal axis t (years) marked 0, 5, 10, 15, 20, 25, 30)

2. M increases as t increases.
3. For $t = 10$, the rate of change in M is 16.5 dollars per year; for $t = 20$, it is 27.3 dollars per year. Your investment will be increasing more rapidly after 20 years than after 10 years.
4. For any fixed value of t, the value of M increases as S increases from 100 to 200 to 400.

8.3

1. (a)

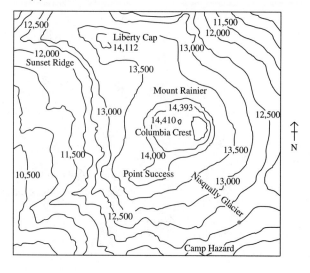

(b) The terrain near Sunset Ridge is steeper, because the contours on its north side are closer together than any of those near Nisqually Glacier.

(c) We would rather climb the west side, where the slope looks more gradual.

2. (a) and (b)

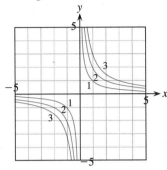

(c) The rate of change in f is greater for the path along $x = 4$. The contours, or level curves, are closer together along the path $x = 4$ than along the path $x = 2$.

8.4

1.

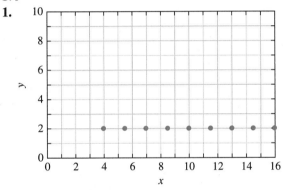

2. $x = 4 + 1.5t$, $y = 2$

3. After 8 seconds

4. $x = 4 + 4.5t$, $y = 2$

5. Tripling the rate triples the distance between consecutive points.

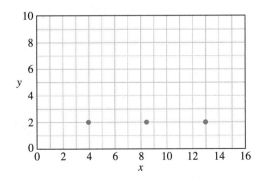

8.5

1.

t	x	y
0	1.0	2.0
1	1.5	3.5
2	2.0	5.0
3	2.5	6.5
4	3.0	8.0
5	3.5	9.5

2. 0.5 unit; 1.5 units

3. Approximately 1.58 units

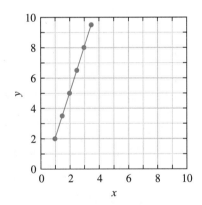

4. $y = 3x - 1$

8.6

1. 20 meters

2. Approximately 2.23 seconds

3. Approximately 10.30 meters

4.

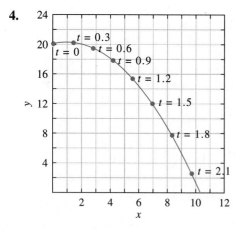

5. $y = 20 + 1.95\left(\frac{x}{4.62}\right) - 4.9\left(\frac{x}{4.62}\right)^2$

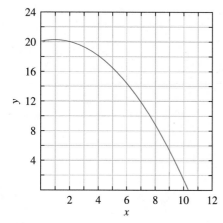

The path looks (and is) exactly the same.

8.7

1. 320 feet

2. $x = 320\cos(s)$, $y = 320\sin(s)$

3. 11.4 minutes

4. $x = 320\cos\left(\frac{2\pi}{11.4}t\right)$, $y = 320\sin\left(\frac{2\pi}{11.4}t\right)$

8.8

1. $\frac{365}{224} \approx 1.63$ revolutions per year

2. $x_h = 67\cos\left(\frac{365}{224} \cdot 2\pi t\right)$,

  $y_h = 67\sin\left(\frac{365}{224} \cdot 2\pi t\right)$

3. $x = 93\cos(2\pi t) + 67\cos\left(\frac{365}{224} \cdot 2\pi t\right)$,

  $y = 93\sin(2\pi t) + 67\sin\left(\frac{365}{224} \cdot 2\pi t\right)$

4.

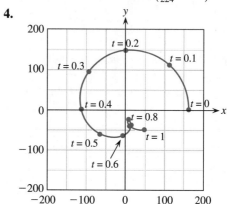

EXERCISES

1. *Skintight.*   How much skin covers your body? One model for estimating your surface area is $A(w,h) = 15.63w^{0.425}h^{0.725}$, a function giving the surface area in square inches of a person who weighs w pounds and who is h inches tall.

  (a) Calculate $A(125,64)$, and write a sentence saying what it means.

  (b) Predict the surface area of a person 5'8" tall who weighs 185 pounds.

  (c) Give a reasonable domain and range for this model if we assume that it works well in predicting body surface area for adults.

2. One of the authors uses the following formula to compute student grades in her precalculus course:

$$G(E_1, E_2, F, P, L) = 0.15E_1 + 0.15E_2$$
$$+ 0.3F + 0.15P + 0.25L$$

where E_1, E_2, F, P, and L are the percentages earned on exam 1, exam 2, the final exam, the projects, and the labs, respectively, and G is the course grade.

  (a) Compute $G(70, 75, 80, 75, 80)$.

  (b) Two students, Carl and Eric, earn exactly the same grades on all exams, However, Carl earns 70% on his projects and 90% on his labs, while Eric earns 90% on his projects and 70% on his labs. If the teacher sticks to her formula, which student will receive the higher grade. Explain. (You can answer this without doing any calculations.)

  (c) Give the domain and the range for this model.

3. Refer to the map of San Francisco in Figure 8.2, page 390, and suppose you ride the cable car along California Street. We can model this situation by setting $y = -0.12$ mile. (Why the minus sign?) Holding y constant at -0.12, we can write the distance d from the cable car's position to the Lafayette Square Park entrance as a function of a single variable, x, the distance of the car from the intersection of California and Gough. Graph and label this function as you did with the examples on page 393 (Figure 8.7).

4. Refer to the map of San Francisco in Figure 8.2, page 390, and suppose you walk the three fifths of a mile from the entrance to Lafayette Square Park down Clay Street to Leavenworth. You know that the diagonal distance between the end of Leavenworth at Fisherman's Wharf and the park is about seven sixths of a mile. Use this to determine the approximate distance on Leavenworth from Clay Street to Fisherman's Wharf.

5. Refer to the map of San Francisco in Figure 8.2, page 390, and imagine yourself walking east on Bay Street ($y \approx 0.8$).
   (a) How rapidly is your distance from the Lafayette Square Park entrance changing when you reach the corner of Bay and Van Ness ($x = 0.2$)?
   (b) Find the rate at which your distance from the park entrance is changing at the moment you reach the corner of Bay and Taylor ($x = 0.7$).

6. The graph of the function $d(x,0) = \sqrt{x^2 + 0^2}$ in Figure 8.8 appears identical to the graph of $f(x) = |x|$.
   (a) Test the suspicion that $d(x,0)$ and $f(x)$ are the same function by evaluating each for several values of x. Include negative values.
   (b) Draw the graph of $f(x) = |x|$, and explain why it is identical to the graph of $d(x,0) = \sqrt{x^2}$

7. Consider arbitrary points $P(x_1,y_1)$, $Q(x_2,y_2)$ in the Cartesian plane (not necessarily in the first quadrant). Using function notation, write the formula for the distance between P and Q as a function D of the four input variables: x_1, y_1, x_2, and y_2.

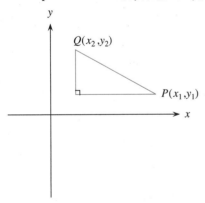

8. In Chapter 5 (page 225), a student tossed a wrapped egg vertically into the air from a height of 20 feet. Suppose the student continues to throw several more eggs, all packaged similarly. For any one of these eggs, its height h above the ground is a function of both its initial velocity v_0 and the elapsed time t since its release. The following table provides data on the relationship between the dependent variable, h (feet), and the independent variables, v_0 (feet per second) and t (seconds). (A height of zero means that the egg is no longer falling.)

(a) Determine $h(5,0.8)$.
(b) Holding v_0 constant, first at 0, then at 10, then at 20 ft/sec, draw three smooth curves showing the relationship between h and t. Use one set of axes, with h on the vertical, t on the horizontal. (Omit points for which h is 0.) Label each curve with its v_0.
(c) Holding t constant at 0.2, at 0.6, and at 1.0 seconds, draw three graphs on the same axes to show the relationship between h and v_0. Label each graph with its corresponding elapsed time.
(d) The function relating h, v_0, and t is given by the formula $h(v_0,t) = 20 + v_0 t - 16t^2$. Use this formula to explain why the graphs in part (b) are parabolas and the graphs in part (c) are lines.

Height of Egg above Ground, $h(v_0, t)$

t \ v_0	0	5	10	15	20
0.0	20.00	20.00	20.00	20.00	20.00
0.2	19.36	20.36	21.36	22.36	23.36
0.4	17.44	19.44	21.44	23.44	25.44
0.6	14.24	17.24	20.24	23.24	26.24
0.8	9.76	13.76	17.76	21.76	25.76
1.0	4.00	9.00	14.00	19.00	24.00
1.2	0.00	2.96	8.96	14.96	20.96
1.4	0.00	0.00	2.64	9.64	16.64

9. The model $A(w,h) = 15.63w^{0.425}h^{0.725}$ from Exercise 1 gives body surface area as a function of weight (pounds) and height (inches).
   (a) Graph the relationship between surface area A and height h for 130-pound adults, using a reasonable interval of h-values. Label the graph $w = 130$. For 130-pound people, does surface area increase or decrease as height increases? How does the graph show this?
   (b) Graph, on the same axes, the relationship between A and h for 175-pound adults, labeling the graph $w = 175$. For a fixed height, does surface area increase or decrease when w changes from 130 to 175? How do the graphs show this?
   (c) The two graphs appear to be lines, but they're not. How you can tell from the formula for A that the curves are not actually linear?

10. Let $F(u,v) = 2u - 3v + 1$.
   (a) Determine whether or not F is a linear function of its two independent variables. Explain how you can decide this from the formula for F.
   (b) Holding v constant at -2, at 0, and at 2, sketch three graphs of $F(u,v)$ on the same axes, showing how F varies with u. Identify each graph.
   (c) Holding u constant at -1, at 1, and at 3, sketch three graphs of $F(u,v)$ on another set of axes, showing how F varies with v. Identify each graph.
   (d) Which would cause a bigger change in the value of F: a change of one unit in u or a change of one unit in v? How did you decide?

11. (a) For the two-variable linear function $F(x,y) = 300 + 2x + 3y$, sketch the graphs of $F(100,y)$, $F(200,y)$, and $F(300,y)$ on the same axes. Label each graph with its constant x-value. Describe the pattern of the graphs.
   (b) The function G is similar to F, but it includes an xy term: $G(x,y) = 300 + 2x + 3y + 0.01xy$. Sketch the graphs of $G(100,y)$, $G(200,y)$, and $G(300,y)$ on the same axes. Label each graph with its constant x-value. Describe the pattern of the graphs.
   (c) The function H is also similar to F, but it includes a quadratic term: $H(x,y) = 300 + 2x + 3y + 0.02y^2$. Sketch the graphs of $H(100,y)$, $H(200,y)$, and $H(300,y)$ on the same axes. Label each graph with its constant x-value. Describe the pattern of the graphs.

12. The pressure-volume-temperature behavior for 1 mole of an ideal gas is given by the function

$$P(T, V) = \frac{0.082057 \cdot T}{V}$$

   where pressure P is measured in atmospheres, volume V in liters, and temperature T in degrees Kelvin (°K). (Add 273 to any Celsius temperature to convert it to degrees Kelvin.)
   (a) Find the pressure exerted by 1 mole of an ideal gas at room temperature (20°C) that is confined to a 3.5-liter container. (Remember to convert the temperature to °K.)
   (b) Sketch a graph of the relationship between pressure and temperature for 1 mole of gas in a 3.5-liter container, and use the graph to determine what happens to the pressure if the temperature of the gas is raised.
   (c) On the same axes used in part (b), sketch a graph of the relationship between pressure and temperature for 1 mole of gas confined to a 1.5-liter container. For each fixed temperature, what is the effect on the pressure of a decrease in volume? How do the graphs show this?

13. Find Turk Street and Beach Street on the topographical map of San Francisco in Figure 8.9. (Both streets run east from Van Ness.) The distance from Beach to Turk looks the same along Polk Street or along Jones Street.
   (a) Which street, Polk or Jones, provides the easier walking route from Beach to Turk? Explain in terms of the altitude function, a.
   (b) Which street (Polk or Jones) do you think would be longer (measured with a tape measure, for example) between Beach and Turk? Explain your answer in terms of the altitude function, a.

14. Look at the topographical map in Figure 8.9. The block of Lombard Street between Hyde and Leavenworth is called "the crookedest street in the world." What practical reason do you think might have caused the city to design the street with so many switchbacks (sharp curves)?

15. Consider the function $H(x,y) = 36 - x^2 - y^2$ with domain $-6 \le x \le 6$, $-6 \le y \le 6$.
   (a) Give the value of the function at $(0,0)$ and at $(2,4)$.
   (b) On the same axes, draw level curves (contours) of the function corresponding to outputs $H = 0$, $H = 12$, $H = 24$, and $H = 36$. Label each contour with its H-value.

16. A dot moves at a constant rate along a line segment from point $P(2,3)$ to point $Q(5,7)$.
   (a) Find the length of the dot's path.
   (b) If the dot starts at P when $t = 0$ seconds and arrives at Q when $t = 10$ seconds, at what rate (units per second) does the dot travel?
   (c) Write a set of parametric equations to describe the dot's motion.

17. Two objects move according to the following parametric equations:

   Object A: $x = -1 + 3t$, $y = -1 + 4t$
   Object B: $x = -1 + 5t$, $y = 3 - 2t$

   (a) Sketch the paths of the two objects over the interval $0 \le t \le 1$.
   (b) By eliminating t, find single equations that describe the path of each object.
   (c) Find the point of intersection of the two paths.
   (d) The paths intersect. Do the objects collide? Explain.

18. The following map shows the section of downtown San Francisco south of Market Street.

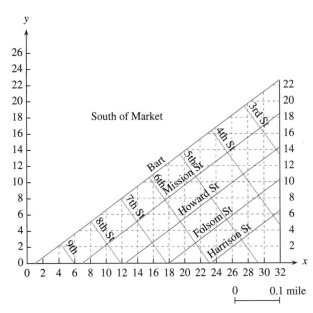

0   0.1 mile

(a) Suppose you are walking down Mission Street from 8th Street on your way to 3rd Street. Approximate the coordinates of the beginning (the intersection of Mission and 8th) and the end (the intersection of Mission and 3rd), and find the length of your walk. If your pace is 2.5 miles per hour, how much time will you need? Write parametric equations that give your location on Mission Street t minutes after you start.

(b) Suppose a friend walks along 5th Street from Harrison to Bart. If the parametric equations $x = 27.2 - 1.68t$, $y = 2.8 + 2.3t$ represent his trip, where t is measured in minutes since he left Harrison, how long will it take him to reach the intersection of Mission and 5th, located at $(21.5, 10.6)$?

(c) Will you and your friend meet?

19. Two dots trace curved paths according to these parametric equations:

Dot A: $x(t) = t^2$, $y(t) = t^4$, $0 \le t \le 1$

Dot B: $x(t) = t$, $y(t) = t^2$, $0 \le t \le 1$

(a) Sketch two graphs, one for each dot, showing the dot's positions at 0.1-unit increments for t. (Use the same scale for both sets of axes.) Do the two dots appear to be moving along the same path? On the basis of your graphs, do they appear to move at the same rate? Explain.

(b) Now use the interval $-1 \le t \le 1$. Are the graphs still the same? Explain.

20. According to Kepler's Second Law of Planetary Motion, the straight line joining a planet to the sun sweeps out equal areas in equal time intervals. Positions A through N on the orbit are determined so that every wedge, shaded or unshaded, has the same area. Thus, the planet's orbital speed varies in such a way that it takes the same amount of time to travel between any two consecutive points on its orbit. (This diagram exaggerates the elliptical orbit.)

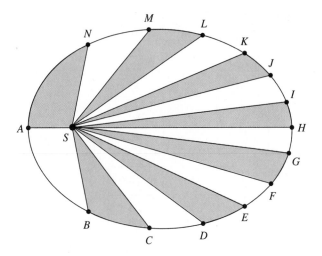

(a) At approximately what position is the planet moving most rapidly? (*Hint*: Consider the distance covered during each time interval.)

(b) At approximately what position is it moving most slowly?

(c) As the planet moves from positions A to B to C to ... to N and back to A, describe how its speed changes. Where is the speed increasing? Where is it decreasing?

21. The table entries were obtained from five different sets of parametric equations. In each case, determine whether the path produced by the equations is linear or curved. If the path is linear, find its slope. If the path is curved, determine whether the path is concave up or concave down.

(a)

t	Δt	$\frac{\Delta y}{\Delta x}$
0	—	—
2	2	5
4	2	7
6	2	11
8	2	13
10	2	15

(b)

t	Δt	$\frac{\Delta y}{\Delta x}$
0	—	—
3	3	5
6	3	5
9	3	5
12	3	5
15	3	5

(c)

t	Δt	$\frac{\Delta y}{\Delta x}$
0	—	—
3	3	9
6	3	8
9	3	7
12	3	6
15	3	5

(d)

t	Δt	$\frac{\Delta y}{\Delta x}$
0	—	—
1	1	3.2
4	3	3.2
9	5	3.2
16	7	3.2
25	9	3.2

(e)

t	Δt	$\frac{\Delta y}{\Delta x}$
0	—	—
1	1	−1
4	3	−3
9	5	−5
16	7	−7
25	9	−9

22. Consider this set of parametric equations:
$x = 2t^3 - 1$, $y = 2t^3 + 1$.

   (a) Using the interval $-2 \le t \le 2$, draw a graph of the path made by these parametric equations. Label the axes x and y; provide scales.

   (b) If a dot were moving along the graph you drew in part (a), would it move at a constant rate? How did you decide?

   (c) Eliminate t from the equations to produce a single equation for the path. Graph this equation, using the same scale as before, and compare the graph to the one in part (a).

23. Show algebraically that each pair of parametric equations represents part of the parabola $x = 1 - y^2$. Sketch the appropriate segment of the parabola and indicate the direction in which a point $P(x,y)$ travels as t increases.

   (a) $x = \cos^2(t)$, $y = \sin(t)$, $-\pi \le t \le \pi$
   (b) $x = 1 - (\ln(t))^2$, $y = \ln(t)$, $\frac{1}{10} \le t \le 10$
   (c) $x = 1 - e^{-2t}$, $y = e^{-t}$, $-1 \le t \le 1$

24. Show algebraically that each pair of parametric equations represents a portion of the circle $x^2 + y^2 = 9$. Sketch the appropriate arc of the circle and indicate the direction in which a point $P(x,y)$ travels as t increases.

   (a) $x = 3\cos(t)$, $y = 3\sin(t)$, $0 \le t \le \pi$
   (b) $x = 3\sin(\pi t)$, $y = -3\cos(\pi t)$, $0 \le t \le \frac{3}{2}$
   (c) $x = \dfrac{3t}{\sqrt{t^2 + 1}}$, $y = \dfrac{3}{\sqrt{t^2 + 1}}$, $-2 \le t \le 2$

25. Refer to the *Calvin and Hobbes* cartoon on page 292. Poor Calvin! His dad's lesson is correct, however, and you are about to supply some of the mathematical details. Assume that Calvin is playing one of those old 45's and that his 8-inch record is turning at a rate of 45 revolutions per minute. Imagine a point on the record label, 1 inch from the center and another point on the rim of the record, 4 inches from the center. Now start the turntable spinning.

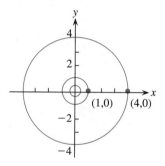

   (a) How fast, in inches per minute, is the point on the record label moving? How fast is the point on the rim moving? Which point is moving faster?

   (b) Write a set of parametric equations for each point, describing its location as the record rotates. Assume that at $t = 0$, the points are in the positions shown and that t is measured in minutes.

26. *Parametric impersonations of a housefly.*   The path of a fly's flight can be filled with loops, turns, and strange twists. The following equations might recreate some of the more memorable flights that you've witnessed. Use a grapher to view the graph produced by each set of equations. Determine whether or not the path eventually folds back upon itself and starts to repeat the pattern. If it does, approximate the smallest t-interval required to complete the pattern. In other words, find the *period* of the pattern.

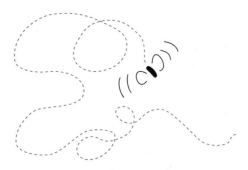

(a) $x = 2\cos(t) + 2\cos(4t)$,
$\quad y = 2\sin(t) + 2\sin(4t)$

(b) $x = t - 2\sin(t)$, $y = 2 - 2\cos(t)$

(c) $x = 2\sin(2t)$, $y = 2\sin(3t)$

(d) $x = 2\cos(t) + 0.75\cos(\frac{4}{3}t)$,
$\quad y = 2\cos(t) - 0.75\cos(\frac{4}{3}t)$

(e) $x = t - \sin(t)$, $y = 1 - \cos(t)$

(f) $x = \cos(t) + 2\cos(0.5t)$,
$\quad y = \sin(t) - 2\sin(0.5t)$

27. In the diagram, point B travels around a circle with center at A, and point C travels around a circle with center at B. In the time it takes point B to make one full revolution about A, point C makes 5.5 complete revolutions about B.

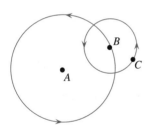

(a) Write parametric equations tracing the path of C relative to an origin located at A if the distance from A to B is two units and the distance from B to C is 1 unit. Use a grapher to help you sketch a graph of the path. (Choose a t-interval wide enough to show one complete pattern.)

(b) If the distance from A to B is increased to four units and the distance from B to C remains one unit, write parametric equations tracing the path of C relative to an origin located at A. Sketch a graph of the path.

(c) How did changing the radius of the inner circle affect the path of C?

28. The graph of any function $f(x)$ can be described parametrically by the equations $x = t$, $y = f(t)$.

(a) If the function $f(x)$ is one-to-one, then it has an inverse. In this case, the graph of $f^{-1}(x)$ can be represented parametrically by the equations $x = f(t)$, $y = t$. Explain why this is true.

(b) Show that the parametric equations $x = t$, $y = t^3$ produce the same graph as the cubing function $f(x) = x^3$. (Use an interval for t that includes both positive and negative values.) Then show that the parametric equations $x = t^3$, $y = t$ produce the same graph as $f^{-1}(x) = \sqrt[3]{x}$.

(c) Find parametric equations to represent the graph of $g(x) = 3^x$. Then find parametric equations to represent the graph of $g^{-1}(x) = \log_3(x)$. Using square scaling, graph the two sets of parametric equations together on the same axes.

8.1 QUILTS

Here you see three traditional American quilt patterns:

Friendship Star 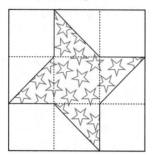   Mohawk Trail 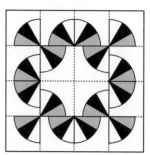   Drunkard's Path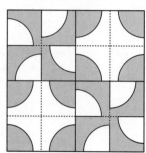

1. Sophie Scott, an avid quiltmaker, decides to incorporate 10-inch squares from all three patterns in a quilt, together with some plain square patches. These last are obviously the easiest to make; in fact she can make eight plain patches in an hour. By contrast, she can make only two Friendship Star squares in an hour. In *hours per square*, what is Sophie's rate for the plain squares? What is her rate for the Friendship Star squares?

2. In the Drunkard's Path pattern, a curved piece is cut out of a small dark $2\frac{1}{2}$-inch square and replaced with an identical light one. This pattern requires much more work because the curves are tricky to sew; it takes Sophie 3 hours to make a 10-inch Drunkard's Path square. The Mohawk Trail pattern is even more difficult, because each curved portion is composed of three separate pieces. She needs 5 hours to make each such 10-inch square. What is her rate, in hours per square, for the Drunkard's Path squares? What is her rate for the Mohawk Trail squares?

The Quilting Function

The length of time that it takes Sophie to complete a quilt depends on the number of squares of each type.

> Let s be the number of Friendship Star squares in the quilt.
>
> Let d be the number of Drunkard's Path squares in the quilt.
>
> Let m be the number of Mohawk Trail squares in the quilt.
>
> Let p be the number of plain squares in the quilt.

3. Using the rates you computed in Questions 1 and 2, write a function $t(s, d, m, p)$ that gives the time required to construct all the squares for the quilt in terms of the number of squares of each type. Is your function linear? Explain, using the algebraic formula for the function to help you decide.

4. You have two sets of values to keep track of: domain variables s, d, m, p, which count the number of squares of each type, and four rates, which tell how long it takes to make a square of a particular type. Let's focus now on the domain variables. Suppose that Sophie plans to make a 60-inch by 80-inch quilt for a queen-size bed. What is the total number of squares she needs?

5. Use that number and all four domain variables to write an equation. (*Hint*: Set your function from Question 3 equal to your answer to Question 4.)

6. The equation you just wrote imposes a restriction on the sum of the domain variables. You could solve that equation for any one of the four variables. For now, solve the equation for d to obtain an expression for d in terms of the other three variables.

7. Go back to the *time* function you wrote and substitute the expression for d. You should now have t as a function of just the three variables p, s, and m.

8. Is the new function linear? Explain your answer in terms of the formula for the function.

9. The Mohawk Trail squares are so difficult that Sophie decides to use only four of them, placed at the very center of the quilt. Use your grapher to see how the dependent variable t varies with p for s-values of 10, 20, and 30.

10. Sketch these three graphs on the same set of axes, labeling the axes, providing scales, and identifying each graph. Use your graphs to complete the assignment.

11. Discuss similarities and differences between the graphs.

12. As p increases, what happens to t? Explain why this makes sense in context.

13. For a given value of p, which of your three graphs has the smallest t-value? (This means that you need to fix the value of p and compare the three graphs with one another at that p-value. You could do this visually by sketching a vertical line that crosses all three of your graphs.) Explain why the answer makes sense in the context of the quilting problem. (This requires a little hard thinking!)

14. Recall that the range of a function is the set of potential output values. Estimate the range of t if we restrict Sophie to 10-inch squares of the four patterns and a 60-inch by 80-inch quilt but impose no other restrictions.

8.2 DIMENSIA

In making containers of various sizes and shapes, manufacturers should understand the relationship between the dimensions they choose for the container and the amount of material needed. Consider, for example, an open cardboard box with a square bottom. In this project, you will analyze how the amount of cardboard used is affected by changing one dimension or the other.

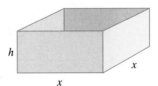

1. Let $S(x,h)$ be the surface area of the box. Write a formula for $S(x, h)$, the relationship between the amount of cardboard material in the box and the dimensions of the box. (Remember that the box has a bottom but no top.)

Holding the Base Constant

2. If the base of the box is a 10-inch square (that is, $x = 10$), rewrite the surface-area function and graph $S(10,h)$, the relationship between the amount of cardboard and the height of the box. What kind of function is $S(10,h)$?

3. How much additional cardboard will be required if the height increases from 2 inches to 3 inches? From 6 inches to 7 inches? From n inches to $n + 1$ inches? Tell how you decided.

4. Now change the size of the base, letting x be 11 inches. Rewrite the formula for S and graph $S(11,h)$.

5. How much additional cardboard will be required with this larger box if the height increases from 2 inches to 3 inches? From 6 inches to 7 inches? From n inches to $n + 1$ inches?

6. Change the size of the base again, letting x be 12 inches. Write the formula for $S(12,h)$, and answer the questions in Question 5 for this box.

7. Sketch the graphs of $S(10,h)$, $S(11,h)$, and $S(12,h)$. Label axes, show scales, and identify each graph.

8. The function $S(x,h)$ has a different rate of change for each fixed value of x.
   (a) For $x = 10$, the rate of increase in surface area is _____ square inches of area per inch of height.
   (b) For $x = 11$, the rate of increase in surface area is _____ square inches of area per inch of height.
   (c) For $x = 12$, the rate of increase in surface area is _____ square inches of area per inch of height.

   As the value of x increases, describe what happens to the *rate of increase* in surface area with respect to height.

Holding the Height Constant

9. Suppose the box has a height of 1 inch. Write the formula for $S(x,1)$, the relationship between the amount of cardboard and the length of a side of the base. View its graph. What type of function is $S(x,1)$?

10. How much additional cardboard would be required to make a box whose base is an 11-inch square instead of a 10-inch square? A 12-inch square instead of an 11-inch square? A 13-inch square instead of an 12-inch square?

11. Suppose you were to change the length of the base from 12.7 inches to 13.7 inches. Without doing any calculations, state whether the additional amount of cardboard would be the same as, more than, or less than the additional amount needed when you changed that dimension from 12 inches to 13 inches. How did you decide?

12. Write the algebraic formulas for $S(x,1)$, $S(x,2)$, and $S(x,3)$. Sketch their graphs, label the axes, showing scales, and identifying each graph.

13. Which of the three graphs appears to be rising most steeply? Calculate the amount of additional cardboard needed when the base dimension changes from 10 inches to 11 inches for boxes of height 1 inch, 2 inches, and 3 inches. In other words, compute three separate *average rates of change*.

14. Compare your two sets of graphs. What can you learn from them about boxmaking?

8.3 WHAT GOES AROUND COMES AROUND

Use square scaling for this project so that circles look like circles and not like ellipses. If you have a choice between *degree mode* and *radian mode*, use radian mode. For all graphing in Questions 1–7, set the time interval as $0 \le t \le 2\pi$ (or approximately $0 \le t \le 6.3$). Assume that time is measured in seconds. If your grapher calls for a time increment, use 0.1 second.

1. Suppose that the motion of an object is governed by the parametric equations

$$x = \cos(t), \qquad y = \sin(t)$$

   Draw a graph of the path of motion. Label the position of the object every 0.5 second. From the spacing of the labeled positions, does it appear that the object is traveling at a constant rate? Explain.

2. The central identity of trigonometry is $\cos^2(t) + \sin^2(t) = 1$. Use this identity to find a single equation, not involving t, to describe the path of the object in (1). What geometric shape does this equation represent?

3. Now alter the basic equations from Question 1 in three ways.
   (a) Add a constant c to the x-equation. Use several values for c, both positive and negative. Experiment until you can describe how the addition of a constant to the x-equation affects the object's path.

(b) Use the original x-equation, but add a constant k to the y-equation. Again, experiment; how does k affect the path?

(c) Add constants to both equations at the same time. How is the path of the object affected?

(d) Does the addition of one or more constants change the rate of travel? Explain.

4. Using this set of parametric equations,

$$x = \cos(t) + 2, \qquad y = \sin(t) + 1$$

write a single equation for the path that does not involve the variable t. What geometric shape does this equation represent?

5. This time, alter the equations in Question 1 by multiplying the x-equation by a positive constant k. How does this modification alter the path of motion? In your investigation, choose at least two values for k that are greater than 1 and at least two that are between 0 and 1. Draw sketches illustrating what you learned.

6. Use the fundamental trigonometric identity in Question 2 to write a single equation, not involving t, for the path of an object moving according to the parametric equations

$$x = 3\cos(t), \qquad y = \sin(t)$$

What geometric shape does this equation represent?

7. What happens to the path of the object if both the x- and y-equations in Question 1 are multiplied by the same positive constant k? Sketch graphs of the paths of motion for several choices of k.

8. Instead of multiplying by a constant, suppose we multiply both equations in Question 1 by the parameter t, as shown:

$$x = t\cos(t), \qquad y = t\sin(t)$$

Using the time interval $0 \le t \le 20$, plot the position of the object at every second. (Remain in square scaling.) Sketch the curve you see.

9. The graph that you have sketched is called the *spiral of Archimedes*. As the value of t increases, does the rate at which the object is moving along the spiral increase, decrease, or stay the same? Explain how you can tell.

8.4 JUST ALGEBRA

1. Solve for the indicated variable.
   (a) $d = \sqrt{x^2 + y^2}$; assume $x \ge 0$ and solve for x.
   (b) $M = Se^{0.05t}$; solve for t.
   (c) $P = \dfrac{kT}{V}$; solve for V.
   (d) $P = \dfrac{kT}{V}$; solve for T.
   (e) $A = 2\pi r^2 + 2\pi rh$; solve for h.
   (f) $A = 2\pi r^2 + 2\pi rh$; assume $r \ge 0$ and solve for r.

2. Each pair of parametric equations describes a function. By algebraically eliminating the parameter, express the function in the form $y = f(x)$. Simplify your answers.
   (a) $x = 6 + 3t,\ y = 8 + 2t$
   (b) $x = 5t,\ y = 3t^2 + 6t - 2$
   (c) $x = t^4,\ y = t^8$
   (d) $x = t^3,\ y = \cos(t)$
   (e) $x = \sqrt[3]{t},\ y = t^2 - t$
   (f) $x = \ln(t),\ y = \ln(t^2)$
   (g) $x = 6 - 3t^2,\ y = 10 + 4t^2$

Bordeaux

Roll out the barrel ...

Preparation

The function $w(x) = 3x - 21$ from Chapter 1 describes how a woman's shoe size depends upon the length of her foot. The function $E(t) = -16t^2 + 20$ from Chapter 5 shows the way in which the height above ground of an egg is determined by the length of time it has been falling.

But many quantities are determined by several inputs rather than a single one. The elevation of a point on the surface of the earth, for example, depends upon two independent variables, longitude and latitude. And the price at the pump of a gallon of gasoline in this country depends upon the wholesale price, the dealer markup, the federal tax rate, and the state tax rate. (Each of these quantities, in turn, depends upon many other factors, but that's another story.)

Write two of your own examples of functions that depend upon two or more independently varying quantities. (You might or might not be able to give formulas for these functions.) We call such relationships **multivariable functions.**

The area of a square depends upon only the length of a side. We write $A(s) = s^2$, the notation $A(s)$ emphasizing the dependence of A upon a single variable s. The area of a rectangle depends upon both the length and the width. We write $A(l, w) = l \cdot w$, emphasizing the dependence of A upon the *two* variables l and w. You are already using such function notation for functions of a single variable; now we are extending it to functions of several variables.

Here are some diagrams that appear inside the front cover of a typical precalculus text:

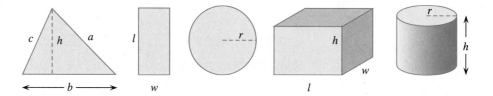

Write the area functions for the triangle, the rectangle, and the circle, expressing them in function notation. How many independent variables appear in each function?

Write the perimeter (circumference) functions for the same shapes. How many variables does each function have?

For the two solids, write the surface area and volume formulas, expressing them in function notation. How many independent variables occur in each?

The Bordeaux Lab

The following excerpt comes from a story in the *New York Times* (March 4, 1990), describing the furor among wine critics over an economist's attempt to quantify the determination of the quality of a particular vintage.

Wine Equation Puts Some Noses out of Joint

Calculate the winter rain and the harvest rain (in millimeters). Add summer heat in the vineyard (in degrees centigrade). Subtract 12.145. And what do you have? A very, very passionate argument over wine.

Professor Orley Ashenfelter, a Princeton economist, has devised a mathematical formula for predicting the quality of red wine vintages in France. And the guardians of tradition are fuming.

... It is widely agreed that weather influences wine quality. What few understand, Ashenfelter argues, is that a mere handful of facts about the local weather tell almost all there is to know about a vintage.*

The Bordeaux Equation, cited in the article, that Professor Ashenfelter uses to predict wine quality is

$$Q = 0.0117\,WR - 0.00386\,HR + 0.6164\,TMP - 12.145$$

where

WR = winter rain (October through March) in millimeters
HR = harvest rain (August through September) in millimeters
TMP = average temperature during growing season (April through September) in degrees Celsius
Q = a number Ashenfelter calls the quality index (the higher the index, the better the wine)

Rewrite the Bordeaux Equation in function notation. How many independent variables are there? What are they?

You have learned to distinguish between linear and nonlinear functions when a single variable is involved. The same concept can be extended to functions of more than one variable. A function having the form

$$f(x_1, x_2, \ldots, x_n) = a_0 + a_1 x_1 + a_2 x_2 + \cdots + a_n x_n,$$

where the a_i are constants and the x_i are the independent variables, is called a linear function of n variables. (Notice that no variable is multiplying or dividing another and that each variable is raised to the first power only.)

Refer to the functions you wrote for the preparation section. Compare with your partners to be sure you are in agreement. Decide which of the functions are linear.

Examine the shoe-size and egg-drop functions: $w(x) = 3x - 21$ and $E(t) = -16t^2 + 20$. Is either one linear? What does its graph look like? How can you tell from the *formula* that one isn't linear? How can you tell from its *graph*?

Now, back to the Bordeaux Equation: To determine Q for a given year, how many inputs are required? Is Q a *linear* function of those variables or not?

Which do you think would contribute more to the quality of a particular vintage: an increase of one degree Celsius in the average temperature during the growing season or an increase of 10 millimeters in the amount of winter rain? Which would you say is more desirable to the wine grower: winter rain or harvest rain? How did you determine your answers to the last two questions?

In 1988, the winter rains were about average, the temperature during the growing season was above average, and August and September were unusually dry. Since the flavor of the wine does not develop fully in the bottle for at least ten years, a definitive testing could not take place before 1998. If Orley Ashenfelter had the opportunity in 1997 to buy the 1988 Bordeaux cheaply, do you think he would do so? Justify your answer.

Suppose that the winter rain and average temperature in another year are about the same as those in 1989 but September of that year is a very rainy month. Does Q increase or decrease with the additional September rainfall? How do you know?

Let's consider the difficulties inherent in drawing graphs for a function of more than one input variable. We are accustomed to plotting the independent variable on the horizontal axis and the dependent variable on the vertical axis. What shall we do if we have more than one independent variable?

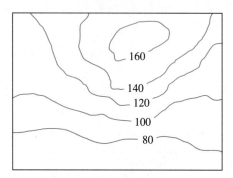

Topographical maps are one solution. They plot altitude as a function of both latitude and longitude by means of **level curves** for different elevations, while the input variables monopolize the horizontal and vertical axes.

The same information could have been presented using different shades of color to indicate different elevations. We see examples of this kind of graphical presentation in an atlas. Even so, these solutions take care of only *two* independent variables. With three or more variables, we need to be either extremely creative (and artistically gifted) or able to take an easier way out.

Perhaps the most straightforward method is to pretend that all of the variables except one are constant and then study what happens to the function when we change just that one variable. In other words, we convert the multivariable problem to a single-variable problem. That's the approach we'll use in this lab.

Suppose that WR is 583 millimeters. Use your grapher to examine how Q varies with HR for three different average temperatures: 15, 18, and 20 degrees Celsius.

Sketch these three graphs on the same set of axes. Label the axes and identify each graph. Be sure that your sketch includes the entire portion of the graph that makes sense in context. (What rainfall amounts might be reasonable for a two-month period? Consider the units in which HR is measured; how large is a millimeter?)

Keep in mind that the purpose of a graph is to convey information. Does your graph give a visual image of how Q depends on HR for various temperatures? Should higher values of TMP be considered as well? Recall what temperatures TMP represents and discuss whether you ought to include any more graphs in your sketch.

Then suppose $TMP = 16$ degrees. Use your grapher to examine how Q varies with WR for three different amounts of harvest rain: 50, 100, and 170 mm.

Sketch these three graphs on another set of axes, again making sure that you show a sensible portion of the graph. Would a person examining your graphs understand what makes the quality of wine go up and what makes it go down?

Recall that the range of a function is the set of output values. Examine your graphs. Do you think the range of this mathematical model could include negative numbers? Explain.

In precalculus, you study various types of scales. The Fahrenheit and Celsius scales give us two different ways to measure temperature. The Richter scale for earthquakes, the decibel scale for sound intensity, and the pH scale for acidity are examples of logarithmic scales. Logarithmic scales are particularly handy in rescaling data whose magnitudes vary widely.

In this lab, we have a scale that presents us with the opposite problem. You might have noticed that the numbers generated by the Bordeaux Equation for realistic temperatures and amounts of rainfall are all in the same ballpark. In other words, most of the Q-values likely to be produced by conditions in a temperate climate lie in a cluster around 3 and 4. That clustering of values makes it difficult to distinguish the quality of one wine from that of another. A Q-value of 4.4, for example, doesn't sound much better than a Q-value of 4.2. To spread out the quality measurements, Ashenfelter treats them as *exponents*—that is, as *logarithms*—whose base is the irrational number e. The scale he uses is actually the *inverse* of a logarithmic scale. Obtain decimal approximations for the numbers $e^{4.4}$ and $e^{4.2}$, and you'll have a sense of how much better a 4.4 wine is than a 4.2 wine.

The constant -12.145 in the Bordeaux Equation was chosen so that when the variables representing winter rain, average temperature, and harvest rain were replaced by their 1961 values, Q turned out to be 4.6052. Why 4.6052? Whip out your calculator again and ask it to approximate $e^{4.6052}$. What's the result? The 1961 vintage was spectacular, and it has become the norm against which all other Bordeaux wines are measured. Do you see why the author of the Bordeaux Equation would want an outcome of 4.6052 for that year?

The ratings in the following table are predictions only, because Bordeaux wine must mature for at least 10 years for its quality to develop fully, and the evaluations were made sooner than that. The table shows a comparison between the results of Orley Ashenfelter's calculations for two types of Bordeaux wines and the ratings of two prominent wine critics.

Year	Ashenfelter (scale of 0 to 100)	Hugh Johnson (scale of 0 to 10)	Robert M. Parker, Jr.
1987	38	3 to 6	Pleasant, soft, clean, fruity
1986	23	6 to 9	Very good, sometimes exceptional
1985	65	6 to 8	Soft, fragrant, very good
1984	33	4 to 7	Austere, mediocre quality
1983	76	6 to 9	Superior to 1981, rarely achieves greatness
1982	56	8 to 10	Most complex and interesting wines since 1961
1981	42	5 to 8	Lacks generosity and richness
1980	28	4 to 7	Light and disappointing

Ashenfelter's calculations, based on his model, suggest that 1989 Bordeaux wines will be the greatest of the century. Would you be willing to invest in wine futures for the 1989 Bordeaux vintages? That is, do you think the wine equation is a good model for determining the quality of a particular vintage? Support your answer with references to the table.

The Lab Report

Functions requiring two or more inputs are very common in fields ranging from weather forecasting to economics. Give three examples of multivariable functions used as models. State what the variables measure. Include at least one linear function.

Explain how you have learned to distinguish linear functions from nonlinear ones when there is more than one input variable. Discuss the difficulties of graphing a multivariable function and describe the method you used to produce graphs for the Bordeaux Equation. Include, as illustrations, the graphs you drew.

Explain how you were able to determine whether each variable had a favorable or unfavorable effect on the quality of the wine, and explain how you could tell whether one variable had a stronger effect than another. Give your group's evaluation of the Bordeaux Equation as a model for predicting wine quality.

Finally, describe the effects you observed when you treated the Q-values as logarithms.

Bézier Curves

Draw your own conclusions

Preparation

A television commercial opens with the image of a car on a computer screen. As you watch, the car's design is modified to give passengers more leg room and then further modified to make the car aerodynamically more efficient. The end result of all this tinkering is, of course, the ideal car, the one that the manufacturer hopes you'll rush out and buy.

With computer-aided design (CAD) software, designers can work interactively with the computer to visualize and then refine their designs for the new models. Because each car design consists of many different curves pieced together, CAD software must enable designers to produce a wide variety of curves on a computer screen.

Imagine that you are part of a team that is in charge of designing next year's model. Sketch a side view of the car you envision.

In another color, shade at least three different curve segments in your design that resemble graphs you have studied this semester. (One of your curves might be a straight line. In mathematics, a line is considered to be one type of curve.) Identify the family of functions or relations to which each of your selected curves belongs.

A computer needs precise mathematical descriptions for all the curves in a design before it can display that design on the screen. In the 1960s, P. Bézier and P. de Casteljau, engineers working in the French automotive industry, independently developed a mathematical method for designing a wide variety of curves. Initially, the results were considered manufacturing secrets; today, **Bézier curves** (the curves produced by this method) are widely used in CAD. (Even though the work of de Casteljau preceded that of Bézier, the latter was published much more widely, and so Bézier gets the credit.) The mathematical descriptions of Bézier curves rely on parametric equations, which allow us to describe curve segments that could not be represented by a function, such as a sideways parabola or a circular loop.

Although you've probably never been part of an automotive design team, you might have used a computer drawing program.* With the curve tool of a drawing program, you can create a variety of curves by clicking on two, three, or four **control points** that you choose. Have you ever wondered how the program uses your control points in determining precisely which curve it will draw? (No, it does not just connect all the points.)

If you specify exactly two control points, the program draws the line segment that connects the two points, a first-degree Bézier curve. For example, suppose that you choose the control points $P_0(3,2)$ and $P_1(5,7)$. Plot these points in a Cartesian (x-y) plane, and then sketch the line segment connecting P_0 to P_1. Any computer drawing program needs a mathematical description in order to draw your first-degree Bézier curve. Several different mathematical formulas can describe this line segment.

- Describe the line segment in function notation. Be sure to restrict the domain of the function so that its graph is precisely the line segment that you've sketched and not the entire line.

---

*Microsoft *Paint* is one example.

- Describe the line segment parametrically: Determine a set of parametric equations that describes uniform (constant rate) motion along the line segment, starting at $P_0(3,2)$ when $t = 0$ and ending at $P_1(5,7)$ when $t = 1$.

You now have two mathematical descriptions for the first-degree Bézier curve determined by control points P_0 and P_1. As you complete the Bézier lab, you will learn the answers to related questions:

- How will the shape of the Bézier curve change if one or two additional control points are chosen?

- Mathematically, how can you describe Bézier curves determined by two, three, or four control points?

The Bézier Lab

In this lab, you'll learn the mathematical rules used by computer drawing software and more sophisticated CAD programs to create Bézier curves of the first, second, and third degree. In addition, you'll use your grapher to investigate the various shapes that these curves can produce.

First-Degree Bézier Curves

A first-degree Bézier curve is completely determined by *two* control points. If the control points are $P_0(x_0, y_0)$ and $P_1(x_1, y_1)$, then the first-degree Bézier curve is simply the line segment joining P_0 and P_1. In the preparation, you wrote two mathematical descriptions for the Bézier curve determined by $P_0(3,2)$ and $P_1(5,7)$. Now we need a mathematical description using parametric equations for the line segment connecting any two arbitrary control points $P_0(x_0, y_0)$ and $P_1(x_1, y_1)$.

As before, we look for a pair of parametric equations to describe motion at a constant rate along the line segment joining $P_0(x_0, y_0)$ and $P_1(x_1, y_1)$, starting at P_0 when $t = 0$ and ending at P_1 when $t = 1$. Recall that parametric equations of the form

$$x(t) = a + bt, \qquad y(t) = c + dt$$

describe motion at a constant rate along a line. Determine the values of the coefficients a and b by substituting $t = 0$ and $t = 1$ into $x(t)$:

$$x(0) = a + b(0) = x_0$$
$$x(1) = a + b(1) = x_1$$

and then solving for a and b in terms of x_0 and x_1. Next, find c and d in terms of y_0 and y_1. Write your mathematical description for the first-degree Bézier curve determined by $P_0(x_0, y_0)$ and $P_1(x_1, y_1)$:

$$x(t) = \underline{\quad} + \underline{\qquad} t \qquad 0 \le t \le 1 \tag{1}$$
$$y(t) = \underline{\quad} + \underline{\qquad} t \qquad 0 \le t \le 1$$

Frequently, the mathematical description for the first-degree Bézier curves determined by the control points $P_0(x_0, y_0)$ and $P_1(x_1, y_1)$ is given in this form:

$$x(t) = (1 - t) x_0 + t x_1 \qquad 0 \le t \le 1 \tag{2}$$
$$y(t) = (1 - t) y_0 + t y_1 \qquad 0 \le t \le 1$$

As you will see later in this lab, this pattern generalizes more easily to higher-degree curves than the pattern in (1). Do the algebra that allows you to express your equations from (1) in the form shown in (2).

Notice that the formulas for the x- and y-coordinates in (2) have the same form: $(1 - t)$ times the coordinate from P_0 plus t times the coordinate from P_1. If we let $B(t)$ represent the point on the line segment at time t, we can summarize the two equations in (2) with a single equation:

$$B(t) = (1 - t)\, P_0 + t\, P_1 \qquad 0 \le t \le 1 \tag{3}$$

This means that $B(t) = (x(t),\, y(t))$, where

$$x(t) = (1 - t)\, x_0 + t x_1 \qquad \text{and} \qquad y(t) = (1 - t)\, y_0 + t y_1$$

Viewed in this form, the first-degree Bézier curve is the result of using the x- and y-coordinates of both control points with two first-degree polynomials in t: $(1 - t)$ and t. Later, we will see how this pattern extends to three or four control points.

Here are two computer-generated first-degree Bézier curves. Coordinate axes have been added so that you can identify the control points.

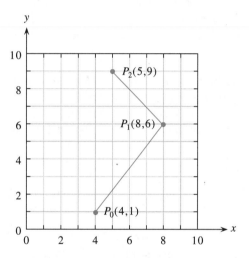

Following the form given in (2), write the mathematical description for $B_0(t)$, the first-degree Bézier curve connecting P_0 to P_1, and $B_1(t)$, the first-degree Bézier curve connecting P_1 to P_2. Then use your grapher to verify that these equations produce the same picture.

Second-Degree Bézier Curves

You just used three control points, but you produced two separate first-degree curves. Now we will generate a single curve controlled jointly by all three points. For comparison, let's use the same points: $P_0(4,1)$, $P_1(8,6)$, and $P_2(5,9)$. We'll let $B(t)$ represent the resulting curve:

$$B(t) = (x(t),\, y(t)) = (1 - t)\, B_0(t) + t\, B_1(t) \tag{4}$$

The difference between form (4) and form (3) is that $B_0(t)$ and $B_1(t)$ are themselves parametric equations rather than simply coordinates of points. In particular,

$$x(t) = (1 - t)(x\text{-coordinate from } B_0) + t\, (x\text{-coordinate from } B_1)$$

Remember that each of those x-coordinates is an entire algebraic expression:

$$= (1 - t)[(1 - t)(4) + t(8)] + t[(1 - t)(8) + t(5)]$$

$$= (1 - t)^2(4) + 2t(1 - t)(8) + t^2(5)$$

$$= (1 - t)^2(x\text{-coordinate of the first control point})$$
$$+ 2t(1 - t)(x\text{-coordinate of the second control point})$$
$$+ t^2(x\text{-coordinate of the third control point})$$

and

$$y(t) =$$

Write out the equation for $y(t)$.

Notice that the formulas for $x(t)$ and $y(t)$ use the x- and y-coordinates of the three control points with three quadratic polynomials in t: $(1 - t)^2$, $2t(1 - t)$, and t^2. The result $B(t)$ is an example of a second-degree Bézier curve.

With the aid of your grapher, view the curve whose mathematical description you completed in (4). The correct curve should begin at $P_0(4,1)$, bend in the direction of $P_1(8,6)$, and end at $P_2(5,9)$.

Every second-degree Bézier curve is determined by three control points: $P_0(x_0, y_0)$, $P_1(x_1, y_1)$, and $P_2(x_2, y_2)$. If, as we did before, we let $B_0(t)$ and $B_1(t)$ be the first-degree Bézier curves connecting P_0 to P_1 and P_1 to P_2, respectively, then we obtain the second-degree Bézier curve by combining $B_0(t)$ and $B_1(t)$ as follows:

$$B(t) = (1 - t)\, B_0(t) + t\, B_1(t) \qquad 0 \le t \le 1 \tag{5}$$

Substitute the correct parametric expressions for $B_0(t)$ and $B_1(t)$, and do the algebra necessary to verify that we can express the combination of parametric equations in (5) as

$$B(t) = (1 - t)^2 P_0 + 2t(1 - t) P_1 + t^2 P_2(t) \tag{6}$$

The second-degree Bézier curve is thus the result of combining the three control points P_0, P_1, and P_2 with the same three quadratic polynomials in t: $(1 - t)^2$, $2t(1 - t)$, and t^2.

Return to the three points you were using earlier: $P_0(4,1)$, $P_1(8,6)$, and $P_2(5,9)$. Let's find out what happens when we change the middle control point. For each change in P_1, write a new mathematical description of the second-degree Bézier curve determined by the original P_0 and P_2 with the new P_1. Then use your grapher to examine the shape of the curve.

- Move P_1 vertically by replacing $(8,6)$ with, say, $(8,11)$ or $(8,0)$.

- Move P_1 to the left or to the right of $(8,6)$.

- Move P_1 diagonally from $(8,6)$.

Describe the shapes you observed. Can a second-degree Bézier curve ever be a line? If so, what must be true of the control points? If not, why not?

Third-Degree Bézier Curves

As you might guess, third-degree Bézier curves are determined by four control points, P_0, P_1, P_2, and P_3. The mathematical description of this curve,

$$B(t) = (1 - t)^3 P_0 + 3t(1 - t)^2 P_1 + 3t^2(1 - t) p_2 + t^3 P_3 \qquad 0 \le t \le 1 \tag{7}$$

is the result of combining four control points and using four cubic polynomials in t. This formula comes from algebraic calculations similar to the ones you used to find $B(t)$ for the second-degree curve, but we won't make you struggle through this one.

Using (7), find a set of parametric equations to describe the Bézier curve determined by the control points $P_0(2,1)$, $P_1(6,5)$, $P_2(4,0)$, and $P_3(10,3)$. With the aid of your grapher, sketch the curve. How many of the control points lie *on* the curve?

Next, investigate the effect of changing the two middle control points, P_1 and P_2. For each change, write the mathematical description of the new third-degree Bézier curve. Then examine the shape of the curve with your grapher. (Notice that for the four given points, the two middle ones lie on opposite sides of the line segment joining the first and last points.)

• Move P_1 higher and P_2 lower than their current positions.

• Move P_2 so that it lies above the direct path from P_0 to P_1.

Continue experimenting until you have a good idea of the variety of shapes possible for a third-degree Bézier curve that starts at $(2,1)$ and ends at $(10,3)$.

Now it's time to test what you have learned. Look at the two sets of control points pictured. Without the aid of a grapher, make a rough sketch of the third-degree Bézier curves determined by these points.

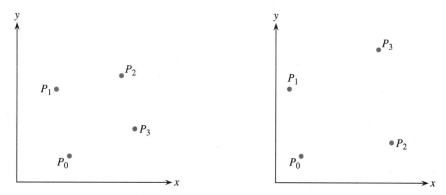

What is the minimum number of control points needed to produce each of the following Bézier curves? Mark the approximate location of each control point.

Let's summarize what you've learned about Bézier curves. First-degree, second-degree, and third-degree Bézier curves are determined by two, three, and four control points, respectively, and are defined by parametric equations that use polynomials. A first-degree Bézier curve consists of the line segment connecting the two points. Each additional control point makes it possible for the curve to bend in another direction. You've been able to create a variety of shapes using only first-, second-, and third-degree Bézier curves. Some sophisticated CAD software allows for Bézier curves of even higher degrees. A click of the mouse adds or moves a control point, allowing designers to experiment with and refine a profile until they are satisfied.

The Lab Report

Discuss what you have learned about Bézier curves.

- Explain how the mathematical descriptions of first- and second-degree Bézier curves arise. Show the missing algebra in the construction of the formulas.

- Provide sketches of several first-, second-, and third-degree Bézier curves. Give mathematical descriptions for the curves you have sketched.

- With a series of sketches, show how a change in one or more of the middle control points of a second- or third-degree Bézier curve alters the shape of the curve.

Conclude your report with a summary of the shapes of curves you could draw using a computer drawing program capable of producing first-, second-, and third-degree Bézier curves. If you have access to such software, feel free to experiment and to include in your report samples of curves you have created.

Supplementary Projects

S.1 SEQUENCES

A sequence, roughly speaking, is a list of numbers written in a specific order. In Chapter 2, you saw many sequences in the form of tables. One of those examples, first encountered in Section 2.1, will serve as an introduction to the topic.

Arithmetic Sequences

Recall the story: The world's population in 1990 was approximately 5.23 billion people and was growing by approximately a quarter of a million people each day. That translated to 91.25 million people per year. The table gives population estimates, in millions of persons, based on these assumptions.

Year	1990	1991	1992	1993	1994	1995
Population	5230	5321.25	5412.50			

1. The population numbers 5230, 5321.25, 5412.50 form the first three terms in a sequence. On the basis of this growth pattern, fill in the next three terms.

   In Chapter 2, you expressed the population P as a function of time t in years since 1990. In ordinary function notation, you would write $P(t)$. Function notation for sequences, however, differs slightly. First, we commonly use one of the letters i, j, k, m, or n for the independent variable. Second, we write the independent variable as a subscript. The independent variable is a nonnegative integer: The domain is, ordinarily, the set of natural numbers, $\{1, 2, 3, \ldots\}$, or the set of whole numbers, $\{0, 1, 2, 3, \ldots\}$.

   In this case, we will use n as the independent variable and will write P_n for the nth term in the sequence. If we let $n = 0$ represent the year 1990, we can use this notation as follows: $P_0 = 5230$, $P_1 = 5321.25$, $P_2 = 5412.50$, and so forth.

2. Write a mathematical expression for the relationship between the term P_{n+1} and the term right before it, P_n.

$$P_{n+1} = \underline{\hspace{4cm}}$$

The formula you just wrote defines a *recursive* relationship, one in which each successive term is expressed as the result of a mathematical operation involving a previous term or terms. In this case, the recursive relationship tells you that you get the next term of the sequence by adding 91.25 to the current term. Whenever a sequence is defined as this one is, by adding a constant quantity to the current term to obtain the next term, it is called an **arithmetic sequence.** The quantity to be added can be positive or negative.

> Subtraction is considered an additive process. Subtracting 5, for instance, is the same as adding −5.

A graphing calculator quickly generates the terms of any arithmetic sequence. Try the following:

- Press 5230 and ENTER.

- Press + 91.25 and ENTER.

- Press ENTER four more times.

449

This sequence of steps generates the sequence in the preceding table.

3. You can also specify the terms of this sequence explicitly. Write an algebraic formula that gives P_n as a function of n.

$$P_n = \text{\underline{\hspace{4cm}}}$$

What type of function is this? (You have written this function before, in Section 2.1. Nothing has changed but the notation.)

Geometric Sequences

The terms in the sequence in Question 1 are based on the assumption that the annual population increase is constant, 91.25 million people per year, regardless of the size of the population. It might be more realistic to assume that the population rises by a constant annual percentage. That way, each annual increase depends on the current population size. Thus, if the population increased by 91.25 million from 1990 to 1991, it grew by $\frac{91.25 \text{ million}}{5230 \text{ million}} = 0.0174$, or 1.74%.

4. The population numbers 5230, 5321.25 form the first two terms in a sequence. Assuming an annual increase of 1.74%, find the next three terms. (Round the results to two decimal places.)

If G_n represents the sequence whose first five terms you have from Question 4, we have $G_0 = 5230$ and $G_1 = 5321.25$.

5. Express the next three terms in sequence notation.

6. Write the relationship between the term G_{n+1} and preceding term, G_n.

$$G_{n+1} = \text{\underline{\hspace{4cm}}}$$

The formula you wrote in Question 6 defines another recursive relationship. In this case, the formula says that you get the next term by multiplying the current term by 1.0174. Whenever a sequence is defined as this one is, by multiplying the current term by a constant to obtain the next term, it is called a **geometric sequence.**

A graphing calculator is handy for generating terms of a geometric sequence, also. Try this:

- Press 5321.25 and ENTER.

- Press * 1.0174 and ENTER.

- Press ENTER three more times.

This sequence of steps generates the sequence for Question 4.

7. You can also specify the terms of this sequence explicitly. Write an algebraic formula that gives G_n as a function of n.

$$G_n = \text{\underline{\hspace{4cm}}}$$

What type of function is this? (You have written this function before, also, in Section 2.2. Again, nothing changes here but the notation.)

Defining a Sequence

There are two ways to define a sequence: recursively or explicitly. In the next two problems, you'll get some practice working with each type of definition.

8. To define a sequence recursively, we specify the recursive relationship and one starting term (or more, if needed, as in part (g)). Write the first six terms of each recursively-defined sequence, and state whether the sequence is arithmetic, geometric, or neither.

(a) $b_{k+1} = 2b_k, b_0 = 1$ (b) $A_n = A_{n-1} + 5, A_1 = -3$

(c) $V_{j+1} = V_j - 300, V_0 = 3000$ (d) $W_{i+1} = 3 + W_i, W_1 = 0$

(e) $V_{i+1} = 3 + 2V_i, V_1 = 0$ (f) $r_m = (-1)r_{m-1}, r_1 = 5$

(g) $F_n = F_{n-1} + F_{n-2}, F_1 = 1, F_2 = 1$ (This is the Fibonacci sequence.)

> The Fibonacci sequence pops up in unexpected places: pine cones, sunflowers, shells, ... You can find a wealth of information on the Internet.

9. To define a sequence explicitly, we write a mathematical expression for the nth term as a function of n and specify the first n-value to use. Write the first six terms of each explicitly defined sequence, and state whether the sequence is arithmetic, geometric, or neither. In each case, use 1 as the initial value for the independent variable.

(a) $a_n = 2 \cdot 3^n$ (b) $W_k = 10 - 2k$

(c) $B_j = \dfrac{1}{2j+1}$ (d) $T_i = \dfrac{3+2i}{4}$

(e) $s_m = m^2$ (f) $R_k = 10^{-k}$

(g) $A_i = \left(-\frac{1}{2}\right)^i$ (h) $h_n = (-1)^n \frac{1}{n}$

10. A certain sequence a_n is defined recursively as follows: $a_{n+1} = a_n + d, a_0 = a$.

  (a) Write the first four terms of this sequence.

  (b) Write the terms of the sequence explicitly. That is, express the nth term, a_n, as a function of n.

  (c) What type of function is this?

11. Another sequence b_n is defined recursively as follows: $b_{n+1} = c \cdot b_n, b_0 = b$.

  (a) Write the first four terms of this sequence.

  (b) Write the terms of the sequence explicitly. That is, express the nth term, b_n, as a function of n.

  (c) What type of function is this?

12. (a) Write a recursive formula that describes the sequence of positive even numbers.

  (b) Write an explicit formula for the positive even numbers; that is, express the nth even number in terms of n.

  (c) Write a recursive formula that describes the sequence of positive odd numbers.

  (d) Write an explicit formula for the positive odd numbers; that is, express the nth odd number in terms of n.

  (e) What type of sequence are these?

Modeling with Sequences

13. Suppose you still have 75 mg of a pain-reducing drug in your system when you take 500 mg more. Your system eliminates 30% of the drug each hour. Let D_n be the amount of the drug in your system n hours after you take the 500-mg dose.

> Reminder: Eliminating 30% means that 70% remains.

  (a) Write the first four terms of the sequence: D_0, D_1, D_2, and D_3.

  (b) What kind of sequence is this?

  (c) Write a recursive definition for D_n.

  (d) Write D_n explicitly as a function of n.

  (e) In how many hours will the drug level in your body drop below 100 mg?

14. You deposit $1000 in a bank account that pays 4% compounded annually. Let A_n be the amount in your account n years after the deposit.

  (a) Write the values of A_0, A_1, A_2, and A_3.

  (b) What kind of sequence is this?

  (c) Write a recursive definition for A_n.

  (d) Write A_n explicitly as a function of n.

  (e) In how many years will your money double?

15. You love music and plan to build up your CD collection. You already have 35 CDs, and you decide to budget for five additional CDs per month. Let N_i be the number of CDs in your collection i months after you make this decision.
   (a) Write the values of N_0, N_1, N_2, and N_3.
   (b) What kind of sequence is this?
   (c) Write a recursive definition for N_i.
   (d) Write N_i explicitly as a function of i.
   (e) After how many months will your collection of CDs surpass 100?

S.2 CREDIT-CARD DEBT AND OTHER LIFE LESSONS

Credit-card companies love to offer credit cards to college students and recent college graduates. In this project, you'll investigate some pitfalls of easily available credit, as well as a way to turn interest rates to your financial advantage. Let's imagine that you have accepted a credit card a few months before graduation. In the time between acquiring the card and getting a job, you amass a debt of $4400.

Parents to the Rescue

In this scenario, your parents agree to lend you the entire $4400, interest-free, provided that you pay them $200 per month until the loan is paid off.

   1. You accept your parents' offer. Let D_n represent the amount of your debt n months after your parents lend you the money.
   (a) Write the first four terms of the sequence and a recursive formula for D_n.
   (b) Write an explicit formula for D_n, that is, D_n as a function of n.
   (c) How long will it take to pay off your loan? Which formula, the recursive formula or the explicit formula, is more helpful in answering this question?

   Credit-card companies bet that a certain percentage of parents will bail out their kids and pay off the debts. Don't get your hopes up, however. Reality is more likely to resemble the next scenario.

Saddled with Debt

The credit-card company charges interest on unpaid balances at an annual rate of 21%. (Think this is high? Check the interest rate on credit cards that you and your friends hold.) You develop a plan to pay off the debt. First, you stop using your card for new purchases. Second, you pay $200 every month toward the balance.

   Let B_k be the balance owed k months after you implement your plan. Clearly, $B_0 = 4400$. But how can we compute B_k for other values of k? Here's a hint:

$$\text{Present Balance} = \text{Previous Balance} + \text{Interest Charge} - \text{Payment}$$

The annual interest rate is 21%. You'll need to convert it to a monthly rate, $\frac{21}{12}\%$, before applying it to the previous balance.

   2. (a) Write a recursive formula for your balance B_k. Simplify the formula.
   (b) Use the formula to calculate the first five terms in the sequence.
   (c) Use a calculator to help you determine how long it will take to pay off the balance. How large is the last payment?
   (d) What is the total amount paid to the credit-card company? How much of this amount represents interest on the debt?

Try the calculator shortcut from Project S.1 for generating the terms.

Getting Your Fiscal House in Order

After paying off the debt, you decide to begin contributing to a retirement fund. You find an annuity with projected annual earnings of 6%, compounded quarterly, or 1.5% every three months. You will make quarterly payments of $600—the exact amount that you had been spending to pay off your credit-card debt.

Let A_m be the balance in the fund after you make the mth payment. Then $A_1 = 600$. Three months later, you make the second deposit. At this time, your balance is

$$A_2 = \text{2nd deposit} + \text{1st deposit with interest} = 600 + 600(1.015) = 1209$$

After six months, you make the third deposit. Now the balance is

$$
\begin{aligned}
A_3 &= \text{3rd deposit} + \text{2nd deposit with 1 quarter interest} \\
&\quad + \text{1st deposit with 2 quarters interest} \\
&= 600 + 600(1.015) + 600(1.015)^2 \\
&= 1818
\end{aligned}
$$

3. **(a)** Compute A_4 and A_5 by continuing the pattern started above. Write out all the terms in each sum; then calculate the sum.
   (b) Generalize the pattern and write a recursive formula for the balance A_m, the balance after m quarterly payments.

Suppose you plan to keep paying into the fund for 30 years, making 121 payments (the initial one at time zero and the final one at the end of 30 years). You'd like to know the projected balance in the annuity, but you certainly wouldn't want to use the formula from Question 3(b) to figure out A_{120}. To do it that way, you would have to compute each of the 119 preceding balances first. There must be an easier way! There is, and it involves a mathematical trick. To see how the trick works, we need to recognize first that the terms you add to obtain A_m are terms of a *geometric* sequence.

Let's consider a generic geometric sequence $a, ar, ar^2, ar^3, ar^4, \ldots, ar^n$. Form a new sequence by summing the terms of the sequence ar^n:

$$
\begin{aligned}
S_0 &= a \\
S_1 &= a + ar \\
S_2 &= a + ar + ar^2 \\
&\;\;\vdots \\
S_n &= a + ar + ar^2 + \cdots + ar^n
\end{aligned}
$$

A sum of terms from a sequence is called a **series.** Because we are summing terms from a geometric sequence, S_n is called a **geometric series.**

We can write a recursive formula showing how the $(k+1)$st term of this new sequence is related to its kth term for any positive integer k:

$$
\begin{aligned}
\text{If } S_k &= a + ar + ar^2 + \cdots + ar^k, \\
\text{then } S_{k+1} &= \underbrace{a + ar + ar^2 + \cdots + ar^k}_{S_k} + ar^{k+1} \\
S_{k+1} &= S_k + ar^{k+1}
\end{aligned}
$$

But we can write S_{k+1} in another way as well, by factoring out an r from all terms except the first:

$$
\begin{aligned}
S_{k+1} &= a + ar + ar^2 + ar^3 + \cdots + ar^k + ar^{k+1} \\
&= a + r[\underbrace{a + ar + ar^2 + ar^3 + \cdots + ar^k}_{S_k}] \\
S_{k+1} &= a + r \cdot S_k
\end{aligned}
$$

This gives us two expressions for S_{k+1}. Equate them and solve for S_k:

$$S_k + ar^{k+1} = a + r \cdot S_k$$
$$S_k - r \cdot S_k = a - ar^{k+1}$$
$$S_k(1 - r) = a[1 - r^{k+1}]$$
$$S_k = \frac{a[1 - r^{k+1}]}{1 - r}$$

This is an important and versatile formula. It allows us to calculate painlessly the sum of a geometric series, as long as we know the initial value, the constant multiplier, and the number of terms in the series. Use this formula in the next two problems.

4. Now back to your annuity. Apply the just-derived formula for the sum of a geometric series to determine A_{120}, the total value of the annuity at the end of 30 years.

5. Suppose you take 500 mg of acetaminophen (the main ingredient in Tylenol) and continue taking 500 mg every 4 hours. About 59% of the drug is eliminated from your body every 4 hours. (That is, 41% remains.) Assume that each dose immediately enters your system. Let a_n be the amount of acetaminophen in your system immediately after you take the nth dose.
  (a) Write a recursive formula for a_n.
  (b) Show that a_n can be viewed as a geometric series.
  (c) Find a_{18}, the amount of acetaminophen in your system after a 3-day period.
  (d) If acetaminophen starts to build up in your body, it could cause liver damage. Calculate a_{19}, a_{20}, and a_{21}. Should you be worried about increasing levels of acetaminophen? Explain.

S.3 JUST ALGEBRA—Sigma Notation

If you add terms in a sequence, you get a series. If you add a lot of terms, the sum could take up more than one line. Mathematicians have devised a shorthand called **sigma notation** that allows them to write sums succinctly. When you see the symbol Σ (the Greek letter *sigma*, or S, for sum), get ready to find a sum.

Here's an example:

$$\sum_{i=1}^{5} 2^i = 2^1 + 2^2 + 2^3 + 2^4 + 2^5$$

The symbol Σ tells you to add some numbers from a sequence. The 2^i says that the terms of the sequence are powers of 2. Which powers of 2? You get the individual terms by replacing the **index** i by, successively, 1, 2, 3, 4, and 5. The statement $i = 1$ at the bottom of the Σ means to start with 1; the 5 at the top says to stop when you reach 5.

1. Express each sum without sigma notation. That is, write out the terms of the sum.

  (a) $\displaystyle\sum_{i=1}^{6} 3i$

  (b) $\displaystyle\sum_{k=2}^{6} \left(\frac{1}{2}\right)^k$

  (c) $\displaystyle\sum_{j=0}^{5} \frac{3^j - 1}{2^j + 1}$

  (d) $\displaystyle\sum_{k=1}^{8} (-2)^k$

  (e) $\displaystyle\sum_{i=1}^{n} (a_{i+1} - a_i)$ Simplify your answer.

2. Rewrite each sum in sigma notation.

(a) $2 + 3 + 4 + \cdots + 100$

(b) $1 + 3 + 3^2 + 3^3 + \cdots + 3^{10}$

(c) $\frac{1}{2} + \frac{1}{3} + \frac{1}{4} + \frac{1}{5} + \frac{1}{6} + \frac{1}{7} + \frac{1}{8} + \frac{1}{9} + \frac{1}{10} + \frac{1}{11} + \frac{1}{12}$

(d) $4 + 6 + 8 + \cdots + 50$

(e) $1 + 3 + 5 + 7 + 9 + 11 + 13 + 15 + 17 + 19 + 21 + 23 + 25$

S.4 ELLIPSES

The **conic sections** are a family of curves so called because they were first studied by the Greeks as the curves that result from cutting or "sectioning" the surface of a cone with a plane. Appolonius of Perga, known as "The Great Geometer," described them in the third century B.C. in his important book, *Conics*. You have studied and graphed the *parabola*, which is the curve that results when the sectioning plane is parallel to an edge of the cone. You are familiar also with the *circle*, which is the curve formed when the cutting plane is parallel to the base of the cone. Figure 1 shows how a parabola and a circle are created with a cone and a plane.

Figure 1 A Parabolic Conic Section and a Circular Conic Section

These two conics are special in the sense that the sectioning (cutting) plane has to have a precise angle relative to the cone. If the sectioning plane intersects the cone in any angle less than the angle formed by the edge of the cone, the resulting curve is an ellipse. See Figure 2.

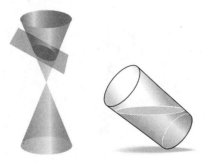

Figure 2 An Elliptical Conic Section and an Elliptical Surface in a Glass of Water

You see an ellipse every time you look at a circle from an angle instead of head-on. For example, the surface of the liquid in a cylindrical glass appears as an ellipse when the glass is tilted. Every circle has exactly the same shape as every other circle, but ellipses vary in shape. Here's why. There is only one angle at which we can slice a cone to create a circle. Circles can be bigger or smaller, but they are all equally round. In contrast, there are infinitely many angles at which we can slice the cone to create an elliptical cross section. The **eccentricity** of an ellipse is a measure of how "out of round" it is.

In this project, you will determine a way to measure the eccentricity of an ellipse, learn how to write and recognize the equation of an ellipse, and draw your own ellipse using string, two tacks, and a pencil.

The Eccentricity of an Ellipse

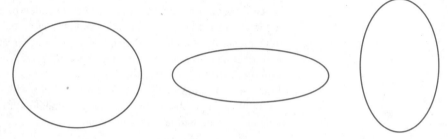

Figure 3   Three Ellipses of Differing Eccentricities

1. Which ellipse in Figure 3 is the most eccentric? How do you know?

You probably remember that a circle is defined as the set of all points in the plane whose distance from a fixed point, called the center of the circle, is a constant, which we call the radius, r. (See Figure 4.) In contrast, an ellipse is defined as the set of all points in the plane for which the *sum of the distances from two fixed points*, called the **foci,** is a constant. (The word *foci* is the plural of the word *focus*.)

> The sum of the two distances, $QC_1 + QC_2$, is constant.

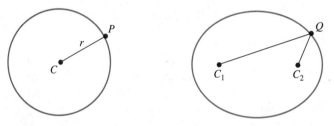

Figure 4   A Circle of Radius r and an Ellipse with Foci C_1 and C_2

In Figure 5, we've placed an ellipse in the Cartesian plane with its center at the origin and its foci at the points $(c,0)$ and $(-c,0)$. We'll designate the vertices of the ellipse—the points at which the curve changes direction—by $(a,0)$ and $(-a,0)$. These are the ends of the **major axis** of the ellipse. The points $(0,b)$ and $(0,-b)$ are the ends of its **minor axis.**[1] Every ellipse is symmetric about both its major and its minor axes.

By definition, the distance from any point (x,y) on the ellipse to the point $(c,0)$, plus the distance from (x,y) to the point $(-c,0)$, is constant. Let's determine that constant.

2. The point $(a,0)$ lies on the ellipse.
  (a) What is its distance from $(c,0)$?
  (b) What is its distance from $(-c,0)$?
  (c) Write the sum of those distances, simplifying the result.

---

[1]In this example, the major, or longer axis is parallel to the x-axis. An ellipse may also be placed in the Cartesian plane with its major axis parallel to the y-axis or on a slant.

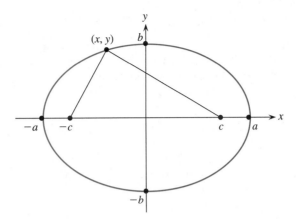

Figure 5 An Ellipse with Foci at $(c,0)$ and $(-c,0)$

Your answer to Question 2(c) provides the key to writing an equation for this ellipse, because that sum is constant for *any* point on the ellipse.

3. In particular, the point $(b,0)$ also lies on the ellipse. Use the Pythagorean theorem to find the distance from $(b,0)$ to $(c,0)$ and from $(b,0)$ to $(-c,0)$. Add them together to get the sum of those distances.

4. Your sum from Question 3 is equivalent to your sum from Question 2(c). Write a mathematical statement expressing that equivalence.

5. Show that your equation in Question 4 is algebraically equivalent to the equation $c^2 = a^2 - b^2$.

The equation $c^2 = a^2 - b^2$ shows how the positions of the two foci are related to the lengths of the axes. Remember it; you'll use it later in this project.

The **eccentricity** e of an elllipse is a measure of how "unround" it is. Eccentricity is defined as a ratio:

$$e = \frac{c}{a}$$

where a is half the length of the major axis and c is the distance from the center of the ellipse to either focus.

6. (a) What happens to the ellipse when $e = 0$?
   (b) What happens when e is close to 1?

In Chapter 8, we mentioned one of Kepler's laws of planetary motion. Johannes Kepler (1571–1630) was a mathematician and astronomer who tried to determine the orbits of four of the inner planets (Mars, Venus, Earth, and Jupiter) using observational data. Like almost all scientists of his time, he believed that the orbits of the planets were circular, because circular motion was held by the philosophers to be perfect. Luckily for Kepler, he chose to work on the planet whose orbit was most "out of round." He realized eventually that the orbit was not a circle with the sun at the center, but an ellipse with the sun at one focus. Here are the eccentricities of the planets' elliptical orbits, computed using modern observation.

Planet	Eccentricity, e
Venus	0.0068
Earth	0.0167
Mars	0.0934
Jupiter	0.0484

7. Which planet did Kepler study? How did you decide?

The Standard Equation of an Ellipse

We can use the distance formula to derive the equation of the ellipse with foci at $(c,0)$ and $(-c,0)$, major axis of length $2a$, and minor axis of length $2b$. (Refer to Figure 5.) By the Pythagorean theorem, the sum of the distance from an arbitrary point (x,y) to the point $(c,0)$ and from (x, y) to $(-c,0)$ is

$$\sqrt{(x-c)^2 + y^2} + \sqrt{(x+c)^2 + y^2}.$$

But we already know from Question 2(c) that the sum of those distances is $2a$, because that is the constant sum that makes this figure an ellipse. This gives us two expressions for the same quantity. Therefore,

$$\sqrt{(x-c)^2 + y^2} + \sqrt{(x+c)^2 + y^2} = 2a.$$

A fair amount of algebra, along with the relationship $c^2 = a^2 - b^2$ that you found in Question 3(c), lets us rewrite the equation in the following, much simpler, form:

Equation of ellipse with center at the origin, major axis on the x-axis

$$\frac{x^2}{a^2} + \frac{y^2}{b^2} = 1$$

Kepler determined that the earth's orbit around the sun is an ellipse with the sun at one focus. The length of half the major axis is $92.957 \cdot 10^6$ miles, and the eccentricity is 0.0167. The **apogee** is the farthest distance from the center of the sun to the center of the earth. The **perigee** is the closest distance.

8. What is the apogee of the earth? What is its perigee? (*Hint*: Draw a sketch.)
9. Imagine the earth's elliptical orbit to be centered at the origin, with its major axis along the x-axis, and write an equation for the orbit.
10. We said before that ellipses come in all shapes and sizes. However, every ellipse with eccentricity $e = 0.0167$ is similar to the earth's orbit.
    (a) Write the equation of an ellipse centered at the origin, with major axis of length 10 (along the x-axis) and eccentricity 0.0167.

Important: Use square scaling.

    (b) Solve this equation for y. Then, using your grapher with square scaling, draw the top half of the ellipse. Why (aside from their philosophical bias in favor of circular motion) do you think that the early astronomers thought that the planets traveled in circular orbits?

Properties of Ellipses—What Is the Exact Shape of an Ellipse?

In Chapter 7, Exercise 7, we talked of the problem one of the authors encountered in making a hole in a metal ceiling plate to fit around a stovepipe, when the ceiling of the room was at a pitch (that is, not parallel to the floor). Exercise 7 determined that 8.94 inches was the length of the major axis of the elliptical hole with minor axis of length 8 inches. Alas, when the carpenter cut a plate with an oval hole of the corresponding dimensions, he found that the hole still did not fit around the stovepipe. Let's see why it didn't fit.

Not every oval is an ellipse. An ellipse is a very particular curve with remarkable properties. A typical oval (a race track, for example) does not satisfy the definition of an ellipse that we gave at the beginning of this project. When the carpenter needed to draw a genuine ellipse on the metal plate, he used a technique that comes right out of the definition of the ellipse. You're about to reproduce what the carpenter did.

Using the desired lengths of the major and minor axes, 8.94 and 8 inches, respectively, the carpenter computed the distance between the two foci. He then placed two thumbtacks that distance apart on either side of the center of what would become the elliptical hole. The thumbtacks became his foci. He connected the tacks with a string 8.94 inches long. Using a pencil and keeping the string taut, he drew an ellipse as in Figure 6. He then cut out the hole with the perfect elliptical shape.

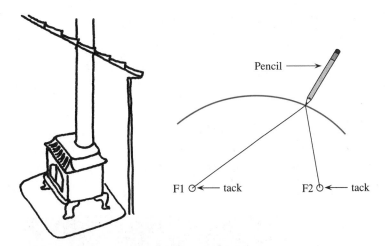

Figure 6 Solution to the Stovepipe Problem

11. **(a)** Using the same ellipse dimensions, minor axis 8 inches, major axis 8.94 inches, determine the distance between the two foci.
    (b) Explain how this technique for drawing an ellipse satisfies the definition that the sum of the distances from any point on the ellipse to the foci is a constant.

12. Now create three of your own ellipses.

    (a) Tack a piece of paper to a bulletin board with two thumbtacks or pushpins. Place the tacks several inches apart. Select a convenient length of string, and tie each end to a tack. Following the directions above, draw an ellipse.
    (b) Move the tacks farther apart and draw another ellipse.
    (c) Move the tacks close together and draw a third ellipse.

13. Describe the relationship between the distance between the foci and the eccentricity of an ellipse.

S.5 WHEN WORLDS COLLIDE—Hyperbolas

On June 14, 2002, an asteroid or irregular "space rock" with diameter varying from 160 to 320 feet came uncomfortably close to the earth. In fact, the asteroid passed inside the moon's orbit, whipping past our planet at 23,000 miles per hour at a distance of 75,000 miles.[2] If this asteroid had struck the earth (see Figure 1), the results would have

Courtesy: NASA

Figure 1 Artist's Depiction of a Major Asteroid Striking the Earth

[2]BBC News Thursday, June 20, 2002, Science/Nature.

been similar to the devastation that occurred in Tunguska, Siberia, in 1908, when 2000 square kilometers of forest were flattened. Just a few months earlier, in March 2002, another asteroid passed the earth, this time at a distance of 288,000 miles, 1.2 times the earth-moon distance. Neither asteroid was observed until after it had passed us by, because each one came from the direction of the sun and thus could not be seen as it approached.

Asteroids (see a sample in Figure 2) travel as do the planets in our solar system: in elliptical orbits with the sun at one focus. However, if an asteroid passes sufficiently close to the earth (and manages to miss it!), it may be deflected by the earth's gravity into a hyperbolic orbit and so escape from our solar system, thus posing no further threat to us here.

Courtesy: NASA

Figure 2   The Asteroid Gaspara

A hyperbola, another example of a conic section, is a curve that is obtained when the sectioning plane intersects both parts, or **nappes,** of the cone. Thus, a hyperbola has two **branches,** one from each nappe. You see a hyperbola when you look at the arcs of light on a wall next to a shaded lamp, as shown in Figure 3. (The wall acts as the sectioning plane.) Just as ellipses come in different shapes, so do hyperbolas, because any one of the infinitely many sectioning planes that intersect both nappes of the cone slices the cone in a hyperbola. As with ellipses, we use **eccentricity** to describe the shape of a hyperbola. The eccentricity of an ellipse tells how "non-round" it is, while the eccentricity of a hyperbola tells how "non-pointy" it is.

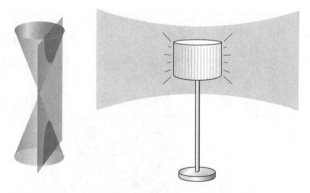

Figure 3   A Hyperbolic Conic Section and a Hyperbolic Shadow on a Wall

In this project, we will study the class of conic sections called hyperbolas: We will determine a way to measure their shape or eccentricity, learn how to write and recognize their equations, and see how to use asymptotes to describe their graphs.

The Eccentricity of a Hyperbola

In Project S.4, "Ellipses," we said that an ellipse can be defined as the set of all points in the plane for which the *sum* of the distances from two fixed points, the foci, is constant. In contrast, a hyperbola is defined as the set of all points in the plane for which the *difference* of the distances from two fixed points, called the **foci,** is constant.

The absolute value of the difference of the two distances, $|PC_1 - PC_2|$, is constant.

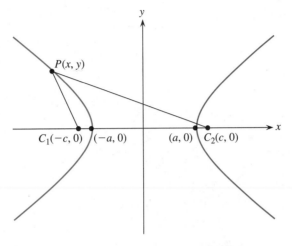

Figure 4 A Hyperbola with Foci C_1 and C_2

In Figure 4, we've placed a hyperbola in the Cartesian plane with its center at the origin and its two foci at the points $(-c,0)$ and $(c,0)$. As with the ellipse, we designate the **vertices** of the hyperbola—the points where the hyperbola changes direction—by $(-a,0)$ and $(a,0)$. The line connecting the vertices is called the **transverse axis** of the hyperbola; it is a line of mirror symmetry for the hyperbola.[3] The point midway between the vertices is called the **center** of the hyperbola. The line perpendicular to the transverse axis and passing through the center of the hyperbola is a second line of symmetry for the hyperbola. By definition, the distance from any point (x,y) on the hyperbola to the point $(-c,0)$ minus the distance from (x,y) to $(c,0)$ is $\pm$ a constant. Let's determine this constant.

1. The point $(a,0)$ lies on the hyperbola.
   (a) What is its distance from $(c,0)$?
   (b) What is its distance from $(-c,0)$?
   (c) Write the positive number that is the absolute value of the difference of these two distances, simplifying the result.

Your answer to Question 1(c) provides the key to writing an equation for this hyperbola, because that difference is constant (ignoring the sign) for any point on the hyperbola. Later in this project, you will write the equation of this hyperbola.

The **eccentricity** e of a hyperbola is a measure of how "non-pointy" the curve is near its vertices. As with the ellipse, the eccentricity of a hyperbola is defined as this ratio:

$$e = \frac{c}{a}$$

where a is the distance from the hyperbola's center to a vertex, and c is the distance from the center to a focus.

---

[3] In this example, the transverse axis is parallel to the x-axis. A hyperbola may also be placed in the Cartesian plane with its transverse axis parallel to the y-axis or on a slant.

2. In any hyperbola, is e less than, greater than, or equal to 1? Justify your answer using the definition of e and the graph in Figure 4.

3. What is the relative position of the foci and vertices of a hyperbola, if e is close to 1?

4. What is the relative position of the foci and vertices of a hyperbola, if e is not close to 1 (say, for example, $e = 5$)?

5. Which of the hyperbolas in Figure 5 is the most eccentric? That is, for which hyperbola is the ratio $\frac{c}{a}$ the greatest?

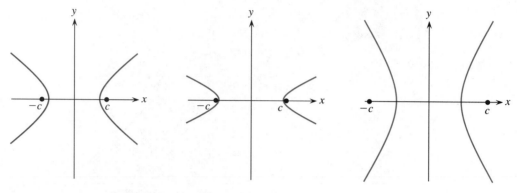

Figure 5 Three Hyperbolas of Differing Eccentricities

Let's go back to our maverick asteroid. Suppose an asteroid is deflected by the sun from its elliptical orbit into a hyperbolic orbit, with the sun at a focus and the earth at the center of the hyperbola. Suppose, also, that observational data allow us to compute that the eccentricity of the orbit is 12.4. (The earth-sun distance is approximately 93 million miles.)

6. Draw a rough sketch of this situation, including the earth, the sun, and one branch of the hyperbola that models the orbit of the asteroid. For simplicity, use a single point to represent the earth and another point for the sun. (In this, we imitate physicists, who imagine the earth and sun to be point masses.)

7. How close is this encounter? That is, how close does the asteroid come to the earth?

The Standard Equation of a Hyperbola

As we did for the ellipse, we can use the distance formula to derive the equation of the hyperbola with foci at $(-c,0)$ and $(c,0)$, center at the origin, and constant difference (and distance between the two vertices) $2a$. (Refer to Figure 4.) As we did with the ellipse, we use the Pythagorean Theorem to write the distance between any point (x,y) on the hyperbola and the point $(c,0)$ and the distance between (x,y) and $(-c,0)$. Then we write the difference of these distances:

$$\sqrt{(x-c)^2 + y^2} - \sqrt{(x+c)^2 + y^2}$$

But we already know that this difference is $2a$, because that is the constant difference that makes this figure a hyperbola. This gives us an equation:

$$\sqrt{(x-c)^2 + y^2} - \sqrt{(x+c)^2 + y^2} = 2a$$

We can use our algebra agility (the algebra gets a bit messy) to rewrite this equation in a form similar to the one we derived for the ellipse:

$$\frac{x^2}{a^2} - \frac{y^2}{c^2 - a^2} = 1$$

Now we introduce the symbol b^2 to represent the quantity $c^2 - a^2$, obtaining the simplified equation

Equation of hyperbola with center at the origin, transverse axis on the x-axis

$$\frac{x^2}{a^2} - \frac{y^2}{b^2} = 1$$

Our equation now looks just like the one for an ellipse, except that the sign between the terms is a minus sign, because it is the *difference* of distances, rather than their sum, that is constant.

8. Write the equation for the hyperbola that models the asteroid's path in Question 4. Let the variables x and y represent distance measured in millions of miles.

9. Solve the equation you wrote in Question 6 for y. Since this involves taking the square root of each side of an equation, you will obtain two functions: One will give you the top halves of the two branches of the hyperbola; the other the bottom halves. Graph both simultaneously using your grapher in the window $-1 \le x \le 15$, $-170 \le y \le 170$, to draw the branch of the hyperbola that models the asteroid's path.

The Asymptotes of a Hyperbola

In deriving the equation of the hyperbola above, we introduced a constant b, saying that $b^2 = c^2 - a^2$. (Careful! This is not the same a^2, b^2, c^2 relationship as the one you used in the equation of the ellipse.) In the equation of the ellipse, the constant b had a geometric meaning; there, b was half the length of the minor axis of the ellipse. (In a way, we can think of a hyperbola as an "inside out" ellipse: Its center is empty, and the hyperbola stretches away forever on either side.) Let's see what information the constant b gives about the graph of a hyperbola.

Rewriting the relationship $b^2 = c^2 - a^2$ as $a^2 + b^2 = c^2$, we see by the Pythagorean theorem that the numbers a, b, and c can be used as the measures of the sides of a right triangle. The right triangle in Figure 6 is drawn so that its base has length a and its hypotenuse has length c, where a is the distance from the center to a vertex, and c is the distance from the center to a focus.

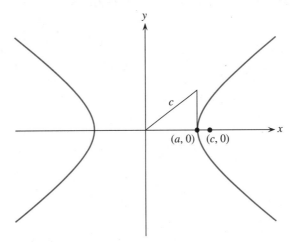

Figure 6   The Hyperbola $\frac{x^2}{a^2} - \frac{y^2}{b^2} = 1$

10. What is the height of the right triangle in Figure 6?

11. Draw a rectangle containing the triangle, with the center of the rectangle at the center of the hyperbola (here, the origin), width $2a$ and height $2b$. Draw the diagonals of the rectangle and continue them to the end of the graph. What appears to be the relationship between the extended diagonals and the graph of the hyperbola?

12. To justify your answer to Question 11, solve the equation of the hyperbola for y. (This will involve taking the square root of both sides of the equation, so make sure you rearrange the equation so that the term containing y^2 is positive.)

13. One form of your answer to Question 12 is $y = \pm\frac{b}{a}x\sqrt{1 - \frac{a^2}{x^2}}$. Remember that for any particular hyperbola, a is constant. What happens to the expression $\frac{a^2}{x^2}$ as $|x|$ becomes large without bound? As $|x|$ approaches infinity, what constant does the expression $\sqrt{1 - \frac{a^2}{x^2}}$ approach?

Look carefully at what remains of your y-equations in Question 13: As $|x|$ approaches infinity, y approaches $\pm\frac{b}{a}x$, a pair of intersecting lines called the **asymptotes** of the hyperbola.

14. Write the equations of the lines through the origin that form the diagonals of the rectangle your drew in Question 11. The equations should agree with the preceding statement and should justify your answer to Question 11, that these lines appear to be asymptotes of the graph of the hyperbola.

How to draw the graph of a hyperbola quickly and painlessly

Your work in Questions 11–14 provides a way to draw a quick sketch of a hyperbola if your grapher's batteries are dead:

- Sketch the rectangle of height $2b$ and width $2a$, centered at the center of the hyperbola.

- Extend the diagonals of the rectangle to draw the asymptotes of the hyperbola.

- Plot the vertices of the hyperbola at $(\pm a, 0)$. (This works only when the center is $(0,0)$.)

- Sketch the graph of the hyperbola, using its symmetry and the asymptotes for guidelines.

15. For the hyperbola $3x^2 - 9y^2 = 27$, determine the foci, vertices, eccentricity, and asymptotes. Draw its graph by hand; then check your work using a grapher. This is one case in which drawing a graph by hand is easier!

S.6 JUST ALGEBRA—Ellipses and Hyperbolas

1. In the "Graph Trek" lab, you learned to shift the graph of a function up or down, left or right. The same principles will allow you to shift the graph of any curve, whether or not it represents a function. In particular, you can shift an ellipse to a new center by subtracting the appropriate constants (positive or negative, as needed) from x and y.
   (a) Write the standard equation of an ellipse whose center is $(0,0)$, with major axis of length 10 and minor axis of length 6.
   (b) Write the equation of the same ellipse, shifting its center to the point $(2,-1)$. Here, you'll need to replace x with $(x - 2)$ and y with $(y + 1)$. Check your equation by solving it for y and graphing the top half with your grapher. Is the center in the right place?
   (c) Write the equation of the same ellipse, shifted now so that its center is the point $(-3,2)$. As before, solve for y, graph the top half, and make sure your ellipse is in the correct position.

2. Now write the general equation for an ellipse with major axis length $2a$ parallel to the x-axis, minor axis length $2b$, and center at (h,k).

3. The graph of $x^2 + y^2 = \frac{1}{4}$ is a circle of radius $\frac{1}{2}$, centered at the origin. Shift that circle down and to the left, so that its center becomes the point $(-\pi, -\sqrt{2})$.

Sometimes the equation of an ellipse is not given in general form (your answer to Question 2) but instead looks like this:

$$Ax^2 + By^2 + Cx + Dy + E = 0$$

where A, B, C, D, and E are constants. (For this equation to represent an ellipse, A and B must have the same sign. If A and B are equal, the graph will be a circle.) To find the coordinates of the center and the lenghts of the major and minor axes, we need to change the form of the equation by completing the square on both the x- and the y-terms.

Completing the Square: See Project 5.3, page 263.

4. Write the equation $2x^2 + y^2 + 4x - 6y - 39 = 0$ in standard form, and tell where the center, foci, and vertices of the associated ellipse are located. We've done the first few steps to get you started:

$2x^2 + 4x + y^2 - 6y = 39$ Arrange x and y terms on one side, constant on the other.

$2(x^2 + 2x \quad) + (y^2 - 6y \quad) = 39$ Factor as needed, making the coefficients of x^2 and y^2 both equal to 1.

$2(x^2 + 2x + 1) + (y^2 - 6y \quad) = 39 + 2$ Add 2 to both sides of the equation to complete the square on x.

Please finish the job.

5. Now you do one from scratch. Write the equation $x^2 + 4y^2 - 3x + 16y + \frac{9}{4} = 0$ in standard form, and tell where the center, foci, and vertices of the associated ellipse are located.

6. Because A and B are equal, this one's a circle: $4x^2 + 4y^2 - 24x - 4y - 63 = 0$. Find its center and its radius.

7. Here's a real challenge. Back in Project S.4, we mentioned that turning the equation

$$\sqrt{(x-c)^2 + y^2} + \sqrt{(x+c)^2 + y^2} = 2a$$

into the equation

$$\frac{x^2}{a^2} + \frac{y^2}{b^2} = 1$$

demands a fair amount of algebra, as well as the identity $c^2 = a^2 - b^2$. Give it a try. Before you finish, you will have squared both sides of the first equation twice, rearranging the terms between squarings. Don't use the identity $c^2 = a^2 - b^2$ until you have succeeded in writing the first equation in a form that is free of square roots. If you solve this problem, your instructor should award you a weightlifting medal for algebra.

8. To see that the graph of the hyperbola $3x^2 - 9y^2 = 27$ (from Question 15 in Project S.5) has no y-intercepts, try setting x equal to 0. What contradiction arises?

9. Complete the square on both x and y to write the equation

$$9x^2 - 72x - 16y^2 - 32y - 16 = 0$$

in the form

$$\frac{(x-h)^2}{a^2} - \frac{(y-k)^2}{b^2} = 1$$

This is the equation of a hyperbola with transverse axis parallel to the x-axis and center shifted to (h,k). Draw a rectangle of width $2a$ and height $2b$ centered at (h,k), and adapt the instructions in Project S.5 to sketch the graph of this shifted hyperbola.

Algebra Appendix

Calvin and Hobbes
by Bill Watterson

In this textbook we assume you have already studied algebra. However, algebra is like any language; it gets rusty unless it's used frequently. The Algebra Appendix is intended not as an algebra text but as a resource, much like a dictionary, to be used to brush up on a technique that you haven't used recently. We don't suggest reading this appendix from start to finish any more than you'd read a dictionary from cover to cover. Use the Appendix Table of Contents on the next two pages, as well as the references in the body of the text, to locate the specific skill you need to review. Each section of the appendix contains practice exercises to make sure you understand that section's techniques. The answers to all the practice exercises are at the end of the appendix.

TABLE OF CONTENTS

A.1 POLYNOMIALS

Polynomials are algebraic expressions that are sums of **terms,** each of which has a real-number **coefficient** and a nonnegative integer power of a variable.

(i) Adding and Subtracting Polynomials

> *Rule*: Add (or subtract) the coefficients of **like terms** (those for which the variables have the same exponent).

Example 1

$$(3x^3 + 2x - 5) + (2x^3 - 4x^2 + 3) = 5x^3 - 4x^2 + 2x - 2$$

Example 2

$$(2x^4 - 3x^3 + x) - (x^4 + 2x^3 - 2x + 5) = x^4 - 5x^3 + 3x - 5$$

(Don't forget to distribute the "−" in front of the second set of parentheses.)

(ii) Multiplying Polynomials

> *Rule*: Multiply each term in one polynomial by each term in the other; then combine like terms.

Example 3

$$(2x + 3)(3x^2 + 2x + 1)$$

$$(2x + 3)(3x^2 + 2x + 1) = 6x^3 + 4x^2 + 2x + 9x^2 + 6x + 3$$

$$= 6x^3 + 13x^2 + 8x + 3$$

Sometimes it is helpful to arrange the multiplication vertically:

$$3x^2 + 2x + 1$$
$$\underline{\quad\quad 2x + 3}$$
$$9x^2 + 6x + 3 \quad\quad \text{First, multiply } 3x^2 + 2x + 1 \text{ by 3, and then by } 2x.$$
$$\underline{6x^3 + \;\; 4x^2 + 2x}$$
$$6x^3 + 13x^2 + 8x + 3 \quad\quad \text{Add terms that have the same degree.}$$

Special Cases

> *Rule*: Multiplying two binomials gives four products. You can remember these as FOIL—first, outer, inner, last.

Example 4

$$(3x + 2)(4x - 6)$$

$$(3x + 2)(4x - 6) = 12x^2 - 18x + 8x - 12 = 12x^2 - 10x - 12$$
$$\text{F} \quad\quad \text{O} \quad\quad \text{I} \quad\quad \text{L}$$

> *Rule*: When the two binomials have the same terms but one difference in sign, the product of the outer terms and the product of the inner terms add to zero: $(a + b)(a - b) = a^2 - b^2$.

Example 5

$$(3x + 2y)(3x - 2y) = (3x)^2 - (2y)^2 = 9x^2 - 4y^2$$

> *Rule*: Squaring a binomial always produces three terms, according to the following pattern:
>
> $$(a + b)^2 = a^2 + 2ab + b^2,$$
>
> the result of applying the FOIL method to the expression $(a + b)(a + b)$.

Example 6

$$(3x - 5y)^2 = (3x)^2 + 2(3x)(-5y) + (-5y)^2 = 9x^2 - 30xy + 25y^2$$

(iii) Factoring Polynomials

Factoring a polynomial means writing it as a product of **factors,** or "unmultiplying" it. Frequently, factoring a polynomial involves a sequence of steps.

Step 1

First, see whether the terms have a factor in common. If so, use the distributive law: $ab + ac = a(b + c)$ to pull out the common factor.

Example 7

Factor $6x^3 - 2x^2$.

$$6x^3 - 2x^2 = 2x^2(3x - 1)$$

Example 8

Factor $16x^2 - 32x - 48$.

$$16x^2 - 32x - 48 = 16(x^2 - 2x - 3)$$ The trinomial can be factored further using the techniques of Example 11.

Step 2

This step is usually determined by the number of terms in the expression to be factored.

Four terms: Try **factoring by grouping,** pulling a common factor out of the first two terms and a different one out of the second two terms.

Example 9

Factor $x^4 + 2x^3 + 6x + 12$.

$$x^4 + 2x^3 + 6x + 12 = x^3(x + 2) + 6(x + 2)$$ The original sum of four terms is now a sum of two expressions. If factoring is going to work, each of these two expressions should have a common factor. Here, it's $(x + 2)$.

$$= (x^3 + 6)(x + 2)$$ Pull out the common factor. You may pull the common factor out on either side.

or

$$= (x + 2)(x^3 + 6)$$

NOTE: You can always check a factoring problem by multiplying the factors and obtaining the original expression.

Example 10

Factor $3x^5 - x^3 - 3x^2 + 1$.

$$3x^5 - x^3 - 3x^2 + 1 = x^3(\underline{3x^2 - 1}) - 1(\underline{3x^2 - 1})$$

Remember, the underlined factors have to match, if factoring by grouping is going to work.

$$= (x^3 - 1)(3x^2 - 1)$$

The expression $x^3 - 1$ can be factored further using the techniques of Example 13.

or

$$(3x^2 - 1)(x^3 - 1)$$

Three terms: See whether the polynomial is the product of two binomials.

Example 11

Factor $6x^2 + 7x - 20$.

The factors, *if we can find them*, will have the form $(ax + b)$ and $(cx + d)$, where b and d will have different signs since their product, -20, is negative. The product of a and c will be 6, and the sum of the inner and outer products must be $7x$. After a lot of trial and error, we obtain

$$6x^2 + 7x - 20 = (2x + 5)(3x - 4)$$

Special Cases

A trinomial of the form $a^2 + 2ab + b^2$ is called a **perfect square** since $(a + b)^2 = a^2 + 2ab + b^2$. These are fairly easy to recognize because the first and third terms of a perfect square trinomial are themselves perfect squares.

Example 12

Factor $4x^2 - 12x + 9$.

$$4x^2 - 12x + 9 = (2x - 3)(2x - 3)$$

The constants 4 and 9 are perfect squares.

$$= (2x - 3)^2$$

Two Terms: If you're sure you have removed all common factors, check to see if you have the difference of two perfect squares, or the sum or difference of two perfect cubes. The patterns to look for are the ones on the left in the following identities:

$$a^2 - b^2 = (a + b)(a - b)$$
$$a^3 - b^3 = (a - b)(a^2 + ab + b^2)$$
$$a^3 + b^3 = (a + b)(a^2 - ab + b^2)$$

NOTE: The sum of two perfect squares does not factor!

Sometimes several factoring techniques can be applied to one polynomial:

Example 13

Factor $3x^4 - 2x^3 - 3x + 2$.

$$3x^4 - 2x^3 - 3x + 2 = x^3(3x - 2) - (3x - 2)$$

First, factor by grouping.

$$= (x^3 - 1)(3x - 2)$$

Now, factor the difference of cubes.

$$= (x - 1)(x^2 + x + 1)(3x - 2)$$

The quadratic $x^2 + x + 1$ does not factor using real numbers.

Example 14

Factor $32x^4 - 2y^4$.

$$32x^4 - 2y^4 = 2(16x^4 - y^4)$$ Always check for a common factor first.

$$= 2(4x^2 + y^2)(4x^2 - y^2)$$ Factor the difference of perfect squares.

$$= 2(4x^2 + y^2)(2x + y)(2x - y)$$ Factor the difference of perfect squares again. (Remember the sum of perfect squares doesn't factor.)

NOTE: If you know one factor of a polynomial, you can use division of polynomials to find another.

(iv) Dividing Polynomials

A polynomial $p(x)$ can be divided by another polynomial $d(x)$, as long as the degree of d is *less than or equal to* the degree of p. In general, the result will not be another polynomial. The process is much like long division of integers.

Example 15

Divide $(10x^3 - 9x^2 - 7x - 3)$ by $(2x - 3)$.

Step 1

$$
\begin{array}{r}
5x^2 \\
2x-3\overline{)10x^3 - 9x^2 - 7x + 3} \\
\underline{10x^3 - 15x^2 } \\
0 + 6x^2
\end{array}
$$

Divide the first term of the divisor into the first term of $10x^3 - 9x^2 - 7x + 3$

Multiply your answer, $5x^2$, by the divisor, $2x - 3$.

Subtract the result from the previous line.

NOTE: The first terms will always have a difference of 0.

Step 2

$$
\begin{array}{r}
5x^2 + 3x \\
2x-3\overline{)10x^3 - 9x^2 - 7x + 3} \\
\underline{10x^3 - 15x^2 } \\
6x^2 - 7x + 3 \\
\underline{6x^2 - 9x } \\
2x + 3
\end{array}
$$

Bring down the remaining terms and repeat the operations in step 1, this time dividing $2x$ into $6x^2$.

Step 3

$$
\begin{array}{r}
5x^2 + 3x + 1 \\
2x-3\overline{)10x^3 - 9x^2 - 7x + 3} \\
\underline{10x^3 - 15x^2 } \\
6x^2 - 7x + 3 \\
\underline{6x^2 - 9x } \\
2x + 3 \\
\underline{2x - 3} \\
6
\end{array}
$$

Repeat the process until the remainder is 0 or has degree *less than* the degree of the divisor.

In this case, we have a remainder of 6. This remainder has *degree* 0, which is less than 1, the degree of the divisor.

We can use this result to write $\frac{10x^3 - 9x^2 - 7x + 3}{2x - 3}$ as $5x^2 + 3x + 1 + \frac{6}{2x-3}$.

Example 16

Divide: $\frac{8x^3-27}{4x^2+6x+9}$.

$$4x^2 + 6x + 9 \overline{)\begin{array}{r} 2x- 3 \\ 8x^3+ 0x^2+ 0x-27 \end{array}}$$

It might be helpful to write in 0 as the coefficient for each missing term.

$$\underline{8x^3+12x^2+18x}$$
$$-12x^2-18x-27$$
$$\underline{-12x^2-18x-27}$$
$$0$$

In this case, the remainder is 0.

Since the division was exact, we can use it to factor $8x^3 - 27$:

$$8x^3 - 27 = (2x - 3)(4x^2 + 6x + 9)$$

Notice that this result fits the pattern $a^3 - b^3 = (a - b)(a^2 + ab + b^2)$.

Practice Problems

In Problems 1 through 3, perform the indicated operations to remove the parentheses, then combine like terms.

1. $(2x^3 - 2x^2 + x) - (3x^2 - 5x^3 + 7)$

2. $(5x^2 + 2x - 3)(2x^2 - 1)$

3. $(4x + 3)(4x - 3)$

4. Use long division to write the rational expression $\frac{2x^4+3x^3+4x}{x^2-1}$ in the form

$$q(x) + \frac{r(x)}{x^2 - 1},$$

where $q(x)$ and $r(x)$ are polynomials and the degree of $r(x)$ is less than 2.

Factor the following polynomials:

5. $x^3 + 4x^2 - 16x - 64$

6. $16x^2 - 24x + 8$

7. $25x^2 - 16$

8. $x^2 + 6x - 16$

A.2 RATIONAL EXPRESSIONS

A rational expression is any quotient (ratio) of two polynomials. The rules for working with rational expressions containing variables are the same as the arithmetic rules for fractions you learned in elementary school. Most of the algebra examples in this section are preceded by an arithmetic example that uses the same technique.

(i) Adding and Subtracting Rational Expressions

Rule: Convert all the individual fractions to equivalent fractions that share the same denominator, then add or subtract numerators. The **lowest common denominator** or **LCD** is the simplest denominator the fractions have in common. Using the LCD usually makes the work easier.

Example 1

Subtract $\frac{5}{6}$ from $\frac{3}{4}$.

$$\frac{3}{4} - \frac{5}{6} = \frac{3 \cdot 3}{4 \cdot 3} - \frac{5 \cdot 2}{6 \cdot 2} = \frac{9}{12} - \frac{10}{12} = \frac{9 - 10}{12} = -\frac{1}{12}$$

Example 2

$$\frac{x+2}{x^2-x} - \frac{1}{x^2+2x}$$

$$\frac{x+2}{x^2-x} - \frac{1}{x^2+2x} = \frac{x+2}{x(x-1)} - \frac{1}{x(x+2)}$$

Factoring the denominators makes finding the lowest common denominator (LCD) easier.

$$= \frac{(x+2)(x+2)}{x(x-1)(x+2)} - \frac{(x-1)}{x(x-1)(x+2)}$$

Convert to equivalent fractions with common denominators.

$$= \frac{x^2+4x+4-(x-1)}{x(x-1)(x+2)}$$

Combine the numerators.

$$= \frac{x^2+3x+5}{1(x-1)(x+2)}$$

Usually, it is more useful to leave the denominator factored.

(ii) Multiplying and Simplifying Rational Expressions

Multiplying

> *Rule*: To multiply two fractions, multiply the numerators and multiply the denominators. You want to cancel any common factors before carrying out the multiplication, so write the numerators and denominators in factored form.

Example 3

Multiply $\frac{10}{49}$ by $\frac{14}{25}$.

$$\frac{10}{49} \cdot \frac{14}{25} = \frac{10 \cdot 14}{49 \cdot 25} = \frac{2 \cdot 5 \cdot 2 \cdot 7}{7 \cdot 7 \cdot 5 \cdot 5} = \frac{2 \cdot 2}{7 \cdot 5} = \frac{4}{35}$$

Example 4

$$\frac{x^2-4}{x^2-3x} \cdot \frac{x^2-5x+6}{x-2}$$

$$\frac{x^2-4}{x^2-3x} \cdot \frac{x^2-5x+6}{x-2} = \frac{(x^2-4)(x^2-5x+6)}{(x^2-3x)(x-2)}$$

Multiply numerators and multiply denominators.

$$= \frac{(x-2)(x+2)(x-2)(x-3)}{x(x-3)(x-2)}$$

Write numerator and denominator in factored form. Cancel the common factors.

$$= \frac{(x+2)(x-2)}{x}$$

It may be more useful to stop here and leave the numerator in factored form.

$$= \frac{x^2-4}{x}$$

Since the original algebraic fraction is not defined at $x=2$ or $x=3$ and the final result is defined at these values, the original algebraic fraction is equivalent to the final result as long as $x \neq 2$ and $x \neq 3$.

Simplifying

Simplifying a rational expression means putting it in **lowest terms**—that is, canceling any common factors that occur in both numerator and denominator.

Example 5

Simplify $\frac{(x^2-4)(x-1)}{(2x^2-x-1)(2-x)}$.

$$\frac{(x^2 - 4)(x - 1)}{(2x^2 - x - 1)(2 - x)} = \frac{(x - 2)(x + 2)(x - 1)}{(2x + 1)(x - 1)(2 - x)}$$

Make sure the numerator and denominator are completely factored so that you can spot any common factors.

$$= \frac{(x - 2)(x + 2)}{(2x + 1)(2 - x)}$$

Wait! What about $(x - 2)$ and $(2 - x)$? Notice that $(2 - x) = -(x - 2)$, so $\frac{x-2}{2-x} = -1$.

$$= \frac{-(x + 2)}{(2x + 1)}$$

These equivalences hold as long as $x \neq 1$ and $x \neq 2$.

(iii) Dividing Rational Expressions

> *Rule*: To divide any expression by a fraction, multiply the number by the reciprocal of the fraction.

Example 6

Divide 4 by $\frac{1}{3}$. Divide $\frac{1}{2}$ by $\frac{3}{4}$.

$$4 \div \frac{1}{3} = 4 \cdot 3 = 12$$

$$\frac{1}{2} \div \frac{3}{4} = \frac{1}{2} \cdot \frac{4}{3} = \frac{1 \cdot 4}{2 \cdot 3} \qquad \text{Cancel first!}$$

$$= \frac{2}{3}$$

Example 7

$$\frac{9x^2 - 1}{x} \div \frac{3x^2 + 2x - 1}{x^2}$$

$$\frac{9x^2 - 1}{x} \div \frac{3x^2 + 2x - 1}{x^2} = \frac{9x^2 - 1}{x} \cdot \frac{x^2}{3x^2 + 2x - 1}$$

Invert the divisor and multiply.

$$= \frac{(3x + 1)(3x - 1)x^2}{x(3x - 1)(x + 1)}$$

Factor the numerator and denominator so that you can cancel.

$$= \frac{x(3x + 1)}{x + 1}$$

The final result holds as long as $x \neq 0$ and $x \neq \frac{1}{3}$.

(iv) Simplifying "Double-Decker" Fractions

Some rational expressions contain fractions in their numerator and/or denominator. We call these monsters "double-decker" fractions.

> *Rule*: Find the lowest common denominator, or LCD, of all the "little fractions" that occur in the double-decker fraction. Then multiply the numerator of the big fraction and the denominator of the big fraction by this LCD. The denominators of all the little fractions will cancel and the result will be a simple fraction.

Example 8

Simplify $\dfrac{-\frac{1}{4}+2}{\frac{1}{3}-\frac{5}{6}}$.

$$\dfrac{-\dfrac{1}{4}+2}{\dfrac{1}{3}-\dfrac{5}{6}} = \dfrac{12\left(-\dfrac{1}{4}+2\right)}{12\left(\dfrac{1}{3}-\dfrac{5}{6}\right)}$$

The LCD of all the little fractions is 12. Multiply the numerator and the denominator of the big fraction by 12.

$$= \dfrac{-3+24}{4-10}$$

Be sure to multiply each term in the numerator and denominator by 12.

$$= \dfrac{21}{-6} = -\dfrac{7}{2}$$

NOTE: The minus sign can be put in the numerator or in the denominator or in front of the fraction.

$$\dfrac{7}{-2} = \dfrac{-7}{2} = -\dfrac{7}{2}$$

Example 9

Simplify $\dfrac{x-\frac{4}{x}}{\frac{2}{x}-1}$.

$$\dfrac{x-\dfrac{4}{x}}{\dfrac{2}{x}-1} = \dfrac{x\left(x-\dfrac{4}{x}\right)}{x\left(\dfrac{2}{x}-1\right)}$$

The LCD of all the little fractions is x. Multiply the numerator and the denominator of the big fraction by the LCD.

$$= \dfrac{x^2-4}{2-x}$$

Multiply each term in the numerator and each term in the denominator by the LCD.

$$= \dfrac{(x-2)(x+2)}{2-x}$$

Factor the numerator and denominator, if possible, to simplify the fraction.

$$= -(x+2)$$

Remember that $\frac{x-2}{2-x} = -1$.

$$= -x-2$$

The final result is equivalent to the original as long as $x \neq 0$ and $x \neq 2$.

Example 10

Simplify $\dfrac{\frac{1}{x-4}-\frac{x}{x^2-16}}{\frac{1}{x+4}-\frac{1}{x}}$.

$$\dfrac{\dfrac{1}{x-4}-\dfrac{x}{x^2-16}}{\dfrac{1}{x+4}-\dfrac{1}{x}}$$

The LCD of all the little fractions is $x(x+4)(x-4)$.

$$= \dfrac{x(x+4)(x-4)\left[\dfrac{1}{x-4}-\dfrac{x}{x^2-16}\right]}{x(x+4)(x-4)\left[\dfrac{1}{x+4}-\dfrac{1}{x}\right]}$$

Multiply the numerator and denominator of the big fraction by the LCD.

$$= \frac{x(x+4) - x^2}{x(x-4) - (x+4)(x-4)}$$

Be sure to multiply *each* term in the numerator and denominator by the LCD. Notice that the two terms in the resulting denominator have a common factor of $x - 4$.

$$= \frac{x^2 + 4x - x^2}{(x-4)(x - (x+4))}$$

Pull out the common factor, $x - 4$.

$$= \frac{4x}{(x-4)(-4)}$$

$$= \frac{x}{-(x-4)}$$

The original complex fraction is equivalent to the final result as long as $x \neq 0$ and $x \neq -4$.

NOTE: This expression can be written $\frac{x}{-(x-4)}$, $\frac{-x}{x-4}$, $-\frac{x}{x-4}$, or $\frac{x}{4-x}$.

Practice Problems

Perform the indicated operations. Be sure the answer is in lowest terms.

1. $\frac{2}{5} \cdot \frac{5}{9}$

2. $\frac{2}{5} - \frac{5}{9}$

3. $\frac{2}{5} + \frac{5}{9}$

4. $\dfrac{\frac{2}{5}}{\frac{5}{9}}$

5. $\frac{x+3}{2x-10} \cdot \frac{x-5}{2x+3}$

6. $\frac{x+3}{2x-10} - \frac{x-5}{2x+3}$

7. $\dfrac{\frac{x-5}{x+5} - \frac{x+5}{x-5}}{\frac{1}{x-5} - \frac{1}{x+5}}$

8. $\dfrac{\frac{1}{(x+h)^2} - \frac{1}{x^2}}{h}$

9. $\frac{x}{x^2-4x+4} - \frac{2}{x^2-4}$

A.3 EXPONENTS AND ROOTS

(i) Definitions and Rules

A positive integer **exponent** or **power** is a convenient shorthand for indicating repeated multiplications of a number b, called the **base.** For example,

$$b^4 = \underbrace{b \cdot b \cdot b \cdot b}_{4 \text{ times}}$$

$$b^n = \underbrace{b \cdot b \cdot b \cdot \cdots \cdot b}_{n \text{ times}}$$

An **nth root** of a number b indicates the operation that is the *inverse* of raising b to the nth power. For example,

$\sqrt[4]{b}$ is the number we raise to the fourth power to get b

$\sqrt[n]{b}$ is the number we raise to the nth power to get b

Here's the relation between roots and powers:

$$\sqrt[n]{b} = a \text{ means } a^n = b$$

Example 1
$$\sqrt[4]{81} = 3 \text{ because } 3^4 = 81$$

$$\sqrt[3]{-27} = -3 \text{ because } (-3)^3 = -27$$

We can extend the definition of *exponent* to any rational number $\frac{m}{n}$ in lowest terms, with m and n integers and $n > 0$:

$$b^0 = 1 \qquad (b \neq 0)$$

$$b^{-n} = \frac{1}{b^n} \qquad (b \neq 0)$$

$$b^{1/n} = \sqrt[n]{b} \qquad (b \geq 0 \text{ if } n \text{ is even})$$

$$b^{m/n} = \sqrt[n]{b^m} = (\sqrt[n]{b})^m$$ These last two expressions indicate that you can raise b to the mth power first or take the nth root of b first, whichever is simpler.

Example 2
$$8^{2/3} = (\sqrt[3]{8})^2 = 2^2 = 4$$ Since 8 is a perfect cube, it might be simpler to take the cube root first.

$$8^{2/3} = \sqrt[3]{8^2} = \sqrt[3]{64} = 4$$ However, this expression can also be simplified by first squaring the 8.

Rules for Exponents

Let a, b, p, and q be real numbers, with $a > 0$ and $b > 0$.

$$b^p \cdot b^q = b^{p+q}$$

$$\frac{b^p}{b^q} = b^{p-q}$$

$$(b^p)^q = b^{pq}$$

$$(a \cdot b)^p = a^p b^p$$

$$\left(\frac{a}{b}\right)^p = \frac{a^p}{b^p}$$

NOTE: There is no "distributive law for exponents." That is, in general,

$$(a + b)^p \neq a^p + b^p.$$

In particular, $(x^{-1} + y^{-1})^{-1} \neq x + y$, and $(x^2 + y^2)^{1/2} \neq x + y$.

Example 3
$$(4x^3 y)^2 (2x^2 y^3) = (16x^6 y^2)(2x^2 y^3) = 32x^8 y^5$$

Example 4
$$\left(\frac{x^3}{x^5}\right)^4 = (x^{-2})^4 = x^{-8} = \frac{1}{x^8}$$

Example 5
$$\frac{(4x^{1/3} y^{1/9})^3}{8y^{4/3}} = \frac{64x^{3/3} y^{3/9}}{8y^{4/3}} = 8xy^{-1} = \frac{8x}{y}$$

Since roots or **radicals** are defined as fractional powers, the rules for exponents apply to radicals as well. Here are the two you'll use most often.

Rules for Roots

$$\sqrt[n]{ab} = \sqrt[n]{a} \cdot \sqrt[n]{b}$$

$$\sqrt[n]{\frac{a}{b}} = \frac{\sqrt[n]{a}}{\sqrt[n]{b}}$$

NOTE: There is no rule for addition of radicals; that is, in general,

$$\sqrt[n]{a+b} \neq \sqrt[n]{a} + \sqrt[n]{b}$$

Example 6

Show that $\sqrt{16+9} \neq \sqrt{16} + \sqrt{9}$.

$$\sqrt{16+9} = \sqrt{25} = 5$$

$$\sqrt{16} + \sqrt{9} = 4 + 3 = 7$$

Since 5 and 7 are not equal, the original radical expressions are not equal.

The rules for roots can sometimes be used to simplify an expression containing a radical.

Example 7

$$\sqrt{4x^3} = \sqrt{4x^2 \cdot x} = \sqrt{4x^2} \cdot \sqrt{x} = 2x\sqrt{x}$$

$$\sqrt[3]{81} = \sqrt[3]{27 \cdot 3} = \sqrt[3]{27} \cdot \sqrt[3]{3} = 3 \cdot \sqrt[3]{3}$$

Example 8

Rationalize the denominator of $\frac{1}{\sqrt{x+h}}$.

NOTE: "Rationalizing the denominator" means rewriting the fraction in order to move the mess (the radical expression) into the numerator.

$$\frac{1}{\sqrt{x+h}} = \frac{1}{\sqrt{x+h}} \cdot \frac{\sqrt{x+h}}{\sqrt{x+h}}$$

We can always multiply the numerator and the denominator of a fraction by the same nonzero quantity.

$$= \frac{\sqrt{x+h}}{x+h}$$

By the definition of square root, $\sqrt{x+h} \cdot \sqrt{x+h} = x+h$.

(ii) Exponent-based Notation

Scientific Notation

We say that a number written in the form $a \cdot 10^n$, where $1 \leq a < 10$ and n is an integer, is written in **scientific notation.**

Example 9

Write 547,000 in scientific notation.

$$547,000 = 5.47 \cdot 10^5$$

Example 10

Write 0.0037 in scientific notation.

$$0.0037 = 3.7 \cdot 10^{-3}$$

Calculator/Computer Shorthand

Calculators and computers use a shorthand version of scientific notation. Instead of indicating a multiplication by 10^n, the calculator will write "E" followed by the power n. *The base is understood to be 10.*

Example 11   Write the numbers in Examples 9 and 10 as they might appear on a calculator.

$$5.47 \text{ E}5$$

$$3.7 \text{ E}^{-3}$$

NOTE: This is entirely different from $(5.47)^5$ and $(3.7)^{-3}$.

(iii) Factoring with Fractional Exponents

Expressions containing variables with fractional exponents are not polynomials, but the techniques from Section A.1(iii) for factoring polynomials can still be used. Being able to factor an expression with fractional exponents is a skill useful for calculus.

Example 12   Factor $5t^{5/2} + 20t^{3/2}$.

$$5t^{5/2} + 20t^{3/2} = 5t^{3/2}(t + 4)$$

Factor out the highest power of t that occurs in each term.

Example 13   Factor $u^{3/2} + 3u^{1/2} + 2u^{-1/2}$.

$$u^{3/2} + 3u^{1/2} + 2u^{-1/2} = u^{-1/2}(u^2 + 3u + 2)$$

The highest power of u that appears in each term is $u^{-1/2}$.

$$= u^{-1/2}(u + 2)(u + 1)$$

Practice Problems

Simplify the expressions in Problems 1 through 3.

1. $(2x)^3 \left(\frac{4x^2}{8x^3}\right)$

2. $(5x^3 y^4) \left(\frac{x}{2y}\right)^4$

3. $\frac{(2x^{1/5} y^{1/3})^3}{6x^{2/5} y^3}$

4. Simplify $(-3y^{-4}x)^2(3^{-2}yx^{-1})$. Express your final answer without negative exponents.

5. Simplify $\frac{4x^3 y^{-2}}{6x^{-1} y^2}$. Express your final answer without negative exponents.

For Problems 6 through 9 write the radical expression using exponents.

6. $(\sqrt[3]{x})^2$

7. $(\sqrt{x})^3$

8. $\frac{1}{\sqrt[4]{x}}$

9. $\frac{1}{\sqrt[4]{x^3}}$

10. Evaluate by hand the following roots.
    (a) $\sqrt{2500}$ **(b)** $\sqrt{0.000016}$ **(c)** $\sqrt[3]{0.008}$

11. Reexpress without a radical $\frac{\sqrt{9x^2}}{\sqrt{16}}$.

12. Use a numerical example to show that, in general,
$(a^2 - b^2)^{1/2} \neq a - b$.

In Problems 13 and 14, write the number using scientific notation and as it might appear on your calculator.

13. 3,965,000,000

14. 0.000000227

In Problems 15 and 16, write the calculator number in ordinary decimal notation.

15. 2.64123 E11

16. 6.34875 E-04

Factor each expression in Problems 17 through 19.

17. $2x^{5/2} - 8x^{1/2}$

18. $y^{7/3} - 25y^{1/3}$

19. $x^{1/2} - 4x^{-1/2} + 3x^{-3/2}$ (*Hint*: The highest power of x appearing in each term is $x^{-3/2}$.)

A.4 LOGARITHMS

(i) Definition and Rules

A **logarithm** is an exponent. In Chapter 5, you learned that a logarithmic function is the inverse of the exponential function with the same base. Usually, logs are defined using this relationship:

$$\log_b(x) = y \text{ means } b^y = x \ (b > 0)$$

NOTE: The logarithm y is an *exponent*.

Example 1

$$\log_2(16) = 4 \qquad \text{because} \qquad 2^4 = 16$$

$$\log_2\left(\frac{1}{4}\right) = -2 \qquad \text{because} \qquad 2^{-2} = \frac{1}{2^2} = \frac{1}{4}$$

The definition and rules for logarithms are discussed in Chapter 5. We repeat them here for convenience.

Rules for Logarithms

Let b, s, and t be positive real numbers with $b \neq 1$:

$$\log_b(s \cdot t) = \log_b(s) + \log_b(t)$$

$$\log_b\left(\frac{s}{t}\right) = \log_b(s) - \log_b(t)$$

$$\log_b(s^x) = x \cdot \log_b(s), \text{ for any real number } x.$$

Example 2

Express $\log_b(\frac{x^2 y^4}{\sqrt{z}})$ in terms of $\log_b(x)$, $\log_b(y)$, and $\log_b(z)$.

$$\log_b\left(\frac{x^2 y^4}{\sqrt{z}}\right) = \log_b(x^2) + \log_b(y^4) - \log_b(\sqrt{z})$$

$$= 2 \cdot \log_b(x) + 4 \cdot \log_b(y) - \frac{1}{2} \cdot \log_b(z)$$

(ii) Change of Base

Calculators and computers generally have built-in logarithm functions for only the bases e and 10. If you want to evaluate a logarithm to another base b, use the conversion formula:

$$\log_b(x) = \frac{\log_a(x)}{\log_a(b)}$$

In particular,

$$\log_b(x) = \frac{\log_{10}(x)}{\log_{10}(b)} \qquad \text{(To convert from base 10 to base } b\text{)}$$

$$\log_b(x) = \frac{\ln(x)}{\ln(b)} \qquad \text{(To convert from base } e \text{ to base } b\text{)}$$

Example 3

Evaluate $\log_2(5)$ in two ways, first using base-10 logs and then using natural logs.

$$\log_2(5) = \frac{\log_{10}(5)}{\log_{10}(2)} \approx \frac{0.69897}{0.30103} \approx 2.32193$$

$$\log_2(5) = \frac{\ln(5)}{\ln(2)} \approx \frac{1.60944}{0.69315} \approx 2.32193$$

Example 4

Let $f(x) = \log_2(x)$. Write f as a function using $\ln(x)$.

Since $\log_2(x) = \frac{\ln(x)}{\ln(2)} = \frac{1}{\ln(2)} \cdot \ln(x) \approx 1.44270 \cdot \ln(x)$, we can write

$$f(x) = \frac{\ln(x)}{\ln(2)} \approx 1.44270 \cdot \ln(x)$$

If the base for a logarithm is not stated explicitly, assume that the base is 10. The expression $\log(N)$ means $\log_{10}(N)$. The expression $\ln(N)$, however, always implies a base of e.

Practice Problems

1. Express $\log_b \left(\frac{\sqrt{xy}}{z^3} \right)$ in terms of $\log_b(x)$, $\log_b(y)$, and $\log_b(z)$.

2. Evaluate $\log_3(8)$ in two ways: using base-10 logs and using natural logs.

3. Let $f(x) = \ln(x)$. Write f as a function using $\log_{10}(x)$. Use your graphing utility to check that both forms for f give the same graph.

© 2003 by Sidney Harris

Papa Newton reading some bedtime equations to little Isaac

A.5 EQUATIONS

In this section, equations are organized by type: linear, quadratic, and higher-degree polynomial equations; those with algebraic fractions; and those containing radical expressions. To solve any equation, you should first decide which type you have because the strategy to use depends upon the type of equation.

If your equation does not fit into any of these categories, you will probably have to use a grapher to obtain a decimal approximation to the solution.

Equations are sometimes used to state a property of real numbers, such as the distributive law, $a(b + c) = ab + ac$, or the fundamental relationship between sine and cosine functions, $\sin^2 x + \cos^2 x = 1$. Such equations are called **identities** because they are true for all values of the variable(s). There's nothing to "solve" in an identity.

The equations in this section are called **conditional equations** because they hold true only under certain conditions—that is, for one or more values of the variable(s). Solving a conditional equation means finding the value(s) that make the equation a true statement. We can always verify a solution to a conditional equation by substituting it into the original equation. A correct solution will produce an identity.

Your first step in solving an equation is to decide which kind of equation you're dealing with because this will determine your method of attack. Once you decide what type equation you have, look in the corresponding section of A.5 for techniques specific to that particular type.

The following general procedures are valid for all equations:

1. Adding or subtracting the same quantity on both sides of an equation does not change the equation's solution.

2. Multiplying or dividing both sides of an equation by the same nonzero quantity *usually* does not change the set of solutions. If that quantity involves a variable, though, extraneous solutions may be introduced or a solution may be lost, so checking all solutions by substitution is very important. This warning applies mainly to equations in Sections (iv) and (v).

(i) Solving Linear Equations

A **linear equation** in one variable is an equation that involves only the first power of the variable and no division by any expression involving the variable.

General Strategy for Solving Linear Equations

Step 1

Get all the terms involving the variable on one side of the equation; get everything else on the other side.

Step 2

Combine like terms, or factor the variable out of all terms.

Step 3

Isolate the variable by dividing both sides of the equation by the coefficient of the variable. (The coefficient may be a constant or an algebraic expression.)

Example 1 Solve $5(3x - 2) + x = 3(2 - x)$ for x.

$$5(3x - 2) + x = 3(2 - x)$$

$15x - 10 + x = 6 - 3x$ To get all the terms involving x on one side of the equation, we need to remove the parentheses by multiplication. (Use the distributive law.)

$15x + x + 3x = 6 + 10$ Step 1: Add $3x + 10$ to both sides of the equation. All terms with x are now on the left-hand side of the equation.

$19x = 16$ Step 2: Combine like terms.

$$x = \frac{16}{19}$$ Step 3: Isolate x by dividing both sides of the equation by 19.

Example 2　　Solve $F = \frac{9}{5}C + 32$ for C.

$$F = \frac{9}{5}C + 32$$

$$F - 32 = \frac{9}{5}C$$

Step 1: Since the term 32 did not involve C, subtract it from both sides of the equation. (Step 2 is unnecessary.)

$$\frac{5}{9}(F - 32) = C$$

Step 3: Isolate C by dividing both sides of the equation by $\frac{9}{5}$. (Notice that division by $\frac{9}{5}$ is equivalent to multiplication by $\frac{5}{9}$.)

Example 3　　Solve $x(2 - y) = y - 1$ for y.

$$x(2 - y) = y - 1$$

$$2x - xy = y - 1$$

In order to get all the terms involving y on one side of the equation, we need to remove the parentheses by multiplication.

$$2x + 1 = y + xy$$

Step 1: All terms involving y are now on the right-hand side of the equation.

$$2x + 1 = y(1 + x)$$

Step 2: Factor y out of all terms on the right-hand side of the equation.

$$y = \frac{2x + 1}{1 + x}$$

Step 3: Isolate y by dividing both sides of the equation by $(1 + x)$.

(ii) Solving Quadratic Equations

An equation is **quadratic** in a variable if it involves only the second power (and possibly the first) of the variable and no division by any expression that includes the variable.

General Strategy for Solving Quadratic Equations

Step 1

Put all terms on one side of the equation and combine like terms to get the equation into the form $ax^2 + bx + c = 0$. (The coefficient a of x^2 cannot be 0 or else the equation wouldn't be a quadratic).

Step 2

Use the quadratic formula to solve for x:

$$x = \frac{-b - \sqrt{b^2 - 4ac}}{2a} \quad \text{or} \quad x = \frac{-b + \sqrt{b^2 - 4ac}}{2a}$$

NOTE: The quadratic formula works for any quadratic equation, although certain quadratic equations (see "Special Cases" and Examples 6 through 8) can be solved more quickly using other methods.

Example 4　　Solve $(x + 1)^2 = -2(x^2 + 6x - 3)$ for x.

$$x^2 + 2x + 1 = -2x^2 - 12x + 6$$

Remove parentheses by multiplication.

$$3x^2 + 14x - 5 = 0$$

Put all terms on the same side of the equation and combine like terms.

$$x = \frac{-14 \pm \sqrt{14^2 - 4(3)(-5)}}{2 \cdot 3}$$

Apply the quadratic formula with $a = 3$, $b = 14$, and $c = -5$.

$$= \frac{-14 \pm \sqrt{256}}{6}$$

$$= \frac{-14 \pm 16}{6}$$

So, $x = \frac{1}{3}$ or $x = -5$.

Example 5

Solve $w(2w + 5) = 9 + w$ for w.

$2w^2 + 5w = 9 + w$ Remove parentheses by multiplication.

$2w^2 + 4w - 9 = 0$ Put all terms on the same side of the equation and combine like terms.

$w = \dfrac{-4 \pm \sqrt{4^2 - 4(2)(-9)}}{2 \cdot 2}$ Apply the quadratic formula.

$= \dfrac{-4 \pm \sqrt{88}}{4} = \dfrac{-4 \pm 2\sqrt{22}}{4}$

$= \dfrac{-2 \pm \sqrt{22}}{2}$ Factor out a 2 from the two terms in the numerator and cancel.

So, $w = \frac{-2 + \sqrt{22}}{2}$ or $w = \frac{-2 - \sqrt{22}}{2}$, approximately 1.345 or -3.345.

Special Cases

Case 1

If the linear term is zero, we can solve the equation directly:

$$ax^2 + c = 0$$

$$ax^2 = -c$$

$$x^2 = \frac{-c}{a}$$

$$x = \pm \sqrt{\frac{-c}{a}}$$

NOTE: Remember that both the positive and negative roots are solutions, and that there are no real-number solutions unless the quantity $\frac{-c}{a}$ is nonnegative.

Example 6

Solve $4x^2 - 7 = 0$ for x.

$4x^2 = 7$ Isolate the term with x^2

$x^2 = \dfrac{7}{4}$ Divide by 4, the coefficient of x^2.

$x = \pm\sqrt{\dfrac{7}{4}} = \dfrac{\pm\sqrt{7}}{2}$ Take the square root of both sides of the equation. Make sure to include both the positive and negative square roots.

$x = \dfrac{\sqrt{7}}{2}$ or $x = \dfrac{-\sqrt{7}}{2}$

Case 2

If after completing step 1 you see a way to factor the quadratic expression, skip the quadratic formula and rewrite the quadratic as the product of two linear factors. Set each factor separately to zero to find one solution from each.

Example 7

Solve $ax^2 + bx = 0$ for x.

$x(ax + b) = 0$ Factor the x from the terms on the left-hand side of the equation.

$x = 0$ or $ax + b = 0$ Zero products must have at least one zero factor. Thus, we can find the solution by setting each factor equal to zero.

$x = 0$ or $x = \dfrac{-b}{a}$

     NOTE: If the constant term is zero, the quadratic equation can always be solved by factoring.

Example 8

Solve $3x^2 + 14x - 5 = 0$ for x.

$(3x - 1)(x + 5) = 0$ Factor the quadratic.

$3x - 1 = 0$ or $x + 5 = 0$ Zero products require at least one zero factor.

$x = \dfrac{1}{3}$ or $x = -5$ Solve each linear equation for x.

     NOTE: If the expression factors easily, this method is a big time-saver. Remember, though, that the quadratic formula *always* works.

(iii) Solving Polynomial Equations of Degree Higher Than Two

Although there is a formula for the solution of any cubic equation, it is cumbersome enough to bring you to tears. If you need to solve a polynomial equation of degree three or higher algebraically, pray that it factors. (If it doesn't, use your grapher to approximate the solutions.)

Example 9

Solve $x^3 - 3 = -x^2 + 2x - 3$ for x.

$$x^3 - 3 = -x^2 + 2x - 3$$

$x^3 + x^2 - 2x = 0$ Bring all terms to one side of the equation and combine like terms.

$x(x - 1)(x + 2) = 0$ Factor and use the zero product property to set each factor equal to 0.

$x = 0$ or $x = 1$ or $x = -2$

Example 10

Solve $z^4 - 5z^2 = -6$ for z.

$$z^4 - 5z^2 = -6$$

$z^4 - 5z^2 + 6 = 0$ Bring all terms to one side of the equation.

$(z^2 - 3)(z^2 - 2) = 0$ Factor.

$z^2 - 3 = 0, \quad z^2 - 2 = 0$ Zero products require at least one zero factor.

$z = \pm\sqrt{3}$ or $z = \pm\sqrt{2}$

Example 11

Solve $3x^5 + 4x^3 + 1 = 5x^3 + 3x^2$ for x.

$$3x^5 + 4x^3 + 1 = 5x^3 + 3x^2$$

$$3x^5 - x^3 - 3x^2 + 1 = 0 \qquad \text{Bring all terms to one side of the equation.}$$

$$x^3(3x^2 - 1) - 1(3x^2 - 1) = 0 \qquad \text{Factor by grouping.}$$

$$(x^3 - 1)(3x^2 - 1) = 0$$

$$(x - 1)(x^2 + x + 1)(3x^2 - 1) = 0 \qquad \text{The difference of two cubes factors.}$$

$$x - 1 = 0, \quad \text{or} \quad x^2 + x + 1 = 0, \qquad \text{Zero products require at least one zero factor.}$$
$$\text{or} \quad 3x^2 - 1 = 0$$

$$x = 1 \quad \text{or} \quad x = \sqrt{\frac{1}{3}} \quad \text{or} \quad -\sqrt{\frac{1}{3}} \qquad \text{The equation } x^2 + x + 1 = 0 \text{ has no real solutions.}$$

(iv) Solving Equations Involving Algebraic Fractions

An **algebraic fraction** is a ratio containing the variable in the denominator. (The variable might also appear in the numerator.)

General Strategy for Solving Equations Involving Algebraic Fractions

Get rid of the fractions by multiplying both sides of the equation by the lowest common denominator (LCD) of all the individual algebraic fractions in the equations. Careful! This technique may introduce extraneous solutions. That is, your new "fraction-free" equation may have more solutions than the original.

> *NOTE*: When you multiply both sides of an equation by an expression that contains the variable, you must check each solution by substituting it into the *original* equation to weed out extraneous roots.

Example 12

Solve $\frac{60n}{r} = \frac{(60+d)n}{r+8}$ for r.

$$r(r + 8) \cdot \frac{60n}{r} = r(r + 8) \cdot \frac{(60 + d)n}{r + 8} \qquad \text{The LCD of the two individual fractions is } r(r + 8).$$

$$(r + 8)60n = r(60 + d)n \qquad \text{Cancel. Notice this equation is linear in } r.$$

$$60nr + 480n = 60nr + dnr \qquad \text{In order to get all the terms involving } r \text{ on one side of the equation, remove the parentheses by multiplication.}$$

$$480n = 60nr + dnr - 60nr \qquad \text{Get all the terms involving } r \text{ on one side.}$$

$$480n = (dn)r \qquad \text{Factor out } r.$$

$$r = \frac{480n}{dn} \qquad \text{Isolate } r \text{ by dividing both sides of the equation by the coefficient of } r.$$

$$r = \frac{480}{d} \qquad \text{Cancel the } n\text{'s. (You might have divided both sides of the equation by } n \text{ in the beginning.)}$$

Example 13

Solve the equation $1 = \frac{1}{x-3} - \frac{3}{x^2-3x}$ for x.

$$x(x-3) \cdot 1 = (x)(x-3) \cdot \left[\frac{1}{x-3} - \frac{3}{x(x-3)} \right]$$

Multiply both sides of the equation by $x(x-3)$, the LCD of the individual fractions.

$$x^2 - 3x = x - 3$$

Be sure to multiply all terms on both sides of the equation by the LCD. Notice that you now have a quadratic equation in x.

$$x^2 - 4x + 3 = 0$$

Get all terms on one side.

$$(x-3)(x-1) = 0$$

Factor the quadratic.

$$x = 3 \quad \text{or} \quad x = 1$$

Be sure to check these in the original equation. Because $\frac{1}{1-3} - \frac{3}{1-3} = 1$, $x = 1$ is a solution. But $x = 3$ is *not* a solution; 3 makes no sense in the original equation because it makes a denominator zero.

$$x = 1$$

Weed out the extraneous root.

(v) Solving Equations Involving Radicals

The techniques in this section apply when the *variable* is under a radical sign. (An equation like $\sqrt{2}x^2 - 3x + \sqrt{5} = 0$ is quadratic in x, and does not belong to this section, even though the real numbers $\sqrt{2}$ and $\sqrt{5}$ involve radicals.)

Strategy for Solving Equations Containing Radical Expressions

If the equation contains a single radical expression, isolate the radical on one side of the equation. Raise both sides of the equation to the power required to remove the radical sign, then solve the resulting equation using the techniques of Sections (i) through (iv).

  If the equation contains three terms and two or more are radicals, you have a real mess on your hands. Isolate one of the radical expressions on one side of the equation; raise both sides of the equation to the power required to remove the radical sign; then simplify the resulting equation as much as possible. If it still contains a radical expression, isolate the radical and perform the whole procedure again.

  CAREFUL: This technique may introduce extraneous solutions. That is, your new "radical-free" equation may have more solutions than the original.

  NOTE: When you multiply both sides of an equation by an expression that contains a variable you must check each solution by substituting it into the *original* equation to weed out extraneous solutions.

Example 14

Solve $\sqrt{t} + 2 = t$.

$$\sqrt{t} + 2 = t$$

$$\sqrt{t} = t - 2$$

Isolate the expression with the radical on one side of the equation.

$$t = t^2 - 4t + 4$$

Square both sides of the equation. Note that the equation has become quadratic.

$$t^2 - 5t + 4 = 0$$

Move all terms to one side of the equation.

$$(t-4)(t-1) = 0$$

Factor.

$$t = 4 \quad \text{or} \quad t = 1$$

Check both solutions in the original equation. Since $\sqrt{4} + 2 = 4$, $t = 4$ is a solution. Since $\sqrt{1} + 2 \neq 1$, $t = 1$ is an extraneous solution.

$$t = 4$$

Weed out the extraneous root.

Example 15

Solve $\sqrt{3x^2 + 5x + 3} - 3 = x$.

$\sqrt{3x^2 + 5x + 3} = x + 3$ Isolate the radical.

$3x^2 + 5x + 3 = x^2 + 6x + 9$ Square both sides of the equation. Now we have a quadratic equation.

$2x^2 - x - 6 = 0$ Move all terms to one side of the equation.

$(2x + 3)(x - 2) = 0$ Factor.

$x = -\dfrac{3}{2}$ or $x = 2$

Check both solutions in the original equation. Since

$$\sqrt{3 \cdot 2^2 + 5 \cdot 2 + 3} - 3 = 2,$$

$x = 2$ is a solution. To check $x = -\frac{3}{2}$, we compute

$$\sqrt{3\left(-\dfrac{3}{2}\right)^2 + 5\left(-\dfrac{3}{2}\right) + 3} = \sqrt{\dfrac{27}{4} - \dfrac{15}{2} + 3} = \sqrt{\dfrac{9}{4}} = \dfrac{3}{2}$$

Since $\frac{3}{2} - 3 = -\frac{3}{2}$, $x = -\frac{3}{2}$ is also a solution.

Example 16

Solve $\sqrt{x + 48} - \sqrt{x} = 4$

$\sqrt{x + 48} = \sqrt{x} + 4$ Isolate one of the radicals.

$x + 48 = (\sqrt{x} + 4)^2$ Square both sides of the equation.

$x + 48 = x + 8\sqrt{x} + 16$

$32 = 8\sqrt{x}$ Isolate the remaining radical.

$4 = \sqrt{x}$

$16 = x$ Square both sides of the equation. Check the solution in the original equation. Since $\sqrt{16 + 48} - \sqrt{16} = 4$, 16 is the solution.

(vi) Solving Systems of Simultaneous Equations

Two or more equations that are to hold true at the same time are called a **system of simultaneous equations.** For work in this course, you will generally not have anything more complicated than a system of two equations in two variables.

General Strategy for Solving a System of Two Equations in Two Variables

Step 1

Solve one of the equations for one of the variables and substitute this expression into the second equation to obtain one equation in one variable.

Step 2

Solve this new equation using the techniques of Sections (i) through (v).

Step 3

Use the solution(s) and either of the original equations to compute the corresponding value for the second variable.

Example 17

Solve the system: $y = 3x^2 - 2x + 1.5$
$$y = 5x - 2.5$$

$5x - 2.5 = 3x^2 - 2x + 1.5$ Step 1: Substitute $5x - 2.5$ for y in the first equation. We now have a quadratic equation in x.

$3x^2 - 7x + 4 = 0$ Step 2: Use the techniques for solving quadratic equations to solve for x. (See A.5 (ii).) Get all terms on one side of the equation.

$(3x - 4)(x - 1) = 0$ Factor.

$x = \dfrac{4}{3}$ or $x = 1$ A zero product requires at least one zero factor.

$y = 5\left(\dfrac{4}{3}\right) - 2.5 = \dfrac{25}{6}$ Step 3: Complete the y-values corresponding to $x = \frac{4}{3}$ and $x = 1$:

$y = 5(1) - 2.5 = \dfrac{5}{2}$ When x is $\frac{4}{3}$, y is $\frac{25}{6}$; when x is 1, y is $\frac{5}{2}$.

$(x, y) = \left(\dfrac{4}{3}, \dfrac{25}{6}\right)$ or $(x, y) = \left(1, \dfrac{5}{2}\right)$ Each solution is a pair of values: one x-value and one y-value.

Example 18

Solve the system: $2x + 3y = 10$
$$4xy = 16$$

$y = \dfrac{16}{4x} = \dfrac{4}{x}$ Step 1: Solve one of the equations for one of the variables. Here we solve the second equation for y.

$2x + 3\left(\dfrac{4}{x}\right) = 10$ Substitute this expression into the first equation. Notice that we now have a fractional equation in x.

$2x^2 + 12 = 10x$ Step 2: Multiply each side of the equation by x to eliminate the fraction. Notice that we now have a quadratic equation in x.

$2x^2 - 10x + 12 = 0$ Get all the terms on one side of the equation.

$x^2 - 5x + 6 = 0$ Divide both sides of the equation by 2.

$(x - 3)(x - 2) = 0$ Factor.

$x = 3$ or $x = 2$ A zero product requires at least one zero factor.

$y = \dfrac{4}{3}$ Step 3: Now compute the corresponding y-values by substituting the solutions for x into $y = \frac{4}{x}$.

$y = \dfrac{4}{2} = 2$ When $x = 3$, $y = \frac{4}{3}$; when $x = 2$, $y = 2$.

$(x, y) = \left(3, \dfrac{4}{3}\right)$ or $(x, y) = (2, 2)$ Each solution is an x-y pair.

Practice Problems

Find the solution sets to the following equations:

1. $4(1 - 5x) = 2(x - 1) + 6x$
2. $\pi r = 4r - 10$
3. Solve $(y + 3)z = 5z - 10$ for z.
4. $x^2 + 2x = 4x + 8$
5. $2x(x + 4) + x = 2x + 4$
6. $2r^2s + 3rs - 5s = 1$ (Solve this equation in two ways: for r in terms of s; for s in terms of r.)
7. $4y^2 - 16 = 0$
8. $-16t^2 + 100t = 0$
9. $y(y^2 + 5y - 4) = 20$
10. $4x^2(x^2 - 1) = x^2 - 1$

11. $\frac{40}{r+10} = \frac{20}{r}$
12. $\frac{x}{x-1} + \frac{2}{3x-1} = \frac{12}{5}$
13. $\sqrt{x + 11} + x = 1$
14. $\sqrt{d^2 - 3d} = \sqrt{2d + 6}$
15. $\sqrt{1 - 3x} + 1 = \sqrt{x + 10}$
16. $\sqrt{2x + 2} + 1 = \sqrt{1 - 6x}$

For Problems 17 and 18, solve the system of equations.

17. $y + 3x = 6$
    $2y - x = 4$

18. $rd = 10$
    $3r = d + 1$

A.6   INEQUALITIES

Solving an inequality is similar to solving an equation, except that we have to pay more attention to signs. The following algebraic manipulations do not change the solution set of an inequality:

1. adding or subtracting the same quantity on each side
2. multiplying or dividing each side by the same positive quantity
3. multiplying or dividing each side by the same negative quantity *and reversing the direction of the inequality sign*

To illustrate these rules, suppose we start with a true statement, such as

$$1 < 5$$

If we multiply both sides by 3, we get

$$3 < 15,$$

which is still a true statement. If instead we multiply both sides of our original statement by -4 *and reverse the direction of the inequality sign*, we get

$$-4 > -20,$$

which yields another true statement.

(i) Interval Notation

Inequalities, if they have solutions, ordinarily have an infinite number of solutions. We cannot list them all, so we need a shorthand for writing them. Interval notation serves this purpose:

$[a, b]$ means "all the real numbers between a and b, including the boundaries a and b"

(a, b) means "all the real numbers between a and b, not including the boundaries a and b"

[a,b) means "all the real numbers between a and b, including a but not including b"

(a,b] means "all the real numbers between a and b, including b but not including a"

($-\infty,b$] means "all the real numbers less than or equal to b"

($a,+\infty$) or (a,∞) means "all the real numbers greater than a"

($-\infty,\infty$) means "all real numbers"

At times, the solution to an inequality cannot be expressed using a single interval, but requires the union ($\cup$) or intersection ($\cap$) of two or more intervals.

[a,b] $\cup$ (c,∞) means "all the real numbers that are either between a and b (inclusive) *or* are greater than c"

[a,b] $\cap$ (c,∞) means "all the real numbers that are between a and b (inclusive) *and* are simultaneously greater than c"

There are two things to note about interval notation. First, the symbol (a,b) looks just like the symbol we use for the ordered pair $x = a$, $y = b$. The meaning of the symbol, however, will be clear from the context in which it is used. Second, the symbols ∞ and $-\infty$ do not represent real numbers, and therefore they are never "included" in an interval. They indicate that an interval is unbounded either at the top or at the bottom (or both).

Example 1

Express each of the following inequalities in interval notation:

$t < -5$	$(-\infty, -5)$
$1.99 \leq v \leq 2.01$	$[1.99, 2.01]$
$y \geq \dfrac{1}{3}$	$\left[\dfrac{1}{3}, +\infty\right)$
$-0.02 < x \leq 0.02$	$(-0.02, 0.02]$
$z < -2$ or $z > 1$	$(-\infty, -2) \cup (1, \infty)$

NOTE: There is no way to write the last inequality as a single interval. It involves two separate sections of the number line.

(ii) Solving Linear Inequalities

All strategies for solving linear equations apply to solving a linear inequality, except that we must be alert to the possible need to reverse the direction of the inequality sign.

Example 2

Solve $4 - 3w < 5w + 16$ for w.

$$4 - 3w < 5w + 16$$

$$4 - 8w < 16 \qquad \text{Subtract } 5w \text{ from both sides.}$$

$$-8w < 12 \qquad \text{Subtract 4 from both sides.}$$

$$w > \frac{12}{-8} \qquad \text{Divide both sides by } -8 \text{ and reverse the direction of the inequality.}$$

The solution can be written as $w < -\frac{3}{2}$ or as $\left(-\frac{3}{2}, \infty\right)$.

(iii) Solving Nonlinear Inequalities

Whenever the inequality involves roots or fractional expressions or higher powers of a variable, algebraic solutions can be very tricky. Difficulties arise when we multiply or divide by an expression containing a variable because that expression could be positive or could be negative, depending upon the value of the variable. We wouldn't know for sure whether or not to reverse the direction of the inequality sign, so we'd have to consider both possibilities.

One way to sidestep the difficulty is to treat the inequality as though it were an equation, and then use the solutions to the equation to divide the real-number line into regions. Each region can be easily tested to see whether it is part of the inequality's solution.

Example 3

Solve $3x^2 - 2x + 1.5 > 5x - 2.5$ for x.

$$3x^2 - 2x + 1.5 > 5x - 2.5$$

$$3x^2 - 7x + 4 > 0 \qquad \text{Move all terms to one side of the inequality.}$$

Next, solve the corresponding equation, and then go back and figure out the solution to the inequality.

$$3x^2 - 7x + 4 = 0 \qquad \text{Replace the inequality sign with an equal sign to get the corresponding equation.}$$

$$(3x - 4)(x - 1) = 0 \qquad \text{Remember that a zero product requires at least one zero factor.}$$

$$x = \frac{4}{3} \quad \text{or} \quad x = 1$$

To find the solutions to the original inequality, use the solutions to the equation to divide the real-number line into three regions:

Test one point from each region in the original inequality.

Region A

Try $x = 0$: $3(0)^2 - 2(0) + 1.5 > 5(0) - 2.5$?   Yes.

(Any number in the region will do. Zero is the most convenient.)

Region B

Try $x = 1.1$: $3(1.1)^2 - 2(1.1) + 1.5 > 5(1.1) - 2.5$?   No.

Region C

Try $x = 2$: $3(2)^2 - 2(2) + 1.5 > 5(2) - 2.5$?   Yes.

If one point from a region makes the inequality true, all points from that region are part of the solution set. The solution, then, consists of Regions A and C, $(-\infty, 1) \cup (\frac{4}{3}, +\infty)$, those numbers that are either less than 1 or greater than $\frac{4}{3}$. We may also write the solution as $x < 1$ or $x > \frac{4}{3}$. It does not include the endpoints 1 and $\frac{4}{3}$ because they make the *equation* true, but not the *inequality*.

Even with clever algebraic manipulations, it is easy to make errors in solving inequalities. Learning to read the appropriate solutions from a graph will help you avoid the most frequent errors, which usually involve the direction of the inequality sign.

Example 4

Solve $\frac{x-1}{2x+3} < \frac{x-2}{x+4}$.

$$\frac{x-1}{2x+3} < \frac{x-2}{x+4}$$ Note that $x \neq -4$ and $x \neq -\frac{3}{2}$

Solve the corresponding equation. Then go back and figure out the solution to the inequality.

$$\frac{x-1}{2x+3} = \frac{x-2}{x+4}$$ Replace the inequality sign with an equal sign to get the corresponding equation.

$$(x+4)(x-1) = (2x+3)(x-2)$$ Multiply both sides by $(2x+3)(x+4)$, the LCD. *(NOTE: This is now a quadratic equation.)*

$$x^2 + 3x - 4 = 2x^2 - x - 6$$ Remove parentheses by mutiplication.

$$x^2 - 4x - 2 = 0$$ Move all terms to one side of the equation.

$$x = \frac{4 \pm \sqrt{16+8}}{2}$$ Use the quadratic formula to find the solutions.

$$x = 2 \pm \sqrt{6}$$ Simplify.

To find the solutions to the inequality, use the solutions to the equation, along with the zeros of the two denominators, to divide the number line into five regions:

| Region A | Region B | Region C | Region D | Region E |

$$\begin{array}{ccccc} & -4 & -\frac{3}{2} \quad 2-\sqrt{6} & & 2+\sqrt{6} \end{array} \longrightarrow x$$

As before, we will test one point from each region in the original inequality.

Region A

Try $x = -5$: $\dfrac{-5-1}{2(-5)+3} < \dfrac{-5-2}{-5+4}$?

$$\frac{6}{7} < 7? \quad \text{Yes.}$$

Region B

Try -2: $\dfrac{-2-1}{2(-2)+3} < \dfrac{-2-2}{-2+4}$?

$$3 < -2? \quad \text{No.}$$

Region C

Try $x = -1$: $\dfrac{-1-1}{2(-1)+3} < \dfrac{-1-2}{-1+4}$?

$$-2 < -1? \quad \text{Yes.}$$

Region D

$$\text{Try } x = 0: \quad \frac{0-1}{2(0)+3} < \frac{0-2}{0+4}?$$

$$-\frac{1}{3} < -\frac{1}{2}? \quad \text{No.}$$

Region E

$$\text{Try } x = 5: \quad \frac{5-1}{2(5)+3} < \frac{5-2}{5+4}?$$

$$\frac{4}{13} < \frac{1}{3}? \quad \text{Yes.}$$

The solution to the inequality consists of regions A, C, and E:

$$(-\infty, -4) \cup \left(-\frac{3}{2}, 2 - \sqrt{6}\right) \cup \left(2 + \sqrt{6}, \infty\right)$$

Isn't it easier to use your grapher to find approximate solutions in this case?

Practice Problems

In Problems 1 through 3, express each inequality in interval notation.

1. $x \geq -3$

2. $y < -1 \quad$ or $\quad y > 2$

3. $8 < z \leq 17$

Find the solutions for each inequality in Problems 4 through 10. Express your answer using interval notation.

4. $3x + 5 > 2(x - 1)$

5. $3(2w + 5) + 3w \leq 5w - 4$

6. $2y^2 - 5y \geq y^2 - 6$

7. $(x + 2)(x - 3) < 3x + 1$

8. $(x - 5)(x + 6)(3x + 6) > 0$

9. $\frac{x-2}{x+4} < \frac{x-1}{2x+3}$ (Compare this problem with Example 4.)

10. $\frac{x}{x+1} > \frac{1-x}{x}$

A.7 ABSOLUTE VALUES

(i) Definition and Properties

Let Q represent any quantity—a number or an algebraic expression. The absolute value of Q, written $|Q|$, is the magnitude of the quantity—that is, its size without regard for its sign. The absolute value of a nonnegative quantity is equal to the quantity itself; the absolute value of a negative quantity is equal to the opposite of the quantity itself. (Note that the opposite of a negative quantity is necessarily positive.)

In mathematical symbols,

$$|Q| = \begin{cases} Q, & \text{if } Q \geq 0 \\ -Q, & \text{if } Q < 0 \end{cases}$$

Example 1

Express each of the following quantities without using absolute-value signs.

$$|5| = 5$$

$$|-2| = -(-2) = 2 \qquad \text{The opposite of } -2 \text{ is } 2.$$

$$|x - 7| = \begin{cases} x - 7, & \text{if } x \geq 7 \\ -(x - 7), & \text{if } x < 7 \end{cases}$$ Since the absolute-value expression contains a variable, we must allow for two possibilities.

You can combine absolute values that are multiplied or divided (but not those that are added or subtracted) according to these rules.

> **Rules for Combining Absolute Values**
>
> $$|a| \cdot |b| = |a \cdot b|$$
>
> $$\frac{|a|}{|b|} = \left| \frac{a}{b} \right|$$

As an illustration of these rules, notice that

$$|x - 3| \cdot |x + 2| = |x^2 - x - 6|$$

and

$$\frac{|x - 3|}{|3 - x|} = \left| \frac{x - 3}{3 - x} \right| = |-1| = 1$$

However, you can't distribute absolute value with addition. In the case of addition, the following rule, called the **triangle inequality,** holds:

$$|a + b| \leq |a| + |b|$$

If we apply the triangle inequality when $a = 5$ and $b = -2$, we get

$$|5 + (-2)| \leq |5| + |-2|$$

Notice that the value of the left-hand side is 3 and that the value of the right-hand side is 7. The resulting inequality

$$3 \leq 7$$

is a true statement.

Absolute values are frequently used to define the distance between two numbers or algebraic expressions.

$$|b - a| = |a - b| = \text{the distance between } a \text{ and } b$$

Interpreted in this manner, the quantity $|x - 7|$ measures the distance between x and 7. So does $|7 - x|$.

Example 2   Find all x-values that make $|x - 7| = 4$.

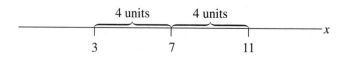

$x = 3$ or   11   These are the x-values that are exactly 4 units from 7.

(ii) Solving Equations Involving Absolute Values

To solve equations involving absolute values, replace expressions containing absolute values with conditional expressions that do not involve absolute values. Every absolute-value expression that contains a variable allows for two possibilities; we must consider both.

Example 3

Find the values of x that make $|2x - 4| = 6$.

By definition $|2x - 4| = \begin{cases} 2x - 4, & \text{if } 2x - 4 \geq 0 \quad \text{(that is, } x \geq 2) \\ -(2x - 4), & \text{if } 2x - 4 < 0 \quad \text{(that is, } x < 2) \end{cases}$

We use 2 to divide the number line into two regions because 2 is where the quantity inside the absolute value changes sign. Then we find the solutions for each region.

Region A $\quad\quad\quad\quad$ Region B
$(-\infty, 2)$ $\quad\quad\quad\quad$ $[2, +\infty)$

$$\underline{}x$$
$$2$$

Find any solutions in region A.

Use $-(2x - 4)$ because $x < 2$ in region A.

$-2x + 4 = 6$

$-2x = 2$

$x = -1$ $\quad$ Since this x-value is in region A, it is part of the solution set.

Find any solutions in region B.

Use $2x - 4$ because $x \geq 2$ in region B.

$2x - 4 = 6$

$2x = 10$

$x = 5$ $\quad$ Since this x-value is in region B, it is part of the solution set.

The complete solution is $x = -1$ or $x = 5$.

Example 4

Solve $|x + 3| = |2x - 4|$.

Rewrite $|x + 3|$ and $|2x - 4|$ without absolute-value signs.

$$|x + 3| = \begin{cases} x + 3, & \text{if } x \geq -3 \\ -(x + 3), & \text{if } x < -3 \end{cases}$$

$$|2x - 4| = \begin{cases} 2x - 4, & \text{if } x \geq 2 \\ -(2x - 4), & \text{if } x < 2 \end{cases}$$

We use -3 and 2 to break the real-number line into three regions because -3 and 2 give the two points where the quantities inside the absolute values change sign.

Region A $\quad\quad\quad$ Region B $\quad\quad\quad$ Region C

$$\underline{}x$$
$$-3 \quad\quad\quad\quad\quad 2$$

Find any solutions in region A.

> Use $-(x + 3)$ and $-(2x - 4)$ because $x < -3$ and $x < 2$ in region A.

$-(x + 3) = -(2x - 4)$

$-x - 3 = 2x + 4$

$x = 7$ Since 7 is not in region A, there are no solutions in region A.

Find any solutions in region B.

> Use $x + 3$ and $-(2x - 4)$ because $x \geq -3$ and $x < 2$ in region B.

$x + 3 = -(2x - 4)$

$3x = 1$

$x = \dfrac{1}{3}$ Since $\frac{1}{3}$ is in region B, it belongs to the solution set.

Find any solutions in region C.

> Use $x + 3$ and $2x - 4$ because $x > -3$ and $x \geq 2$ in region C.

$x + 3 = 2x - 4$

$-x = -7$

$x = 7$ Since 7 is in region C, it belongs to the solution set.

The complete solution consists of two x-values, $x = 7$ and $x = \frac{1}{3}$.

(iii) Solving Inequalities Involving Absolute Values

Solving absolute-value inequalities involves a combination of the techniques for solving absolute-value equations and those for solving ordinary inequalities. We rewrite the problem in a form that contains no absolute-value signs. That means that we must consider two possibilities for each absolute-value expression appearing in the problem. Then we solve each case as though it were an ordinary inequality.

Example 5   Solve $|2x - 4| < 6$.

Compare this problem to Example 3. The number line is divided into the same two regions.

Find any solutions in region A.

> Use $-(2x - 4)$ because $x < 2$ in region A.

$-(2x - 4) < 6$

$-2x + 4 < 6$

$-2x < 2$ We will have to reverse the direction of the inequality.

$x > -1$ For region A, the solutions are all the numbers greater than -1 and simultaneously less than 2.

Find any solutions in region B.

Use $2x - 4$ because $x \geq 2$ in region B.

$$2x - 4 < 6$$

$$2x < 10$$

$$x < 5$$ For region B, the solutions are all the numbers less than 5 but not less than 2.

We combine the solutions from the two regions into a single interval, consisting of all the numbers greater than -1 and less than 5: $-1 < x < 5$, or $(-1, 5)$.

There is an alternative, and simpler, method for solving an inequality having the form

$$|Q| < n,$$

where Q stands for any algebraic expression and n is any positive number. We'll redo Example 5 to illustrate the method:

$|Q| < n$ means that the quantity Q is *between* $-n$ and $+n$

Therefore, $|2x - 4| < 6$ means that the quantity $2x - 4$ is between -6 and 6. This yields a three-part inequality, which can be solved according to the ordinary rules for solving inequalities.

$$-6 < 2x - 4 < 6$$

$$-2 < 2x < 10$$ Add 4 to each part.

$$-1 < x < 5$$ Divide each part by 2.

Example 6   Solve $|2x - 4| > 6$.

The two cases are the same as in Example 5.

Find any solutions in region A.

For $x < 2$, we solve:

$$-(2x - 4) > 6$$

$$-2x + 4 > 6$$

$$-2x > 2$$ We will need to change the direction of the inequality.

$$x < -1$$ For region A, the solutions are all the numbers less than -1 and simultaneously less than 2, or simply the numbers less than -1.

Find any solutions in region B.

For $x \geq 2$, we solve:

$$2x - 4 > 6$$

$$2x > 10$$

$$x > 5$$ For region B, the solutions are all the numbers greater than 5 but not less than 2, or, more simply, the numbers greater than 5.

The solution to the given inequality consists of two separate intervals: it cannot be written with a single interval statement. The solution is $(-\infty, -1) \cup (5, \infty)$.

There is a simpler method for solving this type of inequality, too. Example 6 is an example of an inequality of the form

$$|Q| > n,$$

where Q stands for any algebraic expression and n is any positive number. We'll redo Example 6 to illustrate the method:

$|Q| > n$ mean that the quantity Q is either *less than* $-n$ or *greater than* $+n$.

Therefore, $|2x - 4| > 6$ means that the quantity $2x - 4$ is either less than -6 or greater than 6. We write this as two algebraic inequalities and solve each one separately. Note that there is no legitimate way to do this "in one piece" as we did with Example 5.

$$2x - 4 < -6 \quad \text{or} \quad 2x - 4 > 6$$

$$2x < -2 \quad \text{or} \quad 2x > 10$$

$$x < -1 \quad \text{or} \quad x > 5$$

Example 7

Solve $|x + 3| < |2x - 4|$.

Compare this problem to Example 4. Each absolute-value expression gives rise to two cases and divides the number line into three regions.

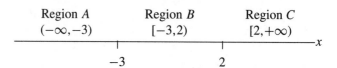

Region A	Region B	Region C
$(-\infty, -3)$	$[-3, 2)$	$[2, +\infty)$

Find any solutions in region A.

Use $-(x + 3)$ and $-(2x - 4)$ because $x < -3$ and $x < 2$ in region A.

$-(x + 3) < -(2x - 4)$

$-x - 3 < -2x + 4$

$\quad x < 7$ Solutions for region A are numbers that are both less than 7 and less than -3. That interval is $(-\infty, -3)$.

Find any solutions in region B.

Use $x + 3$ and $-(2x - 4)$ because $x \geq -3$ and $x < 2$ in region B.

$x + 3 < -(2x - 4)$

$x + 3 < -2x + 4$

$\quad 3x < 1$

$\quad x < \dfrac{1}{3}$ Solutions for region B are numbers between -3 and 2 that are also less than $\frac{1}{3}$. That interval is $[-3, \frac{1}{3})$.

Find any solutions in region C.

Use $x + 3$ and $2x - 4$ because $x > -3$ and $x \geq 2$ in region C.

$$x + 3 < 2x - 4$$

$$-x < -7$$

$$x > 7$$ Solutions for region C are numbers that are both greater than 7 and not less than 2. That interval is $(7, +\infty)$.

Notice that the solutions from regions A and B form a continuous interval, $(-\infty, \frac{1}{3})$. The entire solution, then, can be written $(-\infty, \frac{1}{3}) \cup (7, +\infty)$.

If you have attempted to follow the work of Example 7, you should be thoroughly convinced that, for approximating the solution to a messy inequality, graphs are the way to go!

Practice Problems

Find the solutions to the following equations:

1. $|3x - 9| = 5$

2. $|x + 5| = |3x - 9|$

3. $\frac{|x^2 - 5x + 6|}{|x - 3|} = 10$

4. $\sqrt{(2x - 5)^2} = 6$

Solve the following inequalities:

5. $|3x - 9| < 5$

6. $|3x - 9| \geq 5$

7. $|\frac{1}{x}| < 2$

8. $|2x - 5| < |x + 4|$

A.8   APPLICATIONS OF THE PYTHAGOREAN THEOREM

(i) The Pythagorean Theorem and the Distance Formula

The lengths of the sides of every right triangle satisfy the relationship known as the Pythagorean Theorem.

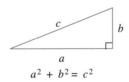

$$a^2 + b^2 = c^2$$

Example 1

Find the height of an equilateral triangle whose side length is 1 foot.

The altitude of the triangle (the line segment representing the height) divides the equilateral triangle into two congruent right triangles. The base of each right triangle is $\frac{1}{2}$ foot long.

Apply the Pythagorean Theorem to one of the right triangles.

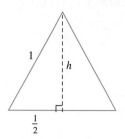

$$h^2 + \left(\frac{1}{2}\right)^2 = 1^2$$

$$h^2 = 1 - \frac{1}{4}$$

$$h^2 = \frac{3}{4}$$

$$h = \frac{\sqrt{3}}{2}$$

The triangle is $\frac{\sqrt{3}}{2}$ foot (approximately 0.87 foot) high.

The Pythagorean Theorem provides a way to find the distance between any two points (x_1, y_1) and (x_2, y_2) in the coordinate plane.

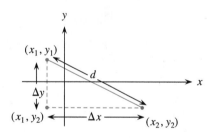

$$d^2 = (x_2 - x_1)^2 + (y_2 - y_1)^2 \qquad \text{by the Pythagorean Theorem}$$

$$d = \sqrt{(x_2 - x_1)^2 + (y_2 - y_1)^2} = \sqrt{(\Delta x)^2 + (\Delta y)^2}$$

Notice that, because the differences are squared, we may subtract the coordinates in either order.

Example 2   Find the distance between the points $(-2, 3)$ and $(7, -4)$.

$$d = \sqrt{(-2 - 7)^2 + (3 - (-4))^2}$$

$$= \sqrt{81 + 49}$$

$$= \sqrt{130}$$

(ii) Completing the Square

To "complete the square" means to take any quadratic expression and rewrite it as a perfect square trinomial added to a constant. The technique is useful for graphing circles and parabolas and is important in traditional calculus courses. You will find another demonstration of the technique in the Project 5.3, page 263, where it is used to graph a parabola.

Recall that an expression in the form

$$x^2 + 2bx + b^2$$

can be written in factored form as

$$(x + b)^2$$

Similarly, $x^2 - 2bx + b^2$ is $(x - b)^2$.

These quadratic expressions are called **perfect square trinomials.** The aim in completing the square is to take an expression such as

$$t^2 + 10t + 14$$

and rewrite it as the equivalent expression

$$(t + 5)^2 - 11$$

Here's the procedure.

Example 3

Find an equivalent expression for $t^2 + 10t + 14$ that contains a perfect square trinomial as one of its terms.

$t^2 + 10t + 14$

$t^2 + 10t + 25 + 14 - 25$ Take half of the 10 and square it. Add *and* subtract the amount from the original expression. (Since $25 - 25 = 0$, we've really only added 0 to the expression.)

$(t + 5)^2 - 11$ Rewrite the first three terms as a perfect square. Combine the rest.

If the coefficient of the quadratic term is something other than 1, there's another step.

Example 4

Complete the square for $3m^2 - 15m - 2$.

$3m^2 - 15m - 2$

$3(m^2 - 5m \qquad) - 2$ Factor a 3 out of the first two terms.

$3\left(m^2 - 5m + \dfrac{25}{4}\right) - 2 - \dfrac{75}{4}$ Take half of 5 and square it, obtaining $\frac{25}{4}$. Add $\frac{25}{4}$ inside the parentheses. Notice that, since $\frac{25}{4}$ is inside the parentheses, we have effectively added $3 \cdot \frac{25}{4}$ or $\frac{75}{4}$ to the expression. So, to balance what we have added, we need to subtract $\frac{75}{4}$.

$3\left(m - \dfrac{5}{2}\right)^2 - \dfrac{83}{4}$ Rewrite the contents of the parentheses as a perfect square. Combine the rest.

(iii) Circles

A circle is the set of points that are the same distance, r, from a point called the *center*. The distance from the center to any point on the circle is the radius. The points on a circle with center (a,b) and radius r satisfy the following equation, which is derived from the distance formula:

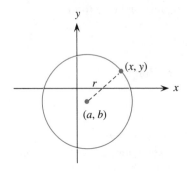

$$(x - a)^2 + (y - b)^2 = r^2$$

Example 5

Describe the circle represented by the following equation.

$5x^2 + 5y^2 + 30x - 10y + 49 = 0$ Our goal is to rewrite this expression in the form $(x - a)^2 + (y - b)^2 = r^2$.

$x^2 + y^2 + 6x - 2y + \dfrac{49}{5} = 0$ Divide both sides of the equation by 5 to change the coefficients of x^2 and y^2 to 1.

$(x^2 + 6x + \quad) + (y^2 - 2y + \quad) = -\dfrac{49}{5}$ Group terms involving x's and y's and then complete the squares.

$(x^2 + 6x + 9) + (y^2 - 2y + 1) = -\dfrac{49}{5} + 9 + 1$ Add 9 and 1 to each side to keep the equation balanced.

$(x + 3)^2 + (y - 1)^2 = \dfrac{1}{5}$

The graph is a circle with center $(-3, 1)$ and radius $\sqrt{\frac{1}{5}}$.

(iv) Ellipses

We can rewrite the basic equation for a circle with center at (a,b) and radius r in the following form:

$$\frac{(x - a)^2}{r^2} + \frac{(y - b)^2}{r^2} = 1$$

When the two denominators on the left-hand side are not equal, as in the following equation, the figure becomes elongated in one direction and we have an ellipse:

$$\frac{(x - a)^2}{c^2} + \frac{(y - b)^2}{d^2} = 1, \text{ where } c \neq d$$

The longest diameter of the ellipse is called its **major axis.** The **minor axis,** perpendicular to the major axis of the ellipse at the center, is the shortest diameter of the ellipse.

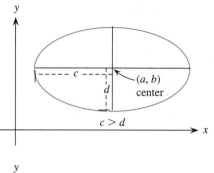

When $c > d$, the major axis is horizontal and the minor axis is vertical.

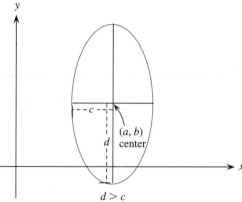

When $d > c$, the major axis is vertical and the minor axis is horizontal.

Note that c is one-half the length of the horizontal axis and d is one-half the length of the vertical axis.

Example 6

Find the equation of an ellipse with center at $(1,3)$ that has a major axis of length 6 parallel to the x-axis and a minor axis of length 4 parallel to the y-axis. Sketch the graph of the ellipse.

$$\frac{(x - 1)^2}{3^2} + \frac{(y - 3)^2}{2^2} = 1$$

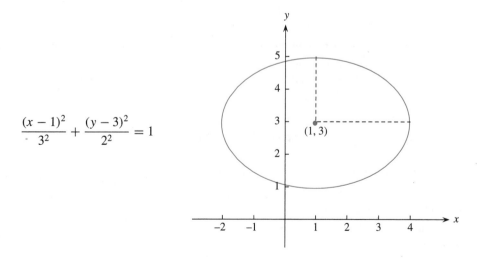

Example 7

Draw the ellipse represented by the equation

$$25x^2 + 50x + 16y^2 - 128y - 119 = 0$$

$$(25x^2 + 50x) + (16y^2 - 128y) = 119$$

Group the x- and y-terms together and move the constant term to the right-hand side of the equation.

$$25(x^2 + 2x) + 16(y^2 - 8y) = 119$$

Factor so that the x^2- and y^2-terms have coefficients of 1.

$$25(x^2 + 2x + 1) + 16(y^2 - 8y + 16) = 400$$

Complete the squares. To Keep the equation balanced, you need to add 25 and 256 to both sides.

$$\frac{(x + 1)^2}{16} + \frac{(y - 4)^2}{25} = 1$$

Divide both sides of the equation by (25)(16) or 400. Express the completed squares as perfect squares.

$$\frac{(x + 1)^2}{4^2} + \frac{(y - 4)^2}{5^2} = 1$$

The ellipse has center $(-1, 4)$. The major axis is parallel to the y-axis and has length 10 units. The minor axis has length 8 units.

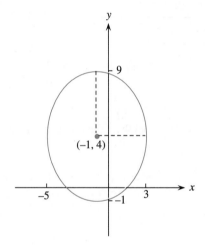

Practice Problems

1. Find the length of the hypotenuse of a right triangle whose legs have lengths 5 and 7.

2. If a right triangle has a hypotenuse of length 10 and a leg of length 6, find the length of the other leg.

3. Find the distance between the points $(-1, 2)$ and $(4, -3)$.

4. Find the perimeter of a triangle made by connecting the points $(1, 5)$, $(2, 1)$, and $(-1, -3)$.

5. Complete the square for $x^2 - 6^x$.

6. Complete the square for $2x^2 - 4x + 9$.

7. Find the equation of a circle with center at $(-\frac{1}{2}, 4)$ and radius 3.

In Problems 8 through 10, determine whether or not the algebraic equation describes a circle. If it does, find the center and radius of the circle.

8. $x^2 + y^2 + 2x - 6y - 6 = 0$

9. $2x^2 - 4x + 3y^2 - 12y = -9$

10. $4x^2 + 4y^2 - 8x + 8y = 0$

11. Find the equation of an ellipse with center at $(-2, 4)$ that has a major axis of length 4 parallel to the y-axis and minor axis of length 2 parallel to the x-axis.

12. Describe the ellipse represented by the equation

$$25x^2 + 9y^2 - 50x + 36y - 164 = 0$$

A.9 FORMULAS RELATED TO GEOMETRIC SHAPES

(i) Two-dimensional Shapes

Rectangle

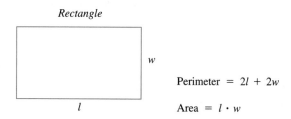

$$\text{Perimeter} = 2l + 2w$$

$$\text{Area} = l \cdot w$$

NOTE: Squares are special cases of rectangles where $l = w$.

Parallelogram

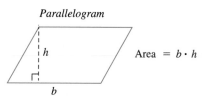

$$\text{Area} = b \cdot h$$

Trapezoid

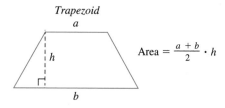

$$\text{Area} = \frac{a + b}{2} \cdot h$$

Triangle

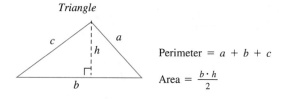

$$\text{Perimeter} = a + b + c$$

$$\text{Area} = \frac{b \cdot h}{2}$$

Circle

$$\text{Circumference} = 2\pi r$$

$$\text{Area} = \pi r^2$$

(ii) Three-dimensional Shapes

Rectangular Solid

$$\text{Surface area} = 2(wh + lh + wl)$$

$$\text{Volume} = lwh$$

Right Circular Cylinder

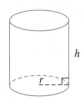

Surface area $= 2\pi r^2 + 2\pi rh$

Volume $= \pi r^2 h$

Sphere

Surface area $= 4\pi r^2$

Volume $= \frac{4}{3}\pi r^3$

Right Circular Cone

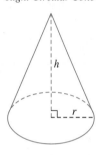

Surface area $= \pi r^2 + \pi r\sqrt{r^2 + h^2}$

Volume $= \frac{1}{3}\pi r^2 h$

Answers to Practice Problems

A.1

1. $7x^3 - 5x^2 + x - 7$

2. $10x^4 + 4x^3 - 11x^2 - 2x + 3$

3. $16x^2 - 9$

4. $2x^2 + 3x + 2 + \frac{7x-1}{x^2-1}$

5. $(x + 4)^2(x - 4)$

6. $8(2x - 1)(x - 1)$

7. $(5x - 4)(5x + 4)$

8. $(x + 8)(x - 2)$

A.2

1. $\frac{2}{9}$

2. $-\frac{7}{45}$

3. $\frac{43}{45}$

4. $\frac{18}{25}$

5. $\frac{x+3}{2(2x+3)}$, $\quad x \neq 5$

6. $\frac{29x-41}{2(x-5)(2x+3)}$

7. $-2x$, $\quad x \neq \pm 5$

8. $-\frac{2x+h}{(x+h)^2 x^2}$, $\quad h \neq 0$

9. $\frac{x^2+4}{(x^2-4)(x-2)}$ or $\frac{x^2+4}{(x+2)(x-2)^2}$

A.3

1. $4x^2$

2. $\frac{5}{16}x^7$

3. $\frac{4x^{1/5}}{3y^2}$

4. $\frac{x}{y^7}$

5. $\frac{2x^4}{3y^4}$

6. $x^{2/3}$

7. $x^{3/2}$

8. $x^{-1/4}$

9. $x^{-3/4}$

10. (a) 50   (b) 0.004   (c) 0.2

11. $\frac{3}{4}x$

12. Suppose $a = 10$ and $b = 2$. Then $(10^2 - 2^2)^{1/2} \neq 10 - 2$. Any other pair of unequal nonzero numbers would do as well.

13. 3.965×10^9, 3.965 E9

14. 2.27×10^{-7}, 2.27 E–7

15. 264,123,000,000

16. 0.000634875

17. $2x^{1/2}(x-2)(x+2)$

18. $y^{1/3}(y-5)(y+5)$

19. $x^{-3/2}(x-1)(x-3)$

A.4

1. $\frac{1}{2}\log_b(x) + \frac{1}{2}\log_b(y) - 3\log_b(z)$

2. $\log_3(8) = \frac{\log(8)}{\log(3)} = \frac{\log(8)}{\log(3)} \approx 1.89$ (In the expression $\frac{\log(8)}{\log(3)}$, base 10 is understood.)

3. $f(x) = \ln(x) = \frac{\log(x)}{\log(e)}$

A.5

1. $\frac{3}{14}$

2. $\frac{10}{4-\pi}$

3. $-\frac{10}{y-2}$

4. $-2, 4$

5. $-4, \frac{1}{2}$

6. $r = \frac{-3s \pm \sqrt{49s^2 + 8s}}{4s}$; $s = \frac{1}{2r^2 + 3r - 5}$

7. ± 2

8. $0, \frac{25}{4}$

9. $-5, \pm 2$

10. $\pm\frac{1}{2}, \pm 1$

11. 10

12. $\frac{11}{21}, 2$

13. -2 (The solution 5 is extraneous.)

14. $6, -1$

15. -1 (The solution $-\frac{15}{4}$ is extraneous.)

16. $-\frac{1}{2}$ (The solution $\frac{1}{8}$ is extraneous.)

17. $x = \frac{8}{7}$, $y = \frac{18}{7}$

18. $r = -\frac{5}{3}, d = -6; r = 2, d = 5$

A.6

1. $[-3, \infty)$

2. $(-\infty, -1) \cup (2, \infty)$

3. $(8, 17]$

4. $(-7, \infty)$

5. $(-\infty, -\frac{19}{4}]$

6. $(-\infty, 2] \cup [3, \infty)$

7. $(2 - \sqrt{11}, 2 + \sqrt{11})$

8. $(-6, -2) \cup (5, \infty)$

9. $(-4, -\frac{3}{2}) \cup (2 - \sqrt{6}, 2 + \sqrt{6})$

10. $(-\infty, -1) \cup (-\frac{1}{\sqrt{2}}, 0) \cup (\frac{1}{\sqrt{2}}, +\infty)$

A.7

1. $\frac{14}{3}, \frac{4}{3}$

2. 1, 7

3. $-8, 12$

4. $-\frac{1}{2}, \frac{11}{2}$

5. $(\frac{4}{3}, \frac{14}{3})$

6. $(-\infty, \frac{4}{3}] \cup [\frac{14}{3}, \infty)$

7. $(-\infty, -\frac{1}{2}) \cup (\frac{1}{2}, \infty)$

8. $(\frac{1}{3}, 9)$

A.8

1. $\sqrt{74} \approx 8.6$

2. 8

3. $5\sqrt{2} \approx 7.07$

4. $3\sqrt{17} + 5 \approx 17.4$

5. $(x-3)^2 - 9$

6. $2(x-1)^2 + 7$

7. $(x + \frac{1}{2})^2 + (y-4)^2 = 3^2$ or
$4x^2 + 4y^2 + 4x - 32y + 29 = 0$

8. circle: center $(-1,3)$, radius 4

9. not a circle, because the coefficients of x^2 and y^2 aren't equal

10. circle: center $(1,-1)$, radius $\sqrt{2}$

11. $\frac{(x+2)^2}{2^2} + \frac{(y-4)^2}{4^2} = 1$

12. The ellipse has center at $(1,-2)$, minor axis of length 6 parallel to the x-axis, and major axis of length 10 parallel to the y-axis.

$$\frac{(x-1)^2}{9} + \frac{(y+2)^2}{25} = 1$$

Trigonometry Appendix

SPECIAL VALUES ON THE UNIT CIRCLE

In Chapter 6, when we found unit-circle coordinates (sine and cosine values, that is), we were limited to cases in which the answer was either 0 or ± 1. In most other cases, either we have an exact value for t (say, 5) but we can find only approximate values for its sine and cosine, or we have exact coordinates (say, $-\frac{12}{13}$ and $\frac{5}{13}$) for a point on the unit circle but we have no way of finding an exact value for t. For those cases, you used a calculator and obtained numbers that were decimal approximations to the actual values. We can, however, find exact sine, cosine, and tangent values for certain special cases, as shown.

t	$\cos(t)$	$\sin(t)$	$\tan(t)$
0	1	0	0
$\frac{\pi}{6}$	$\frac{\sqrt{3}}{2}$	$\frac{1}{2}$	$\frac{1}{\sqrt{3}}$
$\frac{\pi}{4}$	$\frac{\sqrt{2}}{2}$	$\frac{\sqrt{2}}{2}$	1
$\frac{\pi}{3}$	$\frac{1}{2}$	$\frac{\sqrt{3}}{2}$	$\sqrt{3}$
$\frac{\pi}{2}$	0	1	—

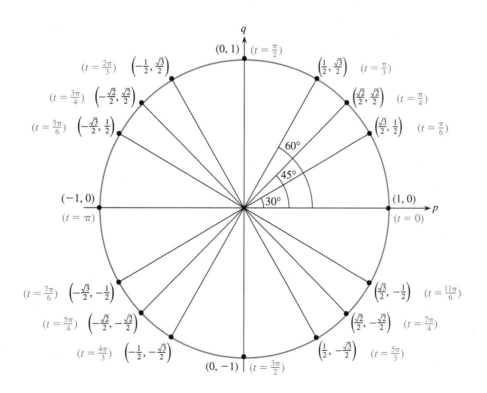

T.2 GRAPHS OF TRIG AND INVERSE TRIG FUNCTIONS

The two fundamental functions derived from the unit circle are the sine and the cosine. Those functions, expressed as ratios, produce four more: the tangent, the cotangent, the secant, and the cosecant. The last three are not nearly as useful in precalculus and calculus as the sine, the cosine, and the tangent are, but we present them here for the sake of completeness.

A. The Sine and the Cosine

For the definitions of these functions, refer to Section 6.2.

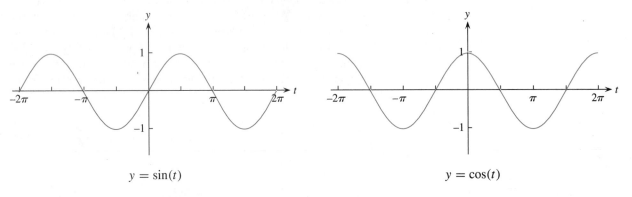

$$y = \sin(t) \qquad\qquad y = \cos(t)$$

The domain for both of these functions is $\mathbb{R}$, the set of all real numbers. Their range is the interval $[-1, 1]$.

B. The Tangent and the Cotangent

Definitions:

$$\tan(t) = \frac{\sin(t)}{\cos(t)} \qquad\qquad \cot(t) = \frac{\cos(t)}{\sin(t)}$$

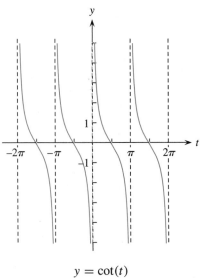

$$y = \tan(t) \qquad\qquad y = \cot(t)$$

Domain: reals except odd multiples of $\frac{\pi}{2}$
Range $(-\infty, \infty)$

Domain: reals except integer multiples of π
Range $(-\infty, \infty)$

C. The Secant and the Cosecant

Definitions:

$$\sec(t) = \frac{1}{\cos(t)} \qquad\qquad \csc(t) = \frac{1}{\sin(t)}$$

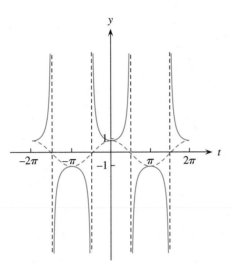

$y = \sec(t)$
($\cos(t)$ shown as a dashed curve)

Domain: reals except odd multiples of $\frac{\pi}{2}$
Range $(-\infty, -1] \cup [1, \infty)$

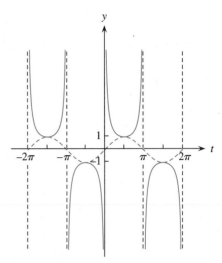

$y = \csc(t)$
($\sin(t)$ shown as a dashed curve)

Domain: reals except integer multiples of π
Range $(-\infty, -1] \cup [1, \infty)$

D. The Inverse Sine, Cosine, and Tangent

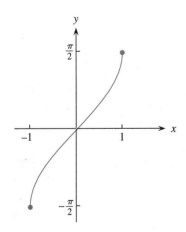

$y = \sin^{-1}(x)$

Domain $[-1, 1]$
Range $\left[-\frac{\pi}{2}, \frac{\pi}{2}\right]$

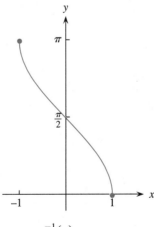

$y = \cos^{-1}(x)$

Domain $[-1, 1]$
Range $[0, \pi]$

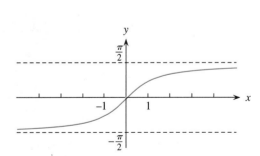

$y = \tan^{-1}(x)$

Domain $(-\infty, \infty)$
Range $\left(-\frac{\pi}{2}, \frac{\pi}{2}\right)$

T.3 DEFINITIONS AND HANDY IDENTITIES

From their definitions, we can see that all six trigonometric functions are interrelated. This section presents some frequently-used relationships. For derivations of these identities and practice in using them, consult any standard trigonometry text.

A. Defining Identities

$$\tan(t) = \frac{\sin(t)}{\cos(t)} \tag{1}$$

$$\cot(t) = \frac{\cos(t)}{\sin(t)} \tag{2}$$

$$\sec(t) = \frac{1}{\cos(t)} \tag{3}$$

$$\csc(t) = \frac{1}{\sin(t)} \tag{4}$$

Also, except for the fact that $\cot(t) = 0$ wherever $\tan(t)$ is undefined,

$$\cot(t) = \frac{1}{\tan(t)} \tag{5}$$

B. Pythagorean Identities

This one is a direct consequence of the way we define the sine and cosine functions on the unit circle:

$$\sin^2(t) + \cos^2(t) = 1 \tag{6}$$

These two are derived algebraically from (6) and from the defining identities:

$$\tan^2(t) + 1 = \sec^2(t) \tag{7}$$
$$1 + \cot^2(t) = \csc^2(t) \tag{8}$$

C. Opposite-Angle or Opposite-Arc Identities

$$\cos(-t) = \cos(t) \tag{9}$$
$$\sin(-t) = -\sin(t) \tag{10}$$
$$\tan(-t) = -\tan(t) \tag{11}$$

The cosine is an *even function*; its graph is symmetric about the y-axis. The sine and the tangent are *odd functions*; their graphs are symmetric about the origin.

D. Sum and Difference Identities

The trigonometric functions are not distributive. In other words, the sine of the sum of two numbers, $\sin(s + t)$, is not equivalent to the sum of their sines, $\sin(s) + \sin(t)$. It is possible, however, to find the sines and cosines of sums of two numbers in terms of functions of the numbers themselves.

$$\sin(s + t) = \sin(s)\cos(t) + \cos(s)\sin(t) \tag{12}$$

$$\sin(s - t) = \sin(s)\cos(t) - \cos(s)\sin(t) \tag{13}$$

$$\cos(s + t) = \cos(s)\cos(t) - \sin(s)\sin(t) \tag{14}$$

$$\cos(s - t) = \cos(s)\cos(t) + \sin(s)\sin(t) \tag{15}$$

$$\tan(s + t) = \frac{\tan(s) + \tan(t)}{1 - \tan(s)\tan(t)} \tag{16}$$

$$\tan(s - t) = \frac{\tan(s) - \tan(t)}{1 + \tan(s)\tan(t)} \tag{17}$$

Identities (12)–(15) are particularly important for calculus and linear algebra. We do not suggest that you memorize them, but rather that you realize that they exist and know where you can find them.

E. Double-Angle or Double-Arc Identities

These follow directly from the "sum" identities when s and t are equal:

$$\sin(2t) = 2\sin(t)\cos(t) \tag{18}$$

$$\cos(2t) = \cos^2(t) - \sin^2(t) \tag{19}$$

$$\tan(2t) = \frac{2\tan(t)}{1 - \tan^2(t)} \tag{20}$$

Identity (19) has two alternative versions that may come in handy:

$$\cos(2t) = 1 - 2\sin^2(t) \tag{21}$$

$$\cos(2t) = 2\cos^2(t) - 1 \tag{22}$$

F. Half-Angle or Half-Arc Identities

These are derived from (21) and (22) by substitution and algebraic manipulation. The "plus or minus" in the formulas doesn't mean that you can pick whichever sign you want or that both are true. The quadrant in which the arc $\frac{t}{2}$ terminates determines the sign to use.

$$\sin\left(\frac{t}{2}\right) = \pm\sqrt{\frac{1 - \cos(t)}{2}} \tag{23}$$

$$\cos\left(\frac{t}{2}\right) = \pm\sqrt{\frac{1 + \cos(t)}{2}} \tag{24}$$

The following versions of $\tan\left(\frac{t}{2}\right)$ come from substitution and algebraic manipulations of several of the previous identities:

$$\tan\left(\frac{t}{2}\right) = \frac{\sin(t)}{1 + \cos(t)} \tag{25}$$

$$\tan\left(\frac{t}{2}\right) = \frac{1 - \cos(t)}{\sin(t)} \tag{26}$$

G. The Law of Sines and the Law of Cosines

These two identities provide efficient methods for determining missing sides or angles in triangles that do not contain an angle of 90°.

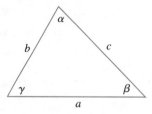

Let α, β, and γ be the three angles of a triangle, and let a, b, and c be the lengths of the sides opposite angles α, β, and γ, respectively, as illustrated.

The Law of Sines

$$\frac{\sin(\alpha)}{a} = \frac{\sin(\beta)}{b} = \frac{\sin(\gamma)}{c} \tag{27}$$

The Law of Cosines

$$c^2 = a^2 + b^2 - 2ab\cos(\gamma) \tag{28}$$

The Law of Cosines is a generalization of the Pythagorean Theorem to triangles that aren't right triangles. With the Law of Cosines, c can be the length of any side of the triangle; it need not be the length of the longest side.

Answers to Odd-Numbered Exercises

Chapter 1

1. The man's foot is longer. Look at Figure 1.3 on page 6: For any given shoe size, the $m(x)$ line lies to the right of the $w(x)$ line, indicating that for a given shoe size (y-value) the man's foot length (x-value) is greater. Or examine the formulas themselves: Because $m(x)$ subtracts a larger number than $w(x)$ does, the foot length would have to be greater to achieve the same output.

3. $10\frac{1}{2}$. If the input and the output are to have the same (unknown) value, we can assign them the same letter and solve an *equation*: $S = 3S - 21$. Notice that S in this case represents both shoe size and foot length, because they have the same numerical value.

5. $y = x^2 + 1$

7. $y = \frac{x}{3} + 1$

9. Answers will vary.

11. (a) $Flip(3) = \frac{1}{3}$, $Flip(\frac{2}{3}) = \frac{3}{2}$,
    $Flip(1000) = \frac{1}{1000} = 0.001$, $Flip(-5) = -\frac{1}{5}$,
    $Flip(0.01) = 100$, $Flip(n+2) = \frac{1}{n+2}$
    (b) $Flip(x) = \frac{1}{x}$
    (c) $Flip(x)$ is undefined for $x = 0$.

13. (a) $G(\frac{1}{2}) = \frac{7}{4}$, $G(-1) = -5$, $G(3) = 3$,
    $G(q^2) = 4q^2 - q^4$
    (b) 4 and 0
    (c) 4 and 0 (Same function; only the letter is changed.)

15. (a) $Quad(-5) = 100 + 21(-5) - (-5)^2 = -30$
    (b) $Quad(-x) = 100 + 21(-x) - (-x)^2$
    $= 100 - 21x - x^2$
    (c) $Quad(z+5) = 100 + 21(z+5) - (z+5)^2$
    $= 180 + 11z - z^2$
    (d) $Quad(z) + 5 = 100 + 21z - z^2 + 5$
    $= 105 + 21z - z^2$
    (e) $1 - Quad(t) = 1 - (100 + 21t - t^2)$
    $= -99 - 21t + t^2$
    (f) $Quad(1-t) = 100 + 21(1-t) - (1-t)^2$
    $= 120 - 19t - t^2$
    (g) $Quad(\frac{1}{x}) = 100 + 21\left(\frac{1}{x}\right) - \left(\frac{1}{x}\right)^2 = 100 + \frac{21}{x} - \frac{1}{x^2}$
    (h) $\frac{1}{Quad(x)} = \frac{1}{100 + 21x - x^2}$
    (i) $Quad(\heartsuit) = 0$ when $100 + 21\heartsuit - \heartsuit^2 = 0$. This is a quadratic *equation*; solve by factoring or with the Quadratic Formula. $\heartsuit = -4$ or 25.

17. In the abstract, the function $y = 3x - 21$ has all real numbers as its domain and range, whereas the shoe-size model $y = 3x - 21$ has a domain restricted to values between $8\frac{2}{3}$ and $11\frac{1}{3}$ and a range limited to numbers between 5 and 13. (You might go further, noting that the only meaningful outputs for the model are the numbers $5, 5\frac{1}{2}, 6, 6\frac{1}{2}, \ldots, 13$. You could also say that the abstract function gives an *exact* relationship between the variables y and x, whereas the shoe-size model gives approximate results.)

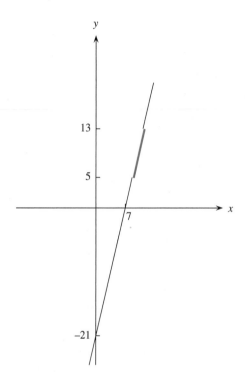

19. (a) $J(0) = 1$
    (b) There are no values for which $J(t) = 0$.
    (c) $J(t) < 0$ for $t < -2$ and for $t > 2$
    (d) Domain: the set of all real numbers except -2 and 2
    (e) Range: the set of all negative numbers and of all numbers greater than or equal to 1; $(-\infty,0) \cup [1,\infty)$

21. (a) From the graph of $x^2 + y^2 = 9$, we see that for every value of x in the interval $(-3,3)$, there are two values of y. A function should give only one.
    (b) Solve the equation for y: $y = \sqrt{9 - x^2}$ and $y = -\sqrt{9 - x^2}$ are the two functions.

23. (a) $Buzz(301) = $ buzz, because 7 divides 301; $Buzz(199) = 199$; $Buzz(71) = $ buzz, because 71 contains the digit 7
    (b) $Buzz(n) = $ buzz for $n = 28, 70, 72, 73, 77, 84, 107$, and infinitely many other integers
    (c) Domain: the set of natural numbers, $\{1, 2, 3, 4, \ldots\}$, a discrete, infinite set
    (d) Range: the discrete set consisting of the word "buzz" as well as all the positive integers that neither contain the digit 7 nor are divisible by 7

25. $\frac{Z(4)-Z(-3)}{4-(-3)} = \frac{20-(-20)}{7} = \frac{40}{7}$

27. $\frac{[5(0)-2(0)^2]-[5(-2)-2(-2)^2]}{0-(-2)} = \frac{0-(-18)}{2} = \frac{18}{2} = 9$, and

$\frac{[5(-2)-2(-2)^2]-[5(0)-2(0)^2]}{-2-0} = \frac{-18-0}{-2} = \frac{-18}{-2} = 9$

29. (a) Increasing on the interval $b < x < c$
   (b) Decreasing on the intervals $x < b$ and $x > c$
   (c) Concave up for all $x < 0$
   (d) Concave down for all $x > 0$

31.

Average Rate of Change	$y = x^2$	$y = x$	$y = \sqrt{x}$
0 to 0.25	0.25	1.00	2.00
0.25 to 0.75	1.00	1.00	0.73
0.75 to 1	1.75	1.00	0.54

(Results have two-place accuracy.)

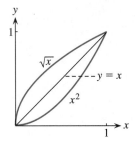

For $y = x^2$, the rate increases and the graph is concave up. For $y = x$, the rate is constant and the graph is a line. For $y = \sqrt{x}$, the rate decreases and the graph is concave down.

33. Sample response: The balance on a loan decreases with time. At first, the balance decreases slowly, because most of the payment covers interest charges. Near the end of the loan, most of the payment serves to reduce the balance. (Other possibilites include height versus horizontal distance for a ball tossed from a window, diameter versus time for a leaky balloon.)

35. Sample response: World population grew with increasing rapidity until the end of the twentieth century, when growth slowed and the population approached a steady state.

37. The independent variable is *driving speed*; the dependent variable is *time required*. An increase in speed means a decrease in time, so the function is decreasing. At low speeds, a small increase in speed saves a lot of time. At high speeds, however, a small increase in speed saves very little time. Therefore, the graph is concave up.

39. The independent variable is *time*, measured in years; the dependent variable is *number of incidents of road rage*. As time passes, the number of incidents rises, so the function is increasing. Each year's increase is greater (same percentage, but of a larger amount) than that of the previous year, so the graph is concave up.

41. The independent variable is *time*, measured in years. The dependent variables are *men's marathon record* and *women's marathon record*. On any time interval, the records either remain constant or drop, so both functions are decreasing, with some constant segments. Over the entire interval shown, the average rate of change of the women's graph is more negative than that of the men's graph. We do not have enough information to determine the curvature of either graph but, based on the statement given, we expect the graphs to intersect.

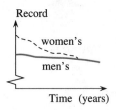

43. The independent variable is *latitude*, measured in degrees north or south of the equator. (It's handy to let the horizontal axis vary from 90°S (the South Pole) to 90°N. (the North Pole), with 0° (the equator) in the middle.) The dependent variable is *hours of daylight on June 21*. For all latitudes from the South Pole to the Antarctic Circle, the function is constant (at 0); for all latitudes from the Arctic Circle to the North Pole, the function is also constant (at 24). For all middle latitudes, the function is increasing: as latitude increases, the daylight hours increase. Without more information from astronomy, we do not know whether the graph for the middle latitudes should be straight or curved.

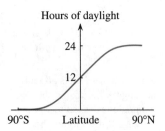

45. The independent variable is *number of hours worked in a week*. The dependent variable is *wages for the week*. Wages rise as hours increase, so the function is increasing. For the first 40 hours, the hourly rate is constant, producing a straight line segment. After 40 hours, the rate rises to a higher constant level, producing another line segment that is steeper than the first.

Wages

40 Hours

Chapter 2

1. (a) The car was moving away from the CBR, because its distance increased as t increased. It appears to be moving at a constant rate because it travels 1.584 feet every 1.8 seconds.
   (b) $d(t) = 2.500 + 0.88t$. Units for 2.500 are *feet*; units for 0.88 are *ft/sec*.

3. (a) $7.50; $157.50
   (b) $T(p) = 0.05p$
   (c) $C(p) = 1.05p$.
   (d) C is proportional to p; the constant of proportionality is 1.05.

5. (a) $R(x) = 50 + 12.5x$. The units associated with 50 are *beats per minute* and with 12.5 are *beats per minute* per *mile per hour*.
   (b) 162.5 beats per minute

7. $y = -3 - \frac{3}{5}x$

9. $s = -3 + t$

11. $r = -2$

13. (a) $ICE(n) = -150 + 0.65n$, where n is the number of cones sold.
   (b) $ICE(0) = -150$. On a day when no cones are sold, the business loses $150.
   (c) $ICE(100) = -85$; if 100 cones are sold, the stand loses $85. $ICE(500) = 175$; if 500 cones are sold, the stand makes a profit of $175.
   (d) 231 cones
   (e) Answers will vary. Sample: Assume that the stand is open for 9 hours and that there are two employees, each needing an average of 1 minute per cone. Then, on a very busy day, they could sell 1080 cones. Domain: the integers in the interval [0, 1080].
   (f) For the domain in (e), the range is those dollar figures from −$150 to $552, rising by 65-cent intervals.

15. Yes. Using any two ordered pairs, $\frac{\Delta y}{\Delta x} = 0.5$.

17. No. From 2 to 5, the rate of change is $-\frac{5}{3}$; from 5 to 12, it is $-\frac{10}{7}$.

19. (a) $F(x) = 5 + \frac{4}{3}x$
   (b) $y = -2 + \frac{4}{3}x$
   (c) $y = 5 - \frac{3}{4}x$

21. (a) A population that undergoes successive doublings in a fixed amount of time, such as 50 years, is growing exponentially.
   (b) $P(x) = 5.5(2^x)$, where P is the population in billions and x is the number of 50-year periods since 1993.
   (c) 100 years; 200 years
   (d) $2^{1/50} \approx 1.014$. This annual growth factor of 1.014 is close to, but slightly lower than the growth factor of 1.0174 given by the model in Section 2.2.
   (e) If this growth pattern were to continue, there would come a time when there would be so many people that a person could not move without bumping into another person. Long before that point, there would be no land left on which to produce food.

23. (a) 216 letters in round 3; 7776 letters in round 5
   (b) $L(n) = 6^n$, where L is the number of letters and n is the number of rounds, $n \geq 1$.
   (c) $L(11) = 362,797,056$, so 11 rounds are enough. At round 10, the sum of all the letters, $L(1) + L(2) + \cdots + L(10)$, is only 72,559,410.

25. (a) $f(t) = 38000(0.75^t)$
   (b) $f(12) \approx 1203.7$, or about 1200 cases
   (c) We have to distinguish between mathematical truth and real life. Mathematically, $f(t)$ is always greater than zero even though, for large values of t, its value is close enough to zero to be negligible. In actuality, with only a small number of cases remaining, public health efforts could be stepped up as they were with smallpox, overriding our mathematical model and eliminating polio.
   (d) $f(-3) = 90074$; approximately 90,000

27. (a) According to the model, there were approximately 31 deer in 1991 and 40 in 1992.
   (b) $P(t) = 24(1.29)^t$, where $P(t)$ gives the number of deer t years since 1990. The growth factor is 1.29.
   (c) The deer population would have reached 240 in a little over 9 years, that is, during 1999.

29. The growth factor is $\left(\frac{34}{12}\right)^{1/3} \approx 1.415$. $B(t) = 12(1.415)^t$ gives the number of bacteria after t hours.

31. $A(t) = 4 \cdot 3^t$

33. $y = 1800\left(\frac{1}{3}\right)^x$

35. $y = 13.2(1.071^x)$

37. The function could be linear; $y = 100 - 3(x - 5) = 115 - 3x$.

39. The function is neither linear nor exponential.

41. (a) $BL(t) = 398900 - 15956t$
   (b) $BD(t) = 398900(0.92)^t$ (Did you remember to *double* the percentage?)
   (c) in about $20\frac{1}{2}$ years (The graphs intersect at $t \approx 20.46$.)

43. $H(x) = 7 - (x - 12) = 19 - x$

45. (a) 83 cents; 83 cents

(b)

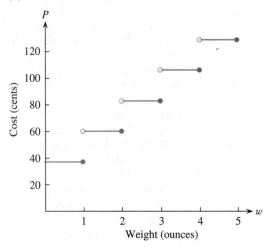

47. (a)

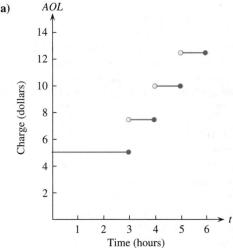

(b)

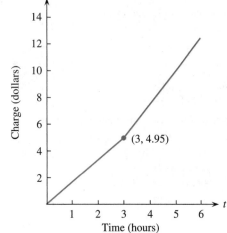

(c) Answers will vary. Sample answer: A user enrolled in the Light Usage Plan probably doesn't use the computer for more than an hour per day, on average. In this case, a reasonable domain would be $0 \le t \le 31$ hours. The corresponding range for the continuous model (b) is $4.95 \le AOL(t) \le 74.95$ dollars. For the step function (a), it is a discrete set of values: $\{4.95, 7.45, 9.95, \ldots, 74.95\}$.

49.

$$S(t) = \begin{cases} 10 + 3t & \text{if } -1 \le t < 0 \\ 10 & \text{if } 0 \le t < 2 \\ 10 - \frac{5}{2}(t-2) = 15 - \frac{5}{2}t & \text{if } t \ge 2 \end{cases}$$

51. (a)

$$w(h) = \begin{cases} 12h & \text{if } 0 \le h \le 40 \\ 480 + 18(h-40) = 18h - 240 & \text{if } h > 40 \end{cases}$$

(b) Sample answer: A reasonable domain is $0 \le h \le 80$ hours; the corresponding range is $0 \le w(h) \le 1200$ dollars.

(c)

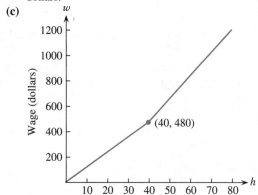

53. (a) 103

(b) The graph has its greatest positive slope during the third 10-day period, when the rate of spread is approximately seven persons per day.

(c) $N(t) = 20 + 3.5t$; $N(t) = 55 + 4.8(t - 10)$; $N(t) = 103 + 7.3(t - 20)$; $N(t) = 176 - 3.6(t - 30)$

(d) $N(t) = \begin{cases} 20 + 3.5t & \text{if } 0 \leq t < 10 \\ 55 + 4.8(t - 10) & \text{if } 10 \leq t < 20 \\ 103 + 7.3(t - 20) & \text{if } 20 \leq t < 30 \\ 176 - 3.6(t - 30) & \text{if } 30 \leq t \leq 40 \end{cases}$

(e) $N(17) = 88.6$, or approximately 89 people.
$N(35) = 158$.

55. (a)

$$IRS(t) = \begin{cases} 398900(0.92)^t & \text{if } 0 \leq t \leq 20.46 \\ 398900 - 15956t & \text{if } 20.46 < t \leq 25 \end{cases}$$

(b)

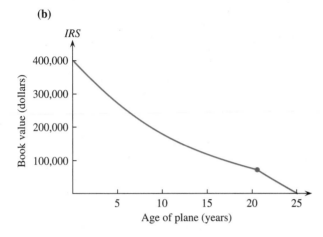

Chapter 3

1. (a) The $\sqrt{x}$ shape has been reflected in the y-axis and in the x-axis: $y = -\sqrt{-x}$.

(b) The family is x^2. The basic parabola was shifted right one unit, reflected in the x-axis, and then shifted up two units: $y = 2 - (x - 1)^2$.

(c) This is the basic sine graph, vertically stretched by a factor of 3 and reflected in the x-axis: $y = -3\sin(x)$ (or reflected in the y-axis—the effect is the same: $y = 3\sin(-x)$).

(d) This is the graph of an exponential function, a^x, that has been shifted one unit down: $y = a^x - 1$. The point $(1,1)$ allows us to solve an equation for the base a: $1 = a^1 - 1$ gives us $a = 2$. The function, then, is $y = 2^x - 1$.

(e) This is the x^3 graph, reflected in the x-axis (or the y-axis — we can't distinguish the effect here) and shifted up one unit: $y = 1 - x^3$.

(f) This is an upside-down x^2 that has been "flattened" by being compressed toward the x-axis: $y = -\frac{1}{36}x^2$. Or, it could have been stretched by a factor of 6 away from the y-axis: $y = -(\frac{1}{6}x)^2$. The expressions are equivalent.

3. (a) $\frac{1}{(x-3)^2}$

(b) $\frac{1}{x^2} + 3$

(c) $-\frac{1}{x^2}$

(d) $\frac{1}{(-x)^2} = \frac{1}{x^2}$

(e) $\frac{1}{3} \cdot \frac{1}{x^2}$

(f) $\frac{1}{(3x)^2} = \frac{1}{9x^2}$

5. (a) $F_1(d) = 1.12d$, $F_2(d) = 1.12(d - 3)$

(b) Shift the graph of F_1 three units to the right to obtain F_2.

(c) $F_2(d) = 1.12d - 3.36$: vertical shift 3.36 units down (3.36 fewer euros)

7. (a) $f \circ g(5) = f(25) = 24$ $f \circ g(x) = x^2 - 1$

(b) $g \circ f(5) = g(4) = 16$
$g \circ f(x) = g(x - 1) = (x - 1)^2$

(c) Yes, the order matters; the two compositions are not equivalent.

(d) The graph of $g \circ f$ is the graph of g shifted one unit to the right. The graph of $f \circ g$ is the graph of g shifted one unit down.

9. $kx^2 = (\sqrt{k}x)^2$. A horizontal compression by a factor of $\sqrt{k}$ gives the same graph.

11. (a) $S(t) = 28170(1.022)^t$, t measured in years since 1990.
$T(t) = 5987(0.982)^t$, t measured in years since 1995.

(b) If t measures *time in years since 1990*, shift Triangle's graph five years to the right: $T(t) = 5987(0.982)^{t-5} = 5987(0.982)^t(0.982)^{-5} \approx 6556(0.982)^t$.

13. (a)

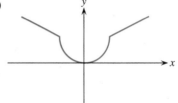

(b)

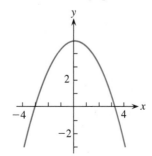

15. (a) Estimate the g-values; use them to estimate f-values. You should obtain, in order, $-3, 3, 5, 3, -3$.

(b) The ordered pairs are $(-4, -3)$, $(-2, 3)$, $(0, 5)$, $(2, 3)$, $(4, -3)$, and the graph is another relative of x^2.

17. The solutions given are not the only possibilities.

(a) If $f(x) = 25x + 7$ and $g(x) = x^{10}$, then $F(x) = g \circ f(x)$.

(b) If $f(x) = 10^x$ and $g(x) = \frac{x}{x^2 - 4}$, then $G(x) = g \circ f(x)$.

(c) If $g(x) = 1.05^x$ and $f(x) = -x$, then $H(x) = g \circ f(x)$.

(d) J is the *product* of two unrelated functions; it cannot conveniently be expressed as a composition. (As one precalculus student observed, "If there's more than one x and they're not all doing the same thing, it's probably not a composition.")

(e) If $f(x) = \sin(x)$, $g(x) = 1 - x^2$, and $h(x) = \sqrt{x}$, then $K(x) = h \circ g \circ f(x)$.

(f) Rather than a composition, L is a quotient of two unrelated functions.

19. (a) $g \circ f(x) = g(\sqrt{x}) = (\sqrt{x})^2 - 3 = x - 3$.

(b) The formula gives -4.

(c) The domain of f is the set of nonnegative real numbers, so something's wrong in part (b).

(d) The domain of $g \circ f$ is the set of nonnegative real numbers, so we should rewrite $g \circ f(x)$ as $x - 3$, $x \geq 0$.

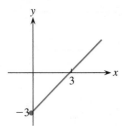

21. (a) $g(h(t)) = \sqrt{\frac{1}{t}}$; $h(g(t)) = \frac{1}{\sqrt{t}}$. These functions are algebraically equivalent, and they have the same graph.

(b) The composition is a reciprocal, so its shape is similar to that of $y = \frac{1}{t}$, but it has only positive numbers in its domain, so the graph has no left branch.

23. (a) $d^{-1}(4) = 1$. When the athlete completes 4 km, 1 hour has elapsed.

(b) $d(6)$ gives the distance traveled in the first 6 hours; $d^{-1}(6)$ tells how much time has elapsed when the athlete passes the 6-km mark.

25. $f^{-1}(x) = \frac{1}{x-3}$

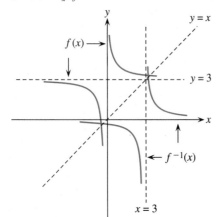

27. (a) $y = -\sqrt{x+2}$

(b) $y = \sqrt{-x+2} = \sqrt{2-x}$

(c) $y = x^2 - 2$, $x \geq 0$. (This is the inverse of $y = \sqrt{x+2}$.)

29. $k \circ h(x) = x$ and $h \circ k(x) = x$, so h and k are inverses.

31. Yes, $B(t)$ has an inverse; it is a strictly decreasing function and therefore is one-to-one. $B^{-1}(1000)$ means the age of the equipment when its value for tax purposes is \$1000. The graph of $B(t)$ shows that in a little more than 2 years, that value has shrunk to \$1000.

33. The reflected graph is a sideways parabola. It fails the vertical line test for a function. The two functions $y = \sqrt{x-3}$ and $y = -\sqrt{x-3}$ together produce this graph.

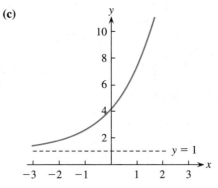

Chapter 4

1. (a) $f(0) = 4$, $f(-2) = \frac{7}{4}$, $f\left(\frac{3}{2}\right) = 6\sqrt{2} + 1$, $f\left(-\frac{1}{3}\right) = \frac{3}{\sqrt[3]{2}} + 1$

(b) Domain: $\mathbb{R}$; range: $f(x) > 1$

(c)

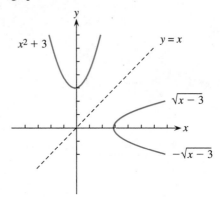

(d)

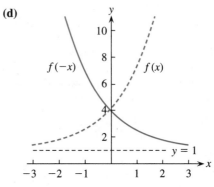

(e) $f(-x) = 3 \cdot 2^{-x} + 1 = 3\left(2^{-1}\right)^x + 1 = 3\left(\frac{1}{2}\right)^x + 1$

3. (a) The function is exponential: $f(x) = 2 \cdot 3^x$

   (b) The function is neither linear nor exponential. A linear or exponential function is either increasing or decreasing but not both. This function first decreases, then increases.

5. (a) $10^{3x}(x^2 - 1) = 0$; $x = \pm 1$

   (b) $x = \frac{1}{2}$

7. The graph shows an exponential function shifted one unit vertically: $y = a^x + 1$. When $x = 2$, $y = 5$. Substitute that information and solve for a: $y = 2^x + 1$.

9. This is the graph of an exponential function that has been stretched vertically by a factor of 2: $y = 2a^x$. When $x = -1$, $y = 1$. Substitute that information and solve for a: $y = 2(2^x)$, or $y = 2^{x+1}$, revealing that we can also consider the given graph to be a left shift of 2^x by one unit.

11. Exponential functions of the form a^x have all positive real numbers as their range, but the range of 1^x is $\{1\}$. Exponential functions have $y = 0$ as a horizontal asymptote, but 1^x has no horizontal asymptote. The graph of a^x is concave up, but the graph of 1^x is a horizontal line.

13. $H(t) = Ca^t + 54$

    $H(0) = Ca^0 + 54 = 160$

    $C = 106$

    $H(20) = 106a^{20} + 54$

    $a = \left(\frac{92-54}{106}\right)^{1/20} \approx 0.95$

    $H(t) = 106(0.95)^t + 54$

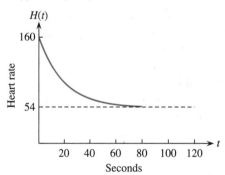

15. (a) Vertical stretch by a factor of 3

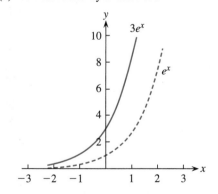

(b) Vertical shift, three units upward

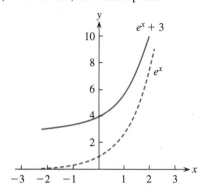

(c) Horizontal compression by a factor of 3

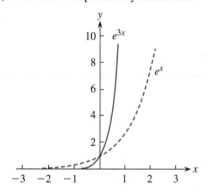

17. Sample answer: Choose 2, 0.5 and −2 as values for A. The table gives corresponding rates of change for the functions Ae^x at $x = 0$ and $x = 1$. (Values are rounded to three decimal places.)

	Slope at $x = 0$	Ae^0	Slope at $x = 1$	Ae^1
$A = 2$	2	2	5.437	5.437
$A = 0.5$	0.5	0.5	1.359	1.359
$A = -2$	−2	−2	−5.437	−5.437

The evidence suggests what is, in fact, true: that the rate of change of the function Ae^x at any particular value x_0 is Ae^{x_0}.

19. (a) $e^x(1 + x) = 0$; $x = -1$

    (b) $2xe^{-x^2}(1 - x^2) = 0$; $x = 0$ or $x = \pm 1$

    (c) $\left(\frac{e^x}{e^{2x}}\right)(3 - (3x - 1)) = 0$; $4 - 3x = 0$; $x = \frac{4}{3}$

21. Locate 0.05 at approximately -1.3 on the power scale, 0.5 at approximately -0.3, 50 at approximately 1.7, and 5000 at approximately 3.7.

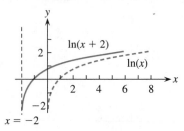

| | 0.05 | | 0.5 | | | 50 | | | 5000 | | | | | |

													x
$\frac{1}{100}$	$\frac{1}{10}$	1	10	100	1000	10,000	100,000	1,000,000	Arithmetic scale
-2	-1	0	1	2	3	4	5	6	Power scale

23. (a) 6 (b) $\frac{3}{2}$ (c) -3 (d) $\frac{1}{3}$ (e) 3.45

25. (a) The x-intercept for $\ln(x+2)$ is $x=-1$; the y-intercept is $y=\ln(2)$. The vertical asymptote is $x=-2$. Shift the graph of $\ln(x)$ two units to the left to obtain the graph of $\ln(x+2)$.

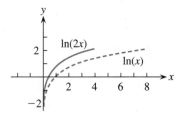

(b) The x-intercept for $\ln(2x)$ is $x=\frac{1}{2}$; there is no y-intercept. The vertical asymptote is $x=0$. Compress the graph of $\ln(x)$ horizontally (toward the y-axis) by a factor of 2 to obtain the graph of $\ln(2x)$.

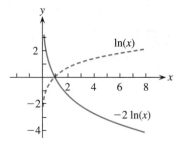

(c) The x-intercept for $-2\ln(x)$ is $x=1$; there is no y-intercept. The vertical asymptote is $x=0$. Stretch the graph of $\ln(x)$ vertically by a factor of 2 and reflect it in the x-axis to obtain the graph of $\ln(2x)$.

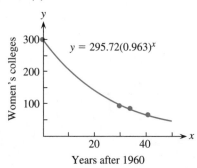

27. This is the graph of a logarithmic function shifted two units to the left: $y=\log_b(x+2)$. We don't know b. When $x=0$, $y=1$. Substitute that information into the formula, and find b: $y=\log_2(x+2)$.

29. This is the graph of a log function reflected in the x-axis: $y=-\log_b(x)$. When $x=5$, $y=-1$. Use that information to find b: $y=-\log_5(x)$.

31. $f(x)=\log_3(x)-2$

33. $f^{-1}(x)=\frac{1}{2}(e^x-1)$

35. $h^{-1}(x)=\ln\left(\frac{x}{10}\right)+1$

37. (a) $x=\frac{\ln(10)}{3}\approx 0.768$

(b) $x=\frac{\log_{10}(e)-1}{2}\approx -0.283$

(c) $e^3\approx 20.086$

(d) $\pm\sqrt{e}\approx \pm 1.649$

39. (a) $\ln\left(\frac{x}{e}\right)=\ln(x)-\ln(e)=\ln(x)-1$

(b) $\log_{10}(0.001x)=\log_{10}(0.001)+\log_{10}(x)=$ $-3+\log_{10}(x)$

(c) $e^{3\ln(x)}=e^{\ln(x^3)}=x^3$

41. (a) $\log_{10}(x^2)+\log_{10}((3x)^{1/2})=\log_{10}(\sqrt{3}x^{5/2})$

(b) $\ln(2t)-4\ln(t^{-1})=\ln(2t)+4\ln(t)=$ $\ln(2t)+\ln(t^4)=\ln(2t^5)$

(c) $\ln(400)+\ln(10^{0.95t})=\ln(400\cdot 10^{0.95t})$

43. (a) $x=\log_{10}(6)\approx 0.778$

(b) $x=\frac{1}{\log_{10}(6)}\approx 1.285$

(c) $x=\frac{\log_{10}(470)+3}{4}\approx 1.418$

(d) $x=\frac{\ln(5/8)}{\ln(3/2)}\approx -1.159$

(e) $x=\sqrt{e}\approx 1.649$

45. Population model: $p(t)=A(1.02)^t$. Find value of t: $A(1.02)^t=2A$; $t=\frac{\ln(2)}{\ln(1.02)}\approx 35$. The population doubles in 35 years, quadruples in 70 years.

47. (a) $x=\frac{\ln(37)}{\ln(5)}$

(b) $\ln(a^x)=\ln(N)$; $x\ln(a)=\ln(N)$; $x=\frac{\ln(N)}{\ln(a)}$. For the base-10 log version, replace $\ln$ with $\log_{10}$ in the calculation.

49. (a) $p=\log_a(M)\longrightarrow a^p=M$ and $q=\log_a(N)\longrightarrow a^q=N$

(b) $a^p a^q=MN$; $a^{p+q}=MN$

(c) $p+q=\log_a(MN)$; therefore, $\log_a(M)+\log_a(N)=\log_a(MN)$

51. (a) $a^q=N$

(b) $(a^q)^p=N^p$; $a^{pq}=N^p$

(c) $\log_a(N^p)=\log_a(a^{pq})=pq$

(d) $\log_a(N^p)=p\cdot\log_a(N)$

53. (a) and (b)

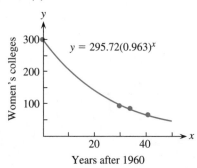

(c) Approximately 139

(d) 3.7 percent

(e) in 2013

55. $(0.95)^t = \frac{1}{2}$; $t \approx 13.5$ years

57. (a) $f(t) = A_0 e^{kt}$; $f(72) = A_0 e^{k \cdot 72} = \frac{1}{2}$;
$k = \frac{\ln(1/2)}{72} \approx -0.0096$; $f(t) = A_0 e^{-0.0096t}$

(b) $f(t) = A_0 (e^{-0.0096})^t \approx A_0 (0.99)^t$; 1% decay per year

59. $y = 2e^{3x}$

61. (a) $\ln(y) = 1 - x$ (b) $y = e^{1-x}$

63. (a)

x	$w(x)$	$w(x+1)/w(x)$
5	151	1.497
6	226	1.460
7	330	1.524
8	503	1.443
9	726	1.339
10	972	

(b) The point of inflection appears to be approximately (11,1248). AIDS in women was spreading most rapidly in 1993, at a rate of approximately 278 new cases per year.

(c) The graph levels off just below 2410.5. According to this model, the maximum number of cases in any one year is 2410.

65. (a) In late 1989 or early 1990, coverage was increasing at an annual rate of about 7.5%.

(b) By 2006, virtually every phone will have an answering machine.

Chapter 5

1. Reflect in t-axis; stretch vertically by a factor of 16; shift up 20 units.

3. $y = 4x^2$ (Stretch x^2 vertically to pass through (1,4) instead of (1, 1).)

5. $t = 2 - r^2$ (Reflect r^2 in r-axis; shift up two units; check the r-intercept.)

7. $q = (p + 1)(p - 2)$ (Create factors from the p-intercepts; check the q-intercept.)

9. $y = \frac{1}{2}(x - 2)^2 - 1$ (Shift x^2 right two units; compress vertically by a factor of 2; shift down one unit.)

11. (a) 27 units
(b) 36 units, at $t = 1$ second;
(c) $t = 3$
(d) Using $t = 2.999$ and $t = 3$, we obtain -35.991; approximately 36 units per second downward.

13. (a) A 37.5-foot square yields the greatest area, 1406.25 ft^2. A function to represent the area is $A(x) = x(75 - x) = 75x - x^2$. The axis of symmetry of that parabola is $x = 37.5$.
(b) The function is $A(x) = x\left(\frac{L}{2} - x\right) = \frac{L}{2}x - x^2$; its axis of symmetry is $x = \frac{L}{4}$. Therefore, the maximum area occurs when each side has the same length, $\frac{L}{4}$.

15. Sections required: $3W + 2L = 300$
Solve for L: $L = 150 - \frac{3}{2}W$
Total area: $A = LW = (150 - \frac{3}{2}W)W$
Area function: $A(W) = 150W - \frac{3}{2}W^2$
Vertex: (50,3750)
Therefore, $W = 50$ feet and $L = 150 - \frac{3}{2}W = 75$ feet gives the maximum area, 3750 ft^2.

17. (a) The graphs intersect where $t \approx 1.61$ and where $t \approx 3.72$. Between those two values, the graph of f lies below the graph of g. Therefore, $f(x) < g(x)$ on the interval $1.61 < x < 3.72$.
(b) $h(x) = 3x^2 - 16x + 18$ is also a quadratic function.
(c) $h(x) < 0$ (below the x-axis) on the interval $1.61 < x < 3.72$.
(d) Because $h(x) = f(x) - g(x)$, solving $h(x) < 0$ means solving $f(x) - g(x) < 0$, or $f(x) < g(x)$, which was the original problem.

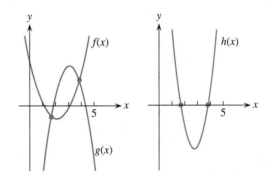

19. (a) Could be. If $a_4 > 0$, the graph will point up on both sides, as this one does. This graph shows one turning point; a degree-4 polynomial has either one turning point or three.
(b) No. The graph of an even-degree polynomial either goes up on both sides or goes down on both sides. This graph could be a polynomial of odd degree.
(c) Could be. The graph goes down on both sides, implying that $a_4 < 0$; there are four zeros and three turning points, consistent with a polynomial of degree 4.
(d) No. For $x > 0$, this graph appears to approach the x-axis and level off; it cannot be the graph of a polynomial function.

21. The sharp "corner" at the origin says that $|x|$ is not a polynomial.

23. Graph goes down on the left, up on the right. Zeros are $0, \pm\frac{1}{\sqrt{2}}$.

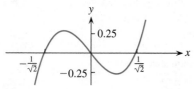

25. Graph goes up on the left, down on the right. Zeros are $-\frac{3}{2}, 0, \frac{1}{2}$.

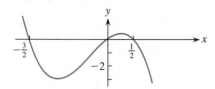

27. **(a)** $p(-1) = 9(-1)^3 + 5(-1)^2 - 6(-1) - 2$
$\qquad = -9 + 5 + 6 - 2 = 0$

(b) $p(x) = (x + 1)(9x^2 - 4x - 2)$

(c) Zeros of $9x^2 - 4x - 2$ are $\frac{2\pm\sqrt{22}}{9} \approx -0.30, 0.74$. Those, together with -1 are the zeros of $p(x)$.

(d)

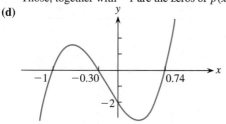

29. $y = (x - 1)^3 + 1$

31. $y = \sqrt{x + 2}$ (not a polynomial!)

33. **(a)**

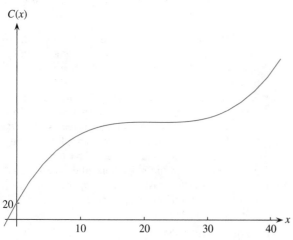

(b) The fixed cost is given by $C(0)$: $20.

(c) By zooming in on the graph of $C(x)$, we see that the function actually decreases slightly from $x = 20$ to $x \approx 23$ before it rises again. $C(20)$ and $C(25)$ are equal: $120. This model says that the cost of making 21, 22, 23, or 24 pots is *less* than the cost of making 20.

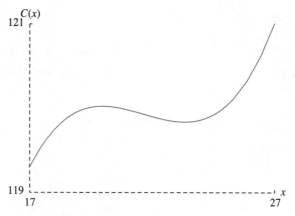

(d) $10.94, $3.66, −$0.04, $8.42

(e) As x increases from 0 to 20, the rate of change in C is positive but decreasing. At $x = 20$, the rate is zero. Between $x = 20$ and $x = 23$, the rate is slightly negative, but it goes back to zero again just after $x = 23$. From $x = 24$ on, the rate is positive and increasing.

35. **(a)** at $t = 15$, $-1.625 \frac{\text{miles per hour}}{\text{second}}$; at $t = 26$, $0.52 \frac{\text{miles per hour}}{\text{second}}$.

(b) The first rate is negative, because at $t = 15$, the car's speed is decreasing. The other is positive, because at $t = 26$, the car's speed is increasing.

(c) The speed was changing more rapidly at $t = 15$; $|-1.625| > |0.52|$. The graph of $s(t)$ is steeper at $t = 15$ than at $t = 26$.

37. **(a)**

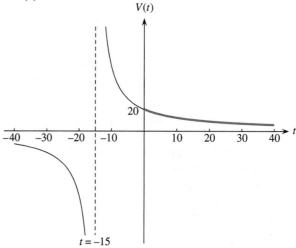

(b) The domain of the model is a continuous set, because time and velocity are continuous.

(c) The engine was cut off at $t = 0$; $v(0) = 20$ feet per second.

(d) Solve the equation $\frac{300}{15+t} = 10$; $t = 15$ seconds. On the graph, we see that when $v(t) = 10$, $t = 15$.

(e) At $t = 5$, approximately -0.75 feet per second per second; at $t = 10$, approximately -0.48 feet per second per second. (Negative acceleration means that the velocity is decreasing; the boat is slowing down.) From $t = 5$ to $t = 10$, the magnitude, or absolute value, of the acceleration decreased.

39.

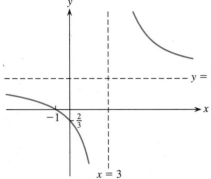

41.

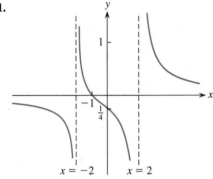

43. $y = \frac{x+2}{x+1}$. Relative of $\frac{1}{x}$; x-intercept at -2 puts $(x + 2)$ into numerator; vertical asymptote at -1 puts $(x + 1)$ into denominator. An algebraically equivalent version, $y = \frac{1}{x+1} + 1$, shifts the graph of $\frac{1}{x}$ left 1 and up 1.

45. $y = \frac{1}{2}(x + 2)(x + 1)(x - 1)(x - 2)$. Graph shows a polynomial function of even degree with four zeros. Create the polynomial from its zeros; then check the y-intercept. Multiply the whole function by $\frac{1}{2}$ so that the y-intercept is correct.

47. $y = \frac{x^2-4}{x^2-2}$. Vertical asymptotes at $\pm\sqrt{2}$ put $(x + \sqrt{2})(x - \sqrt{2})$ into the denominator. Zeros at ± 2 put $(x + 2)(x - 2)$ into the numerator. Check the y-intercept algebraically. Check the horizontal asymptote algebraically.

49. The domain of g is $1 \leq r \leq 59.7$, where r is the distance in earth-radii from the center of the earth. The spaceship position varies from the earth's surface ($r = 1$) to the moon's surface ($r = 59.7$).

Chapter 6

1. $\cos(-2468\pi) = 1$, $\sin(-2468\pi) = 0$, $\cos\left(\frac{5\pi}{2}\right) = 0$, $\sin\left(\frac{5\pi}{2}\right) = 1$

3. (a)

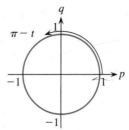

Halfway around, then back counterclockwise t units

(b)

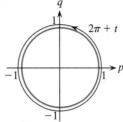

One complete revolution plus an arc of length t

(c)

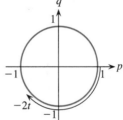

Twice as far as arc t in the counterclockwise direction

5. In degree mode, $\sin(90) = 1$, $\cos(90) = 0$. In radian mode, $\sin(90) \approx 0.89$, $\cos(90) \approx -0.45$. Ninety degrees is one-quarter of the way around the unit circle, at the point $(0,1)$; hence, the cosine and sine values are 0 and 1, respectively. An arc of length 90 radians wraps around the circle approximately $14\frac{1}{3}$ times, ending in quadrant II, where the sine value is positive and the cosine value is negative.

7. (a) The hypotenuse is one unit long because it is a radius of the unit circle.

(b) Angle α contains $45°$, because $\frac{\pi}{4}$ is one eighth of the entire circle, which contains $360°$. Because the sum of the angles of a triangle is $180°$, angle β also contains $45°$.

(c) This is an isosceles triangle because the base angles are equal. Therefore the legs of the right triangle are of equal length.

(d) Solving $x^2 + x^2 = 1$, we obtain $x^2 = \frac{1}{2}$ or (because the lengths are positive) $x = \frac{1}{\sqrt{2}}$.

(e) Each leg of the triangle has length $x = \frac{1}{\sqrt{2}}$. But the horizontal length is the cosine value for $\frac{\pi}{4}$, and the vertical length is the sine value. Therefore, $\cos\left(\frac{\pi}{4}\right) = \frac{1}{\sqrt{2}}$, and $\sin\left(\frac{\pi}{4}\right) = \frac{1}{\sqrt{2}}$.

9. (a) $OA = OB = 1$, as both are radii of the unit circle.

(b) The arc representing $\frac{\pi}{3}$ is one sixth of the unit circle, or $60°$, leaving $30°$ remaining in the first quadrant. So $\angle AOC$ measures $30°$; by symmetry, so does $\angle BOC$. Their sum, $\angle BOA$, measures $60°$.

(c) Because $OA = OB$, $\triangle AOB$ is isosceles, and angles CAO and CBO are congruent. Their sum is $120°$ (because $\angle BOA$ measures $60°$), so they each measure $60°$. A triangle with three equal angles is equilateral: $AB = OA = OB = 1$.

(d) AC is half of AB, so $AC = \frac{1}{2}$.

(e) Apply the Pythagorean theorem to $\triangle AOB$ to find OC: $(OC)^2 + \left(\frac{1}{2}\right)^2 = 1$; $OC = \frac{\sqrt{3}}{2}$.

(f) AC and OC give the coordinates of point A: $\left(\frac{1}{2}, \frac{\sqrt{3}}{2}\right)$.

Therefore, $\cos\left(\frac{\pi}{3}\right) = \frac{1}{2}$ and $\sin\left(\frac{\pi}{3}\right) = \frac{\sqrt{3}}{2}$.

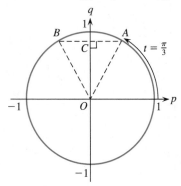

11. $\tan(-t) = \frac{\sin(-t)}{\cos(-t)} = \frac{-\sin(t)}{\cos(t)} = -\tan(t)$

13. $t = \pi + 2n\pi$

15. $t = \frac{\pi}{4} + n\pi$

17. Because the arc of length t terminates in the second quadrant, $\cos(t)$ and $\tan(t)$ are both negative; $\cos(t) = -\frac{\sqrt{7}}{4}$, $\tan(t) = -\frac{3}{\sqrt{7}}$.

19. Because the terminal point of the arc lands in the fourth quadrant, $\cos(t)$ and $\sec(t)$ are positive, while $\tan(t)$, $\csc(t)$, and $\cot(t)$ are negative; $\cos(t) = \frac{\sqrt{5}}{3}$, $\tan(t) = -\frac{2}{\sqrt{5}}$, $\sec(t) = \frac{3}{\sqrt{5}}$, $\csc(t) = -\frac{3}{2}$, $\cot(t) = -\frac{\sqrt{5}}{2}$.

21. (a) (a,b) and (d,e)
(b) $(a,0)$ and (c,e)
(c) (b,d)

23. (a) $g \circ f(x) = \cos^2(x)$, which means $(\cos(x))^2$
(b) $C(x) = f(g(h(x)))$, where $f(x) = x^2$, $g(x) = \sin(x)$, and $h(x) = 3^x$.

25. (a)

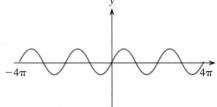

Graph of $y = \sin(t)$

(b)

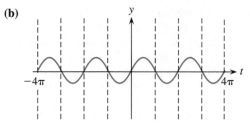

Vertical asymptotes where $\sin(t) = 0$

(c)

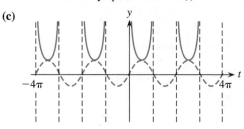

Positive portions of $\csc(t)$

(d)

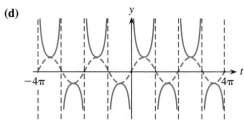

Complete graph of $\csc(t)$

(e) $\csc(t) = \frac{1}{\sin(t)}$; since the numerator of this fraction is never zero, the value of $\csc(t)$ is never zero.

(f) The graph of $\csc(t)$ has odd symmetry (symmetry about the origin).

(g) Domain: $t \neq n\pi$; range: $y \geq 1$ or $y \leq -1$

27. (a) $t = \frac{\pi}{4}$
(b) Approximately 0.707
(c) Approximately -0.707
(d) At $t = \frac{\pi}{4}$, the graphs appear equally steep but slanting in opposite directions, the sine function increasing while the cosine function decreases. The slopes are equal in magnitude but opposite in sign.

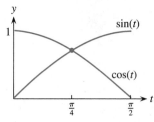

29. The identity $\sin^2(t) + \cos^2(t) = 1$ holds for all values of t, including algebraic expressions involving another variable. In the first case, think of the expression $x^3 - 4$ as representing t; in the second case, t is the expression $\sqrt[3]{x}$. Both functions are equivalent to the constant function $y = 1$.

31. $y = 10 \sin(8t) - 5$; y-intercept: -5

33. $y = \sin\left[\frac{\pi}{3}(t + 2)\right]$; y-intercept: 0.87

35. Solid curve: $\sin(2x)$; dashed curve: $2\sin(x)$; dotted curve: $\sin(\frac{1}{2}x)$; $a = \frac{\pi}{2}$; $b = \frac{3\pi}{2}$; $c = 4\pi$

37. **(a)** Sample answer: The pedals are revolving 60 times per minute (once per second). If time is measured in seconds, the period is one second. Assume that at $t = 0$, the reflector is 11 inches above the road.

(b) $h(t) = 7\sin(2\pi t) + 11$

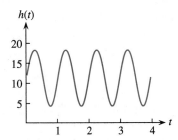

39. **(a)** $T = 2\pi\sqrt{\frac{39.375 \text{ in.}}{384 \text{ in./sec}^2}} \approx 2\pi\sqrt{0.102539 \text{ sec}^2} \approx 2.0$ seconds

(b) Sample answer: If the pendulum is closest to the left side of the case when $t = 0$, then its distance (in.) from the left side as a function of time (seconds) is $d(t) = 2.25\sin[\pi(t - 0.5)] + 6$ or, equivalently, $-2.25\cos(\pi t) + 6$.

(c) On the basis of the model in part (b), the pendulum is moving most rapidly away from the left side when it passes through the center, at $t = 0.5$ seconds. Its approximate speed at that moment is

$$\frac{d(0.5 + 0.01) - d(0.5 - 0.01)}{0.02} \approx 7 \text{ in./sec}$$

Every time the pendulum reaches the center of its swing, it moves at approximately 7 in./sec.

41. $y = 4\cos\left(\frac{1}{3}t\right)$ is one possible formula.

43. $y = 2\sin\left(t - \frac{\pi}{4}\right)$ is one possible formula.

45. $y = 6\cos\left(\frac{\pi}{12}t\right) + 6$ is one possible formula.

47. $\frac{\tan(0.01) - \tan(-0.01)}{0.02} \approx 1$, equal to the slope of the sine function at $t = 0$

49. Graph of $3\cos\left(3t + \frac{\pi}{4}\right) = 3\cos\left[3\left(t + \frac{\pi}{12}\right)\right]$

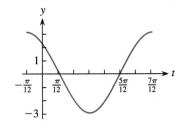

51. One possibility is $y = \cos\left(t - \frac{\pi}{2}\right)$.

53. From a calculator, $\cos^{-1}(0.44) \approx 1.115$. The unit circle shows that $t \approx -1.115$ is another solution. All solutions, correct to three decimal places, are given by $t = \pm 1.115 + 2n\pi$.

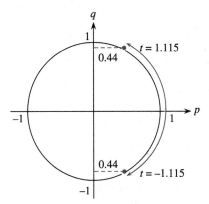

55. $t = \pm\frac{1.231}{4} + \frac{2n\pi}{4} = \pm 0.308 + \frac{n\pi}{2}$

57. $t = 0.554 + \frac{n\pi}{2}$

59. True for all values of t. (The statement is an identity.)

61. $t = 1.373 + n\pi$

63. Solve the equation $1 - \sin\left(\frac{\pi}{2}t\right) = 1.25$:

$$t = \begin{cases} -0.161 + 4n \\ 2.161 + 4n \end{cases}$$

On $-4 \le t \le 4$, there are four solutions: -1.839, -0.161, 2.161, and 3.839.

65. A possible formula is $f(t) = 4\cos\left(\frac{\pi}{5}t\right) + 1$. Solve the equation $4\cos\left(\frac{\pi}{5}t\right) + 1 = 0$. The t-intercepts pictured are $a \approx 2.902$ and $b \approx 7.098$.

67. **(a)** At midnight, the tip of the hour hand is farthest from the floor. This fact suggests a cosine model: $H(t) = 4\cos\left(\frac{\pi}{6}t\right) + 86$.

(b) $H(4) = 4\cos\left(\frac{2\pi}{3}\right) + 86 = 84$ inches

(c) 3:00 or 9:00

(d) 2:31 or 9:29

69. Graphs of $y = \cos(t)$ and $y = t$ indicate exactly one solution: $t \approx 0.74$.

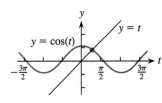

Chapter 7

1. $AC = \frac{14.1}{\cos(22°)} \approx 15.2$; $BC = 14.2 \tan(22°) \approx 5.7$;
$\angle C = 90° - 22° = 68°$

Check: $14.1^2 + 5.7^2 \approx 15.2^2$

3. (a)

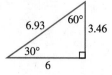

(b) Not possible to solve without one side length.

(c)

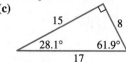

5. (a) Approximately 17 feet.
(b) Answers will vary. Allowing 4 feet of exit space at the end, the ramp can be approximately 13.4 feet.
(c) For the ramp in part (b), the rise-to-run ratio is $\frac{1}{2}$, and its angle of elevation is approximately 46°.
(d) Answers will vary. A switchback would allow a longer ramp.

7. (a) $\theta = \tan^{-1}(\frac{6}{12}) \approx 26.57°$; $a = \frac{8}{\cos(26.57°)} \approx 8.94$ in.
(b) By similar triangles, the "rise" for the small triangle is 4 inches. By the Pythagorean theorem, $a^2 = 4^2 + 8^2$, $a \approx 8.94$ inches.

9. (a) 30 feet
(b) 60 feet

11. Sample answers shown.

(a)

(b)

(c)

For part (c), we distorted the two similar rectangles from part (a) into two nonsimilar parallelograms.

13. 4528 feet of vertical distance

15. $\angle B = 72°$; $AB \approx 118.6$; $BC \approx 171.1$

17. $HI \approx 21.9$; $\angle H \approx 20.4°$; $\angle I \approx 14.6°$

19. For $\triangle M_1 NO$, $\angle N \approx 63.2°$, $M_1 O \approx 11.7$
For $\triangle M_2 NO$, $\angle N \approx 16.8°$, $M_2 O \approx 3.8$

21. By the Law of Cosines, $\angle S = \cos^{-1}\left(\frac{7^2 - 13^2 - 19^2}{-2(13)(19)}\right) \approx 13.2°$.
Similarly, $\angle T \approx 141.8°$, $\angle U \approx 25°$.

23. $\angle DAB = 180° - 53.5° = 126.5°$
$\angle BDA = 180° - 28° - 126.5° = 25.5°$
$AD = \frac{1000 \sin(28°)}{\sin(25.5°)} \approx 1090.5$ ft.
$h = 1095.5 \sin(53.5°) \approx 877$ feet, as before.

25. 200 feet

27. (a) 120°   **(b)** −225°   **(c)** 338°

29. (a) $\frac{\pi}{6}$ **(b)** 6   **(c)** $-\frac{\pi}{15}$

31. (a) $\frac{3\pi}{2}$ units; $\frac{\pi}{4}$ radians
(b) At the point $(-3,0)$; $\frac{1}{2}$; $\frac{\pi}{2}$
(c) $\frac{1}{3}$; $\frac{2\pi}{3}$

33. Exercise 32 provides a hint. Dad is talking about two different kinds of velocity. A point on the edge of the record *is* traveling faster (in inches per minute) than a point on the label, but the two points have the same *angular* velocity (revolutions per minute).

35. (a) 85°, in Quadrant IV
(b) 20°, in Quadrant I
(c) 32°, in Quadrant III

37. −40° and 320° (or −40° plus any integer multiple of 360°)

39. 165° and −195° (or 165° plus any integer multiple of 360°)

Chapter 8

1. (a) $A(125,64) = 15.63(125)^{0.425}(64)^{0.725} \approx 2481$. A 125-pound person who is 64 inches tall has approximately 2481 square inches of skin.
(b) $A(185,68) \approx 3062$ square inches
(c) The domain could be $100 \le w \le 300$ pounds, $58 \le h \le 80$ inches. The corresponding range would be $2101 \le A \le 4231$ square inches. (Answers will vary.)

3. We need the minus sign because California Street is south of Clay Street, our x-axis.

$$d(x,-0.12) = \sqrt{x^2 + (-0.12)^2} = \sqrt{x^2 + 0.0144}$$

(Distances d and x are measured in miles.)

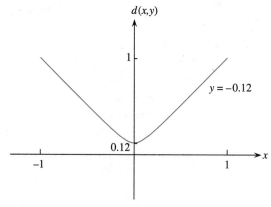

5. We compute the rate of change of $d(x,0,8) = \sqrt{x^2 + 0.8^2}$ when $x = 0.2$ and when $x = 0.7$.
(a) Approximately 0.24 mile (of total distance) per mile (of change in the x-direction)
(b) Approximately 0.66 mile (of total distance) per mile (of change in the x-direction)

7. $D(x_1, y_1, x_2, y_2) = \sqrt{(x_1 - x_2)^2 + (y_1 - y_2)^2}$

9.

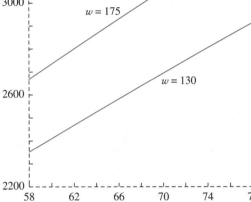

(a) For 130-pound people, body surface increases as height increases. We know this by the positive slope of the $w = 130$ graph.

(b) For a fixed height, surface area increases or decreases when w changes from 130 to 175. We know this because the $w = 175$ graph is higher on the coordinate plane than the $w = 130$ graph.

(c) The formula for A is exponential in both of its variables, w and h. The graph of an exponential function is curved, even though the bending may be so slight as to be imperceptible to our eyes.

(b)

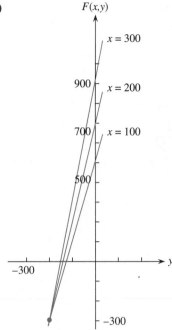

$G(100, y) = 500 + 4y$
$G(100, y) = 700 + 5y$
$G(100, y) = 900 + 6y$
Three lines meet at the point $(-200, -300)$.

(c)

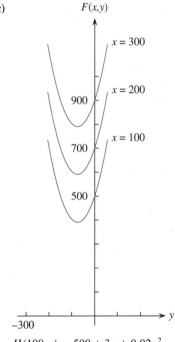

$H(100, y) = 500 + 3y + 0.02y^2$
$H(100, y) = 700 + 3y + 0.02y^2$
$H(100, y) = 900 + 3y + 0.02y^2$
The graphs are parabolas with axis of symmetry $x = -75$ and 200 vertical units apart.

11. (a)

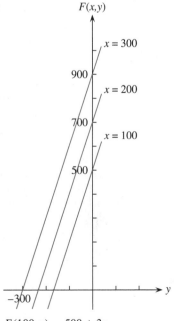

$F(100, y) = 500 + 3y$
$F(100, y) = 700 + 3y$
$F(100, y) = 900 + 3y$
The graphs are equally spaced parallel lines.

13. (a) Polk Street provides an easier walking route than Jones Street, because Polk crosses fewer level curves, indicating smaller changes in elevation.

(b) Jones Street would be longer, because of the additional vertical distance that must be covered.

15. (a) $H(0,0) = 36$; $H(2,4) = 16$

(b)

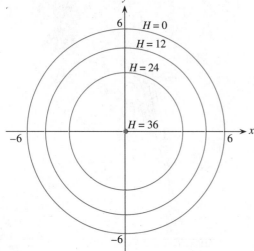

17. (a)

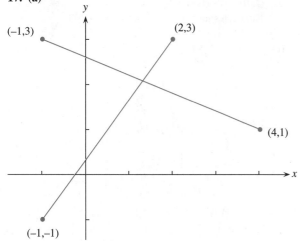

(b) $y = \frac{4}{3}x + \frac{1}{3}$ and $y = -\frac{2}{5}x + \frac{13}{5}$

(c) $\left(\frac{17}{13}, \frac{27}{13}\right)$

(d) Object A reaches the point $\left(\frac{17}{13}, \frac{27}{13}\right)$ at $t = \frac{10}{13}$. Object B reaches the same point earlier, at $t = \frac{6}{13}$. The objects do not collide.

19. (a) Both dots appear to move along the same path. They move at different rates, however, because their positions for t-values other than 0 and 1 are not the same.

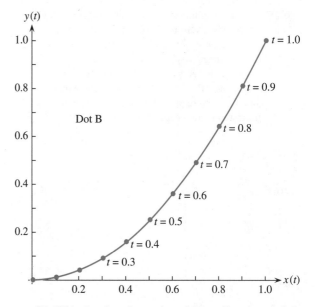

(b) Widening the t-interval would not affect the graph for dot A, whose path would trace a portion of the parabola $y = x^2$ backward from the point $(1,1)$ to the origin as t goes from -1 to 0, and retrace that path from the origin to $(1,1)$ as t goes from 0 to 1. But it would cause dot B to trace out more of the parabola $y = x^2$, going from the point $(-1,1)$ through the origin to the point $(1,1)$ as t goes from -1 to 1.

21. (a) Curved; concave up, because $\frac{\Delta y}{\Delta x}$ increases as t increases.

(b) Linear; slope is 5.

(c) Curved; concave down, because $\frac{\Delta y}{\Delta x}$ decreases as t increases.

(d) Linear; slope is 3.2.

(e) Curved; concave down, because $\frac{\Delta y}{\Delta x}$ decreases as t increases.

23. (a) $1 - y^2 = 1 - \sin^2(t)$
$= \cos^2(t)$
$= x$

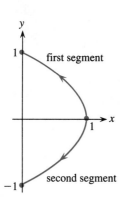

first segment

second segment

(b) $1 - y^2 = 1 - (\ln(t))^2$
$= x$

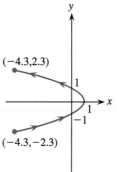

$(-4.3, 2.3)$

$(-4.3, -2.3)$

(c) $1 - y^2 = 1 - (e^{-t})^2$
$= 1 - e^{-2t}$
$= x$

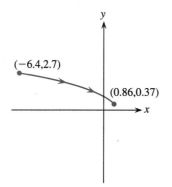

$(-6.4, 2.7)$

$(0.86, 0.37)$

25. (a) Figure the two circumferences: In one revolution, the inner point travels 2π inches while the outer point travels 8π inches. So at 45 revolutions per minute, the inner point travels at 90π inches per minute while the outer point travels at 360π inches per minute. The outer point moves four times as fast.

(b) The respective amplitudes will be 1 and 4, but the equations must have the same period, because each point makes a complete revolution in $\frac{1}{45}$ minute. Therefore, t minutes corresponds to $45 \cdot 2\pi$ radians. The equations are $x(t) = \cos(90\pi t)$, $y(t) = \sin(90\pi t)$ for the inner point and $x(t) = 4\cos(90\pi t)$, $y(t) = 4\sin(90\pi t)$ for the outer point.

27. (a) $x(t) = 2\cos(t) + \cos(5.5t)$
$y(t) = 2\sin(t) + \sin(5.5t)$

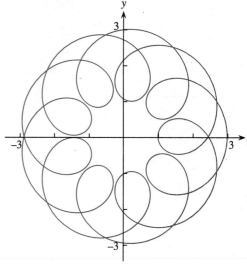

(b) $x(t) = 2\cos(t) + \cos(5.5t)$
$y(t) = 2\sin(t) + \sin(5.5t)$

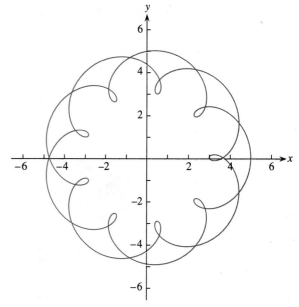

(c) Enlarging the radius of the inner circle makes the entire pattern larger (10 units in diameter rather than 6) and makes the inner loops much smaller.

Index

Fundamental Classes of Functions

- Constant: $f(x) = b$; b any constant
- Linear: $f(x) = b + mx$; b and m constants
- Exponential: $f(x) = C \cdot a^x$; C and a constants, $C \neq 0$, $a > 0$, $a \neq 1$
- Logarithmic: $f(x) = \log_a(x)$; a constant, $a > 0$, $a \neq 1$
- Quadratic: $f(x) = ax^2 + bx + c$; a, b, c constants, $a \neq 0$
- Polynomial, of degree n: $f(x) = a_n x^n + a_{n-1} x^{n-1} + \cdots + a_1 x + a_0$; a_i constant, $a_n \neq 0$, and n a nonnegative integer
- Rational: $f(x) = \frac{p(x)}{q(x)}$; p and q polynomials, $q(x)$ not the zero function
- Sinusoidal: $f(x) = A \sin(B(x + C)) + D$ or $f(x) = A \cos(B(x + C)) + D$; A, B, C, D constants, $A \neq 0$

A Brief Glossary

Abstract function: a purely mathematical functional relationship, devoid of context

Amplitude: for a sinusoid, half the difference between the highest value and the lowest

Asymptote, horizontal: a horizontal line that the graph of a function, in the long run, approaches

Asymptote, vertical: a vertical line $x = a$ that the graph of a function begins to resemble, as the input values approach the number a

Base: for an exponential function, the growth factor

Circular function: one of the six functions (sine, cosine, tangent, secant, cosecant, and cotangent) defined in terms of the unit circle

Composition: performing two or more functions in sequence, the output of one becoming the input to the next

Concave up: characterized by increasing rates of change (the curve "holds water")

Concave down: characterized by decreasing rates of change (the curve "spills water")

Continuous variable: one whose values can change without interruption or gap

Contour: for a two-input function, a curve along which the output value is constant

Decreasing: becoming less as the independent variable increases

Dependent variable: the output of a function, usually represented on the vertical axis of a graph

Discrete variable: one whose values are separated from each other by a measurable amount; not continuous

Domain: the set of allowable inputs to a function

Even Symmetry: symmetry about the y-axis; $f(-x) = f(x)$

Function: a relationship between the domain and the range, such that each element of the domain determines one element of the range

Growth factor: in an exponential function, the amount by which the initial value is multiplied for each one-unit increase in the input

Increasing: becoming greater as the independent variable increases

Independent variable: the input to a function, usually represented on the horizontal axis of a graph

Inflection point: a point on a graph at which the concavity changes

Input variable: the independent variable in a function

Intercept: a point at which a graph touches or crosses an axis

Inverse function of a function f: the function that undoes the operation(s) performed by f

Level curve: same as *contour*

Logarithm: an exponent, when considered as the output of a function; the power to which a fixed positive number must be raised to equal the input variable

Logistic: exponential growth in a restricted environment

Mathematical model: a function when used to represent or approximate an actual situation

Multivariable function: one whose value depends upon two or more input variables

Natural exponential function: the exponential function whose base is e; e^x

Natural logarithm: the logarithm whose base is e; $\ln(x)$

Odd symmetry: symmetry about the origin; 180-degree rotational symmetry; $f(-x) = -f(x)$

One-to-one: characterizing a function that has exactly one input for each output

Output variable: the dependent variable in a function

Parameter: (1) in a function, a value that is constant for that particular function; (2) in a set of parametric equations, the independent variable

Parametric equations: a set of functions with the same input variable, whose outputs become the coordinates of a curve

Period of a function: the interval in which a periodic function completes its pattern once

Periodic function: one whose output values repeat at regular intervals

Piecewise function: one whose domain is divided into two or more intervals, and which is defined differently on each interval

Proportional: for two variables, related to each other through multiplication by a constant

Radian: an arc of a circle equal in length to the radius of the circle; an angle at the center of the circle subtended by such an arc, 2π radians corresponding to 360 degrees

Range: the set of possible outputs of a function

Real number: any number that can be represented as a point on the real number line

Reciprocal: the multiplicative inverse of a number; "one over" the number

Rate of change: for a function, the ratio of a change in the output to the corresponding change in the input

Root of an equation: a solution to the equation; a number that makes the equation true

Slope: for a line, the constant rate of change

Step function: a piecewise function whose outputs consist of constant-valued segments

Turning point: a point at which a function changes from increasing to decreasing, or vice-versa

Unit circle: a circle, centered at the origin, whose radius is 1 unit

Variable: a quantity that can take on more than one value

Zero of a function: an input value for which the output is zero; an x-intercept